B/NA

Kinetics of Ordering and Growth at Surfaces

NATO ASI Series

Advanced Science Institutes Series

A series presenting the results of activities sponsored by the NATO Science Committee, which aims at the dissemination of advanced scientific and technological knowledge, with a view to strengthening links between scientific communities.

The series is published by an international board of publishers in conjunction with the NATO Scientific Affairs Division

| A | Life Sciences | Plenum Publishing Corporation |
| B | Physics | New York and London |

C	Mathematical and Physical Sciences	Kluwer Academic Publishers Dordrecht, Boston, and London
D	Behavioral and Social Sciences	
E	Applied Sciences	

F	Computer and Systems Sciences	Springer-Verlag
G	Ecological Sciences	Berlin, Heidelberg, New York, London,
H	Cell Biology	Paris, and Tokyo

Recent Volumes in this Series

Series B: Physics

Kinetics of Ordering and Growth at Surfaces

Edited by
Max G. Lagally

University of Wisconsin–Madison
Madison, Wisconsin

Plenum Press
New York and London
Published in cooperation with NATO Scientific Affairs Division

Proceedings of a NATO Advanced Research Workshop on
Kinetics of Ordering and Growth at Surfaces,
held September 18-22, 1989,
in Acquafredda di Maratea, Italy

Library of Congress Cataloging-in-Publication Data

NATO Advanced Research Workshop on Kinetics of Ordering and Growth at
    Surfaces (1989 : Acquafredda di Maratea, Italy)
    Kinetics of ordering and growth at surfaces / edited by Max G.
Lagally.
        p.   cm. -- (NATO ASI series. B, Physics ; v. 239)
    "Proceedings of a NATO Advanced Research Workshop on Kinetics of
Ordering and Growth at Surfaces, held September 18-22, 1989, in
Acquafredda di Maratea, Italy."
    "Published in cooperation with NATO Scientific Affairs Division."
    Includes bibliographical references and index.
    ISBN 0-306-43702-3
    1. Thin films--Surfaces--Congresses. 2. Order-disorder models-
-Congresses. 3. Epitaxy--Congresses. 4. Surfaces (Physics)-
-Congresses. 5. Surface chemistry--Congresses.   I. Lagally, Max G.
II. North Atlantic Treaty Organization. Scientific Affairs
Division. III. Title. IV. Series.
QC176.84.S93N37 1989
530.4'275--dc20   90-47499
                                                           CIP

© 1990 Plenum Press, New York
A Division of Plenum Publishing Corporation
233 Spring Street, New York, N.Y. 10013

Printed in the United States of America

PREFACE

This volume contains the papers presented at the NATO Advanced Research Workshop on "Kinetics of Ordering and Growth at Surfaces", held in Acquafredda di Maratea, Italy, September 18-22, 1989.

The workshop's goal was to bring together theorists and experimentalists from two related fields, surface science and thin-film growth, to highlight their common interests and overcome a lack of communication between these two communities.   Typically surface scientists are only concerned with the microscopic (atomic) description of solids within one monolayer of the surface.   Thin-film growers are usually considered more empirical in their approach, concerned primarily with the "quality of their product", and have not necessarily found it useful to incorporate surface science understanding into their art. This workshop aimed to counter at least in some measure these stereotypes.   Its focus was on generating dialogue on the fundamental structural and kinetic processes that lead to the initial stages of film growth, from both the surface science and crystal growth perspectives.   To achieve this, alternate days emphasized the view of surface science and thin-film growth, with considerable time for discussion, a format that appeared to succeed well.

The success of the workshop is in large measure due to the efforts of the organizing committee, L. C. Feldman, P. K. Larsen, J. A. Venables, and J. Villain, whose advice on the constitution of the program was invaluable.   In addition, I would like to thank in particular Poul Larsen, the Associate Director, for his considerable efforts in acting as treasurer, arranging for the conference site, and taking care of all local arrangements.   The Workshop was made possible through the generous support of the NATO Scientific Affairs Division, Brussels, Belgium and of the U.S. Office of Naval Research, Physics Program, which supported travel by U. S. participants. Finally I would like to thank Mrs. Lynn Neis, who single-handedly prepared both the conference program and the manuscript for this book in camera-ready form.   The latter was completed from diskettes of every conceivable format sent by the authors.   The high visual quality of the book is the result of her efforts; possible remaining errors are the responsibility of the editor and authors.

About one week before this Workshop, Professor N. Cabrera died in Madrid, Spain. He was one of the early great figures in the field of crystal growth and a continuing proponent of fundamental studies in the areas represented by this workshop. We would like to dedicate this volume to his memory.

Max G. Lagally
Department of Materials Science
  and Engineering
University of Wisconsin
Madison, WI

INTRODUCTION

This book contains papers summarizing the "state of the art" of growth and ordering processes at surfaces, from the perspective of both the surface scientist and the vapor-phase crystal grower. It contains chapters that present latest research results befitting of a workshop, as well as ones that review more traditional results and methods with the aim of providing the reader wishing an introduction to this subject a sufficient overview and extensive reference lists from which to proceed to a more detailed understanding. The book is divided into three parts. Part I covers those topics that primarily relate to kinetic processes in two dimensions. In this part also a variety of techniques are discussed including the novel applications of scanning tunneling microscopy (STM) and thermal-atom beam scattering for growth and ordering studies. The introductory chapter provides a summary of the essential processes in two-dimensional ordering. In addition, the chapter on "Molecular Kinetics of Steps" describes the application of the well known decoration technique for the investigation of surface kinetics processes.

Part II provides an up-to-date look at the microscopic processes involved in epitaxial growth. It begins with a summary chapter on the use of diffraction for investigating all manner of finite-size and time-dependent effects in epitaxy. Part II has three main topical concentrations: GaAs epitaxy, Si epitaxy, and theoretical developments. In particular, two chapters discuss the application of STM to epitaxial growth. It is evident that much understanding of growth mechanisms can be quickly obtained with STM, even if the few studies that have so far been performed are to a large degree still at the qualitative level.

Part III discusses the morphological evolution of surfaces and films grown on them. Morphological evolution includes both the dynamic roughening and the smoothening of structures at the surface, issues of considerable theoretical interest at present, as well as of technological concern, because the roughness of interfaces affects many properties of devices based on multiple layers. Part III begins with two chapters reviewing equilibrium properties, respectively surface melting and roughening. It then proceeds to several theoretical chapters on surface smoothening and on the development of morphology during growth. These chapters are followed by discussions of kinetic processes involved in cluster formation at surfaces and finally by chapters on applications dealing with strained layers and multilayer films.

Crystal growth has been and continues to be a dynamic area of research, with a large potential for applications to technologically useful devices. The potential for applications has been proven time and again. In the modern era of micro and nano-engineered materials, it is important to continue to improve our understanding of the kinetic processes responsible for growth at the most microscopic level. This book aims to provide a useful reference and source of stimulation for researchers in this and related fields, as well as a starting point and guide for those who may be interested in gaining a better fundamental knowledge about modern crystal growth.

CONTENTS

II. EPITAXIAL GROWTH

III. MORPHOLOGICAL EVOLUTION IN SURFACES AND FILMS

KINETICS OF ORDERING AND GROWTH IN 2-D SYSTEMS

Ole G. Mouritsen

Department of Structural Properties of Materials
The Technical University of Denmark
Building 307, DK-2800 Lyngby, Denmark

ABSTRACT

An overview is given of the lessons learnt from using simple two-dimensional kinetic lattice models to study the kinetics of ordering processes and late-stage growth in atomic and molecular overlayers adsorbed on solid surfaces. The general phenomenology and the classical theories of late-stage non-equilibrium ordering processes in condensed matter are reviewed within a framework in which the ordering process is seen as a pattern formation process producing random interfaces. The growth is then controlled by the dynamics of the random network of interfaces. The fundamental questions of the dynamics involve the form of the growth law, the value of the growth exponent, the dynamical scaling property of the non-equilibrium structure factor, and a possible universal classification of ordering kinetics. Evidence from kinetic lattice model calculations is presented in favor of a universal description of late-stage ordering kinetics in terms of algebraic growth laws with exponents which only depend on whether or not the order parameter is a conserved quantity. Other variables, such as temperature, ordering degeneracy, details of the interaction potential and the dynamical model, as well as additional conservation laws, appear to be irrelevant. However, it is pointed out that crossover effects in the irrelevant variables and in time may well veil the asymptotic growth behavior. Moreover, imperfections, such as impurities and vacancies, may change the algebraic growth law into a logarithmic one. It will be emphasized that two-dimensional overlayers adsorbed on solid surfaces constitute a particularly suitable class of systems for studying fundamental aspects of ordering kinetics in two dimensions since these systems provide a richness of ordering symmetries and degeneracies. The specific examples dealt with include O/W(112), O/W(110), H/Fe(110), O/Pd(110), N_2/Graphite, (N_2 + Ar)/Graphite, ad N-doped O/W(112).

INTRODUCTION

Ordering and growth processes are very common and general phenomena in nature.[1-6] They involve as diverse phenomena as formation of snowflakes, ordering in alloys, and quark confinement in the inflationary universe. The kinetics and dynamics of the ordering and growth processes are highly non-linear and irreversible phenomena and hence involve

Kinetics of Ordering and Growth at Surfaces
Edited by M. G. Lagally
Plenum Press, New York, 1990

fundamental questions in non-equilibrium thermodynamics. In condensed matter in particular, ordering and growth phenomena have striking manifestations in connection with non-equilibrium phase transitions. Moreover, ordering processes in solids are of prime interest within materials science and technology.

Ordering and growth processes in condensed matter often proceed in a seemingly disorderly fashion,[2] leading to complex pattern formation and random networks of interfaces between the different ordered and growing domains. It is the dynamics of this random network of interfaces and their instabilities and fluctuations which hold the key to understanding and describing the ordering kinetics.

Surface science has proved to be an extremely rich testing ground for theoretical ideas of the equilibrium properties of 2-D phase transitions.[7] The reason for this, in the case of adsorbed overlayers for example, is that the competition between the adatom-adatom interactions and adatom-substrate potential leads to a very rich phase behaviour, possibly even more rich than in 3-D bulk matter. This richness has led to an extremely fruitful interplay between experimental and theoretical research. A similar fruitful interplay is now anticipated regarding the non-equilibrium and kinetics of surface phenomena. Whereas the experimental study of time-dependent ordering processes in 3-D systems is well developed,[3,9-12] only recently have detailed time-resolved studies been conducted for surface systems.[13-19] These studies are so far limited to the late-stage ordering processes of oxygen chemisorbed on tungsten surfaces.

The theoretical description of non-equilibrium phase transitions and ordering kinetics is known to be a very difficult problem.[1] It is therefore not surprising that a substantial part of our present theoretical knowledge about such phenomena stems from computer-simulation studies of a variety of models. Despite the apparent complexity of the pattern-formation processes involved in the ordering, as well as the disorderly way in which the ordering processes proceed, and despite the obvious lack of conventional symmetry properties, it has in recent years become clear that a description is possible using the concept of dynamical scaling symmetry. Moreover, there are strong indications that some universal principle is operative and that the associated universality classes are mainly determined by whether or not the order parameter is a conserved quantity.

In this brief overview a status will be given of the field of ordering kinetics in 2-D systems with particular emphasis on the results obtained from computer-simulation studies of two-dimensional kinetic lattice models. The focus is on the late-time kinetics, i.e. the time regime which is dominated by the competition between large domains of different types of ordering.

GENERAL PHENOMENOLOGY: DYNAMICS OF RANDOM NETWORKS OF INTERFACES

1.   Domain Growth

The prototypical situation we consider here is one in which a system with a phase transition is thermally quenched below the phase transition temperature. Immediately after the quench the system will be far from its equilibrium state characterized by the temperature to which the quench is performed. The system will then undergo a spontaneous non-equilibrium phase transition or an ordering process. This leads

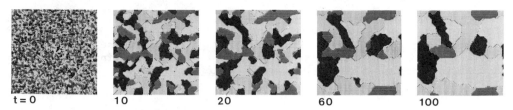

t = 0　　　　10　　　　20　　　　60　　　　100

Fig. 1　　Snapshots of the evolving domain pattern in a four-fold
degenerate (p = 4) ordering process with non-conserved order
parameter, e.g. O/W(110). Results are shown for quenches from
a high to a very low temperature below the ordering transition.

initially to nucleation and growth of small clusters of ordered domains
and later to a coarsening of the domains. Eventually the system will
approach its uniformly ordered equilibrium state. In the case where the
ordered phase is degenerate, i.e. when the system can order in several
thermodynamically equivalent ordered structures, the approach to
equilibrium is governed by the competition between the different ordered
domains which are nucleated simultaneously. In equilibrium, one of these
domains has taken over the whole system. The occurrence of several
domains leads to a network of boundaries separating domains of different
order, cf. Fig. 1. It is the dynamics of this network and the associated
thermodynamic forces which govern the evolution towards equilibrium.

The scenario described above accounts in general for
non-equilibrium spontaneous ordering processes and covers such general
phenomena as phase separation, spinodal decomposition, and order-disorder
transitions. Particular systems include binary alloys, superfluids,
physisorbed and chemisorbed overlayers, magnetic systems, and polymer
mixtures. We shall specifically consider the process of late-time domain
growth (or grain growth) in systems with order-disorder transitions.

2.　　Some Fundamental Questions

Obviously the type of pattern-formation processes described above
(cf. Fig. 1) leads to order in a disorderly fashion. The processes are
subject to random fluctuations and they are chaotic in the sense that the
specific pattern formed at a given time is intimately determined by the
random fluctuations in the initial state before the quench. In fact, the
basic mechanism for the evolution of the patterns may be said to be the
repeated amplification of the noise in the initial state. Despite the
apparent chaotic behaviour, ensemble-averaged properties, e.g. the
domain-size distribution function, of the patterns in Fig. 1 will have
perfectly well-defined time-dependent values which do not depend on the
noise of the initial state.

The pattern-formation phenomena we are considering here are
associated with irreversible growth processes caused by non-equilibrium
conditions. A number of general questions arise for these
pattern-formation phenomena: (i) Can they be described quantitatively in
space and time? (ii) If so, how can they be described? (iii) Do they
arise from simple physical principles? (iv) Can they be modelled by
simple models? (v) Are there any universal principles operative?
Experimental and theoretical studies of the ordering kinetics in 2-D
systems may provide some answers to these fundamental questions. The

answers will have to involve the nature of the growth law, the form of the non-equilibrium scattering function, the possible existence of a dynamical scaling function, and the possible relevant variables for a universal classification of ordering kinetics. Candidates for such variables include the temperature, the ordering degeneracy, the interaction potential, the conservation laws, as well as randomness (impurities, vacancies, and imperfections).

DOMAIN GROWTH KINETICS

1. Ordering Degeneracy

An important parameter in the domain growth problem is the ordering degeneracy, i.e. the number, p, of thermodynamically equivalent types of ordered domains. In the case of grain growth in the formation of a polycrystalline aggregate after quenching a liquid, the value of p is very large, in principle infinity, corresponding to the continuum of ways in which the nucleating crystallite can orient its principal crystal axes in space. For other problems, such as order-disorder transitions in binary alloys, the value of p is some small and finite number. In the case of adsorbed molecular or atomic monolayers on solid surfaces, the experimentalist has the option of choosing a variety of values of p by appropriate choices of the adsorbant and the substrate symmetry. As an example, $p = 2$ for O/W(112),[14,18] $p = 2$ or 3 for H/Fe(110)[20] and O/Pd(110),[21] $p = 4$ for O/W(110),[15,16,19,22] $p = 6$ for N_2/Graphite,[23,24] and $p = 8$ for $(N_2 + Ar)$/Graphite.[25]

In Fig. 2 are shown typical domain-boundary configurations at intermediate times for a series of models with different values of p. Obviously, the larger the value of p, the more compact the domains become. For low values of p, the domain pattern is very convoluted and percolation-like.

2. Algebraic Growth Laws and Dynamical Scaling

One may anticipate from the evolution of the domain pattern, cf. Fig. 1, that the ordering problem is characterized by some time-dependent linear length scale, r(t). This scale may be derived quantitatively from the domain-size distribution function, P(R,t), in which R(t) measures the linear domain size, e.g. as defined by $R \sim V^{1/d}$, where V is the domain volume (area) and d is the spatial dimension ($d = 2$). At a given time, P(R,t) has a peak at some value of R, and this peak moves towards larger values of R as time elapses. From the function P(R,t) we obtain the first measure of a time-dependent linear length scale, the average domain size, R(t)

$$R(t) = \int_0^\infty R(t)P(R,t)dR, \qquad (1)$$

where P(R,t) is appropriately normalized, $\int P(R,t)dR = 1$.

In a high-resolution real-time study of ordering dynamics it is not possible to measure the length scale directly but only indirectly via the dynamical structure factor, $S(q,t)$. Appropriate powers, $k_m(t)^{-1/m}$, of the time-dependent moments

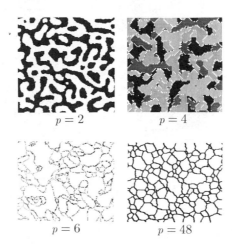

$p = 2$ $p = 4$

$p = 6$ $p = 48$

Fig. 2   Typical domain-boundary configurations at intermediate times as
found during a thermal quench from a very high temperature to a
very low temperature below the ordering transition. p denotes
the number of degenerate ordered domains. For p = 6 and p = 48
only the network is shown. The results are obtained from Monte
Carlo simulations on two-dimensional models with symmetries
corresponding to O/W(112), O/W(110), N_2/Graphite, and to
two-dimensional grain growth.

$$k_m(t) = \sum_{\vec{q}} |\vec{q}|^m \, S(\vec{q},t)/\sum_{\vec{q}} S(\vec{q},t) \qquad (2)$$

of the structure factor have dimensions of length. Similarly the
intensity at the Bragg point, $q = q_o$,

$$L(t) = [S(\vec{q}_o,t)/N]^{1/2} \qquad (3)$$

may be used as a length-scale measure. In Eq. (3), N is the number of
scattering centers (atoms).

Finally, the excess domain-boundary energy[26]

$$\Delta E(t) = E(t) - E(T), \qquad (4)$$

which measures the non-equilibrium energy of the total domain-boundary
network relative to the equilibrium energy at the quench temperature,
$E(T) = E(t \rightarrow \infty)$, is related to the total interface perimeter, $\Delta E(t)^{-1}$,
and hence the average domain radius.

Obviously, a simple linear time-dependent length scale is a rather
primitive and far from complete description of the structure and
evolution of the patterns in Fig. 1. A fuller characterization could
involve the curvature distribution function or similar more intimate

measures of the evolving morphology. However, such quantities would be difficult to monitor experimentally. Nevertheless, the time evolution of domain patterns is subject to a remarkable scaling symmetry which provides a rather exhaustive description of the growth process. This may be anticipated from Fig. 1 in which the pattern at some given time has a similar appearance as an appropriately magnified portion of a pattern at an earlier time. This phenomenon may be called pictorial scaling: the pattern evolves in a self-similar fashion. In other words, the pattern formation is invariant with respect to a temporal change in length scale. It hence has a temporal scaling symmetry.

The above statements about a temporal scaling symmetry can be put in a more rigorous form by introducing the dimensionless time-dependent variable x . In the case of the domain-size distribution function, dynamical scaling then implies

$$\tilde{P}(x) = R(t)\overline{P}(R,t) , \qquad (5)$$

where $P(x)$ is the scaling function. The appropriate choice of scaling variable would in this case be $x = R(t)/\overline{R}(t)$. In the case of the structure function the scaling property may be expressed as

$$\tilde{S}(x) = k_m(t)^{d/m}S(\vec{q},t) , \qquad (6)$$

with the scaling variable $x = k_m(t)^{-1/m}|q|$. The statement of Eq. (6) is that the structure factor only depends on time via the characteristic time dependence of the linear length scale. The regime of dynamical scaling is expected to apply only after some transient time t_o. Similarly, corrections to scaling are expected. The temporal scaling symmetry implies that there is only one length scale in the problem, i.e.

$$r(t) \sim \overline{R}(t) \sim \Delta E(t)^{-1} \sim k_m(t)^{-1/m} . \qquad (7)$$

The explicit time dependence of the linear length scale is often analyzed in terms of an algebraic growth law

$$r(t) \sim t^n , \qquad (8)$$

where n denotes the kinetic growth exponent. Evidence for the validity of Eq. (8) will be provided below.

3.   Phenomenological Theories of Late-Time Growth and the Role of Conservation Laws

In characterizing the non-equilibrium dynamics of an ordering process it is important to specify the conservation laws in effect,[27] in particular whether the order parameter is a conserved quantity or not.

In 1962 Lifshitz[28] presented a phenomenological theory of late-time interface motion in systems with non-conserved order parameter. The theory assumes that the driving force is determined by the interface curvature. Later, Allen and Cahn[29] presented a field theoretical treatment of the problem by assuming that the driving force is proportional to the gradient square of the order parameter at the interface. Although assigning very different driving forces to the growth, the growth-law exponent resulting from these two approaches turns out to be the same and we shall therefore here present the simpler Lifshitz argument.

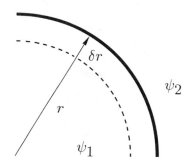

Fig. 3   Moving interface between two domains of order ψ_1 and ψ_2.

With reference to Fig. 3 the driving force free energy for the interface between two domains of order ψ_1 and ψ_2 is assumed to be

$$\delta F = [F^\circ(\psi_1) - F^\circ(\psi_2)]\delta r + C\kappa\delta r ,\qquad (9)$$

where $\kappa \sim r^{-1}$ is the interface curvature and C is an effective diffusion constant. F° is the bulk free energy. When the two domains are thermodynamically equivalent, only the interface term survives. By inserting δF into the diffusion equation, one obtains

$$\frac{\delta r}{\delta t} \sim \frac{\delta F}{\delta r} \sim \frac{C}{R} ,\qquad (10)$$

which leads to the growth law

$$r(t) \sim t^{1/2} \text{(nonconserved order parameter)} .\qquad (11)$$

Hence the growth exponent is n = 1/2 for curvature-driven growth. The theoretical argument presented above assumes implicitly that there are only two different domains, p = 2. It is not obvious how this can be generalized to p > 2 in which case vertices and the topology of the boundary network will come into play.

In the case of conserved order parameter, which is the case of spinodal decomposition, a phenomenological theory for a two-component system also exists. A version of this theory is due to Lifshitz and Slyozov[30] who found also in this case an algebraic growth law but with a different exponent value

$$r(t) \sim t^{1/3} \text{(conserved order parameter)} ,\qquad (12)$$

i.e., n = 1/3. The assumptions underlying Eq. (12) are that the growth is limited by diffusion through the domains and that the domain pattern is described by a single length scale r(t). This growth law may be thought of as the leading term in the solution of the interface diffusion equation[31]

$$\frac{\delta r}{\delta t} \sim \frac{C(\kappa)}{r} , \qquad\qquad (13)$$

in which the effective diffusion constant is now a function of the interface curvature, i.e., $C(\kappa) \sim \kappa + O(\kappa^2)$. The physical reason for the lower exponent value in this case is that the moving interface has to comply with the constraint of conserved order which implies long-range diffusion processes of material. The original Lifshitz-Slyzov theory was restricted to low volume fractions. However, Huse[31] has recently shown that the theory can be generalized to apply for arbitrary volume fractions of the two phases.

Since the ordering processes in both the above cases are driven by purely thermodynamic forces, the growth exponents n are not dependent on the spatial dimension. They are hence superuniversal.

A situation may arise in which the order parameter is not conserved but at the same time coupled to another field which is. This will frequently be the case in chemisorbed overlayers with sublattice ordering and conserved coverage (density) but non-conserved order parameter. Provided the equilibrium densities of the two types of ordering are the same,[32] there is no need for long-range diffusion during the interface motion and the Lifshitz-Allen-Cahn growth law will apply, Eq. (11). If this is not the case, the Lifshitz-Slyozov law, Eq. (12), is expected to be valid at late times.[32]

It is interesting to note that there is a conceptual difference between chemisorbed overlayer systems and e.g. physisorbed overlayers and other open systems for which the density (or the total coverage) is not conserved during the ordering process. For the chemisorbed systems the density is constant although the order parameter of the particular sublattice ordering is not. Only for (1 x 1)-ordering is the order parameter and the density the same quantity. Hence, for physisorbed overlayers, growth can always proceed by reduction of interface, whereas in chemisorbed systems the interface motion must involve processes controlled by adatom diffusion.

THEORETICAL MODELS OF ORDERING KINETICS

Theoretical model studies of the late-time ordering kinetics have proceeded by mainly using (i) phenomenological field-theoretical equations of motion, (ii) cellular automata and cell model dynamics, and (iii) simple statistical mechanical kinetic lattice models. The properties of the various theoretical models are obtained almost exclusively by numerical techniques. Specifically, the kinetic lattice models are usually studied by means of Monte Carlo computer-simulation techniques. In this section we shall mainly be concerned with the kinetic lattice model description and only briefly refer to the methods and results used and obtained from phenomenological modelling.

1.   Kinetic Lattice Models

Two things are required as necessary ingredients of a kinetic lattice model: (i) a microscopic interaction model (i.e. a Hamiltonian, H) defined on a lattice with certain variables coupled by certain interaction strengths) and (ii) a dynamical principle (or a so-called move class). The Hamiltonian needs to govern a phase transition and it provides the ordering symmetry of the problem. The move class embodies the various conservation laws in effect. It may seem superfluous to supplement the Hamiltonian with a dynamical principle since the Hamiltonian should provide its own dynamics. However, in many cases this is not so, e.g. in the case of the Ising model which has no non-trivial equation of motion. Moreover, in the cases where the Hamiltonian furnishes an equation of motion it is not possible to solve this equation (e.g. by molecular-dynamics techniques) under far-from equilibrium conditions. The dynamic principle imposed is therefore usually of a stochastic character and is formulated in terms of an appropriate master equation.

The computer-simulation techniques used are based on Metropolis Monte Carlo importance-sampling methods[33] by which an ensemble of successive microstates of the system is created, Ω_1, Ω_2,..., Ω_M. The transition between two successive states is governed by the transition probability

$$p(\Omega_i \to \Omega_{i+1}) = \tau^{-1} \exp \left[- \frac{H(\Omega_{i+1}) - H(\Omega_i)}{k_B T} \right], \tag{14}$$

where T is the temperature to which the quench is performed and τ is the relaxation time typical of the transition. The integer index for the microstate defines a time parameter, t, which is usually measured in units of Monte Carlo steps per site of the lattice (MCS/S). This time parameter refers to the time defined in the appropriate master equation to which the transition probabilities, Eq. (14), pertain. It is not obvious that the time parameter of the stochastic dynamics is related in a simple fashion to a real physical time scale. However, a simple non-rigorous argument based on an Arrhenius-type single-particle excitation scheme suggests that there is a linear relation. Furthermore, there is now some evidence that, as far as relaxation out of non-equilibrium is concerned, the Monte Carlo time parameters of different move classes are linearly related as long as the updating algorithms are built on spatially local decision criteria.[34,35] Further theoretical work is needed to settle this question definitively.

The thermal quenches are performed by initiating the system in a disordered configuration corresponding to a high-temperature state and then suddenly changing the temperature to a low value below the transition temperature.

The advantages of using a computer-simulation approach to domain-growth kinetics is that it is simple and easy to implement, it takes faithful account of the topology of the boundary network and the local features of the growth, and it gives direct access to the evolution of the microstructures. The drawbacks include the restriction to finite systems and finite times, as well as the fact that the method is of a purely numerical nature which may suffer from crossover and parameter effects.

Two-dimensional Ising models and their isomorphic lattice gas analogs constitute an important class of simple kinetic lattice models from which seminal information on non-equilibrium ordering kinetics and growth has been obtained. Of particular importance is the two-dimensional spin-1/2 Ising model with nearest-neighbor interactions (nn) defined on a square lattice by the Hamiltonian

$$H = - J \sum_{i,j}^{nn} \sigma_i \sigma_j, \tag{15}$$

9

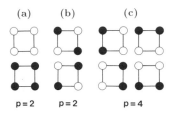

(a)　　(b)　　　　(c)

p=2　　p=2　　　　p=4

Fig. 4　　Schematic representation of the ordered states of the square-lattice Ising model.　(a) (1 x 2) ferromagnetic ordering (p = 2).　(b) (2 x 1) simple antiferromagnetic ordering (p = 2).　(2 x 1) super-antiferromagnetic ordering (p = 4) corresponding to $\alpha > 1/2$ in Eq. (16).

where $\sigma_i = \pm1$. Ferromagnetic and antiferromagnetic ordered states are stabilized for J>0 and J<0, respectively, cf. Figs. 4a and b. Both these types of order are two-fold degenerate, i.e. p = 2. A higher degeneracy may be obtained by extending Eq. (15) to include next-nearest-neighbor (nnn) interactions

$$H = - J \left[\overset{nn}{\underset{i,j}{\Sigma}} \sigma_i\sigma_j + \alpha \overset{nnn}{\underset{i,j}{\Sigma}} \sigma_i\sigma_j \right] , \qquad (16)$$

which for $\alpha > 1/2$ lead to four-fold degenerate, p = 4, super-antiferromagnetic ordering, cf. Fig. 4c.　Important thermodynamic variables for the different kinds of order in the Ising models include the antiferromagnetic order parameter

$$\psi = N^{-1} \underset{j}{\Sigma} \sigma_j \exp (i\vec{q_o} \cdot \vec{r_j}) , \qquad (17)$$

corresponding to sublattice modulation q_o, and the ferromagnetic order parameter (the density or coverage)

$$\rho = N^{-1} \underset{j}{\Sigma} \sigma_j . \qquad (18)$$

　　　　The dynamical principles imposed on Ising models are usually of two stochastic types: (i) Glauber-type single-site spin-flip dynamics, $\sigma_i \rightarrow -\sigma_i$, which conserves neither ψ nor ρ, and (ii) Kawasaki-type two-site spin-exchange, $\sigma_i \leftrightarrow \sigma_j$, i $\neq$ j, which conserves ρ but not ψ. The two types of dynamics can be used separately or in combination.

　　　　Hence, the simple Ising models in Eqs.(15) and (16) provide a unique testing ground for ordering dynamics and its possible dependence on ordering degeneracy (p), temperature, details of the interaction potential (α), and the conservation laws for two different quantities.

　　　　Another simple model which has proved useful in studying ordering kinetics is the two-dimensional p-state ferromagnetic Potts model defined by the Hamiltonian

$$H = - J \sum_{i,j} \delta_{\sigma_i \sigma_j} \tag{19}$$

with $J > 0$ and $\sigma_i = 1,2,\ldots,p$. This model has a p-fold degenerate ground state and may hence for low values of p serve as a model of rare gas ordering on Graphite[7] and for high values of p as a model of grain growth in two dimensions.[36-38] Possible realistic modifications of Eq. (19) involve an anisotropic grain-boundary potential.[39,40] Kawasaki dynamics will conserve the order parameter of Eq. (19) whereas Glauber dynamics will not.

A number of authors[41-45] have proposed an extension of the computer-simulation approach for studying the ordering governed by kinetic lattice models to include a Monte Carlo renormalization-group scheme. Such a scheme, which has been applied both to the Potts model and the kinetic Ising model for conserved as well as non-conserved order parameter, is based on a matching condition for the average domain sizes calculated in two different systems of the same size but at different levels of renormalization:

$$\bar{R}(N,m,t) = \bar{R}(Nb^d, m + 1, t'). \tag{20}$$

When the time rescaling factor, $\Delta(T,t) = t'/t$, is constant it follows that $\Delta = b^{1/n}$ from which the kinetic growth exponent may be determined. A main and unresolved problem with this approach is that Δ may depend on b,T, as well as t.

2. Phenomenological Models

The phenomenological description takes its starting point in a set of time-dependent continuum field variables for the order parameter whose time evolution is governed by a non-linear Langevin equation[1]. In general this equation takes the form

$$\frac{\partial \psi(\vec{r},t)}{\partial t} = \Gamma(\vec{r}) \frac{\delta F(\psi(\vec{r},t))}{\delta \psi(\vec{r},t)} + \xi(\vec{r},t), \tag{21}$$

where $\Gamma(\underline{r})$ is a kinetic function, F is a coarse-grained ϕ^4 free-energy functional

$$F[\psi(\vec{r},t)] = \frac{1}{2} \int d\vec{r} \ [V[\psi(\vec{r},t)]\psi + C(\nabla\psi)^2], \tag{22}$$

with

$$V[\psi(\vec{r},t)] = - \alpha\psi + \frac{b}{2} \psi^3 , \tag{23}$$

and $\xi(\underline{r},t)$ is a delta-correlated noise term satisfying the fluctuation-dissipation theorm

$$<\xi(\vec{r},t)\xi(\vec{r}',t') > = 2k_B T\Gamma(\vec{r})\delta(\vec{r} - \vec{r}')\delta(t - t'). \tag{24}$$

In the case of a non-conserved order parameter (Model A in the classification scheme of critical dynamics[27]) the kinetic function is a constant phenomenological parameter, $\Gamma(\vec{r}) = D$. In the case of a conserved order parameter (Model B[27]) the kinetic function is

$$\Gamma(\vec{r}) = - M\nabla^2, \tag{25}$$

where M is a mobility parameter. In the absence of noise, Eq. (21) is in this case equivalent to the so-called Cahn-Hilliard equation which is a model of spinodal decomposition.[1] The equations of motion are solved by numerical integration. It has recently been found, somewhat surprisingly, that the noise term does not influence the late-stage growth kinetics and the scaling function.[46]

A computationally efficient way of dealing with phenomenological equations for ordering dynamics has recently been proposed by Oono and Puri[47] who by a cell dynamics approach basically replace V($\underline{r}$,t) in Eq. (22) by

$$V[\psi(\vec{r},t] \sim \tanh \psi(\vec{r},t) , \tag{26}$$

which has a fix-point structure similar to that of Eq. (23).

3.   Summary of Results Obtained from Theoretical Model Studies

In this section we briefly review the results obtained from theoretical model studies of late-stage ordering kinetics in 2-D systems. Since there is a general agreement between the results of the different theoretical approaches, we shall describe these results jointly and later in Sec. VI return to a detailed description of the specific results obtained from two-dimensional kinetic lattice models of relevance for ordering kinetics on surfaces.

In the case of a non-conserved order parameter it has been found that the structure factor exhibits dynamical scaling[20,22,35,43,48-57] and that the late-time growth law is that of the Lifshitz-Allen-Cahn theory,[20-22,35,38,40-43,47,48,50-53,55-65] Eq. (11), with kinetic growth exponent value, n = 1/2. This result holds independent of the ordering degeneracy, p, independent of additional conservation laws, independent of details of the interaction potential, and independent of temperature.

In the case of a conserved order parameter, which has turned out to be a much more difficult case to settle, there has been some controversy regarding the nature of the growth law. However, the most reliable data currently available[31,44,46,47,66-69] provide strong evidence in favor of dynamical scaling in the late-time regime and an asymptotic growth law as predicted by the Lifshitz-Slyozov theory, Eq. (12), with kinetic growth exponent value n = 1/3. Again this result is found to be independent of details of the dynamical model (e.g. whether it is the Langevin model or the kinetic Ising model). For this case, less is known about the possible dependence on the ordering degeneracy.[70] Very recently[67] some new information has been obtained about dynamical scaling and ordering dynamics in tricritical systems with coupled order parameters, one of which is conserved and the other one is not (Model C[27]). This information is consistent with n = 1/3.

Finally, it should be mentioned that the few available theoretical model studies of late-stage ordering kinetics in 3-D systems,[38,71,72]

suggest that the exponents for both the non-conserved and conserved order parameter cases are the same as in two dimensions supporting the conjecture that n is superuniversal.

In summary, our current knowledge about late-stage ordering kinetics therefore suggests the possibility of a universal classification scheme with basically two universality classes. Whether a particular dynamical system falls in one class or the other should depend on whether or not the order parameter is a conserved quantity. Other variables, such as temperature, ordering degeneracy, details of the interaction potential and the dynamical model, as well as additional conservation laws, appear to be irrelevant.

In reality, the picture of a universal classification scheme of ordering kinetics is, unfortunately, not as clear as outlined above. This is true of many theoretical model studies (as well as a variety of experimental studies). The reasons for this are manifold. On the theoretical side the picture is often obscured by crossover effects in either temperature[37,57,65,73] or in some model parameter.[37,38,56,57,63,73,74] Experimentally, the effective growth exponent may be influenced by similar effects in addition to restrictions in observation time and effects due to imperfections and impurities.[17] We shall return to such crossover effects below.

In conclusion, the universality in domain-growth kinetics is not a fully settled matter and further theoretical work is needed to clarify the influence of crossover effects as well as the role of corrections to dynamical scaling.

SOME SPECIFIC EXAMPLES

Two-dimensional chemisorbed and physisorbed overlayers on solid surfaces constitute a particularly suitable class of systems for studying fundamental aspects of ordering kinetics in 2-D systems since these overlayers display a very rich phase behavior with phases of a great variety of ordering symmetries and degeneracies. They are attractive to model by computer-simulation techniques due to the low dimensionality and the possibility of using a lattice model formulation. Below we shall review the existing information on some selected overlayer systems and discuss the kinetic lattice model results in relation to experimental data where available. It should be generally noted that most of the microscopic models dealt with in these studies do not represent faithfully the interaction potential of the specific materials but mostly focus on the symmetry properties of the pertinent dynamical model.

1.   p = 2 : O/W(112)

The W(112)-surface presents itself to the adsorbing oxygen atoms as a rectangular substrate, cf. Fig. 5a. The adsorbed overlayer has a phase of (2 x 1)-ordering, Fig. 5b, which is two-fold degenerate, i.e. p = 2. The simplest possible model of the half-monolayer coverage system and its order-disorder transition is the nearest-neighbor antiferromagnetic Ising model, Eq. (15), with J < 0 and ordered states as shown in Fig. 4b. As remarked above, this model does not, however, represent all the details of the interactions in the real system, specifically it lacks further neighbor interactions and does therefore not provide a quantitative description of the equilibrium phase diagram. Nevertheless, it is in the correct dynamical class.

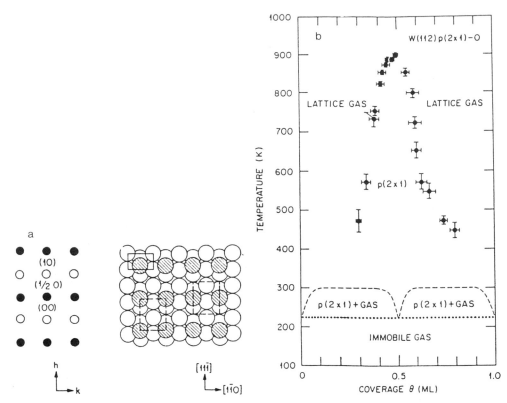

Fig. 5   LEED pattern of O/W(112) at half-monolayer coverage and
real-space illustration of the rectangular mesh of oxygen
adsorption sites on W(112).   (b): Phase diagram of O/W(112).
[According to Refs. 75 and 14.].

The ordering kinetics of this model (as well as that of a
two-dimensional XY-model with uniaxial anisotropy[54,57] which is in the
same dynamical class) is well understood[41-43,47,48,50,58,61] and is known
to be described by the Lifshitz-Allen-Cahn growth law, Eq. (11), for
non-conserved order parameter.   This is true for conserved as well as
non-conserved density.   Moreover, the model exhibits dynamical scaling
for the structure factor as well as the domain-size distribution
function, Eq. (5), as shown in Fig. 6.

Experimentally, the Lifshitz-Allen-Cahn growth law has been
confirmed in an early experimental LEED study[14] of O/W(112) for
half-monolayer coverage. Furthermore, a recent high-resolution LEED
study[18] of the same system provided strong evidence in favor of dynamical
scaling of the structure factor.

2.   p = 4: O/W(110)

Part of the phase diagram of O/W(110) is shown in Fig. 7.   This
diagram contains a (2 x 1) oxygen ordered overlayer phase.   The
degeneracy of this phase is still in dispute and depends on whether the
oxygen atoms sit in the short-bridge sites (p = 4) or in the three-fold

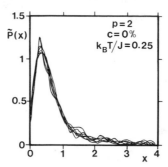

Fig. 6   Dynamical scaling function, Eq. (5), for the domain-size distribution function for the p = 2 antiferromagnetic Ising model with non-conserved order parameter quenched to a temperature k_BT/J = 0.25. Results are shown for times t = 79,98,154,194,244,307,386, and 486 MCS/S. [According to Ref. 35].

hollow sites[19] (p = 8). We shall here examine some model results which base themselves on a model[49,51,56,63,74] which assumes p = 4.

The model we shall be concerned with is the nearest- and next-nearest neighbor antiferromagnetic Ising model, Eq. (16), with coupling-strength ratio, α > 1/2, which supports a p = 4-fold degenerate (2 x 1)-phase, cf. Fig. 4c. In zero magnetic field, and for

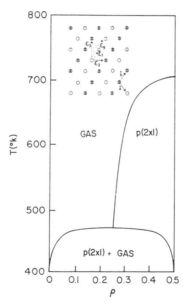

Fig. 7   Phase diagram of 0/W(110) up to half-monolayer coverage. The insert shows the possible oxygen ordering in the (2x1) phase. [According to Ref. 49].

15

non-conserved order parameter this model describes the ordering process in (2 x 1)-phases of O/W(110) at half monolayer coverage. We shall first focus on this particular coverage and then later return to cases of other coverages.

In Fig. 1 is shown a series of typical microconfigurations as they evolve in time for the model with $\alpha = 1$ using Glauber dynamics which conserves neither the order parameter nor the density. That the growth process obeys the Lifshitz-Allen-Cahn law is demonstrated in Fig. 8 using different measures of length scale. Data obtained for other temperatures, cf. Fig. 9, show the same result, demonstrating that the growth exponent is independent of temperature. As the critical temperature is approached, however, the effective value of the growth exponent becomes suppressed due to critical slowing-down which has to be corrected for[51] in terms of the dynamic critical exponent z. Figure 10 shows that the structure factor exhibits dynamical scaling at late times. At very low temperatures, one observes in this model a crossover to freezing-in behavior for $\alpha < 1$, cf. Fig. 11. The reason for this is that, whereas for $\alpha \geq 1$ there is always some admixture of domain walls along the lattice axis and along the diagonals (which support boundary-kink migration through vertices even at zero temperature), cf. insert of Fig. 12, for $\alpha < 1$ only the antiphase boundaries prevail, cf. Fig. 12. Antiphase boundaries alone lead to a frozen-in tiling structure at very low temperatures. In this tiling structure, domain-boundary motion is an activated process. This type of behavior is likely to be restricted to a singular point, $T = 0$, although it will appear as smeared out in a finite-time, finite system-size calculation. A comprehensive compilation of results for the kinetic exponent n is given in Fig. 12 for different values of T and α.

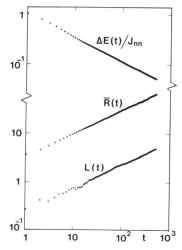

Fig. 8   Log-log plot vs time of length scales L(t) [Eq. (3)], $\overline{R}(t)$ [Eq. (1)], and excess energy $\Delta E(t)$ [Eq. (4)] for the p = 4 Ising antiferromagnet with non-conserved order parameter and density for a quench to $T = 0$ in the case of $\alpha = 1$. The time is in units of MCS/S. [According to Ref. 63].

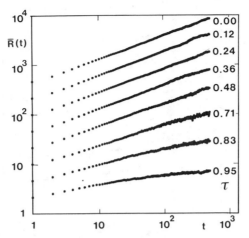

Fig. 9   Log-log plot of the average linear domain size, $\overline{R}(t)$ (in
arbitrary units) vs time for different reduced quench
temperatures, $\tau = T/T_c$, for the p = 4 Ising antiferromagnet
with non-conserved order parameter and density and $\alpha = 1$.
[According to Ref. 63].

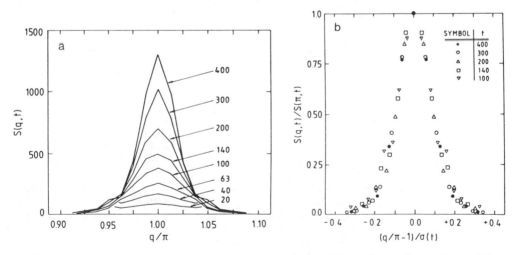

Fig. 10   Structure favor and dynamical scaling function, Eq. (6),
constructed using the first moment as a length-scale measure.
The data refer to the p = 4 Ising antiferromagnet with
non-conserved order parameter and density, (2x1) ordering and
$\alpha = 1$.   The quench temperature is $k_B T/J = 1.33$.   [According to
Ref. 51].

17

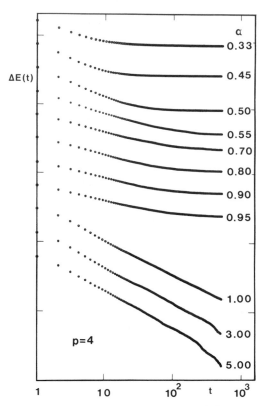

Fig. 11 Log-log plot of the excess energy $\Delta E(t)$, Eq. (4), vs time for quenches to zero temperature of the $p = 4$ Ising model with non-conserved order parameter and density in the case of varying coupling strength ratio α. The time is in units of MCS/S. [According to Ref. 63].

We then turn to the case of imposing a conservation law for the density,[51,56] ρ in Eq. (18), while still not conserving the order parameter, ψ, in Eq. (17). The density is conserved by using Kawasaki spin-exchange dynamics, and exchange between nn as well as nnn sites are allowed with equal probability, i.e. $\delta \equiv [\nu_{nnn}/(\nu_{nn} + \nu_{nnn})] = 1/2$, where the ν refer to the hopping frequencies. A selection of data from quenches of this model to different temperatures is shown in Fig. 13. Again we find, for different length-scale measures and for different temperatures, that the classical growth law holds with $n = 1/2$. Hence, the additional conservation law for the density (which is not an order parameter for the antiferromagnet) does not influence the growth exponent.[32] The dynamical scaling property of this model is demonstrated in Fig. 14 which also shows that the small-distance structure of the domain pattern is well described by a Porod-type law,

$$\tilde{P}(x) = F_2(x) \sim x^{-\omega} , \tag{27}$$

for large x, where $\omega = d + 1$.

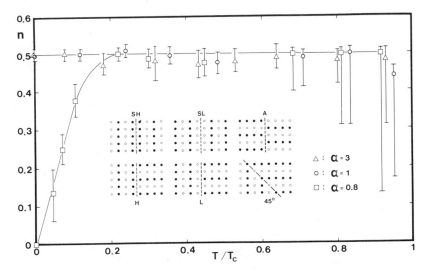

Fig. 12   Growth exponent, n, as a function of reduced quench
temperature, T/T_c, where T_c is the critical temperature, for
different values of the parameter α in the four-fold degenerate
(p = 4) Ising antiferromagnet, Eq. (16), with non-conserved
order parameter and density. The inset shows the six possible
different types of domain walls in this model: SH (superheavy),
SL (superlight), A (antiphase), H (heavy), L (light), and 45°
walls. [According to Ref. 63].

It should be noted that some parameter effects have been
discovered[51,56] in the present model at low temperatures in the case
where the exchange frequency between nnn sites is zero or very small,
i.e. $\delta \ll 1$. In that case, the effective growth exponent becomes
temperature-dependent with a crossover to freezing-in behavior at T = 0
and a levelling out of n being around 0.35 at higher temperatures.[51]
This finding was explained[51] as a manifestation of Lifshitz-Slyozov
kinetics since an observation of excess density in the domain walls
suggested that long-range diffusive processes were limiting the growth.
However, the finding[56] that a short-range nnn exchange removes this
effect led to a reinterpretation of the δ = 0 data as influenced by
finite-time and finite-size effects revealing T = 0 as a singular point.
Figure 15 demonstrates that the freezing-in at zero temperature is
restricted to very low frequencies of nnn jumps and for less than 5% nnn
exchange the Lifshitz-Allen-Cahn growth law is recovered, even at zero
temperature.

Before turning to a comparison with experimental results it should
be pointed out that the above model results are not restricted to p = 4
Ising models. A computer-simulation study[57,76] of a 2-D square-lattice
XY antiferromagnet with cubic anisotropy with non-conserved order
parameter and p = 4 also leads to dynamical scaling and n = 1/2. The
continuous nature of the single-site variables permits formation of
'soft' domain boundaries which at very low temperatures seem to lead to a
new universality class[57,65] with n = 1/4, cf. Fig. 16. For finite
temperatures there is, however, a distinct crossover to the
Lifshitz-Allen-Cahn growth law with n = 1/2, cf. Fig. 16. All these data
corroborate to the concept of universality in domain-growth kinetics.

19

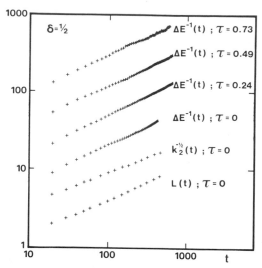

Fig. 13   Log-log plot vs time of length scales (in arbitrary units) for
quenches of the four-fold degenerate (p = 4) Ising
antiferromagnet, Eq. (16), with conserved density and
non-conserved order parameter. The data refer to quenches from
infinite temperature to different reduced temperatures,
$\tau = T/T_c$, below the critical point. [According to Ref. 56].

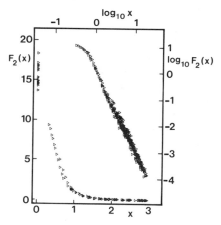

Fig. 14   Zero-temperature dynamical scaling function, $F_2(x) = S(x)$ in
Eq. (6), constructed by using the second moment of the
structure factor to define a length-scale measure. Only data
for times t > 60 MSC/S are shown. The data are shown in linear
as well as log-log plots. [According to Ref. 56].

The computer-simulation results presented above for the p = 4 Ising
antiferromagnet with non-conserved order parameter and conserved density
may be important for interpreting experimental data, specifically for the
ordering dynamics of O/W(110) at half-monolayer coverage, which has the

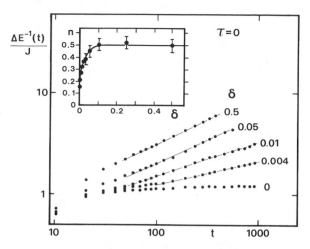

Fig. 15   Inverse excess energy $\Delta E^{-1}(t)$ vs time for zero-temperature
quenches for different values of the relative jump frequency,
δ, for the p = 4 Ising antiferromagnet with non-conserved order
parameter and conserved density. The inset shows the effective
kinetic growth exponent n as a function of δ. The time is in
units of MCS/S. [According to Ref. 56].

same symmetry as the present model and the same conservation laws. A LEED
study[16] found dynamical scaling of the structure factor and an
anomalously low exponent value, n $\simeq$ 0.28, which in the light of the model
calculations (cf. Figs. 12 and 15) can be interpreted as a non-asymptotic
value caused by a low-temperature condition or a low nnn exchange
frequency.[56] Such an interpretation is supported by the fact that in the
experiments[16] the overlayer is prepared in a glassy frozen-in state at
low temperatures and then quenched up in temperature to a fairly low
reduced temperature, $\tau = T/T_c \sim 0.3$. Furthermore, for O/W(110) the
interaction constants are known[77] and from these we have $\alpha < 1$ and then
estimate δ to be rather small. In the experimental paper[16] it was ruled
out that the low exponent value observed is due to impurities since the
average domain size at which the growth slows down is dependent on
temperature. Instead it was suggested that a large ground state
degeneracy (p = 8) might be responsible for the slow growth. However, on
the basis of the results quoted above we argue that the exponent value is
not expected[38] to depend on the value of p. It should also be noted that
a recent computer-simulation study of a similar model of this problem has
suggested that the low exponent value may be accounted for effectively by
allowing for a high rate of deexcitation of a precursor state into the
chemisorbed state.[74]

    If we now turn to quenches of O/W(110) at sub-half-monolayer
coverages, a completely different situation arises. When the system is
quenched into the coexistence region of (1 x 1) gas and a (2 x 1)
oxygen-ordered phase, the following scenario will occur: First islands of
(2 x 1) ordering will form and grow and eventually a phase separation
process will take place between the two phases. Since these two phases
have different densities, the phase separation process is controlled by
long-range diffusion and the situation might therefore at late times,

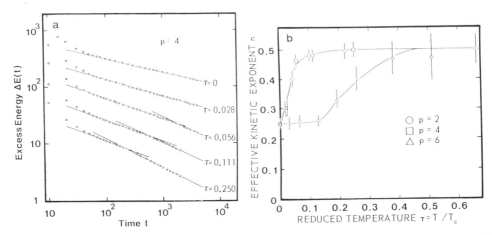

Fig. 16 (a): Excess energy, $\Delta E(t)$, vs time for a $p = 4$ XY antiferromagnet with non-conserved order parameter being quenched to different reduced temperatures, $\tau = T/T_c$, below the ordering transition temperature. The solid lines denote the effective growth laws. The time is in units of MCS/S. (b): Effective growth exponents for soft-wall models with different ground-state degeneracy. The $p = 4$ model refers to the data in (a). The $p = 6$ model is the model of herringbone order in N_2/Graphite, see below [According to Ref. 57].

where growth is limited by diffusion, be accounted for by the Lifshitz-Slyozov theory. It should be pointed out, however, that the applicability of this theory for $p > 2$ is unclear. The pioneering computer-simulation study[49] within kinetics of ordering on surfaces was in fact devoted to this situation at coverage $\rho = 1/4$ by considering the ordering process in a lattice gas model with a range of interaction extending to the third-nearest neighbors on a square lattice using Kawasaki dynamics to conserve the density. Extremely slow growth kinetics was encountered as described by an effective growth exponent value $n \simeq 0.15$. Experimentally, the sub-half-monolayer coverage situation of O/W(110) has also been studied by LEED.[16,19] Also in this case the structure factor is found to exhibit dynamical scaling and the kinetic exponent derived from the Bragg peak intensity, cf. Eq. (3), is found to be $n \simeq 0.28$. Super-half-monolayer coverages have been studied as well[19], $\rho = 0.65$, at which the equilibrium is a coexistence between a (2 x 1) and a (2 x 1) ordered phase. Also these two phases have different density. The experimental value of the kinetic exponent is found to be[19] $n \simeq 0.2$. Both these exponent values are substantially below the Lifshitz-Slyozov value $n = 1/3$. Since it is not clear whether or not the Lifshitz-Slyozov theory applies for $p > 2$, it is at the moment difficult to interpret these experimental observations theoretically.

3. **p = 2,3: H/Fe(110)**

Viñals and Gunton[20] have performed a Monte Carlo study of the ordering kinetics of the (2 x 1) and (3 x 1) ordered phases of H chemisorbed on Fe(110). These two phases, which have $p = 2$ and $p = 3$

respectively, cf. Fig. 17(a), occur in the phase diagram of the adsorbate system for coverages 1/2 and 2/3, respectively. The results of the simulations, which are based on a density-conserving Kawasaki dynamics and non-conserved order parameter, show that the growth in the (2 x 1) phase follows the Lifshitz-Allen-Cahn growth law with exponent n = 1/2. Furthermore, the structure factor obeys dynamical scaling. For quenches into the (3 x 1)-phase a much slower growth mode is observed[20] and it was not found possible to characterize this model by a simple algebraic growth law within the time regime investigated although the data are consistent with n ≈ 0.2 - 0.3. Moreover, evidence of anisotropic growth was found. It is likely that these results may be influenced by crossover effects due to the particular choice of the Kawasaki exchange mechanism,[20] in line with our findings above for the p = 4 Ising model.

So far no experimental study has been reported for the ordering kinetics of hydrogen chemisorbed on Fe(110).

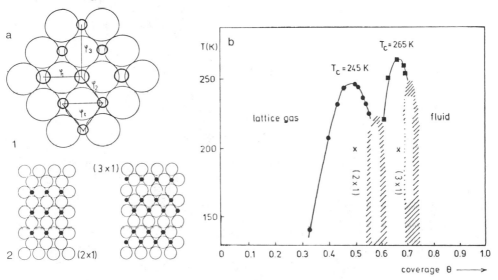

Fig. 17   (a): Interactions of the H/Fe(110) model and the adsorption sites of Fe(110) shown together with the two types of stable hydrogen ordering at coverages 1/2 (2 x 1) and 2/3 (3 x 1). (b): Phase diagram of H/Fe(110). [According to Ref. 20].

4.   p = 2,3: O/Pd(110)

The ordering kinetics of the (3 x 1) phases, cf. Fig. 18a, in the phase diagram of oxygen chemisorbed on Pd(110), Fig. 18b, has been analyzed by Ala-Nissila and Gunton[21] using an axial next-nearest-neighbor Ising model (ANNNI model). This is a particularly interesting system since it displays a wetting transition in analogy with the p = 4 (2 x 2) phases of the same model.[52,55,62] For both conserved and non-conserved density (and non-conserved order parameter) the ordering kinetics was found[21] to be well described by the Lifshitz-Allen-Cahn growth law, n = 1/2, everywhere within the (3 x 1) phase. The growth behavior was found[21] to be rather anisotropic with a generalized form of anisotropic scaling to hold. No experimental data is available for the ordering kinetics of O/Pd(110).

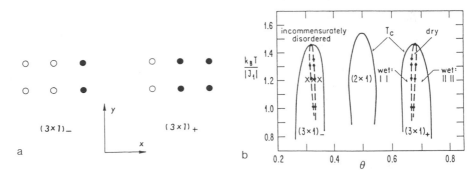

Fig. 18   (a): (3x1)-ordered structures of O/Pd(110) corresponding to coverages 1/3 and 2/3.   (b): Schematic phase diagram of O/Pd(110).   [According to Ref. 21].

5.   p = 6: N_2/Graphite and CO/Graphite

N_2 or CO physisorbed on Graphite undergoes an orientational ordering transition at low temperatures. Below the transition temperature the registered ($\sqrt{3}$ x $\sqrt{3}$)30° commensurate phase orders orientationally into a (2 x 1) herringbone structure as shown schematically in Fig. 19a. This phase has p = 6 degenerate ordered domains. The effective Hamiltonian governing the herringbone ordering can be shown to be that of classical planar quadrupoles on a triangular lattice.[78] The domain-growth kinetics of the herringbone ordering has been studied via an anisotropic planar rotor model[78]

$$H = K \sum_{i>j} X_i X_j \cos (2\psi_i + 2\psi_j - 4\theta_{ij}), \ K > 0 \tag{28}$$

using Monte Carlo simulation.[23,79] In Eq. (28), ψ_i are rotor variables, $0 \le \psi_i \le \pi$, arrayed on a triangular lattice, and θ_{ij} are the directional angles between the centers of a pair of nearest-neighbor rotors. The single-site occupation variables are denoted X_i, X_i = 0,1. For the herringbone structure, X_i = 1 for all i.

The growth kinetics of this model for non-conserved order parameter has been shown to be described by an algebraic growth law[23,79] and the structure factor is found to exhibit dynamical scaling.[79] Despite some controversy[23,79] regarding the value of the growth exponent at zero temperature it seems now clear that at finite temperatures[25,57] the growth exponent is n = 1/2 as expected for Lifshitz-Allen-Cahn kinetics, cf. also Fig. 16b.

Although the herringbone ordering transition has been studied extensively[80] for N_2/Graphite and CO/Graphite systems, no quantitative data are currently available for the ordering kinetics. Finally, it could be mentioned that there are a number of other adsorbate systems with positionally ordered overlayer structures of the same symmetry as the herringbone structure, e.g. the (2 x 1) phase of O/Ir(111).

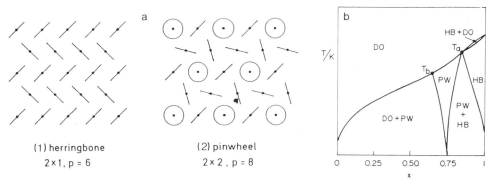

(1) herringbone
2×1, p = 6

(2) pinwheel
2×2, p = 8

Fig. 19   (a): Herringbone (2x1)-structure of N_2/Graphite or CO/Graphite
           and pinwheel (2x2)-structure of (3:1) (N_2+Ar)/Graphite.   (b):
           Schematic phase diagram (N_2+Ar)/Graphite, PW = pinwheel, HB =
           herringbone, DO = disordered phase.   [According to Refs. 25 and
           78].

6.   **p = 8: $(N_2 + Ar)$/Graphite and (CO + Ar)/Graphite**

     An interesting situation arises when an additional component is
introduced in the anisotropic planar rotor model, Eq. (28), in a way
which leads to a coupling to the herringbone order.   The simplest
situation is that of introducing a rotationally inert impurity or a site
vacancy ($<X> < 1$ in Eq. (28)).   In the experimental systems, N_2/Graphite
or CO/Graphite, this may be realized by mixing in an appropriate rare
gas, such as Ar or Kr.   Under appropriate circumstances,[81,82] this leads
to a stabilization of the so-called pinwheel structure, cf. Fig. 19a, in
which the diatomic molecules are orientationally ordered and there is an
additional compositional ordering of the two species into a (2 x 2)
superstructure.   Such   a   structure   has   indeed   been   observed
experimentally[80] in (CO + Ar)/Graphite which incidentally is the only
two-dimensional system in which chemical (compositional) alloy-ordering
has ever been observed.   A schematic phase diagram of the annealed
site-diluted anisotropic planar rotor model is shown in Fig. 19b.   This
diagram is not the correct phase diagram for N_2 + Ar or CO + Ar mixtures
since it does not take the interactions between the two species properly
into account.[82]   Nevertheless, it reproduces a (2 x 2) pinwheel phase
with the correct symmetry properties.   The degeneracy of this phase is p
= 8.   Other adsorbate models with the same symmetry properties include
positional ordering in the (2 x 2)-phases of O/Ni(111) and CF_4/Graphite.
No time-resolved experimental data exists for the ordering kinetics in
any of these systems.

     The ordering kinetics of the pinwheel structure has been studied by
Monte Carlo computer simulation techniques[25] which neither conserve the
pinwheel order parameter nor the orientational order parameter but
conserve the total composition.   The results of this study at a
composition $<X> = 3/4$, cf. Fig. 19b, is that the growth law is
effectively algebraic but the exponent value is temperature-dependent
with some tendency to level out at $n \simeq 0.3$ at temperatures around $0.6T_c$.
The growth behavior seems to be complicated by the fact that the
herringbone and the pinwheel phases are energetically degenerate locally,

and herringbone domains therefore appear as transients at early times and in the walls between the pinwheel domains at late times. Since the two phases have different density, long-range diffusive processes may be rate-limiting and the Lifshitz-Slyozov theory might be applicable, as suggested by the numerical results for the effective kinetic exponent. The data are, however, not of sufficient accuracy to settle the matter. It is interesting to note, however, that a study[25] of a version of the model in which the vacancies are immobile being fixed to the (2 x 2) superlattice positions (implying p = 2) leads at finite temperatures to the Lifshitz-Allen-Cahn growth law, n = 1/2. Hence, diffusion of the vacancies seems to play an important role. Further testimony to this conjecture comes from preliminary calculations[83] of the ordering kinetics of (2 x 2) phases of the p = 4 Ising model, Eq. (16), with 50% dilution. These calculations give strong evidence in favor of a Lifshitz-Slyozov mechanism being operative (see also the statements in Ref. 51 p. 582 about a similar situation of (2 x 2)-phases on the triangular lattice).

7.   Impurity Effects : N-doped O/W(112)

It is well known in materials science that randomness, such as impurities, vacancies and other types of imperfections, have dramatic effects on the mobility of grain boundaries and interfaces. The important concepts in such problems are the grain-boundary velocity relative to the diffusivity of the impurity (vacancy), as well as the impurity's interaction with the grain boundary: If the impurity dispersion is static, the growth will eventually become pinned; if the impurity diffusivity is very fast compared to the grain-boundary velocity, the boundary mobility will not be affected by the impurities. Consequently, randomness is expected also to have dramatic effects on the ordering kinetics of surface systems[84] which often are contaminated and suffer from various heterogeneities.

A number of computer-simulation studies have been carried out on the ordering dynamics in two-dimensional models with quenched randomness, e.g. the random-field Ising model and Ising and Potts models with quenched site dilution.[85] All these studies suggest a dramatic slowing down (possible logarithmic) of the growth due to the randomness. Very few computer-simulation studies have been reported on the effect of annealed site dilution.

Recently[35], the ordering kinetics in the p = 2 and p = 4 Ising antiferromagnets with (2 x 1) ordering has been studied in the presence of an annealed site dilution. For the dilute systems, a distinct crossover is found at late times to an effectively logarithmic growth behavior similar to that seen in systems with quenched randomness. These results apply to both types of ordering suggesting that the effects on ordering kinetics of vacancy and impurity diffusion do not depend on the ordering degeneracy.

In a high-resolution LEED experiment Zuo et al.[17] recently studied the ordering kinetics of an oxygen monolayer chemisorbed on W(112) doped with various amounts of nitrogen impurities. The symmetry of the oxygen ordering is that of the p = 2 model mentioned above. Zuo et al.[17] found a distinct crossover from algebraic growth in the pure oxygen monolayer (described by the Lifshitz-Allen-Cahn law) to a progressively slower growth mode as the impurity content is increased. This slow growth mode was, within the scatter of the data, found to be consistent with a logarithmic growth law, R(t) $\sim$ lnt. Zuo et al.[17] interpreted their finding within the theoretical framework of the random-field Ising model, i.e. in terms of quenched randomness. However, as we have recently pointed out,[35,84] surface diffusion data for O and N on tungsten surfaces in the pertinent temperature range suggest that the diffusion of the nitrogen impurities is similar to that of oxygen. Therefore, our model predictions should apply when we interpret the mobile nitrogen impurities as a site dilution which does not participate in the oxygen ordering and does not couple directly to the order parameter (which is required by the random-field Ising model). Therefore, we reinterpret the LEED results as a manifestation of slow, possible logarithmic growth, due to annealed randomness.

The ordering kinetics of oxygen on W(110), which has the symmetry of the p = 4 model, has not been studied systematically in the presence of impurities. It has been ruled out experimentally that the observed late-time slowing down of the growth in supposedly pure oxygen monolayers is due to impurities.[19]

CONCLUSIONS

In this brief overview we have discussed the current theoretical knowledge about the kinetics of ordering and growth in two-dimensional systems. Most of this knowledge stems from computer-simulation studies of kinetic lattice models and from simple phenomenological theories. We have pointed out that it is fruitful to consider the far-from-equilibrium ordering processes in degenerate systems as processes which involve the formation of a complex, random network of interfaces between the different ordered domains. It is the physics of these interfaces, in particular the diffusion processes along and across these interfaces, which control the mobility and the overall kinetics. We have demonstrated that such complex pattern-formation phenomena, cf. Figs. 1 and 2, can indeed by described in space and time and that the concept of dynamical scaling symmetry becomes a seminal one for this description. Furthermore, these patterns and their kinetics can be derived from simple principles and may be described by means of very simple physical models. Finally, some universal principle seems to be operative. This universality refers to the asymptotic late-time growth behavior as being described in terms of algebraic growth laws with exponent values which only depend on whether or not the order parameter is a conserved quantity. Other variables, such as temperature, ordering degeneracy, details of the interaction potential and the dynamical model, as well as additional conservation laws, appear to be irrelevant. This is a very remarkable universality. However, we have also pointed out that crossover effects may well veil the asymptotic behavior. Moreover, imperfections and various types of randomness, such as impurities and vacancies, may change the algebraic growth law into a much slower growth mode (possibly logarithmic). The theory of the whole field of ordering kinetics is still in a very unsatisfactory state and what seems to be a natural approach, a renormalization-group-type theory, has so far had only limited success.

A substantial amount of experimental evidence exists which supports the universality in ordering kinetics. Most of this experimental work refers to three-dimensional systems[3] and only a small number of studies have been reported for two-dimensional systems and surfaces.[8] The results of some of these studies have been discussed in the present overview in relation to theoretical predictions and pertinent model calculations. In view of the strong current interest in non-equilibrium phenomena and dynamical systems this situation is probably going to change in the near future and more experimental studies on kinetics of ordering and growth at surfaces will be conducted as the time-resolution of dynamic surface experimentation improves.

Acknowledgements

The work described in this paper was supported by the Danish Natural Science Research Council under Grants Nos. 5.21.99.72 and 11-7785, and the Danish Technical Research Council under Grant No. 16-4296.K. Many of the results described herein were obtained in collaboration with a number of colleagues and I wish to acknowledge the stimulating interaction with Hans C. Fogedby, Peter Jivan Shah, Eigil Praestgaard, Per-Anker Lindgaard, Claus Jeppesen, og Henrik Flyvbjerg.

REFERENCES

1. For two excellent reviews in the field of non-equilibrium phase transitions and ordering phenomena, see e.g., J. D. Gunton, M. San Miguel, and P. S. Sahni, in Phase Transitions and Critical Phenomena, C. Domb and J. L. Lebowitz, eds. Vol. 8 (Academic Press, New York, 1983) p. 267 and H. Furukawa, Adv. Phys. 34, 703 (1985).
2. O. G. Mouritsen, Ann. N.Y. Acad. Sci. (in press).
3. Dynamics of Ordering Processes in Condensed Matter, S. Komura and H. Furukawa, eds. (Plenum Press, New York, 1988).
4. D. Kessler, J. Koplik, and H. Levine, Adv. Phys. 36, 255 (1988).
5. J. S. Langer, Science 243, 1150 (1989).
6. Random Fluctuations and Pattern Growth, H. E. Stanley and N. Ostrowsky eds. (Kluwer Academic Publishers, 1989).
7. M. Schick, Surf. Sci. 125, 94 (1983).
8. K. Heinz, in Kinetics of Interface Reactions, M. Grunze and H. J. Kreuzer eds. (Springer Verlag, 1987), p. 202.
9. S. Komura, K. Osamura, H. Fujii, and T. Takeda, Phys. Rev. B31, 1278 (1985).
10. S. Katano, M. Izumi, R. M. Nicklow, and H. R. Child, Phys. Rev. B38, 2659 (1988).
11. S. E. Nagler, R. F. Shannon, C. R. Harkless, M. A. Singh, and R. M. Nicklow, Phys. Rev. Lett. 61, 718 (1988).
12. R. F. Shannon, C. R. Harkless, and S. E. Nagler, Phys. Rev. B38, 9327 (1988).
13. M. G. Lagally, G.-C. Wang, and T.-M. Lu, CRC Crit. Rev. Solid St. and Mat. Sci. 7, 233 (1978).
14. G.-C. Wang and T.-M. Lu, Phys. Rev. Lett. 50, 2014 (1983).
15. P. K. Wu, J. H. Perepezko, J. T. McKinney, and M. G. Lagally, Phys. Rev. Lett. 51, 1577 (1983).
16. M. C. Tringides, P. K. Wu, and M. G. Lagally, Phys. Rev. Lett. 59, 315 (1987).
17. J.-K. Zuo, G.-C. Wang, and T.-M. Lu, Phys. Rev. Lett. 60, 1053 (1988).
18. J.-K. Zuo, G.-C. Wang, and T.-M- Lu, Phys. Rev. B39, 9432 (1989).
19. P. K. Wu, M. C. Tringides, and M. G. Lagally, Phys. Rev. B39, 7595 (1989).
20. J. Viñals and J. D. Gunton, Surf. Sci. 157, 473 (1985).
21. T. Als-Nissila and J. D. Gunton, Phys. Rev. B 38, 11418 (1988).
22. K. Binder, Ber. Bunsenges. Phys. Chem. 90, 257 (1986).
23. O. G. Mouritsen, Phys. Rev. B28, 3150 (1983).
24. J. D. Gunton and K. Kaski, Surf. Sci. 144, 290 (1984).
25. O. G. Mouritsen, Phys. Rev. B32, 1632 (1985).
26. K. Binder and D. Stauffer, Phys. Rev. Lett. 33, 1006 (1974).
27. P. C. Hohenberg and B. I. Halperin, Rev. Mod. Phys. 49, 435 (1977).
28. I. M. Lifshitz, Sov. Phys. JETP 15, 939 (1962) [J. Eksptl. Theoret. Phys. (USSR) 42, 1354 (1962)].
29. S. M. Allen and J. W. Cahn, Acta Metall. 27, 1085 (1979).

30. I. M. Lifshitz and V. V. Slyozov, J. Phys. Chem. Solids 19, 35 (1961).
31. D. A. Huse, Phys. Rev. B34, 7845 (1986).
32. W. W. Mullins and J. Vinals, Acta Metall. 37, 991 (1989).
33. O. G. Mouritsen, Computer Studies of Phase Transitions and Critical Phenomena (Springer-Verlag, Heidelberg, 1984).
34. A. N. Burkitt and D. W. Heermann, Europhys. Lett. (in press).
35. O. G. Mouritsen and P. J. Shah, Phys. Rev. B40, 11445 (1989) P. J. Shah and O. G. Mouritsen, Phys. Rev. B. April 1 (1990).
36. D. J. Srolovitz, M. P. Anderson, G. S. Grest, and P. S. Sahni, Scr. Metall. 17, 241 (1983).
37. J. Viñals and M. Grant, Phys. Rev. B36, 7036 (1987).
38. G. S. Grest, M. P. Anderson, and D. J. Srolovitz, Phys. Rev. B38, 4752 (1988).
39. G. S. Grest, D. J. Srolovitz, and M. P. Anderson, Phys. Rev. Lett. 52, 1321 (1984).
40. C. Jeppesen, H. Flyvbjerg, and O. G. Mouritsen, Phys. Rev. B40, 9070 (1989).
41. J. Viñals, M. Grant, M. San Miguel, J. D. Gunton, and E. T. Gawlinski, Phys. Rev. Lett. 54, 1264 (1985).
42. G. F. Mazenko, O. T. Valls, and F. C. Zhang, Phys. Rev. B31, 4453 (1985).
43. S. Kumar, J. Viñals, and J. D. Gunton, Phys. Rev. B34, 1908 (1986).
44. C. Roland and M. Grant, Phys. Rev. B39, 11971 (1989).
45. J. Viñals and J. D. Gunton, Phys. Rev. B33, 7795 (1986).
46. T. M. Rogers, K. R. Elder, and R. C. Desai, Phys. Rev. B37, 9638 (1988).
47. Y. Oono and S. Puri, Phys. Rev. B38, 434 (1988); S. Puri and Y. Oono, Phys. Rev. B38, 1542 (1988).
48. P. S. Sahni, G. Dee, J. D. Gunton, M. Phani, J. L. Lebowitz, and M. Kalos, Phys. Rev. B24, 410 (1981).
49. P. S. Sahni and J. D. Gunton. Phys. Rev. Lett. 47, 1754 (1981).
50. K. Kaski, M. C. Yalabik, J. D. Gunton, and P. S. Sahni, Phys. Rev. B28, 5263 (1983).
51. A. Sadiq and K. Binder, J. Stat. Phys. 35, 517 (1984).
52. K. Kaski, T. Als-Nissila, and J. D. Gunton, Phys. Rev. B31, 310 (1985).
53. A. Milchev, K. Binder, and D. W. Heermann, Z. Phys. B63, 521 (1986).
54. O. G. Mouritsen, Phys. Rev. Lett. 56, 850 (1986).
55. T. Als-Nissila, J. D. Gunton, and K. Kaski, Phys. Rev. B37, 179 (1988).
56. H. C. Fogedby and O. G. Mouritsen, Phys. Rev. B37, 5962 (1988).
57. O. G. Mouritsen and E. Praestgaard, Phys. Rev. B38, 2703 (1988).
58. S. A. Safran, P. S. Sahni, and G. S. Grest, Phys. Rev. B28, 2693 (1983).
59. P. S. Sahni, D. J. Srolovitz, G. S. Grest, M. P. Anderson, and S. A. Safran, Phys. Rev. B28, 2705 (1983).
60. G. S. Grest and D. J. Srolovitz, Phys. Rev. B30, 5150 (1984).
61. E. T. Gawlinski, M. Grant, J. D. Gunton, and K. Kaski, Phys. Rev. B31, 281 (1985).
62. T. Als-Nissila, J. D. Gunton, and K. Kaski, Phys. Rev. B33, 7583 (1986).
63. A. Høst-Madsen, P. J. Shah, T. V. Hansen, and O. G. Mouritsen, Phys. Rev. B36, 2333 (1987).
64. G. N. Hassold and D. J. Srolovitz, Phys. Rev. B37, 3467 (1988).
65. T. Castan and P.-A. Lindgaard, Phys. Rev. B40, 5069 (1989).
66. J. G. Amar, F. E. Sullivan, and R. D. Mountain, Phys. Rev. B37, 196 (1988).
67. A. Chakrabarti, J. B. Collins, and J. D. Gunton, Phys. Rev. B38, 6894 (1988).
68. T. M. Rogers and R. C. Desai, Phys. Rev. B39, 11956 (1989).

69. E. T. Gawlinski, J. D. Gunton, and J. Viñals (preprint).
70. G. S. Grest and P. S. Sahni, Phys. Rev. B30, 226 (1984).
71. M. K. Phani, J. L. Lebowitz, M. H. Kalos, and O. Penrose, Phys. Rev. Lett. 45, 366 (1980).
72. R. Toral, A. Chakrabarti, and J. D. Gunton, Phys. Rev. Lett. 60, 2311 (1988).
73. O. G. Mouritsen, H. C. Fogedby, and E. Praestgaard, in Ref. 3, p. 133.
74. H. C. Kang and W. H. Weinberg, Phys. Rev. B38, 11543 (1988).
75. G.-C. Wang and T.-M. Lu, Phys. Rev. B31, 5918 (1985).
76. P.-A. Lindgaard, H. E. Viertiö, and O. G. Mouritsen, Phys. Rev. B38, 6798 (1988).
77. W. Y. Ching, D. L. Huber, M. G. Lagally, and G.-C. Wang, Surf. Sci. 77, 550 (1978); E. D. Williams, S. L. Cunningham, and W. H. Weinberg, J. Chem. Phys. 68, 4688 (1978).
78. O. G. Mouritsen and A. J. Berlinsky, Phys. Rev. Lett. 48, 181 (1982).
79. K. Kaski, B. Kumar, J. D. Gunton, and P. A. Rikvold, Phys. Rev. B29, 4420 (1984).
80. H. You, S. C. Fain, S. Satija, and L. Passell, Phys. Rev. Lett. 56, 244 (1986); H. You and S. C. Fain, Phys. Rev. B34, 2840 (1986).
81. A. B. Harris, O. G. Mouritsen, and A. J. Berlinsky, Can. J. Phys. 62, 915 (1984).
82. E. J. Nicol, C. Kallin, and A. J. Berlinsky, Phys. Rev. B38, 556 (1988).
83. O. G. Mouritsen and P. J. Shah (unpublished).
84. O. G. Mouritsen and P. J. Shah in these proceedings.
85. See e.g. the references in Ref. 84.

GROWTH KINETICS OF WETTING LAYERS AT SURFACES

K. Binder

Institut für Physik, Universität Mainz
Staudinger Weg 7
D-6500 Mainz, Federal Republic of Germany

ABSTRACT

Monte Carlo simulation of lattice gas models for the wetting transitions in systems with short range forces are described. A nearest-neighbor simple cubic lattice with nonconserved "Glauber dynamics" is used, applying a slab geometry (LxL cross section). It is shown that the growth proceeds in two stages: for short times t, the thickness of the wetting layer at an initially nonwet wall increases proportional to the logarithm of the time; for $t \gg L^2(\ln L)^2$ the thickness increases proportional to $t^{1/2}/L$. Generalizations to other systems are briefly discussed. Also two-dimensional growth of a wetting film at surface steps is considered, considering "terraces" of an LxM geometry with M≫L as substrate for adsorption. The steps may act like a "boundary field" at the adsorbed layer. If this field couples quadratically to the order parameter only, rather than wetting one observes the film to spontaneously break up in a one-dimensional sequence of ordered domains.

INTRODUCTION

Understanding the dynamical laws which control the growth kinetics of layers on a substrate is of interest for wetting phenomena,[1-5] thin film growth, and epitaxy.[6] In the present work, Monte Carlo simulations of the dynamics of wetting phenomena in simple Ising models[7-9] will be discussed: these models are far too simplified to be applicable to any real system, but they are nevertheless useful for testing some aspects of more general theories.[10-14]

One aspect which is present in any Monte Carlo simulation[15] is the use of a <u>finite system geometry</u>. However, such finite size effects are not only a limitation of computer simulation, but often are indeed present in experimental work on adsorption phenomena. E.g., the smearing of the delta-function singularity associated with the latent heat of the first-order solid-gas transition of oxygen monolayers adsorbed on grafoil can be accounted for quantitatively[16] by the appropriate scaling theory developed in the context of Monte Carlo simulations.[17] In the same spirit, the finite size effects discussed here may also be of experimental relevance.

Kinetics of Ordering and Growth at Surfaces
Edited by M. G. Lagally
Plenum Press, New York, 1990

In this paper, we consider two different physical situations: (i) a solid substrate surface of linear dimensions LxL is in equilibrium with a gas of adatoms at chemical potential difference $\delta\mu \sim k_BT \ln(p*/p)$, p being the gas pressure and p* its value at liquid-vapor coexistence. For small (but positive) $\delta\mu$ thermal equilibrium would require a rather thick wetting layer, whose thickness Z_0 diverges to infinity when $\delta\mu \to 0$. This wetting layer results due to attractive interactions between the adatoms and the surface. Suppose now that initially (for times t < 0) $\delta\mu$ is very large, so that the thickness of the liquid layer is very small, and that at t = 0 we put $\delta\mu \to 0$ instantaneously. Then the thickness of the wetting layer will steadily grow with time, $Z_0(t \to \infty) \to \infty$. We are interested in understanding the growth law $Z_0(t)$, as well as the time-dependence of the density profile $\rho(Z,t)$, Z being the distance from the surface. Figure 1 shows Monte Carlo results for a simple cubic lattice gas model with nearest neighbor attractive interactions, that will be analyzed further in the third section. (ii) While the above situation corresponds to a growth of a three-dimensional film on a two-dimensional substrate, we also consider growth of two-dimensional monolayers starting out as a one-dimensional row. This case is physically relevant for adsorption on stepped surfaces,[8,9] see Fig. 2. For suitable parameters ϵ_1, ϵ_L the adsorption starts preferentially at one boundary of the terrace, and in fact for the nearest-neighbor model a second order wetting transition can be located exactly.[18] For this problem one can again ask how the width x(t) of an adsorbed wetting layer along a step grows with time if one starts with an initially empty terrace. Of course, one can also consider the inverse problem: suppose initially the terrace (or part of it) is covered with an adsorbed layer but thermodynamic parameters are chosen such that the equilibrium

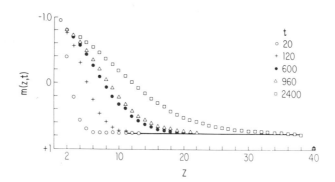

Fig. 1   Time evolution of the ensemble averaged density profile $\rho(Z,t)$ or, equivalently, magnetization profile $m(Z,t)$ of the corresponding Ising magnet, $m(Z,t) = 1 - 2\rho(Z,t)$, for 8x8x40 lattices at a temperature $k_BT/J = 4$, "exchange interaction" J_s in the surface plane being $J_s = 0.5J$. Boundary "magnetic fields" are chosen $H_1 = -5J$, $H_{40} = 17.5J$ (i.e., the surface layer Z = 1 is strongly attractive for adatoms, the other boundary layer at Z = 40 is strongly repulsive). In the direction parallel to the surface periodic boundary conditions are used. At time t = 0 $\delta\mu$ is changed from $\delta\mu \to -\infty$ (i.e., "bulk magnetic field" $H \to -\infty$) to $\delta\mu = 0$ (H = 0). Time t is measured in units of Monte Carlo steps per site. Each Monte Carlo step physically means that one considers a lattice site for a random condensation or evaporation of an adatom. The solid line shows the bulk magnetization. From Mon et al.[7]

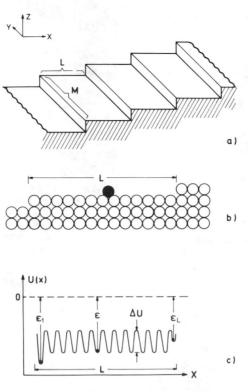

Fig. 2   a) Schematic view of a regularly stepped surface, where steps a
          distance L apart in the x-direction run parallel to each other a
          distance M in the y-direction, to form a "staircase" of LxM
          terraces, on which adsorption can take place. b) Cross section
          through one terrace of width L.   Open circles represent
          substrate atoms, full circle represents an adsorbate atom. c)
          Corrugation potential corresponding to the geometry of case b).
          We assume that the substrate creates a lattice of preferred
          sites, at which adatoms can be bound to the surface with an
          energy ϵ.   In the rows adjacent to the terrace boundaries,
          however, we assume in general different binding energies ϵ_1, ϵ_L,
          which correspond to the Ising magnet terminology to the
          "boundary magnetic fields" $H_1 = J - (\epsilon_1-\epsilon)/2$, $H_L = J - (\epsilon_L-\epsilon)/2$.
          The energy barrier ΔU separates neighboring preferred sites.
          From Albano et al.[9]

coverage is very low.   Then one can study how $x(t \rightarrow \infty)$ decreases to a
very small value (see Fig. 3 for an example where the system evolves to a
final state with a small boundary excess coverage $\theta_{ex} = \Sigma_\ell (\theta_b - \theta_1)$,
while the bulk coverage θ_b there is less than one percent only).

      In the next section we summarize the main theoretical predictions of
Lipowsky[10] and others, while the third section describes the computer
simulation results on the growth of three-dimensional wetting layers.[7]

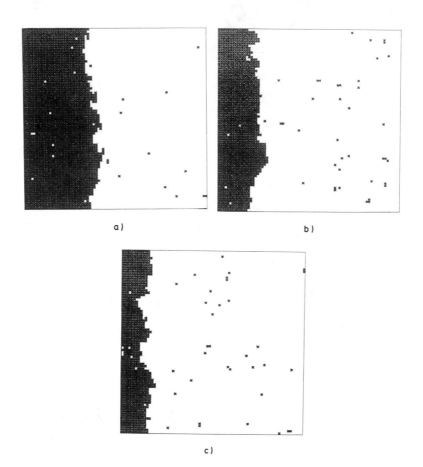

a)

b)

c)

Fig. 3 Snapshot pictures of a terrace of size L = M = 80 with
H_1 = -0.5J, H_L = 0.5J and zero bulk field, at a temperature
T = 0.66T_c (T_c being the critical temperature of the
two-dimensional Ising model), where in the initial stage the
coverage $\theta(x,t)$ = 1 in all rows from x = 1 to x = 40, while
$\theta(x,t)$ = 0 in all rows from x = 41 to x = 80. Adatoms are shown
as black stars (adsorption sites form a square lattice). Times
shown are t = 5000 MCS(a), 6000 MCS(b) and 7000 MCS(c). From
Albano et al.[8]

PHENOMENOLOGICAL THEORY

We consider a state off the wetting transition in the wet region.
Then the thickness $Z_o(t)$ of the wetting layer can be expressed in terms
of a dynamical scaling assumption[10] similar to dynamic scaling laws in
domain growth,[19]

$$Z_0(t) = (\delta\mu)^{\beta_S} \tilde{Z}(\xi_{/\!/}^{-z} t), \tag{1}$$

where $\xi_{/\!/} \sim \delta\mu^{-\nu_{/\!/}}$ is the correlation length for interfacial fluctuations, z a dynamic exponent, and β_S the static exponent of the thickness (or surface excess coverage, respectively). For an adsorbed layer exchanging adatoms with the gas, the order parameter (i.e., the local density) is not a conserved quantity, and in this case[10,14,20] z = 2. With respect to the static exponents β_S, $\nu_{/\!/}$, different cases need to be distinguished: (i) the surface forces are of short range, dimensionality d = 3. Then β_S = 0 (i.e. a logarithmic variation results), $\nu_{/\!/} = 1/2$. (ii). The surface forces are of short range, dimensionality d = 2 (e.g., wetting at steps). Then β_S = -1/3, $\nu_{/\!/}$ = 2/3. (ii). The surface forces are nonretarded or retarded van der Waals forces, d = 3: this means that even if the interface is a distance Z apart it feels a potential energy $U(Z) \sim Z^{-2}$ or Z^{-3}) respectively.[21] Then β_S = -1/3 or -1/4, respectively, and $\nu_{/\!/}$ = 2/3 or 5/8, respectively.[10] Now the asymptotic growth law for $\delta\mu \to 0$ is simply found from Eq. (1) assuming a power law for the scaling function $Z(X)$ for small arguments X such that $\delta\mu$ cancels out, i.e., $Z(X) \sim X^{-\beta_S/\nu_{/\!/}z}$ and hence[10]

$$Z_0(t) \sim t^{-\beta_S/(\nu_{/\!/}z)} . \tag{2}$$

While for the short range case we thus obtain $Z_0(t) \sim \ln t$ (d = 3) and $Z_0(t) \sim t^{1/4}$ (d = 2), the growth law for non-retarded or retarded van der Waals forces in d = 3 is $Z_0(t) \sim t^{1/4}$ or $Z_0(t) \sim t^{1/5}$, respectively. Even slower growth results when the density is conserved and the growth thus is diffusion-limited: then z = 4 in Eqs. (1,2) implies that the growth exponent - $\beta_S/(\nu_{/\!/}z)$ takes one half of its value in the corresponding nonconversed case.

All these results refer to linear dimensions L of the substrate of infinite extent (or M, respectively in Fig. 2). We now consider the effect of finite size.[7,22] Then at late times a quasi-one-dimensional motion takes over, where the interface behaves like a rigid object of mass L^2 (d = 3) or M (in Fig. 2), and performs random hops backward and forward in the direction normal to the boundary, with a diffusion constant D inversely proportional to its "mass", e.g., one predicts[7,22,23]

$$Z_0(t) \sim (Dt)^{1/2} \sim L^{-1}t^{1/2} \ (d = 3) \ \text{or} \ M^{-1/2}t^{1/2} \ (d = 2), \ t \to \infty . \tag{3}$$

These growth laws take over at times later than a crossover time $t^*(L)$, which can be estimated matching both expressions Eqs. (2), (3), which yields,[23] with $Z_0(t^*) = Z_0^*$

$$t^* \sim L^{2/[1+2\beta_S/(\nu_{/\!/}z)]} , \ Z_0^* \sim L^{-2\beta_S/[\nu_{/\!/}z+2\beta_S]} \tag{4}$$

for d = 3. For the short range case where β_S = 0 there is a logarithmic correction $t^* \approx L^2(\ln L)^2$, $Z_0^* \approx \ln L$, while for non-retarded or retarded van der Waals forces the result would be $t \approx L^4$, $Z_0^* \approx L$, or $t^* \approx L^{10/3}$, $Z_0^* \approx L^{2/3}$. Thus for non-retarded van der Waals forces this growth law Eq. (3) only sets in when the thickness $Z_0(t)$ of the adsorbed film equals the lateral dimension, while in the other cases this occurs for rather thin films already. For the two-dimensional short range case, $t^* \sim M^2$, $Z_0^* \sim M^{1/2}$.

Finally we draw attention to the fact that the discussion so far refers to cases only where the initial nonwet state is unstable. If the wetting transition is of first order, mean field theory also predicts a region where this state is metastable. This occurs for a Ginzburg-Landau theory of wetting with short range forces, with a free energy[14,24]

$$\frac{\Delta F}{k_B T} = \int_0^\infty d\varsigma \ \left\{ \frac{1}{2} \left[\frac{\partial}{\partial \varsigma} \mu \right]^2 - \mu^2 + \frac{1}{2} \mu^4 \right\} - \frac{h_1}{\gamma} \mu_1 - \frac{1}{2} \frac{g}{\gamma} \mu^2 , \tag{5}$$

where $\mu(\varsigma)$ results from rescaling the magnetization m(z) and ς the rescaled distance z such that constants in the bulk part of the free energy are eliminated, while the surface correction is expanded quadratically in terms of $\mu_1 = \mu$ ($\varsigma = 0$), h_1/γ and g/γ being some coefficients. Figure 4 shows the resulting surface phase diagrams,[14] indicating two quenching experiments which start out in the nonwet state of the surface and bring the system to a state where the surface is wet. While the growth of the wetting layer starts immediately if the quench has crossed the "surface spinodal",[14,24] $h_{1s}^{(2)}$, for final states between the wetting transition line and this surface spinodal the decay of the metastable nonwet state starts by the formation of a droplet of approximately sphere-cap shape (Fig. 5). It has been found[14] that the formation free energy ΔF^* of such a droplet is proportional to $\Delta F \sim (\Delta h_1)^{-2}$, Δh_1 being the distance of the metastable state from the first-order wetting transition.

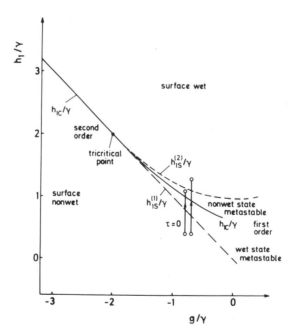

Fig. 4   Surface phase diagram of the Ginzburg-Landau model plotted in terms of the coefficients h_1/γ and g/γ. For $g/\gamma < -2$ critical wetting occurs, for $g/\gamma > -2$ first order wetting. In the latter region, mean field theory predicts metastable wet and non-wet regions limited by the two surface spinodal lines $h_{1s}^{(1)}$, $h_{1s}^{(2)}$, respectively. Also two quenching experiments at scaled time $\tau = 0$ are indicated. From Schmidt and Binder[14].

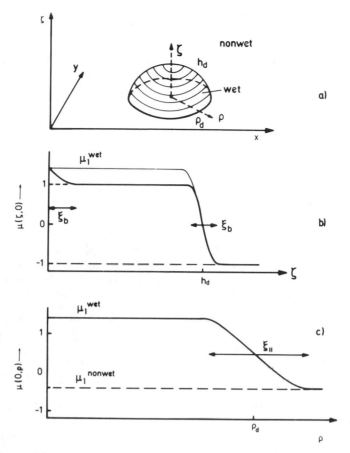

Fig. 5 Metastable non-wet states of the surface decay by formation of wet critical droplets which approximately have the shapes of sphere caps of height h_d and radius ρ_d attached to the surface (a). Microscopically the droplet is described by an order parameter profile $\mu(\zeta,\rho)$. The figure shows special symmetric cuts through this profile: (b) shows the order parameter variation with ζ at $\rho = 0$ and (c) shows the order parameter variation with ρ at the surface ($\zeta = 0$). These figures indicate the precise definitions of h_d and ρ_d. Note that the interfacial width ξ_b (correlation length in the bulk) is adsorbed in the units of Eq. (5) while in case (c) a nontrivial length $\xi_{//}$ appears, the parallel correlation length at the surface in the metastable non-wet phase. For states very close to the first order wetting transition $\rho_d \gg \xi_{//}$ and then $\mu(0,\rho_d) \approx \mu_1^{wet}$, the equilibrium value of the order parameter at the surface in its wet state. From Schmidt and Binder[14].

MONTE CARLO SIMULATION OF THE GROWTH OF WETTING LAYERS IN THE ISING MODEL[7]

Simulations were carried out for simple cubic LxLxD systems with D = 40 layers and two free surface LxL layers, and periodic boundary conditions in the two remaining directions. L was varied from L = 8 to L = 24, while for a study of the corresponding static wetting phase

diagrams much larger systems were used.[25,26] The model Hamiltonian is
($S_i = \pm 1$: Ising spins)

$$H = J\Sigma\ S_iS_j - J_s\Sigma\ S_iS_j - H\Sigma\ S_j - H_1\ \Sigma\ S_i - H_D\ \Sigma\ S_k ,$$

$$\underset{\substack{<i,j> \\ \text{bulk}}}{} \quad \underset{\substack{<i,j> \\ \text{surfaces}}}{} \quad \underset{\substack{j \\ \text{top} \\ \text{surface}}}{} \quad \underset{\substack{k \\ \text{bottom} \\ \text{surface}}}{} \tag{6}$$

where the bulk field $H = 0$, $J_s/J = 0.5$, $H_1/J = -5$, $H_D/J = 17.5$. The precise values of the boundary fields are not crucial but are chosen to ensure the formation of a simple interfacial profile which matches smoothly into the bulk magnetization as the bottom surface is approached (Fig. 1). These parameters also put the system in the wet phase far from the critical wetting transition (which occurs at[25] $H_{1c}/J \approx -1.35$). Note also that the bulk system with $J/k_BT = 0.25$ is far above the roughening transition temperature ($J/k_BT_R \approx 0.41$).[27] Thus, the interface in this study is rough and the dynamics is not expected to be dominated by the layering transitions.[26,28]

The initial configuration is always a uniformly ordered state, modeling a bare substrate. Standard single spin-flip updates with sites chosen at random are used.[15] As the simulation progresses, a wet layer begins to form and grows, producing a density profile which evolves with "time" t. The profile is recorded and ensemble averaged over many samples to obtain m(Z,t), see Fig. 1. The number of samples ranges from 60 (for L = 24) to 3000 (for L = 8), and times up to t = 5760 MCS/spin were studied. We define the layer thickness $Z_0(t)$ from $m(Z_0(t)) = 0$, to obtain the results of Fig. 6. For short times t, the expected behavior $Z_0(t) \sim \ln t$ is indeed observed, while for $t > t^*$ Eq. (3) is verified (Fig. 7).

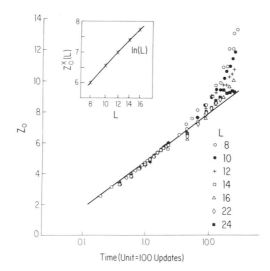

Fig. 6   Ensemble-averaged position of the interface $Z_0(t)$ vs. ln(t) for LxLx40 lattices and various linear dimensions as indicated in the figure. The size dependence of the crossover position $Z_0*(L)$ is given in the inset. From Mon et al.[7]

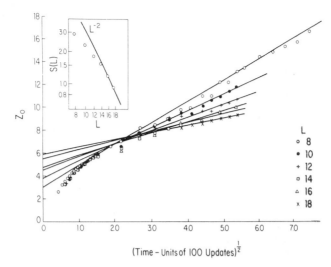

Fig. 7   Ensemble-averaged position of the interface $Z_0(t)$ vs. $\sqrt{t}$ for
LxLx40 lattices.   The size dependence of the slope is roughly
consistent with a dependence slope $\sim 1/L$ as required in Eq. (3).
From Mon et al.[7]

      As pointed out by Mon et al.[7] the probability distribution for the
interface position depends on the system linear dimension only through
the diffusion constant $D \sim L^{-2}$, which implies a lack of self-averaging[29]
- this is responsible for the huge sample-to-sample fluctuations.
Therefore the study of the growth of wetting layers by simulations is
computationally very demanding, and due to the need of much computer time
no study of other situations described by different sets of parameters
has yet been performed. We emphasize, however, that the model Eq. (6)
exhibits wetting transitions of both second and first order, tricritical
wetting, prewetting as well as layering. Figure 8 presents sections of
its phase diagram.

WETTING AT SURFACE STEPS

      Simulations were carried out for the model Eq. (6) in a LxM geometry
and the special case $J_s = J$ but various choices of the two boundary
fields H_1 (top boundary) and H_L (bottom boundary), cf Fig. 2. While for
the special case where H_1 and H_L have the same magnitude but competing
signs, close enough to T_c and/or for strong enough boundary fields an
interface running parallel to the steps always is present, see Fig. 9, a
different behavior occurs when H_1 and H_L have the same sign or differ in
absolute magnitude - then the first-order transition of the adsorbed
strip at the terrace from low coverages, $\theta_b^{(1)}$, to high coverage, $\theta_B^{(2)}$,
does not occur for $\delta\mu = 0$ but rather at a shifted value $H^* \sim \delta\mu^* \sim$
$1/L$.[32,33] For $L \to \infty$ the wetting phase diagram can be obtained exactly[18]
and controls also the behavior of small strips of finite width L when H_1
$= -H_L$ (Fig. 10). The wetting transition line is given by[18]

$$\exp\left[\frac{2J}{k_BT}\right]\left\{\cosh\left(\frac{2J}{k_BJ}\right) - \cosh\left(\frac{2H_1}{k_BT}\right)\right\} = \sinh\left(\frac{2J}{k_BT}\right). \qquad (7)$$

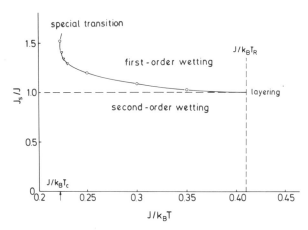

Fig. 8   Surface phase diagram in the plane of variables J_s/J and J/k_BT, exhibiting the line of tricritical wetting transitions, separating the region of first-order wetting (above this line) from second-order wetting (below this line). The tricritical wetting line at J/k_BT_c ends in the so-called "special transition"[30,31] (or surface-bulk multicritical point), and at lower temperature it ends at the roughening temperature, where layering transitions replace wetting[28]. The estimate of T_R was taken from Mon et al.[27] From Binder et al.[26]

Figure 9 shows the increasing roughness of the interface as the temperature is raised.

An additional complication arises due to the fact that infinite strips of width L are quasi-one dimensional systems which cannot maintain true spontaneous long range order. Consider e.g. the model of Eq. (6) where all fields vanish, $H = 0$, $H_1 = H_L = 0$. This system for finite width L does not develop infinite-range order at any nonzero temperature, but rather the strip is always broken up in domains (Fig. 11). The size of these domains in the direction parallel to the step is very large, namely[35] $\xi_D \sim \exp(2L\sigma_{int}/k_BT)$, where σ_{int} is the interfacial free energy of the Ising model. Near T_c we have[34] $\xi_D \sim L$, however.

If H_1 and H_L differ from zero, we still expect a behavior qualitatively similar to Fig. 11 to occur in the nonwet part of the phase diagram, Fig. 10, for fields H near $H^*(\delta\mu \approx \delta\mu^*)$. Close to the wetting transition we expect a well-defined interface to occur which is sometimes bound to the top boundary and sometimes to the bottom boundary, while in between it crosses the terrace from one boundary to the other. In this fashion, a gradual transition from the domain wall geometry of Fig. 11 to that of Fig. 9 is possible.

For such strips of finite width, the dynamics of wetting hence is influenced by the dynamics of the domain pattern, which exists here in thermal equilibrium: the domain walls crossing the terrace such as shown in Fig. 11 will randomly diffuse back and forth along the strip with a diffusion constant proportional to 1/L, analogous to the behavior discussed in the previous section. A Monte Carlo investigation of the dynamics of these domains is in progress.[8] At this point, we also draw attention to recent work of Stauffer and Landau[40] who have observed a $t^{1/4}$ growth law of a initially straight interface between bulk two-dimensional phases in the Ising model.

40

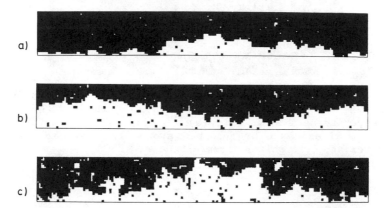

Fig. 9   Snapshot pictures of a system with L = 24, M = 288 at H = 0,
          H_1 = -3J, H_L = -3J, at the timestep t = 24000 MCS/site of a
          Monte Carlo simulation, and three temperatures: T = $0.68T_c$ (a),
          T = $0.78T_c$ (b), and T = $0.88T_c$ (c). Sites taken by adsorbed
          atoms are shown in black, empty sites are left white. From
          Albano et al.[9]

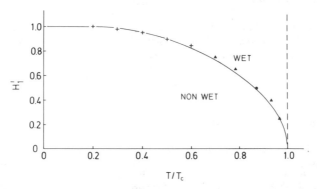

Fig. 10 Wetting phase diagram for the square nearest-neighbor Ising
          model (Eq. 6) with a free surface where a boundary field
          $H'_1 \equiv H_1/J$ acts. The solid curve is the numerical solution of
          Eq. (7), while the points result from various extrapolations of
          Monte Carlo data obtained from strips with L ≤ 24. From Albano
          et al.[9]

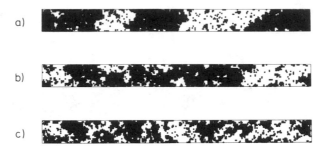

Fig. 11 Snapshot pictures of a system with L = 24, M = 288 at H = H_1 = H_L = 0 and the timestep t = 18000 MCS/site at the temperatures T = 0.95 T_c (a), T = 1.0 T_c (b) and T = 1.05 T_c (c). Sites taken by adsorbed atoms are shown in black, empty sites are left white. From Albano et al.[34]

CONCLUSIONS

The growth of wetting layers at surfaces with time is predicted[10,11] to occur by much slower growth laws than the growth laws controlling the kinetics of ordering or phase separation in the bulk.[19,36,37] The type of growth law depends on the nature of the forces.[10] So far, computer simulations have been able to verify the logarithmic growth law predicted for short range forces;[7,38] in addition, interesting new types of finite size effects have been found. While these size effects clearly are relevant to properly analyze the simulations, it is unclear whether they have any counterparts in an experiment.

At stepped surfaces the growth of adsorbed monolayers also proceeds via wetting at steps. There the problem is complicated due to the fact that in strips of finite width a true long range order is not expected, rather the strip breaks up into domains. We also note that related phenomena also can occur in models where interfacial wetting is predicted, see e.g. Ref. 39 for a study of (3x1) order on the centered rectangular lattice. E.g., by changing parameters of the system a heavy wall between domains of type A, C may get unstable and split into two light walls with a growing B domain in between. It would be interesting to study as well the dynamics of such phenomena by simulations.

One problem which we have disregarded completely so far is the conversion of the Monte Carlo time units to the physical time scale. Kang and Weinberg[41] have recently studied this question for lattice gas models of surface diffusion in adsorbed monolayers. As expected, they find that for the standard choices of transition probabilities[15] a nontrivial temperature - and coverage - dependent time scale conversion factor arises, when one compares the results to a model with a microscopically more realistic choice of transition probability arising from a thermally activated motion. However, they suggest that for domain growth even the asymptotic exponent depends on the type of transition probability. This is very puzzling and disagrees with the idea of "dynamic universality",[19] according to which asymptotic exponents of growth laws should not depend on microscopic details of the considered motions. Clearly, this problem does deserve further attention also with respect to the growth of wetting layers.

42

It is hoped that the present work stimulates the experimental search for such growth phenomena in adsorbed films and adsorbed monolayers.

Acknowledgements

This paper is based on work done in collaboration with K. K. Mon and D. P. Landau (Ref. 7) and with E. V. Albano, D. W. Heermann and W. Paul (Refs. 8, 9). It is a pleasure to thank them for a fruitful and stimulating interaction.

REFERENCES

1. For general reviews on wetting, see D. E. Sullivan and M. M. Telo da Gama, in Fluid Interfacial Phenomena, C. A. Croxton ed. (Wiley, New York 1986), and Ref. 2-5.
2. S. Dietrich, in Phase Transitions and Critical Phenomena, vol. XII (C. Domb and S. L. Lebowitz, eds.) p. 1 (Academic, New York, 1988).
3. E. H. Hauge, in Fundamental Problems in Statistical Physics VI (E. G. D. Cohen ed.) p. 65 (North Holland, Amsterdam 1985).
4. P. G. de Gennes, Rev. Mod. Phys. 55, 825 (1985).
5. M. E. Fisher, J. Chem. Soc., Faraday Trans 2, 82, 1569 (1986); J. Stat. Phys. 34, 667 (1984).
6. J. A. Venables, G. D. T. Spiller, and M. Hanbrücken, Rep. Progr. Phys. 47, 399 (1984).
7. K. K. Mon, K. Binder and D. P. Landau, Phys. Rev. B 35, 3683 (1987).
8. E. V. Albano, K. Binder, D. W. Heermann, and W. Paul, to be published.
9. E. V. Albano, K. Binder, D. W. Heermann, and W. Paul, Surface Science 223, 151 (1989).
10. R. Lipowsky, J. Phys. A18, L585 (1985).
11. R. Lipowsky and D. A. Huse, Phys. Rev. Lett. 52, 353 (1986).
12. M. Grant, K. Kaski and K. Kankaala, J. Phys. A20, L571 (1987).
13. M. Grant, Phys. Rev. B37, 5705 (1988).
14. I. Schmidt and K. Binder, Z. Phys. B67, 369 (1987).
15. K. Binder and D. W. Heermann, Monte Carlo Simulation in Statistical Physics: An Introduction (Springer, Berlin 1988).
16. R. Marx, Phys. Rev. B40, 2585 (1989).
17. M. S. S. Challa, D. P. Landau, and K. Binder, Phys. Rev. B34, 1841 (1986).
18. D. B. Abraham, Phys. Rev. Lett. 44, 1165 (1980); J. Phys. A21, 1741 (1988).
19. A. Sadiq and K. Binder, J. Stat. Phys. 35, 517 (1984).
20. G. Forgacs, H. Orland and M. Schick, Phys. Rev. B31, 7434 (1985).
21. I. E. Dzyaloshinskii, E. M. Lifshitz and L. P. Pitaevskii, Adv. Phys. 10, 165 (1961).
22. D. M. Kroll and G. Gompper, Phys. Rev. B39, 433 (1989).
23. In Ref. 7 it was erroneously written $Z_0(t) \sim L^{-2}t^{1/2}$ and thus the estimates quoted there for $t^*(L)$ are in error.
24. H. Nakaniski and P. Pincus, J. Chem. Phys. 79, 997 (1983).
25. K. Binder and D. P. Landau, Phys. Rev. B37, 1745 (1988).
26. K. Binder, D. P. Landau, and S. Wansleben, Phys. Rev. B40, 6971 (1989).
27. E. Bürkner and D. Stauffer, Z. Phys. B53, 241 (1983); K. K. Mon, S. Wansleben, D. P. Landau, and K. Binder, Phys. Rev. Lett. 60, 708 (1988).
28. R. Pandit, M. Schick and M. Wortis, Phys. Rev. B26, 5112 (1982).
29. A. Milchev, K. Binder and D. W. Heermann, Z. Phys. B63, 521 (1986).
30. K. Binder, in Phase Transitions and Critical Phenomena, Vol. VIII, (C. Domb and J. L. Lebowitz, eds.) p. 1 (Academic, New York 1983).

31. K. Binder and D. P. Landau, Phys. Rev. Lett. $\underline{52}$, 318 (1984), D. P. Landau and K. Binder, Phys. Rev. B (1990, in press).

32. M. E. Fisher and H. Nakanishi, J. Chem. Phys. $\underline{75}$, 5875 (1981); H. Nakanishi and M. E. Fisher, J. Chem. Phys. $\underline{78}$, 3279 (1983).

33. E. V. Albano, K. Binder, D. W. Heermann, and W. Paul, J. Chem. Phys. $\underline{91}$, 3700 (1989).

34. E. V. Albano, K. Binder, D. W. Heermann, and W. Paul, Z. Physik $\underline{B77}$, 445 (1989).

35. M. E. Fisher, J. Phys. Soc., Japan Suppl. $\underline{26}$, 87 (1969).

36. K. Binder and D. W. Heermann, in <u>Scaling Phenomena in Disordered Systems</u>, (R. Pynn and A. Skjeltorp eds.) p. 207 (Plenum, New York 1985).

37. K. Binder, in <u>Materials Science and Technology, Vol. 5: Phase Transformation in Materials</u> (P. Haasen, ed.) (VCH Verlagsges., Weinheim, in press).

38. Z. Jiang and C. Ebner, Phys. Rev. $\underline{B36}$, 6976 (1987).

39. I. Sega, W. Selke, and K. Binder, Surface Sci. $\underline{154}$, 331 (1985).

40. D. Stauffer and D. P. Landau, Phys. Rev. $\underline{B39}$, 9650 (1989).

41. H. C. Kang and W. H. Weinberg, J. Chem. Phys. $\underline{90}$, 2824 (1989).

THE EFFECTS OF MOBILE VACANCIES AND IMPURITIES ON THE

KINETICS OF ORDERING AT SURFACES

Ole G. Mouritsen and Peter Jivan Shah

Department of Structural Properties of Materials
The Technical University of Denmark
Building 307, DK-2800 Lyngby, Denmark

ABSTRACT

The kinetics of the ordering processes in two-dimensional lattice models with annealed vacancies (inert impurities) and non-conserved order parameter is studied as a function of vacancy concentration by means of Monte Carlo temperature-quenching simulations. The models are Ising antiferromagnets with couplings leading to two-fold degenerate as well as four-fold degenerate (2 x 1)-ordering. The growth for the pure systems is found to be described by a power law with the classical growth exponent, n = 1/2. For the dilute systems there is a distinct crossover at late times to a much slower growth behavior similar to that found in systems with quenched randomness. These results apply to both types of ordering suggesting that the effects on ordering kinetics of vacancy diffusion and annealed randomness do not depend on the symmetry of the order parameter. The results of the model study are relevant for the interpretation of experiments on ordering in impure overlayers on solid surfaces. The finding of a crossover from an algebraic growth law for the pure system to a much slower growth mode in the dilute system is in accordance with recent high-resolution LEED experiments on the oxygen ordering on W(112)-surfaces doped with nitrogen.

INTRODUCTION

The ordering kinetics in condensed matter systems is known to be strongly influenced by imperfections, such as site-impurities[1-5], vacancies,[6-11] random couplings[12,13] or random fields,[14-25] and second-phase particles.[26] All these different circumstances impose some kind of randomness with which the ordering process has to comply. This leads to effects which may slow down the growth and eventually pin it, or possibly even lead to a completely different growth mode. Two intensively studied cases are that of random fields[14,15,16-25] and that of quenched (immobile) impurities and vacancies.[1-9] In both cases the growth is found to be slowed down dramatically and there is theoretical evidence that the growth mode becomes logarithmic at late times.[5,18]

The case of annealed (mobile) vacancies or impurities and their effects on ordering kinetics has been studied much less.[7-9] This case is that of a system with at least two components and it often involves phase separation dynamics,[27] depending on the actual phase diagram, i.e. the microscopic interactions with and among the mobile imperfections. Similar to the case of quenched impurities[5] or random fields,[18] the present case is interesting because it may implicate activated processes by which the impurities interact with the moving domain boundary.[9] This proves to have a dramatic effect on the growth behavior. The cases of annealed and quenched impurities (or vacancies) differ from the case of the random-field Ising model[5] in that the randomness in the annealed case does not couple directly to the local order parameter as is the case in the random-field Ising model. This is an important distinction from the point of view of non-equilibrium dynamics.[5].

Two-dimensional chemisorbed and physisorbed molecular overlayers on solid surfaces constitute a particularly suitable class of systems for studying fundamental aspects of ordering dynamics since these systems provide a richness of ordering symmetries and degeneracies.[28] They are attractive to model by computer-simulation techniques, but it is unfortunately very difficult to obtain reliable time-resolved experimental data from them.[29,30] Moreover, the kinetics of ordering and growth on surfaces may be strongly influenced by surface inhomogeneities, such as steps and impurities. Only very recently has the first systematic experimental work been reported on the growth kinetics of a chemisorbed overlayer in the presence of impurities.[1]

A MODEL STUDY

In this paper we report on a computer-simulation study,[11] carried out in the simplest possible setting, of the ordering kinetics in a two-dimensional kinetic lattice model[28] with mobile vacancies (inert impurities). The model is the site-diluted square-lattice spin-1/2 Ising antiferromagnet with nearest-neighbor (NN) and next-nearest-neighbor (NNN) interactions described by the Hamiltonian

$$H = J \left\{ \sum_{i>j}^{NN} \sigma_i \sigma_j + \alpha \sum_{i>j}^{NNN} \sigma_i \sigma_j \right\} , \tag{1}$$

with $\sigma_i = 0, \pm 1$. The value $\sigma_i = 0$ is associated with vacant sites. The global vacancy concentration is c. The parameter α measures the ratio between NNN and NN coupling strengths. The lattice has free boundaries. We shall focus on the antiferromagnetic case, J > 0. For the pure system, c = 0, and for α < 1/2, this model leads to simple (2 x 1) antiferromagnetic order which is two-fold degenerate (p = 2), cf. Fig. 1. For α > 1/2, the order is of (2 x 1) super-antiferromagnetic type which is four-fold degenerate (p = 4), Fig. 1. We have investigated the model,[11] Eq. (1), for α = 0 and α = 1. We shall in the following refer to these two cases as the 'p = 2' model and the 'p = 4' model. The transition temperatures in the thermodynamic limit of the two models for c = 0 are $k_B T_c/J \approx 2.27$ and 2.10, respectively. In the case of annealed site dilution,[27] c > 0, the system is effectively a two-component system which will phase separate at sufficiently low temperatures. It is important to notice that annealed site-diluted models for any concentration, in contrast to models of quenched site dilution, will display cooperative behavior and ordering, provided the temperature is sufficiently low.[27] In the very dilute range this implies segregation of spins into ordered islands, i.e. phase separation.

The ordering processes of the model are assumed to be governed by a microscopic dynamics which involves spin-flip excitations as well as vacancy diffusion. The spin-flip excitations involve single sites and they do not conserve the antiferromagnetic order parameter. The vacancy

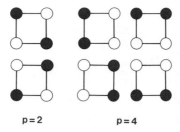

p=2   p=4

Fig. 1   Schematic representation of the two types of (2x1)
            antiferromagnetic ordering of the Ising model of Eq. (3).
            (a): $\alpha < 1/2$, two-fold degenerate (p = 2) antiferromagnetic
            ordering.   (b): $\alpha > 1/2$, fourfold-degenerate (p = 4)
            super-antiferromagnetic ordering.

diffusion involves exchange of a spin and a vacancy at nearest or
next-nearest neighbor sites and a possible simultaneous spin flip of the
spin involved.   This process conserves the global vacancy concentration.
The time-evolution of the model is simulated by Monte Carlo techniques.[28]
The time is measured in units of Monte Carlo steps per site (MCS/S).

RESULTS

        In Figs. 2 and 3 are shown snapshots of typical configurations as
they evolve in time for different concentrations for both the p = 2 and
the p = 4 models.   One observes from this figure that the vacancies
localize themselves at the domain boundaries at a very early time.
Simultaneously, vacancies trapped inside the domains tend to precipitate.
Large precipitates are found when small domains within large domains
disappear, leaving their trapped vacancies behind.   Similar to
observations made for the pure system, finite-size effects at late times
may lead to formation of metastable 'slab' configurations.   For small
degrees of dilution, the late-time growth process proceeds via
intermittency of migration of the vacancies along the domain-boundary
network towards the surface and curvature-driven domain growth.   The
accumulation of vacancies in the domain boundaries leads to a screening
of the direct domain-domain interactions and a slowing down of the
growth.

        In Figs. 4a and 5a are shown log-log plots of the results for the
time dependence of the average domain size, $\bar{R}(t)$, for the two models in
the case of quenches to k_BT/J = 0.25 for different vacancy
concentrations.   It is noted that the algebraic growth law

$$\bar{R}(t) \sim t^n \qquad\qquad (2)$$

holds for c = 0 with the expected[28] classical Lifshitz-Allen-Cahn value
n = 1/2.   As the system gets diluted, the growth slows down and there is
a crossover from an algebraic growth law at early times to a much slower
growth mode at late times.   At no stage does the growth become pinned,
not even at zero temperature. The crossover point occurs earlier in time

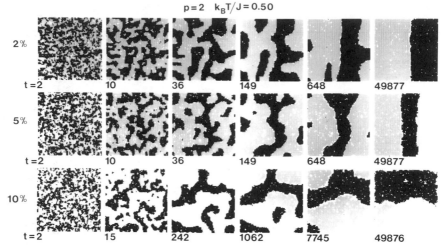

Fig. 2   Snapshots of microconfigurations for the p = 2 model as they
         evolve in time t (in units of MCS/S) after quenches to a
         temperature k_BT/J = 0.540 for vacancy concentrations c = 2,
         5, and 10%.   The two types of (2 x 1) antiferromagnetically
         ordered domains are indicated by grey and black regions.   The
         snapshots refer to lattices with 100 x 100 sites.

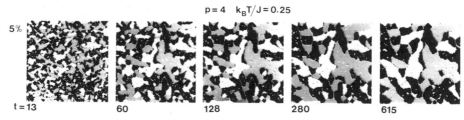

Fig. 3   Snapshots of microconfigurations for the p = 4 model on a 200
         x 200 lattice for c = 5% as they evolve in time t (in units
         of MCS/S) for quenches to a temperature k_BT/J = 0.25.   The
         four types of (2 x 1) antiferromagnetically ordered domains
         are indicated by black and three different grey-toned
         regions.

the larger c is. In Figs. 4b and 5b the same data are reanalyzed in a
semi-logarithmic plot which shows that the dilute systems at late times
have an ordering kinetics consistent with a logarithmic growth law

$$\overline{R}(t) \sim \ln t \, .$$ (3)

The same statement holds for other quench temperatures. We cannot exclude
that this slow growth mode is not a crossover from the
Lifshitz-Allen-Cahn law to a slower but still algebraic growth behavior.

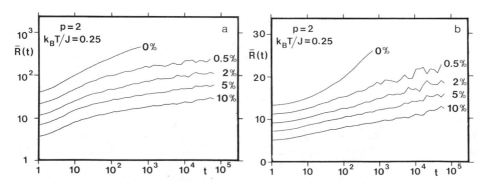

Fig. 4   (a): Log-log plot of the average domain size, $\bar{R}(t)$, vs time t
(in units of MCS/S) for quenches to a temperature k_BT/J =
0.25 for different vacancy concentrations in the p = 2 model.
For the sake of clarity, the various data sets have been
appropriately translated along the vertical axis.   (b): Same
data as (a) in a semilogarithmic plot.

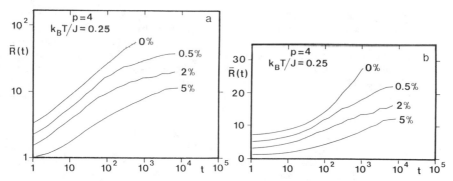

Fig. 5   (a): Log-log plot of the average domain size, $\bar{R}(t)$, vs t, (in
units of MCS/S) for quenches to a temperature, k_BT/J = 0.25,
for different vacancy concentrations in the p = 4 model.   For
the sake of clarity, the various data sets have been
appropriately translated along the vertical axis.   (b): Same
data as (a) in a semilogarithmic plot.

49

COMPARISON WITH EXPERIMENTS

Annealed randomness in systems undergoing ordering processes is realized in a number of different experimental systems. We shall here refer to Zuo et al.'s[1] high-resolved LEED study of oxygen ordering in a monolayer chemisorbed on W(112). In this case the oxygen atoms have an ordering with the same symmetry as the (2 x 1)-structures of the p = 2 model. In the presence of nitrogen impurities a crossover from Lifshitz-Allen-Cahn algebraic growth to a much slower growth is observed.[1] This result was interpreted[1] as a case of the random-field Ising model, i.e. in terms of quenched randomness. However, surface diffusion data[31] including prefactor and activation energy for O and N on tungsten surfaces in the pertinent temperature range (800-900K) suggest that the diffusion of the nitrogen impurities is similar to that of oxygen. Therefore, our model predictions should apply when we interpret the mobile nitrogen impurities as a site dilution which does not participate in the oxygen ordering and does not couple directly to the order parameter (which is required by the random-field Ising model).[5] Therefore, we reinterpret the LEED results as a manifestation of slow, possible logarithmic growth, due to annealed randomness.

The ordering dynamics of oxygen on W(110), which has the symmetry of the p = 4 model, has not been studied systematically in the presence of impurities. The experimentally observed late-time slowing down of the growth in supposedly pure oxygen monolayers has been ruled out experimentally as being due to impurities.[30]

CONCLUSIONS

We have in the present paper reported on a Monte Carlo computer-simulation study[11] of the influence on the ordering kinetics of annealed dilution. Two-fold (p = 2) as well as four-fold (p = 4) degenerate ordering has been studied. The ordering kinetics in the case of a non-conserved order parameter has been investigated as a function of temperature and degree of dilution. Our main result is that, independent of the temperature and the ordering degeneracy, there is a distinct crossover from an algebraic growth law, $\bar{R}(t) \sim t^{1/2}$, to a much slower growth mode as the system becomes diluted. These results are in accordance with recent experimental studies of the ordering kinetics in impure overlayers chemisorbed on solid substrates.

Another important result is the finding of vacancy accumulation in the domain interfaces. As the vacancies accumulate in the domain boundaries, the direct domain-domain interactions become screened or even decoupled, implying that the curvature-driven pure-system Lifshitz-Allen-Cahn growth mechanism becomes ineffective. This may provide a possible explanation of the logarithmic growth behavior in the dilute system since the growth then proceeds via an evaporation-condensation mechanism which is an activated process, and the time it takes for two domains to merge depends on the size of the domains. Since the order parameter is not conserved in this process, it seems that the Lifshitz-Slyozov theory[32] should not apply. However, since we are basically dealing with a phase separation problem this theory may still apply and it should be pointed out that we can not exclude on the basis of the present set of data that the growth at late times crosses over to the Lifshitz-Slyozov growth law

$$\bar{R}(t) \sim t^{1/3} . \tag{4}$$

We emphasize that there is at present no theory for the slow growth behavior in the dilute system and wish to remark that, although this behavior is also described by an effectively logarithmic growth law, it is not covered by either the theory of the random-field Ising model[17,18] or the theory by Huse and Henley[5] for a quenched dilution.

Acknowledgements

The work described in this paper was supported by the Danish Natural Science Research Council under Grants Nos. 5.21.99.72 and 11-7785, and the Danish Technical Research Council under Grant No. 16-4296.K. Illuminating discussions with Kurt Binder during the workshop are gratefully acknowledged.

REFERENCES

1. J.-K. Zuo, G.-C. Wang, and T.-M. Lu, Phys. Rev. Lett. $\underline{60}$, 1053 (1988).
2. D. J. Srolovitz, M. P. Anderson, G. S. Grest, and P. S. Sahni, Acta Metall. $\underline{32}$, 1429 (1984).
3. G. S. Grest and D. J. Srolovitz, Phys. Rev. $\underline{B32}$, 3014 (1985).
4. D. J. Srolovitz and G. S. Grest, Phys. Rev. $\underline{B32}$, 3021 (1985).
5. D. A. Huse and C. L. Henley, Phys. Rev. Lett. $\underline{54}$, 2708 (1985).
6. D. Chowdhury, M. Grant, and J. D. Gunton, Phys. Rev. $\underline{B35}$, 6792 (1987).
7. O. G. Mouritsen, Phys. Rev. $\underline{B32}$, 1632 (1985).
8. G. F. Mazenko and O. T. Valls, Phys. Rev. $\underline{B33}$, 1823 (1986).
9. D. J. Srolovitz and G. N. Hassold, Phys. Rev. $\underline{B35}$, 6902 (1987).
10. T. Ohta, K. Kawasaki, A. Sato, and Y. Enomoto, Phys. Lett. $\underline{A126}$, 93 (1987).
11. O. G. Mouritsen and P. J. Shah, Phys. Rev. $\underline{B50}$, 11445 (1989), P. J. Shah and O. G. Mouritsen Phys. Rev. B, April 1 (1990).
12. J. H. Oh and D.-I. Choi, Phys. Rev. $\underline{B33}$, 3448 (1986).
13. A. E. Jacobs and C. M. Coram, Phys. Rev. $\underline{B36}$, 3844 (1987).
14. D. P. Belanger, A. R. King, and V. Jaccarino, Phys. Rev. $\underline{B31}$, 4538 (1985).
15. H. Yoshizawa, R. A. Cowley, G. Shirane, and R. Birgeneau, Phys. Rev. $\underline{B31}$, 4548 (1985).
16. M. Grant and J. D. Gunton, Phys. Rev. $\underline{B29}$, 1521 (1984).
17. J. Villain, Phys. Rev. Lett. $\underline{52}$, 1543 (1984).
18. G. Grinstein and J. F. Fernandez, Phys. Rev. $\underline{B29}$, 6389 (1984).
19. E. T. Gawlinski, K. Kaski, M. Grant, and J. D. Gunton, Phys. Rev. Lett. $\underline{53}$, 2266 (1984).
20. E. T. Gawlinski, S. Kumar, M. Grant, J. D. Gunton, and K. Kaski, Phys. Rev. $\underline{B32}$, 1575 (1985).
21. S. R. Anderson and G. F. Mazenko, Phys. Rev. $\underline{B33}$, 2007 (1986).
22. D. Chowdhury and J. D. Gunton, J. Phys. A: Math. Gen. $\underline{19}$, L1105 (1986).
23. M. Grant and J. D. Gunton, Phys. Rev. $\underline{B35}$, 4922 (1987).
24. D. A. Huse, Phys. Rev. $\underline{B36}$, 5383 (1987).
25. S. R. Anderson, Phys. Rev. $\underline{B36}$, 8435 (1987).
26. M. Hillert, Acta Metall. $\underline{36}$, 3177 (1988).
27. R. B. Stinchcombe, in: Phase Transitions and Critical Phenomena, C. Domb and M. S. Green eds. (Academic, New York, 1983) Vol. 8, p. 151.
28. For a recent review on the kinetics of ordering and growth in 2-D systems, see the article by O. G. Mouritsen in this volume.
29. K. Heinz, in: Kinetics of Interface Reactions, M. Grunze and H. J. Kreuzer eds. (Springer Verlag, New York, 1987) p. 202.
30. P. K. Wu, M. C. Tringides, and M. G. Lagally, Phys. Rev. $\underline{B39}$, 7595 (1989).

31. A. Polak and G. Ehrlich, J. Vac. Sci. Technol. 14, 407 (1977); A. G. Naumovets and Y. S. Vedula, Surf. Sci. Rep. 4, 365 (1985).
32. I. M. Lifshitz and V. V. Slyovov, J. Chem. Phys. Solids 19, 35 (1961).

ATOMIC BEAM SCATTERING STUDIES OF ORDERING AT SURFACES

Klaus Kern and George Comsa

IGV/KFA, P.O. Box 1913
D-5170 Jülich, Federal Republic of Germany

ABSTRACT

The scope of this workshop is to bring the "growth" people and the "surface science" people closer; to make them interact. This contribution tries to parallalize this scope by discussing experimental aspects of the interaction between the substrate surface and the growing film. The focus is on the structure and dynamics of rare gas films growing on close packed metal substrates as investigated by high-resolution He scattering.

The influence of the substrate on the film growth can be easier understood by realizing the crucial role played by the first adsorbed monolayer. Indeed, on the one hand, the properties and in particular the structure of this first layer are decisively influenced by the substrate. On the other hand, the nature of the film growth depends directly on the structure of this first monolayer; e.g. layer-by-layer growth can hardly take place if the structure and the lattice constant of the first monolayer deviate markedly from the equilibrium structure and lattice constant of the bulk material.

The influence of the substrate on the structure and dynamics of the first adlayer proceeds via the strength and lateral corrugation of the holding potential and - in real life - also via the always present defects. The influence of the lateral corrugation of the holding potential on the structure - in particular on the lattice constant and orientation - of the first adsorbed full monolayer will be emphasized. The controversial issue concerning the height of the lateral corrugation is also discussed.

INTRODUCTION

The first monolayer of a film growing on a substrate plays an essential role in heteroepitaxy. Indeed, on the one side of the layer, it has to cope with the surface layer of the substrate, which is made of atoms of a different species, i.e. with different electronic properties, and which in general has a different structure (lattice constant and even symmetry) than the layer would have as a part of its own bulk crystal. As a consequence the properties and in particular the structure of the first adlayer differs into a lesser or larger extent from the properties of the

Kinetics of Ordering and Growth at Surfaces
Edited by M. G. Lagally
Plenum Press, New York, 1990

outermost layer of its own bulk crystal. The nature and magnitude of this difference determine primarily the nature of the film growth on the other side of the adlayer.[1,2] The nature of the atoms on the growing side being the same, even minute structural misfits and the ensuing stress may forbid layer-by-layer epitaxial growth.

Thermal He-scattering proves to be a particularly adequate probe to investigate the various properties of this first monolayer as well as those of the growing film.[3] This is due to features like exclusive sensitivity to the outermost layer, nondestructiveness, high momentum resolution in diffraction, very high energy resolution of energy loss and last but not least outstanding sensitivity for minute impurities and defects. In section 2 the principles and application of He-scattering to adlayer investigation are outlined.

The heteroepitaxial growth is known to be a very complex process. In order to be able to determine important details and to reach clear conclusions concerning main aspects of the process we have chosen to look at simple systems: rare gas films on a close packed metal surface (Xe, Ar, Ne on Pt(111)). The lowest energy surface of the rare gas crystals has the same symmetry as the Pt(111)-surface and the corrugation of the (111) surface of fcc metals is the lowest compared to other orientations. Some of the significant results will be illustrated for the Xe/Pt(111) system; in particular the monolayer statics and dynamics and the multilayer growth.

THERMAL He ATOMS AS A PROBE OF THIN FILM GROWTH AND DYNAMICS

There are several ways of using thermal atom scattering to study the proper- ties of surfaces and adsorbed layers. In this chapter we will outline the different types of surface scattering, emphasizing the features that make He atoms to a particularly appropriate probe of thin film growth and dynamics.

Figure 1 shows schematically a He-atom-surface scattering experiment. A highly monochromatic beam of thermal He-atoms ($\Delta\lambda/\lambda \leq 1\%$) is generated in a high pressure supersonic expansion and collimated to a few tenths of a degree by a series of specially shaped collimators, i.e. so called skimmers. Depending on the source temperature, the wave-length of the He atoms ranges between He 0.3 and 2.0Å; typical fluxes are of the order 10^{19} He atoms/sec·sr. The He atoms scattered from the crystal surface into a well defined solid angle element Ω are detected by an electron impact ionization mass spectrometer. The scattered intensity can be either measured energy integrated or energy resolved. In the latter case, the scattered beam is divided into pulses by a pseudo random (PR) chopper and the times of arrival of He-atoms at the detector are analyzed. Since the initial energy of the He-atoms and the geometry of the experiment are known, the energy transfer during the scattering can be calculated. For experimental details of the spectrometer we refer the reader to Ref. 4.

The attractivity of thermal He-atoms as probe particles for the analysis of the structure and dynamics of surfaces and thin films is inherently associated with the following aspects:

1.   The Exclusive Surface Sensitivity of the Method

At large separations the He-atom interacts with the electrons of the surface via induced dipol-dipol interactions. These long range dispersion forces are attractive and decay as z^{-3} with z being the distance between

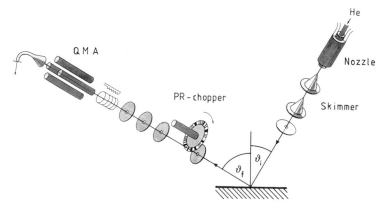

Fig. 1 Schematic arrangement of a He-surface scattering experiment.

He-atom and surface. Upon a closer approach, the electron orbitals of the He-atom overlap with the valence electrons of the surface atoms causing a strong repulsion, which is usually assumed to be proportional to the electron density $\rho(\underline{r})$ of the solid surface. A reasonable Ansatz for the atom surface potential thus writes:[5]

$$V(x,y,z) = V_0 \exp[-\kappa(z-\xi(x,y))] - \frac{C_3}{(z+z_0)^3} , \tag{1}$$

with the constants V_0 and z_0, the so called "softness parameter" κ, and the van der Waals-coefficient C_3. In this simple model, the periodic part of the potential given by the modulation $\xi(x,y)$ of the electron density $\rho(\underline{r})$ and reflecting the discrete symmetry of the surface in the x-y plane, only appears in the repulsive part of the interaction potential. The classical turning point of the He-atom is a few angstroms above the ion cores of the surface, i.e. the He-atoms do not penetrate into the selvedge. Accordingly the information sampled by the scattered He-atom is exclusively determined by the outermost surface layer.

2.   The Complete Nondestructiveness of the Method

He being a noble gas at very low thermal energies ($\sim$ 10-200 meV), He-atom scattering is a completely nondestructive surface probe. This property is of particular importance in the investigation of weakly bound and sensitive adlayer structures, like physisorbed films with binding energies less than $\sim$300 meV.[3]

3.   Energy - Wavelength Matching

The de Broglie wavelength of the thermal He-atoms generated in the super-sonic expansion is comparable to the interatomic spacing and the translational energy is close to the energy of collective surface excitations (e.g. surface phonons). This favorable energy-wavelength matching allows the spectroscopy of static as well as dynamic properties of solid surfaces and thin films.

The matching of the He-atom wavelength with typical interatomic distances of surfaces gives rise to interference effects in the scattering which can be detected by scanning the scattering angle or the wavelength. In such a diffraction experiment no energy analysis is usually done, i.e. the differential cross section $d\sigma/d\Omega$ is measured.[6] In some special experiments the energy analysis of the scattered He-atoms may be used to separate the true elastic scattering.[7] For diffraction from a periodic two dimensional array atoms the structure factor does not depend on the momentum transfer perpendicular to the surface (we are dealing here with diffraction rods!) and the differential cross-section denotes

$$\frac{d\sigma}{d\Omega} \propto S(Q", \hbar\omega = 0) = \sum_m \delta (Q" - G"_m) \ . \tag{2}$$

The terms $Q" = (Q_x, Q_y)$ and $G_m" = (G_{mx}, G_{my})$ are the momentum exchange vector and the reciprocal lattice vector in the surface plane of the two-dimensional structure, respectively. Thus, the Laue-condition for diffraction from a 2D-lattice reads

$$Q" = G"_m \ . \tag{3}$$

The momentum-resolution of the He-diffractometer is determined by the angular opening of the He-beam and of the detector and by the monochromaticity of the He-wave. The instrument used in the authors laboratory, designed to measure also dynamical surface properties in the same experiment (which is achieved at the expense of very high diffraction capabilities) has a momentum resolution of ~ 0.01 Å^{-1}.

In Fig. 2 we show as an example a He-diffraction scan from a complete Xe-monolayer adsorbed on Pt(111) ($\theta_{Xe} \approx 0.42$ Xe-atoms per Pt substrate atoms); a well behaved diffraction pattern with sharp Bragg-peaks is observed. This diffraction scan has been measured in a fixed scattering geometry $\theta_f + \theta_i = 90°$ by rotating the Pt-crystal around an axis perpendicular to the scattering plane; the angle on the abscissae scales to a momentum transfer scale by the relation $Q" = 5.723$ $(\sin\theta_f - \cos\theta_f)$. The diffraction pattern characterizes a hexagonal densely packed 2D-Xe crystalline solid with a nearest neighbor distance of 4.33Å. In the inset of Fig. 2 the orientational structure (with respect to the Pt-substrate) of the Xe-monolayer is characterized in an azimuthal diffraction profile, which is obtain by rotating the Pt-crystal around its surface normal at a fixed polar angle corresponding to a Bragg-position. The symmetric peak-doublet centered along the $\bar{\Gamma} \bar{K}_{Pt}$ direction of the substrate surface characterizes a Novaco-McTague phase rotated $\pm$ 3.3° degrees off the natural R 30° orientation of submonolayer Xe-films (see Chapter 3.2).

The energy of thermal He atoms is comparable with the energy of most atomic motions of surfaces and overlayers.[8] Thus, in a scattering experiment the He-atom may exchange an appreciable part of its energy with the surface. This energy can be measured in time-of-flight experiments with a resolution of ~ 0.3 meV. The range of energy transfer that can be covered by thermal He atoms is limited at the low end by the present maximum resolution of ~ 0.3 meV and at the upper end by the nature of the scattering mechanism. The interaction time of thermal He atoms with the surface being of the order of 10^{-13} sec, the upper limit for observable collective excitations is about 40 meV.

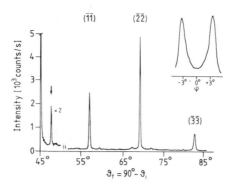

Fig. 2   Polar and azimuthal (inset) He-diffraction scan of a complete Xe
          monolayer on Pt(111); He wavelength λ_i = 1.098 Å, surface
          temperature 25 K.   The polar scan is taken along the Γ M_{Xe}
          azimuth.

The   kinematics   of   inelastic   He-scattering   is   governed   by
conservation of momentum and energy. Because of the loss of vertical
translational invariance at the surface, only the momentum parallel to
the surface is conserved in the scattering process,

$$K_f = K_i + Q'' + G''_m \tag{4}$$

$$\hbar\omega = \hbar^2\, k_f^2/2m - \hbar^2 k^2_i/2m \,, \tag{5}$$

where Q'' and $\hbar\omega$ are the phonon momentum and energy respectively, m is the
mass of the He-atom, k_i (k_f) and K_i (K_f) are the wave vectors and their
components parallel to the surface of incident (scattered) He-atoms, and
G''_m is a reciprocal-lattice vector. Depending on the sign of ω, a phonon
is annihilated (ω>0) or created (ω<0) during the scattering process.
Thus, by inelastic He-atom scattering from surfaces we can measure the
dispersion relations of the collective excitations point by point by
observing where in the (Q'',$\hbar\omega$) space the He-atoms undergo one-phonon
annihilation or one-phonon creation events.[8]   Formally, what we are
measuring in this experiment is the double differential cross section

$$\frac{d^2\sigma}{d\Omega_f dE_f} \propto \sum_{\vec{Q},j}{}'' \; |\vec{e}\,(Q'',j) < \Psi_G^{f*} |\nabla V(Q'',z)\, |\Psi_G^{i} > |^2 \; x \; |n^\pm| \; x$$

$$\delta\,(\vec{K_f}-\vec{K_i}-\vec{Q}'') \; x \; \delta\,(E_f - E_i - \hbar\omega\,(Q'',j)) \,, \tag{6}$$

which describes the exchange of a single phonon of wavevector Q'',
frequency ω(Q'',j) and polarization e(Q'',j).   The Bose factor for
annihilation (-) or creation (+) of a phonon, respectively, i.e., the
phonon occupation number, is denoted by $n^\pm$.

Figure 3 shows a series of typical He-energy loss spectra which have
been measured under identical scattering conditions from Xe-films on
Pt(111) 1,2,3, and 25 monolayers (ML) thick. By varying the scattering
conditions complete phonon dispersion curves for each film can be plotted
(see Ref. 9).

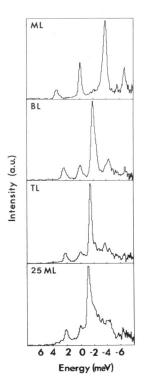

Fig. 3 He energy loss spectra measured in the $\bar{\Gamma}$ $\bar{M}_{Xe}$-azimuths of Xe-films on Pt(111) 1, 2, 3, and 25 monolayers thick. All spectra are taken under identical scattering conditions, i.e., primary beam energy 18.4 meV and incident angle θ_i = 42°C.

These dispersion plots reveal a layer-by-layer evolution of the dynamical film properties with thickness. The monolayer film is characterized by a dispersionless Einstein-oscillator. The vibrational modes of multilayer films show dispersion across the Brillouine-zone, the amount of dispersion increases with the film thickness. Eventually, 25 ML-films 'exhibit a well developed Rayleigh-wave, characteristic for a semi-infinite crystal.

There are two significant features evidenced by the energy loss spectra which are particularly useful in the investigation of film growth. Due to the high resolution, the energy losses (gains) of 1,2 and 3ML films, measured under identical scattering conditions, can be clearly discriminated. This allows the straightforward determination of the completion of the first and of the second monolayers within a few percent. This kind of information is very valuable also in the study of the thermodynamics of physisorbed films.[10] The other feature is the diffuse elastic peak ($\Delta E = 0$), its intensity being known to be a sensitive measure for the presence of surface defects. In Fig. 3 the diffuse elastic intensity decreases with film thickness, which is a direct proof of the layer-by-layer growth of the film.

The energy transfer between a He-atom and a surface is not restricted to collective excitations such as phonons. The He-atom can exchange energy with individual atoms diffusing on the surface. Diffusive motions of adatoms are associated with a continuum of low energy excitations and the interaction of He-atoms with this continuum gives rise to a broadening δE of the elastic peak. For low energy He-scattering from a two-dimensional fluid adlayer, Levi et al.[11] have shown that the width of the "quasielastic" peak is related to the diffusion coefficient D via $\sigma E = 2hDQ^2$. Frenken et al. have recently used this technique to study the self diffusion of Pb adatoms on the Pb(110) surface.[11]

Another remarkable way to use He scattering for the study of adsorbed layers is based on the large (~ 100 $\mathring{A}^2$) total cross-section Σ for diffuse He scattering of isolated adsorbates.[12] This large cross-section is attributed to the long-range attractive interaction between the adatom and the incident He atom, which causes the He atoms to be scattered out of the coherent beams. The remarkable size of the cross-section, 4-6 times the geometrical size, A, of the adsorbate allows the extraction of important information concerning the lateral distribution of adsorbates, mutual interactions between adsorbates, dilute-condensed phase transitions in 2D, adatom mobilites, etc., simply by monitoring the attenuation of one of the coherently scattered beams.[12] This technique also allows the detection of impurities (including hydrogen!) in the permill range, a level hardly attainable with almost all other methods.

The possibility to investigate the lateral distribution of adsorbates, in particular the dilute-condensed phase transition in 2D, is based upon the large difference between the cross-section for diffuse scattering, Σ, and the geometrical size, A, of the adsorbates. The degree of overlap of the cross-sections Σ at a certain adsorbate coverage, θ, which determines the He-reflectivity, depends on the nature of the lateral distribution of the adsorbate. For instance, as long as the adsorbates form a lattice gas, the He-reflectivity depends on θ as

$$I/I_o = (1-\theta)^{\Sigma n_g} , \qquad (7)$$

with I_o and I the intensities of the specular beam scattered from the clean and the adsorbate covered surface, respectively, and n_s the density of adsorption sites. At low coverages Eq. (7) becomes

$$I/I_o \simeq 1- \Sigma n_s \theta . \qquad (7')$$

On the other hand, when islanding starts, i.e., at the 2D gas → 2D solid + 2D gas transition, the He-reflectivity is determined by the much smaller geometrical size, A, of the adatoms in the 2D condensed phase:

$$I/I_c \simeq 1 - An_s (\theta-\theta_c) , \qquad (8)$$

with I_c being the specular intensity at the critical coverage θ^c, where condensation sets in. A comparison of eqs. (7) and (8) shows that the slope of the specular He-intensity I versus coverage θ is expected to change dramatically in the ratio Σ/A ($\simeq 4-6$) when condensation sets in. This is illustrated in Fig. 4 for a Pt(111) surface at 54 K exposed to Kr at P_{Kr} 2.1×10^{-9} mbar.

The sudden slope change in Fig. 4 obviously marks directly the onset of islanding. The critical coverage θ_c corresponds to the 2D vapor pressure of Kr on Pt(111) at 54 K. From measurements of θ_c at different temperatures, the 2D latent heat of vaporization is obtained under the

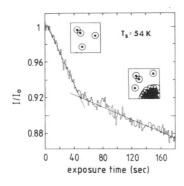

Fig. 4   Attenuation of the specularly reflected He-beam from Pt(111) upon exposure to Kr (P_{Kr} = 2.1 x 10^{-9} mbar) at T_s = 54 K.

assumption that the 2D gas is nearly perfect.[13] In Table I the 2D latent heats of vaporization obtained in this way for Xe, Kr, and Ar on Pt(111) are listed.[3] They represent actually the lateral energy between the adlayer atoms.[12,14]

In addition, at lower temperatures where $\theta_c \to 0$ Eq. (8) can be used in a straightforward manner to determine the sticking probability. Figure 5 shows the same type of measurement as Fig. 4 but at 25 K. The linear drop of the specular intensity with exposure, indicates a coverage-independent sticking coefficient. A simple kinetic calculation gives a value $s_0 \approx 0.7$. Only close to completion of the first monolayer (determined by phonon-spectroscopy to be at an exposure of 7.2 x 10^{-6} mbar . sec) the curve deviates from Eq. (8) and eventually levels off.

MONOLAYER FILMS AND STATICS OF COMPETING INTERACTIONS

The properties of monolayer films on solid surfaces, i.e. quasi two dimensional (2D) systems, are quite often different from those of the three dimensional (3D) bulk material. For instance, a one monolayer solid Xe phase melts at a substantial lower temperature than the bulk Xe-solid; on the basal plane of graphite the monolayer Xe-film has a 2D gas-liquid-incommentsurate solid triple point at T_t^{2D} = 0.61 T_t^{3D}.[15] Upon increasing thickness, the film properties gradually approach three dimensional behavior.

Table 1
Lateral Interaction Energies for Rare Gases Adsorbed
on Pt(111)

	Xe	Kr	Ar
e_1 (meV)	43	26	17

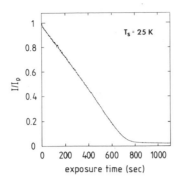

Fig. 5   Same as Fig. 4 but at a surface temperature of T_s = 25 K and a
Kr pressure P_{Kr} = 8.4 x 10^{-9} mbar.

What makes the difference between a monolayer film and a film, say
25 monolayers thick, which shows almost bulk behavior? First, the average
atom in a bulk film has twelve nearest neighbors (when condensing in a
fcc lattice) while an atom in a monolayer film has only six of the same
kind. Thus, thermal fluctuations are more important in monolayer than in
bulk films. Second, real adsorbent surfaces are structured, providing a
periodic potential relief which interferes with the lattice structure of
the overlayer, inducing modulations in the latter.[16] The properties of
the first monolayer are certainly influenced most dramatically by the
substrate surface. The thicker the grown film the smaller the <u>direct</u>
influence of the substrate on the properties of the remote film layers.
However, we have to take into account that the film growth depends
directly on the structure of this first monolayer. We will thus first
discuss the static interaction of substrate and adlayer in the monolayer
regime.

1.   <u>Substrate Corrugation and Commensurability</u>

Atoms adsorbed on a periodic substrate can form ordered structures.
These structures may be either in or out of registry with the structure
of the substrate. It is convenient to describe this ordering by relating
the Bravais lattice of the adlayer to that of the substrate surface.
Park and Madden[17] have proposed a simple vectorial criterion to classify
the structures. Let a_1 and a_2 be the basis vectors of the adsorbate and
b_1 and b_2 those of the substrate surface; these can be related by

$$\begin{bmatrix} \vec{a}_1 \\ \vec{a}_2 \end{bmatrix} = F \begin{bmatrix} \vec{b}_1 \\ \vec{b}_2 \end{bmatrix} \qquad\qquad (9)$$

with the matrix

$$F = \begin{bmatrix} F_{11} & F_{12} \\ F_{21} & F_{22} \end{bmatrix} ; \qquad\qquad (10)$$

a_1 x a_2 and b_1 x b_2 are the unit cell areas of the adlayer and substrate
surface, respectively; det F is the ratio of the two areas. The relation

between the two ordered structures is classified by means of this quantity as follows:

i)   det F = integer

the structure of the adlayer has the same symmetry class as that of the substrate and is in registry with the latter; the adlayer is termed commensurate.

ii)   det F = irrational number

the adlayer is out of registry with the substrate; the adlayer is termed incommensurate.

iii) det F = rational number

the adlayer is again in registry with the substrate. However, whereas in i) all adlayer atoms are located in equivalent high-symmetry adsorption sites, here only a fraction of adatoms is located in equivalent sites; the adlayer is termed high-order commensurate.

In Fig. 6, we show a simple one-dimensional model illustrating this classification. The periodicity of the substrate surface is represented by a sinusoidal potential of period b and the adlayer by a chain of atoms with nearest neighbor distance a.

Assuming that the structural mismatch between adlayer and substrate is not too large ($\leq$10-15%), the nature of the adlayer ordering on the substrate is determined by the relative interaction strength e_1/V_c which is the ratio of e_1, the lateral adatom interaction in the layer, to V_c, the modulation of the adsorbate-substrate potential parallel to the surface. When the diffusional barrier Vc is large compared to the lateral attraction, commensurate structures will be formed. On the other hand, when the lateral adatom interactions dominates, incommensurate structures will be favored. Only when the competing interactions are of comparable magnitude, may both registry and out of registry structures be stabilized by the complex interplay of these interactions.

Before discussing some examples of structural monolayer phases and their mutual transitions, it seems useful to comment shortly on the substrate corrugation V_c. Remember: V_c corresponds to the modulation along the surface of the bottom of the adsorbate-surface potential well. Its magnitude has been often underestimated because it has been correlated with the corrugation felt by a thermal He atom when scattering at the surface. With respect to the corrugated felt by an adsorbed Xe atom migrating along the surface there are two fundamental differences: the scattering atom feels the corrugation of the repulsive potential and not that of the bottom of the attractive well and the polarizability of He is about 20 times smaller than that of Xe.

It has been shown by Steele,[18] that the physisorption potential of rare gas atoms on a crystal surface is represented, in a good approximation, by a Fourier expansion in the reciprocal lattice vectors G of the substrate surface

$$V(r,z) = V_o(z) + V_{mod}(r) = V_o(z) + \sum_G V_G(z) \exp (iGr) , \qquad (11)$$

evidencing nicely the lateral modulation of the adsorption energy. Here z (> 0) is the distance of the adatom perpendicular to the surface and r the coordinate parallel to the surface. $V_o(z)$ is the mean potential

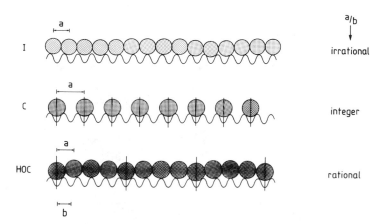

Fig. 6 One-dimensional model for adsorbed monolayers. The substrate is represented by a sinusoidal potential of period b and the adlayer by a chain of atoms with lattice constant a. C-commensurate, HOC-high order commensurate, and I-incommensurate.

energy of an adatom at a distance z, V_G the principal Fourier amplitude. At low temperatures $V_o(z)$ can be replaced by its harmonic approximation:

$$V_o(z) \simeq V_o + V_o'' (z-z_o)^2 . \tag{12}$$

Owing to the rapid convergence of the Fourier series, the second term is usually already one order of magnitude smaller than the first order term, and for the basal plane of graphite Gr(0001) and the fcc(111) surface we obtain

$$\vec{V(r)} = V_o(z) + V_G \{\cos(2\pi s_1) + \cos(2\pi s_2) + \cos 2\pi(s_1+s_2)\} \tag{13}$$

respectively. Here s_1 and s_2 are the dimensionless coordinates of the atoms in the substrate surface unit cell (see Fig. 7).

If we assume a 12-6 Lennard-Jones pair potential to represent the interaction between various atoms of the adsorbate/substrate system, the corrugation in the unit cell of a particular surface is entirely determined by the ratio σ/a and by the magnitude of the binding energy V_o, with σ being the Lennard-Jones diameter of the adatom and a being the nearest neighbor distance in the substrate surface. In this model, the energy difference between adsorption of a Xe-atom in a hollow and in a bridge position, V_{H-B}, amounts to $\sim 0.02\ V_o$ for the graphite (0001) surface and to $\sim 0.01\ V_o$ for the (111) surfaces of Pd, Pt and Ag (similar atomic radii).[18] It is this energy difference between desorption in a hollow site and in a bridge site which represents the corrugation of the potential which determines the activation barrier for the diffusive motion of adatoms and thus ultimately also whether, at a given natural misfit between adlayer and substrate, commensurate structures appear. Taking the actual binding energies for Xe-adsorption on these surfaces,[10,14] the Xe-corrugation is calculated to ~ 3 meV (Gr), ~ 3 meV (Pd, Pt) and ~ 2 meV (Ag).

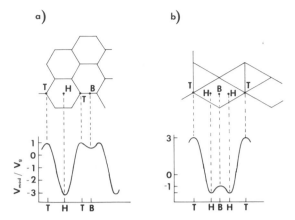

Fig. 7   The variation of the well depth of the rare gas/graphite (0001) (a) and rare gas/Pt (111) (b) potential in the plane parallel to the substrate surface, in the framework of a simple Lennard-Jones model (see text).

The two-body Lennard-Jones potential is only a crude approximation of the interaction between a rare gas atom and a solid surface. While for the rare gas-graphite interaction the inclusion of many body terms, in particular three body forces, significantly improves the physisorption potentials, this approach seems to be insufficient in the case of metal surfaces. Drakova et al.[19] demonstrated recently in a self consistent Hartree-Fock calculation, that the corrugation of the rare gas-transition metal surface potential is substantially enhanced by the hybridization between occupied rare gas orbitals and empty metal d-orbitals. For the interaction of a Ne-atom with the (110) surface of Cu, Ag and Pd, these authors calculated the corrugation of the short bridge site, V_{H-SB} to increase from 0.04 V_0 (Cu) over 0.08V_0 (Ag) to 0.27 V_0 (Pd). Based on the Pd(110) value we can estimate the bridge corrugation V_{H-B} of the (111) surfaces of Pd and Pt to about 0.07-0.08 Vo and that of Ag to about 0.02 Vo" These large values are consistent with the semiempirical Xe-Pd potential, recently developed by Girard and Girardet[20] in order to explain the experimentally observed face specificity of the binding energy and of the induced dipole moment. For Pd(111) these authors evaluated $V_{H-B} \simeq V_0$, and for the Pd(110) surface they calculated $V_{H-SB} \simeq 0.35\ V_0$.

A similar trend also holds for the graphite-rare gas interactions; including anisotropic interactions between the adatom and each carbon atom Vidali and Cole[21] calculate, for example, a corrugation V_{H-B} 0.04 V_0 $\simeq$ = 6 meV for the Xe-atom, which is twice the original value given by Steele. This corrugation seems to be fairly consistent with various experimental results.

The large rare gas-d-metal corrugation has been confirmed recently in experiments and in molecular dynamics studies of Xe adsorption on Pt(111). From the Xe-coverage dependence of the isosteric heat of adsorption, Kern et al.[22] determined the corrugation of the Xe/Pt(111) potential to $\sim$ 30 meV, i.e. about 10% of the binding energy (V_0 = 277 meV in the limit of zero coverage[10]). A somewhat lower value of $\sim$ 10 meV was needed in a molecular dynamics simulation by Black and Janzen[23] in order to stabilize the commensurate $\sqrt{3}$ Xe-phase which has been observed experimentally (see below). Note, that these relatively larger values are compatible with values of diffusion barriers determined in direct experiments on other close packed surfaces as for instance, for Xe on W(110) yielding 47 meV[24].

2. Monolayer Phases of Xe on Pt(111)

As already stated, when the lateral adatom interaction and the substrate corrugation are comparable, the adsorbed monolayer may form various ordered structures, commensurate as well as incommensurate phases allowing the investigation of the corresponding CI-transition as a function of coverage or temperature. In view of the lateral interaction energy e_1 = 43 meV of Xe on Pt(111) and the similar value of the corrugation $V_c \approx$ 30 meV, this adsorption system should be a model system to test the ordering phenomena in two dimensional systems with competing interactions. The Xe-Xe and Xe-Pt interactions favor Xe adsorption in the three fold hollow sites of the Pt(111) surface. Below coverages of $\theta_{Xe} \approx$ 0.33 (Xe adatoms per Pt-substrate atoms) and in the temperature range 60-99 K the xenon condenses in a ($\sqrt{3}$ x $\sqrt{3}$)R30° commensurate solid phase (Fig. 8).[25] This phase leads to very sharp He-diffraction peaks, characteristic for coherent Xe-domains about 800 A in size. As the Xe-coverage is increased above 0.33 the Xe-structure undergoes a transition from the commensurate $\sqrt{3}$ structure (C) to an incommensurate striped solid (SI) phase with superheavy walls (for details we refer the reader to Ref. 26). This weakly incommensurate solid is able to accommodate more Xe atoms than the commensurate phase by consisting of regions of commensurate domains separated by a regularly spaced array of striped denser domain walls. The domain walls have been found to be no sharp interfaces but relaxed broadened regions of increased density, 45 Xe-interrow distances wide (FWHM). Increasing coverage causes the commensurate domains to shrink and brings the walls closer together. The domain walls are thus a direct consequence of the system efforts to balance the competition between the lateral Xe-Xe and the Xe-Pt interactions. The C-SI transition can also be induced by decreasing the temperature below $\sim$ 60 K at constant coverage ($\theta_{Xe} \leq$ 0.33); the driving force for this temperature induced CI-transition are anharmonic effects.[27]

The usual measure for the incommensurability of an I-phase is the misfit m = $(a_C - a_I)/a_C$ where a_C is the lattice parameter of the commensurate phase and a_I that of the incommensurate structure. For striped I-phases, the misfit has of course uniaxial character, being defined only along the direction perpendicular to the domain walls. Quantitative measurements[26] of the misfit during the C-SI transition of Xe on Pt(111) have revealed a power law of the form m = $1/\ell \propto$ $(1-T/T_c)^{0.51\pm0.04}$, i.e., the distance between nearest neighbor walls ℓ scales with the inverse square root of the reduced temperature. This square root dependence is the results of an entropy mediated repulsion between meandering nearest neighbor walls and is in accord with theoretical predictions.[28]

With increasing incommensurability the domain wall separation becomes progressively smaller until at a critical misfit of $\sim$ 6.5% the Xe domain wall lattice spontaneously rearranges from the striped to the hexagonal symmetry in a first order transition.[29] A further increase of the incommensurability by adding more and more Xe eventually results in an adlayer rotation to misalign itself with the substrate in order to minimize the increasing strain energy due to the defect concentration. This continuous transition starting at a rotated phase (HIR) follows a

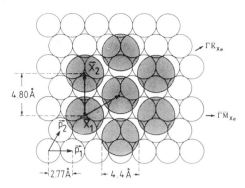

$(\sqrt{3} \times \sqrt{3})R\,30°-Xe\,/\,Pt\,(111)$

Fig. 8 Geometry of the commensurate $(\sqrt{3}x\sqrt{3})$ R30° Xe-monolayer on Pt(111).

power law $\phi \propto (m-0.072)^{1/2}$ starting at a critical separation between nearest neighbor walls $\ell_c \simeq 10$ Xe-interrow distances[29] (Fig. 9a).

Novaco and McTague[30] have shown that these adlayer rotations for monolayers far from commensurability are driven by the interconversion of longitudinal stress into transverse stress. These authors also showed that the rotational epitaxy involves mass density waves (MDW)[also known as static distortion waves (SDW)], i.e., there exists a periodic deviation of the position of monolayer atoms from their regular lattice sites. Indeed, it is the combination of rotation and small displacive distortions of the adatom net which allows the adlayer to minimize its total energy in the potential relief of the substrate. In a diffraction experiment, these mass density waves should give rise to satellite peaks.

Fuselier et al.[31] have introduced an alternative concept to explain the adlayer rotation: the "coincident site lattice". They pointed out that energetically more favorable orientations are obtained for rotated high-order commensurate structures. The larger the fraction of adatoms located in high-symmetry, energetically favorable sites, the larger the energy gain and the more effective the rotated layer is locked. It turns out that the predictions of the coincident site lattice concept for the rotation angle versus misfit agrees well with the Novaco-McTague predictions.

The experimental results do not allow so far to decide whether the Novaco-McTague mechanism involving MDW or the "coincident in lattice concept involving HOC structures, or may be both have the determining role in driving the adlayer rotation. In particular, no mass density wave satellites have been observed in electron and x-ray diffraction experiments from rotated monolayers[32] so far. In He diffraction scans of rotated Xe monolayers on Pt(111) we have, however, observed satellite peaks (see the peak marked by arrow, Fig. 2) at small Q-vectors. Originally we assigned these peaks to a higher-order commensurate superstructure[9]. However, Gordon[33] pointed out that these satellites could be due to the MDW. Here, we will show that both MDW as well as commensurate buckling satellites are present in the rotated Xe monolayers

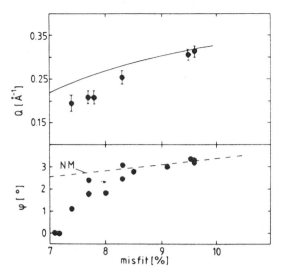

Fig. 9 Rotation angle ϕ of hexagonal incommensurate rotated Xe-monolayer on Pt(111) as a function of misfit m (bottom) and dispersion of the mass density wave satellites with m (top). The solid line is Gordon's relation (see text).

on Pt(111). The distinction between the two types of satellites is straightforward. As pointed out by Gordon, the wave vector, Q, of the MDW satellites should be subject to the following relation:

$$Q \approx (8\pi/a_{Xe}^R) \ m/\sqrt{3}) \ (1+m/8) \ , \tag{14}$$

with m the misfit, and a_{Xe}^R the lattice constant of the rotated Xe layer. For a not too large misfits, this MDW satellite should appear in the same direction as the principal reciprocal lattice vector of the Xe layer, i.e., in the $\bar{\Gamma} \bar{M}_{Xe}$ direction. On the other hand, according to its particular structure (Fig. 10b), the commensurate buckling should have its maximum amplitude in the $\bar{\Gamma} \bar{K}_{Xe}$ direction. Moreover, these commensurate buckling satellites should only be present at the particular coverages where a certain high-order commensurability becomes favorable, in the present case at monolayer completion (m = 9.6%) (see also Ref. 9), whereas the MDW satellites should be present in the entire misfit range where the Xe layer is rotated (7.2%-9.6%).

In Fig. 9b we show the dispersion of the MDW-satellites deduced from a series of diffraction scans like in Fig. 2, taken in the $\bar{\Gamma} \bar{M}_{Xe}$-direction, and compare them with Gordon's prediction for the MDW given above. The data follow qualitatively the predicted dependency; the agreement becomes quantitative at misfits ≥ 8%. The reason for the better agreement at large misfits is due to the fact that Gordon's analysis of the MDW (similar to Novaco-MacTague's model calculations) have been performed in the linear response approximation of the adsorbate-substrate interaction; this approximation is only justified at larger misfits, where the adlayer topography corresponds rather to a weakly modulated uniform layer than to a domain wall lattice.[34]

In Fig. 10(a) we show scans like in Fig. 2 but now measured in the $\bar{\Gamma} \bar{K}_{Xe}$ direction at small Q for rotated Xe layers of misfits 8% and 9.6%. At variance with the scans in the $\bar{\Gamma} \bar{M}_{Xe}$ direction (Fig. 2), a satellite peak is observed only for the complete Xe monolayer (m = 9.6%). Being present only at a particular misfit this peak does not originate from a MDW but from the buckling of a HOC-structure. The location of this satellite peak at Q = 0.28 Å^{-1} corresponds to a buckling period of 23Å and can be ascribed to a high-order commensurate structure shown in Fig. 10(b) and described in detail in Ref. 9.

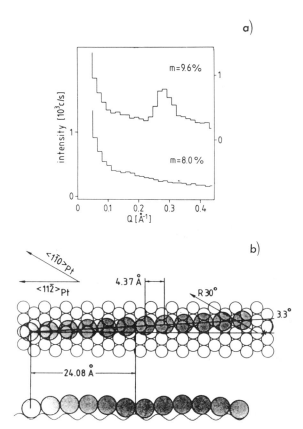

Fig. 10 a) Polar He-diffraction scan of rotated Xe monolayers on Pt(111) taken along the $\bar{\Gamma} \bar{K}_{Xe}$ azimuth at two misfits. b) Plane and side view of a 3.3° rotated Xe-domain at the misfit m = 9.6%.

MONOLAYER FILMS AND DYNAMICS OF COMPETING INTERACTIONS

When adsorbing a layer of atoms (for example rare gas atoms) on a substrate surface we are not dealing only with static interaction effects which determine the structure of the adlayer but also with dynamical interactions between collective excitations (for example phonons) of adlayer and substrate.

The phonon spectrum of a crystal surface consists of two parts: The bulk bands, which are due to the projection of bulk phonons onto the two-dimensional Brillouine zone of the particular surface, and the

68

specific surface phonon branches[8]. A surface phonon is defined as a localized vibrational excitation with an amplitude which has wavelike characteristics parallel to the surface and decays exponentially into the bulk, perpendicular to the surface. In Fig. 11a we display as an example the dispersion curves of the clean Pt(111) surface along the Γ M_{Pt} azimuth[35]. The solid lines, termed S_1 and S_2, are specific surface phonon branches. These true surface modes only exist outside or in gaps of the projected bulk phonon bands of equivalent symmetry. The mode MS is an example of a 'surface resonance' which exists inside the bulk bands. Of particular interest is the lowest frequency mode S_1 below the transverse bulk band edge. In this mode, the atoms are preferentially vibrating in the plane defined by the surface normal and the propagation direction, i.e. in the sagittal plane. This wave is the famous Rayleigh wave.

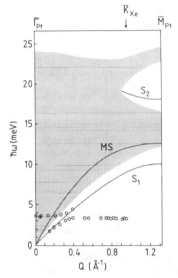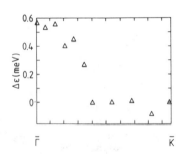

Fig. 11 Left: Surface phonon spectrum of the clean Pt(111) surface along the $\overline{\Gamma}$ $\overline{M}_{Pt}$-azimuth (solid lines-discrete surface phonon branches, shaded regions-projected bulk bands) and of a Xe-monolayer adsorbed on this surface (dots). Right: Measured line width $\Delta\epsilon$ = $[(\delta E)^2 - E_I^2]^{1/2}$, with δE the FWHM of the photon peak and E_I the intrinsic fundamental broadening (here E_I = 0.32 meV) of the Xe-monolayer phonon.

Adsorbing a layer of densely packed atoms on the substrate surface adds three additional phonon modes to the system; two in-plane modes and one mode with polarization perpendicular to the surface. The frequency and dispersion of these modes is governed by the "spring constants" which couple the adatoms laterally and to the substrate. In the following we concentrate on the perpendicular mode. In physisorbed systems, the electronic ground state of the adsorbate is only weakly perturbed upon adsorption. The physisorption potential is rather flat and shallow, i.e. the spring constant of the vertical adatom-motion is weak and the corresponding phonon frequencies are low (dots in Fig. 11a).

Being a dynamical system, the adlayer modes can couple to substrate phonon modes of the same polarization. The main effects of the phonon mediated coupling between adlayer and substrate, the mode hybridization and life time shortening can be also seen in Fig. 11 for the system Xe/Pt(111).

At the Brillouin zone boundary the adlayer mode lying well below the substrate modes the influence of the substrate-adlayer coupling is negligible; neither dispersion nor linewidth of the adlayer mode are influenced by the substrate phonons. At the zone center in the whole region near Γ, however, where the adlayer mode overlaps the bulk phonon bands of the substrate a substantial linewidth-broadening, i.e. lifetime-shortening, is obvious (Fig. 11b). The excited adlayer modes decay by emitting phonons into the substrate; they become leaky modes. The most dramatic coupling effect, is a dramatic hybridization splitting around the crossing between the adlayer mode and the substrate Rayleigh wave obvious in Fig. 11a. A detailed quantitative account of these dynamical coupling effects has been given recently in Refs. 36 and 37.

MULTILAYER FILM GROWTH

1. <u>Multilayer Growth Mode and Monolayer Structure</u>

The mode of nucleation and initial growth of thin films is a matter of long standing interest. The growth-mode of a film is usually classified according to the morphology.[38] Frank-van der Merwe growth, type-1 growth: with increasing coverage the film forms a sequence of stable uniform layers, i.e. the film grows layer-by-layer. Volmer-Weber growth, type-3 growth: small 3D-cluster are nucleated directly on top of the bare substrate. An intermediate situation is the Stranski-Krastanov growth, type-2 growth: the deposit growths initially in a few (often only one) uniform layers on top of which 3D-islands are formed.

The important parameters which determine the equilibrium configuration of an heteroepitaxial system are the film-substrate interaction strength (not the corrugation as in the case of monolayer phases) and the lateral adatom-interaction in the film. Usual measures of these quantities are the isosteric heat of adsorption (which is identical to the quantity V_0 defined in Eq. (11)) and the lateral adatom attraction e_1. Based on simple thermodynamics,[39] it was argued that uniform layer growth should be observed if the adsorbate-substrate interaction V_0 is stronger than the lateral film interaction, e_1, i.e. for so called strong substrates. Weak substrates, on the other hand, should favor a three-dimensional cluster growth. Experiments designed to test these predictions[40] showed, however, that at low temperatures Frank-van der Merwe growth is restricted to a very narrow intermediate range of substrate strengths (see Fig. 12). Both, small as well as large V_0/e_1 values resulted in a nonuniform growth.

A more transparent and general criterion to analyze the type of growth to be expected results from the thermodynamic treatment given by Ernst Bauer many years ago.[41] The criterion for layer-by-layer (Frank van der Merwe) growth has been given recently by Bauer and van der Merwe in a more explicit form for an n-layer film:[42]

$$\gamma_{f(n)} + \gamma_{i(n)} - \gamma_s \leq 0 , \qquad (15)$$

with γ_s and $\gamma_{f(n)}$ the surface energies of the semi-infinte substrate and the n-monolayer film and $\gamma_{i(n)}$ the interfacial energy. Equation (15) is rigorously fulfilled in the trivial case of homoepitaxy, when $\gamma_{f(n)} = \gamma_s$ and $\gamma_{i(n)} = 0$. In the heteroepitaxial case the obvious condition is that $\gamma_f < \gamma_s$. The inequality has to be large enough to fulfill Eq. (15) because in general the interfacial energy has no reason to be negative

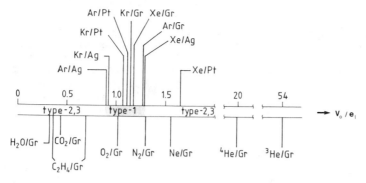

Fig. 12 Multilayer film growth modes of various adsorbate/substrate systems ordered on the scale of substrate strengths. Systems showing Frank-van der Merwe growth are marked by bars above the center line.

and, in addition, because substrate and film bulk have at least a different natural lattice constant and thus the first monolayer has in general an "unnatural" structure (at least an "unnatural" lattice constant); thus in general $\gamma_{f(1)} \geq \gamma_f$ and/or $\gamma_{i(1)}$ positive and non-negligible.

Bauer and van den Merwe[42] have included in $\gamma_{i(n)}$ the n-dependent strain energy of the film. In order to emphasize the role of the structure, Eq. (15) may be also written for each layer of the film as

$$\gamma_{f(n)} + \gamma_{i(n,n-1)} - \gamma_{f(n-1)} \leq 0 , \qquad (16)$$

with $\gamma_{i(n,n-1)}$ the interfacial energy between the n-th layer and the (n-1)-layer film and thus e.g. for the second monolayer

$$\gamma_{f(2)} + \gamma_{i(2,1)} - \gamma_{f(1)} \leq 0 . \qquad (17)$$

Because the surfaces of the n and n-1 film consist of the same atom species, $\gamma_{f(n)} \approx \gamma_{f(n-1)}$ with $\gamma_{f(n)}$ slightly smaller if the n-th layer has a more "natural" structure than the (n-1)th layer, which it certainly tends to have. However, if the structures of the two layers (n and n-1) differ, the interfacial energy $\gamma_{i(n,n-1)}$ becomes important and of course positive and Eq. (16) is not fulfilled. We may thus conclude that even if $\gamma_f < \gamma_s$ fulfilled, but the structure of the first monolayer differs appreciably from that of its own bulk, 3-D growth sets in above the first monolayer (Stranski-Krastanov mode). The set in of this mode may be retarded a few monolayers by long range influence of the substrate, but still there will be no layer-by-layer growth.

The rather restrictive condition for layer-by-layer growths that the structure (symmetry and lattice constant) of the first monolayer should be almost identical to that of the own bulk, explains also the reintrant non-wetting illustrated in Fig. 12. Indeed, on strong substrates the first adsorbed monolayer is in general compressed beyond the density of the close packed plane of its own bulk and thus no layer-by-lower growth takes place.

71

This behavior is consistent with a recent molecular dynamics simulation of Grabow and Gilmer,[44] the result of which is summarized in a 'phase' diagram shown in Fig. 13. The deposition on a fcc(100) substrate surface only occurs in a layer-by-layer mode for strong substrates $V_o/e_\ell \geq 1$ with negligible structural misfit $m \approx 0$ (thick line). The importance of the structural mismatch is also evident in the phase boundary between Stranski-Krastanov and Volmer-Weber growth which shifts substantially to larger substrate strengths with increasing misfit.

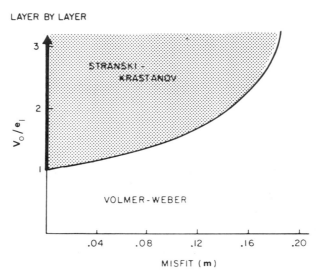

Fig. 13 "Phase" diagram of multilayer growth on a fcc(100) surface in the substrate strength V_o/e_ℓ-misfit plane, according to molecular dynamics calculations of Grabow and Gilmer.[44]

As will be shown below (section 5.2) the peculiar case of Xe/Pt(111), a strong substrate case growing yet layer-by-layer can be explained in the same frame. Indeed, the Xe monolayer is not compressed beyond its natural bulk lattice constant (4.34 A at 25 K) simply because it becomes locked into a HOC-phase with exactly this lattice constant.

2.   Epitaxial Multilayer Growth of Xe on Pt(111)

The He energy loss spectra shown in Fig. 3 contain besides lattice dynamics information also direct information concerning the growth characteristics of the Xe multilayers. This information is in the diffuse elastic peak, i.e., in the peak at zero energy exchange. This peak originates from scattering at defects and its intensity is a sensitive measure of surface disorder. From the comparison with spectra taken from surfaces of known disorder, we can infer that the monolayer is well ordered and the multilayers even better. For the 25 ML thick film in Fig. 3, the diffuse elastic peak has nearly vanished. This shows that the 25 ML film is very flat, and thus that Xe on Pt(111) exhibits complete wetting. This goes along with the layer-by-layer evolution of the surface phonon dispersion discussed in detail in Ref. 9 and 36.

The structure of the Xe multilayers has been characterized by measuring polar and azimuthal He-diffraction scans shown in Fig. 14. As already emphasized earlier at monolayer completion the Xe monolayer on 3.3°. Pt(111) is a Novaco-McTague rotated layer with rotation angle $\phi = \pm$ 3.3°. The azimuthal plots in Fig. 14 show that all consecutive layers growing on top of the first layer are likewise rotated by $\phi = \pm$ 3.3°. Lattice constant and average domain size, as deduced from the polar diffraction and average domain size, are also unchanged with increasing film thickness, i.e., $a_{Xe} = 4.33 \pm 0.03$ Å and average domain size ≈ 300 Å, respectively. Thus, the consecutive layers grow epitaxially on the preceding ones. Within experimental confidence there is no misfit between the nearest neighbor distances in the monolayer ($a_{Xe}^{R-ML} = 4.33$ Å, T = 25K) and in the bulk Xe ($a_{Xeb} = 4.34$ Å, T = 25 K). As already mentioned this structural compatibility, which leads to an unstrained layer-by-layer growth, appears to be a direct result of registry forces. Indeed Xe/Pt(111) being a "strong substrate" system, the monolayer lattice parameter would be expected, in the absence of registry forces,

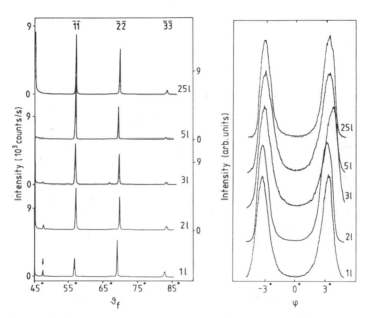

Fig. 14 Polar (left) and azimuthal (right) He diffraction scans of Xe-films of indicated thickness.

to be compressed well beyond the bulk value. However, as shown in chapter 3.2, at misfits of 9.6% the rotated monolayer locks into an energetically favorable high order commensurate structure. It is this high order commensurate locking of a fraction of adatoms which, by counterbalancing the tendency of the strong substrate to compress the monolayer lattice beyond the bulk value, allows for an unstrained layer-by-layer growth.

Note that the small peak at low Q-values in Fig. 14 is a MDW satellite already shown in Fig. 2. The very gradual disappearance of the mass density waves with increasing film thickness, emphasizes that the influence of the substrate on adsorbed multilayer films extends over several layers.

REFERENCES

1. M. Bienfait, Surf. Sci. <u>162</u>, 411 (1985).
2. J. G. Dash, Physics Today, Dec. 1985, 26.
3. K. Kern, P. Zeppenfeld, R. David, and G. Comsa, J. Vac. Sci. Technol. <u>A6</u>, 639 (1988).
4. R. David, K. Kern, P. Zeppenfeld and G. Comsa, Rev. Sci. Instr. <u>57</u>, 2771 (1986).
5. E. Zaremba and W. Kohn, Phys. Rev. <u>B 15</u>, 1769 (1977).
6. T. Engel and K. H. Rieder, <u>Springer Tracts in Modern Physics</u> (Springer, Berlin, 1982), Vol. 91.
7. P. Zeppenfeld, K. Kern, R. David and G. Comsa, Phys. Rev. Lett. <u>62</u>, 63 (1989), A. M. Lahee, J. R. Manson and J. P. Toennies, Ch. Wöll, Phys. Rev. Lett. <u>57</u>, 471 (1986).
8. K. Kern and G. Comsa, Adv. Chem. Phys., K. P. Lawley, ed. (Wiley, New York, 1989 p. 211).
9. K. D. Gibson and S. J. Sibener, Faraday Discuss. Chem. Soc. <u>80</u>, 203 (1985); K. Kern, R. David, R. L. Palmer and G. Comsa, Phys. Rev. Lett. <u>56</u>, 2823 (1986).
10. K. Kern, R. David and G. Comsa, Surf. Sci. <u>175</u>, L669 (1986).
11. J. W. M. Frenken, B. J. Hinch and J. P. Toennies, Surf. Sci. <u>211/212</u>, 21 (1989), A. C. Levi, R. Spadacini and G. E. Tommei, Surf. Sci. <u>182</u>, 504 (1982).
12. G. Comsa and B. Poelsema, Appl. Phys. <u>A38</u>, 153 (1985).
13. B. Poelsema, L. K. Verheij and G. Comsa, Phys. Rev. Lett. <u>51</u>, 2410 (1983)
14. J. Unguris, L. W. Bruch, E. R. Moog and M. B. Webb, Surf. Si. <u>109</u>, 522 (1981).
15. J. P. McTague, J. Als Nielsen, J. Bohr and M. Nielsen, Phys. Rev. <u>B25</u>, 7765 (1982).
16. K. Kern and G. Comsa, in <u>Physics and Chemistry of Solid Surfaces VII</u>, R. Vanselow and R. F. Howe, eds. (Springer, Berlin, 1988) p. 64.
17. R. L. Park and H. H. Madden, Surf. Sci. <u>11</u>, 188 (1968).
18. W. Steele, Surf. Sci. <u>36</u>, 317 (1973).
19. D. Drakova, G. Doyen and F. V. Trentini, Phys. Rev. <u>B32</u>, 6399 (1985).
20. C. Girard and C. Girardet, Phys. Rev. <u>B36</u>, 909 (1987).
21. G. Vidali and M. W. Cole, Phys. Rev. <u>B29</u>, 6736 (1984).
22. K. Kern, R. David, P. Zeppenfeld and G. Gomsa, Surf. Sci. <u>195</u>, 353 (1988).
23. J. Black and A. Janzen, Surf. Sci. <u>217</u>, 199 (1989).
24. J. R. Chen and R. Gomer, Surf. Sci. <u>94</u>, 456 (1980).
25. K. Kern, R. David, R. L. Palmer and G. Comsa, Phys. Rev. Lett. <u>56</u>, 620 (1986).
26. K. Kern, R. David, P. Zeppenfeld, R. L. Palmer and G. Comsa, Solid State Comm. 62, 361 (1987).
27. M. B. Gordon and J. Villain, J. Phys. <u>C18</u>, 3919 (1985).
28. V. L. Pokrovsky and A. L. Talapov, Sov. Phys. JETP <u>51</u>, 134 (1980).
29. K. Kern, Phys. Rev. <u>B35</u>, 8265 (1987).
30. A. D. Novaco and J. P. McTague, Phys. Rev. <u>B19</u>, 5299 (1979).
31. C. R. Fuselier, J. C. Raich and N. S. Gillis, Surf. Sci. <u>92</u>, 667 (1980).
32. C. G. Shaw, S. C. Fain and M. D. Chinn, Phys. Rev. Lett. <u>41</u>, 955 (1978), K. L. Damico et al., Phys. Rev. Lett. <u>53</u>, 2250 (1984).
33. M. B. Gordon, Phys. Rev. Lett. <u>57</u>, 2094 (1986).
34. H. Shiba, J. Phys. Soc. Jpn. <u>4R</u>, 211 (1980).
35. K. Kern, R. David, R. L. Palmer, G. Comsa and T. S. Rahman, Phys. Rev. <u>B33</u>, 4334 (1986).
36. K. Kern, P. Zeppenfeld, R. David and G. Comsa, Phys. Rev. <u>B35</u>, 886 (1987), B. M. Hall, D. L. Mills, P. Zeppenfeld, K. Kern, U. Becher

and G. Comsa, Phys. Rev. <u>B40</u>, 6326 (1989)..

37. B. M. Hall, D. L. Mills and J. Black, Phys. Rev. <u>B32</u>, 4932 (1985).

38. J. A. Venables, G. D. T. Spiller and M. Hanbücken, Rep. Prog. Phys. <u>47</u>, 399 (1984).

39. D. E. Sullivan, Phys. Rev. <u>B20</u>, 3991 (1979).

40. M. Bienfait, J. L. Senguin, J. Suzanne, E. Lerner, J. Krim and J. G. Dash, Phys. Rev. <u>B29</u>, 983 (1984).

41. E. Bauer, Z. Kristallogr. <u>110</u>, 372 (1958).

42. E. Bauer and J. H. Van der Merwe, Phys. Rev. <u>B33</u>, 3657 (1986).

43. R. J. Murihead, J. G. Dash and J. Krim, Phys. Rev. <u>B29</u>, 6985 (1984).

44. M. Grabow and G. H. Gilmer, Surf. Sci. <u>194</u>, 333 (1988).

HELIUM ATOM SCATTERING STUDIES OF SINGLE UNCORRELATED AND CORRELATED DEFECTS ON SURFACES

A. Lock, B. J. Hinch and J. Peter Toennies

Max-Planck-Institut für Strömungsforschung
Bunsenstr. 10
D - 3400 Göttingen, Federal Republic of Germany

ABSTRACT

Helium atom scattering (HAS) from crystal surfaces is unique among all surface science probes in that it is sensitive to only the outermost surface layer, entirely non-destructive and applicable to all types of surfaces: insulators, semiconductors, and metals. With the recent advent of improved energy resolution (1%) and sensitivity (dynamic range of 10^6) HAS is becoming increasingly useful in the study of defects on surfaces as well as for studying growth processes.

First the basic HAS techniques are reviewed and compared with LEED. Then several examples of high resolution studies of disordered surface structures will be presented. Recent work has concentrated on studying randomly distributed point or linearly extended defects of low concentration. The large angle diffuse (incoherent) elastic scattering from isolated step edge defects on otherwise smooth surfaces, e.g. metallic fcc(001) and (111) planes, yields information on the concentration, orientation and structure of single steps (Pt(111), Al(111), Cu(111) and Ni(001)). In accord with the unique sensitivity of HAS to step edges the diffraction patterns from periodically stepped surfaces show a rich structure. A careful kinematic analysis of the stepped (332) aluminum surface reveals extensive facetting. The temperature induced changes in step concentration, roughening and facetting have also been studied.

Finally very recent results on the epitaxial growth of Pb on Cu(111) have revealed an unexpected intensity nodulation indicating oscillations in the stability of the layers related to a quantum size effect. Further improvements in sensitivity and resolution will make even more detailed studies of the structure and growth on insulators, semi-conductors and metal surfaces possible.

INTRODUCTION

The first helium atom scattering experiments from surfaces were carried out in 1929 by Stern.[1] He observed broad diffraction peaks in scattering from the (001) cleavage faces of LiF. These experiments, following two years after the electron diffraction experiments of

Davisson and Germer,[2] provided the first evidence for the wave nature of atomic particles. In the following years low energy electron diffraction (LEED) became a powerful technique for surface studies.[3] Analogous experiments with atoms were hampered by problems related to producing nearly monoenergetic beams of atoms. This was achieved by either using a mechanical velocity selector of the Fizeau type[4] or a grating monochromator[5] which not only made the apparatus very complicated but also reduced the beam flux making measurements difficult. This changed with the advent of nozzle beams introduced by Becker and coworkers in the 1950's.[6] In 1977 Toennies and Winkelmann[7] discovered that a quantum effect peculiar to ^{4}He is responsible for an extraordinary strong adiabatic cooling of ^{4}He to temperatures in the expanding gas of the order of $T = 10^{-3}$ K. As a result these beams have a velocity half width of $\Delta v/v \propto (T/T_0)^{1/2} \approx 10^{-2}$, where $T_0 = 80K$ is typically the temperature of the gas in the source. Thus monoenergetic beams with very high fluxes (10^{20} atoms/sr·sec) became available for surface studies. Helium atoms are especially suitable since they interact only very weakly with all solid surfaces and appreciable condensation will only occur at temperatures below about 4K. Helium atoms cannot penetrate the crystal surfaces because of the strong Pauli exchange repulsion between the electrons of their outer shell and those of the solid at distances of about 3Å from the surface atomic layer. Neither can they damage the crystal because of their low kinetic energies of about 20 meV. Figure 1 compares the typical trajectories of electrons with those of He atoms in a surface scattering experiment. The interaction of He atoms usually only involves a single collision. The theoretical analysis is thus considerably simplified compared to that required for LEED where multiple interactions play a big role.

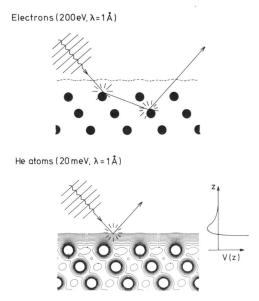

Electrons (200eV, λ=1Å)

He atoms (20 meV, λ=1Å)

Fig. 1   Comparison of the different types of interactions of electrons and helium atoms in scattering from a single-crystal surface. Electrons can penetrate the surface and undergo multiple scattering before leaving the surface, whereas helium atoms, because of their much greater cross sections, interact in a single collision with the outer electrons of the surface via a physisorption potential with a well minimum at about $z \simeq 3Å$.

Figure 2 shows a simple comparison of the characteristic experimental features of a monoenergetic electron source with a helium atom beam source. It is interesting to note that the helium beam fluxes are five orders of magnitude greater than for electrons. This advantage is, however, lost at the mass spectrometer detector, which has a sensitivity of about 10^{-5}. Thus the overall signals are about the same for helium and electron scattering. The helium atom beam energies can be varied from about 5 meV to over 80 meV by changing the source temperatures from 30K to 350K. The energetic half-width of the incident beam then goes from about 0.08 meV up to about 1 meV depending also on the source pressure. Figure 3 shows a typical helium atom scattering (HAS) apparatus.[8] The apparatus has many differential pumping stages between source and detector in order to reduce the helium background gas in the detector to about 10^{-15} Torr ($\approx$10 He atoms/cc). At this pressure the background count rate is about 10 ions/sec compared to an estimated signal for the incident beam of over 10^8 counts/sec. Thus in principle measurements over a dynamic range of 10^7 are possible. In the apparatus of Fig. 3 the detector is mounted at the end of a long flight tube to provide for a high (1%) time-of-flight resolution. To change scattering angles it is possible to either rotate the detector flight tube, which is connected to the scattering chamber by a bellows, or keep the detector fixed and rotate the target, which is the preferred mode of operation. This type of high-resolution helium scattering apparatus is now in operation or in construction at several laboratories: Genova (Boato), Madrid (Miranda), Nancy (Mutaftschiev), Paris (Lapujoulade), Cambridge (Allison), Seibersdorf, Austria (Semerad), Berlin (Rieder), Köln (Neuhaus), Jülich (Comsa), Göttingen (Toennies), Karlsruhe (von Blanckenhagen), Murray Hill (Doak), Tallahassee (Skofronick), Chicago (Sibener), Iowa (Stassis), Providence (Greene), Pennsylvania State (Frankl), Seattle (Engel), where the names in parenthesis indicate the senior scientists in each of the laboratories.

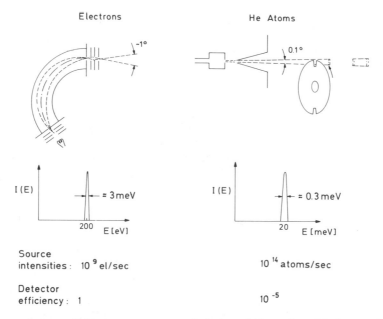

Fig. 2 Comparison of beam sources and intensities for high-resolution electron and helium atom scattering.

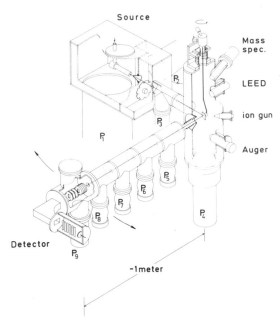

Source

Mass spec.

LEED

ion gun

Auger

P_2

P_3

P_1

P_5

P_6

P_7

P_4

P_8

Detector

P_9

~1meter

Fig. 3   Typical apparatus used for elastic and inelastic helium atom
scattering experiments. Nine differential pumping stages are
used between the beam source (pump P_1) and the detector (pump
P_9) to reduce the partial pressure of helium in the mass
spectrometer. The angular distributions are usually measured by
rotating the crystal and keeping the detector fixed. In the
apparatus shown the detector can be rotated around the
scattering chamber as well.

Helium atom scattering has previously been extensively used to
study the periodic structures of clean and adsorbate covered structures.
This work is described in several extensive reviews.[9] Inelastic helium
atom scattering is also an important method for studying surface phonon
dispersion curves of clean and adsorbate covered surfaces.[10] In this
review we will describe some recent studies of the scattering of helium
atoms from single defects and from stepped surfaces. Our emphasis here
will be on the phenomena and the basic experimental observations and less
on the details of the theoretical analysis. The following section will
first deal with scattering experiments from single randomly distributed
defects. Next we will describe studies from periodically stepped
surfaces and show how helium scattering makes it possible to identify
directly the actual facets on an annealed vicinal surface.

HELIUM ATOM SCATTERING FROM SINGLE DEFECTS

Before discussing scattering from defects we present some data on
the diffraction of helium atoms from the smooth face of a metal surface.
Figure 4 shows a typical angular distribution with a fixed angle
$\theta_i + \theta_f = \theta_{S_D} = 90°$ for scattering from the densely packed (111) surface of
platinum, which is an fcc crystal.[11] In this figure and elsewhere the
incident angle has been converted to parallel momentum transfer by the

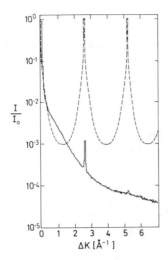

Fig. 4   Comparison of the angular distribution for scattering of
         electrons (----, estimated) and helium atoms (———, measured)
         from a Pt(111) surface along the <112> azimuth.   The incident
         helium beam wavevector is 7.83Å^{-1} (E_i = 32 meV).

simple expression

$$\Delta K = k_i(\sin \theta_f - \sin \theta_i) ,$$ (1)

assuming only elastic scattering.   The wave vector in this particular
experiment was k_i = 7.83Å^{-1} (E_i = 32 meV) corresponding to a source
temperature of $T_s \approx$ 130 K.   Note that the intensity is plotted on a log
scale and extends over five orders of magnitude.   The first-order
diffraction peak is clearly visible even though its intensity is only
10^{-3} of the incident beam intensity, I_o.   The second-order diffraction
peak is much weaker and has an intensity of only 10^{-5} I_o.   The sharply
decaying background is due to scattering from defects and phonons.   The
diffraction intensities can be simply analyzed using the Eikonal
approximation if it is assumed that the He-surface interaction potential
is a hard repulsive wall with a sinusoidal corrugation with the
periodicity of atoms of the surface layer.   Such an analysis reveals that
the effective corrugation amplitude is only 10^{-3}Å, in good agreement with
a calculation of the expected electronic density distribution at the
surface.[11]   This weak diffraction intensity contrasts greatly with that
expected from LEED which is shown qualitatively in the same figure for
comparison.   In the early days of helium surface scattering this weak
diffraction intensity was considered to be a serious disadvantage of the
method, but as we show here, it is, in fact, of considerable advantage in
studying defects.   Very similar results are obtained for the (111), (100)
fcc faces and for the (110) and (100) bcc faces of metal single crystals,
all of which are very smooth as probed by thermal helium beam scattering.
Only on the furrowed (110) face of an fcc crystal do scattering
measurements perpendicular to the rows show considerable diffraction
intensities.   Thus most of the metal surfaces act as nearly perfect
mirrors for helium atoms.   This extreme flatness of metal surfaces is
commonly explained by the surface smoothing effect of the conduction
electrons, first proposed by Smoluchowski.[12]

In 1984 Poelsema, Palmer and Comsa showed that even small amounts of adsorbed CO molecules on the surface of Pt(111) lead to a considerable attenuation of the specular intensity with an effective integral cross section of about 120Å^2.[13] This cross section is far greater than typical gas kinetic cross sections which are about 30Å.[2] The large size can be explained by realizing that the long-range attractive potential of the CO molecules will also effect the trajectories of the helium atoms. Thus the helium atoms passing through this weak potential field will be deflected sufficiently so that they are no longer detected in the specular direction as illustrated in Fig. 5. Similar large cross sections are known from crossed-molecular-beam scattering experiments where the cross sections are referred to as total integral cross section.[14] As a result of the large cross sections, adsorbate concentrations of about 10^{-4} monolayers lead to 1% attenuation of the beam, which is easily detectable. This effect is very useful for studying adsorbate surface concentrations. Subsequently these same authors showed that step edges had effective cross sections for scattering in a plane perpendicular to the edges of about 12Å.[15] Unfortunately the integral cross sections are difficult to interpret in terms of the structure of the adsorbates or steps.

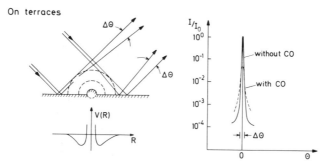

Fig. 5   Intensity attenuation by single adsorbates. The schematic diagram shows the trajectories of helium atoms scattered from adsorbed CO molecules on a smooth metal surface. The long-range attractive potential leads to small deflections of the atoms out of the specular beam.

More direct information on the physical features of the adsorbates is contained in the wide-angle distribution of atoms scattered by the defects out of the specular beam. In 1987 Lahee, Manson, Toennies and Wöll were able to observe oscillations in the wide-angle scattering of helium from single molecules.[16] Figure 6 shows the angular distribution for scattering from CO molecules adsorbed on Pt(111).[17] At small parallel wavevector transfers ΔK a small amount of attenuation of the beam is visible, but beyond $\Delta K = 3\text{Å}^{-1}$ an increase in intensity is observed with three distinct maxima. The CO surface coverage was estimated to be only $\theta = 0.05$ monolayers. At this temperature and coverage CO is known not to form islands and there is evidence that the single CO molecules are distributed randomly over the surface.[18] Thus the observed scattering can be attributed to an incoherent superposition of the diffraction patterns from single CO molecules.

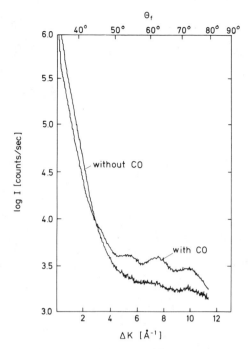

Fig. 6　Effect of CO adsorption on the angular distribution of the total intensity measured at 10° away from the <112> azimuth. The CO exposure was 0.5 Langmuir corresponding to about 5% of a monolayer. The incident helium atom wavevector was $k_i = 9.7\text{Å}^{-1}$.

　　　　The theory for the observed wide-angle scattering reveals that in fact there are two contributing scattering amplitudes as shown in Fig. 7. One amplitude f_1 describes the forward Fraunhofer diffraction from the molecules. This is reflected back by the smooth metal surface into angles near the specular. The other contribution f_2 is due to the uniform back-scattering from the short-range repulsive potential of the adsorbed CO molecules. This can be approximated by assuming a hard hemispherical boss. The observed intensity, which is proportional to $|f_1+f_2|,^2$ contains interference terms between the two contributions. Numerical scattering calculations for a realistic potential predict that the spacing of the undulations is inversely proportional to the diameter of the hard boss.[19] The diameter of the CO molecules determined by a best fit of the data of Fig. 6 yields a value of 2.4Å, which agrees remarkable well with that predicted for the He-CO gas phase potential surface. It is important to realize that this experiment measures the form factor for scattering from a single "defect" directly.

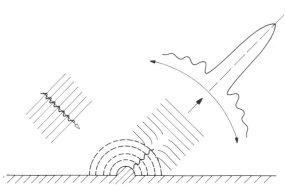

Fig. 7 Diffraction from a single chemisorbed molecule. Schematic
diagram showing the interaction of an incident wave packet with
an adsorbed atom or molecule on a smooth metal surface. The
forward Fraunhofer diffraction with amplitude f_1 is reflected
from the surface and interferes with the backscattered waves f_2.

Similar angular distributions were observed for stepped surfaces
with a small concentration of steps produced by a special sputtering
treatment.[18] These angular distributions were also shown to provide
information on the width of the repulsive disturbance produced by a
single step. In this case the scattering theory is, in principle, very
formidable because of the difference in height of the two terraces on the
both sides of the step. As shown by Hinch however the problem is greatly
simplified if the finite coherence length of the beam is accounted for in
the scattering calculations.[20] Such calculations reveal that the angular
distributions, which as discussed above are equivalent to form factors,
are, in fact, quite different for scattering from an upward or downward
step. This is illustrated with a hard wall model calculation in
Fig. 8.[21] The greater intensity on the large-angle side where the
undulations appear can be explained by the classical deflection of the
atoms from the sloped portion of the steps. This "focussing" of
intensity is referred to as a surface rainbow effect.[22] As a result of
the large asymmetry in the form factors it is possible to attribute
scattering at $\Delta K < 0$ to downhill steps and scattering at $\Delta K > 0$ to uphill
steps. It is perhaps surprising to see that the distributions for up and
down steps are mirror symmetric with respect to each other. For a more
realistic soft potential with an attractive well small deviations from
this symmetry are expected.

We are now in a position to understand the angular distributions
measured from a sputtered surface. Figure 9 shows a typical angular
distribution taken at 225K for a Cu(111) surface that had been sputtered
at 300K for 1 hour (2.0 μA cm^{-2}, Ne$^+$, 1 keV) and allowed to stand for a
further hour at 225K.[23] Under these room temperature sputtering
conditions we expect the defects to coalesce to produce randomly
distributed steps normal to the [11$\bar{2}$] and [$\bar{1}\bar{1}2$] directions. Since the
surface was initially a smooth (111) surface they should have equal

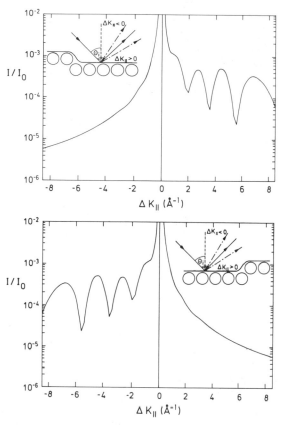

Fig. 8   Orientation dependence of diffraction from steps.   Normalized
intensities calculated for a density of 1/150Å of steps at an
incident wavevector of k_i = 9.1Å^{-1}.   The step parameters were
h = 2.09Å, width 5.3Å and angle ϵ = 43°.

densities in order to maintain on average the low-index face.   The
orientation of the step edges can easily be ascertained by rotating the
azimuth of the crystal.   For rotations of 5° out of <112> the
oscillations largely disappear in accord with the idea that the observed
undulations are concentrated in directions normal to the step edge
extension.   Shown at the top of Fig. 9 is a billiard ball model of the
step structure corresponding to the diffraction undulations seen on the
corresponding sides of the lower panel.   The model reveals that the
uphill and downhill steps on a (111) surface actually have different
structures even though both have close packed rows of atoms at the edge.
The shapes of the undulations on each side were fitted to a hard wall
potential shown in Fig. 10.   Since the step heights are known from fitted
to a hard wall potential shown in Fig. 10.   Since the step heights are
known from the bulk crystallography it was necessary to fit only the
widths W and the angles of inclination.   The results of this study and
other similar experiments are summarized in Table I.[23]   For the Cu(111)
and Pt(111) steps only small differences are observed for steps normal to
the [11$\bar{2}$] and [$\bar{1}\bar{1}$2] directions.   The step widths vary from 5.3 to 7.0Å
with increasing step height.   Since the widths increase with step height
it is perhaps not so surprising that the angles of inclination do not
change too much from one metal to another for the (111) face.

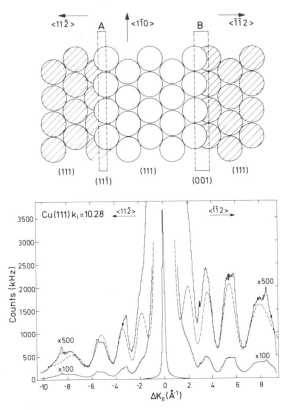

Fig. 9   Diffuse helium scattered intensity oscillations observed along the [11$\bar{2}$[and [$\bar{1}\bar{1}$2] direction from a stepped Cu(111) structure. The incident wave vector was 10.28Å^{-1}. The dashed curves are best-fit calculations. The top part shows a ball model illustrating the difference in the structure of the oppositely oriented steps.

Table I
Parameters for Step Edges

Material	Face	Height h[Å]	Direction Normal to Steps	Width W[Å]	Angle ϵ[deg]
Al	(111)	2.34	<11$\bar{2}$> <$\bar{1}\bar{1}$2>	7.0	37
Cu	(111)	2.09	<11$\bar{2}$> <$\bar{1}\bar{1}$2>	5.4 5.3	42 43
Ni	(001)	1.76	<100)	5.7	34
Pt	(111)	2.27	<11$\bar{2}$> <1$\bar{1}$2>	6.6	38

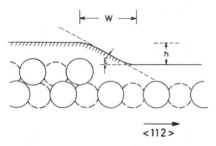

<112>

Fig. 10 Hard-wall corrugation function at a single step edge. The width
w, height h and steepest angle of inclination ϵ are shown.

Figure 11 shows an application for this new diagnostic technique to
a study of the effectiveness of successive annealing treatments at
different temperatures in healing out step edge defects. The crystal was
first sputtered as described above and heated to the first temperature
setting of 250K for two minutes and then cooled down to 225K, after which
the angular scan was measured. The measuring time for each scan was

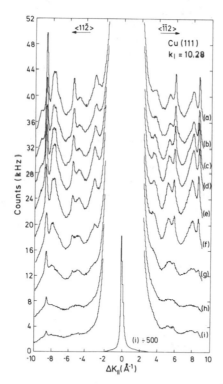

Fig. 11 A succession of diffuse scattering measurements from a sputtered
Cu(111) surface at various crystal temperatures: (a) 225K, (b)
250K, (c) 275K, (d) 300K, (e) 325K, (f) 350K, (g) 375K, (h) 400K
and (i) 425K. The central specular peak of scan (i) is shown,
reduced by a factor of 500. All other scans, (a) → (h), are
shifted by 2 kHz with respect to one another. Incident wave
vector, k_i = 10.28 Å^{-1}.

about 12 minutes. This procedure was repeated for successively larger temperatures and in each case the amplitude of the diffraction oscillations is observed to decrease. After an anneal at 400K the undulations reached a minimum indicating an almost complete healing of the surface.

DIFFRACTION FROM REGULARLY STEPPED SURFACES

The previous experiments suggest that helium atom scattering is also an ideal method for studying periodically stepped surfaces. The first studies of this type were carried out by Lapujoulade and colleagues[24] and by Comsa and coworkers.[25] These experiments revealed a large number of closely spaced diffraction peaks due to diffraction from the step edges. However, the dynamic range of these experiments was only about a factor 10 to 100 and consequently some of the finer structures may have been overlooked. Our group has recently embarked on an extensive study of stepped aluminum surfaces.[26] Aluminum was chosen because of its nearly free electron nature making it accessible to theoretical treatments and because Al is a particularly effective inelastic scatterer and therefore especially well suited for a study of surface phonons at step edges. This work will be described elsewhere.[27]

Since most previous studies of stepped surfaces have been carried out with LEED it is instructive to compare the expected diffraction intensities in an HAS experiment with those of a LEED experiment. Figure 12 shows the positions of the atoms on a typical stepped surface and defines some of the notation to be used below.

The differential intensity is given in both experiments by

$$I = | \sum_i f_i \, e^{i\Delta k . \rho_i} |^2 , \tag{2}$$

where f_i is the atomic form factor, Δk the momentum transfer and ρ_i the position of each atom. For simplicity we neglect scattering from atoms below the topmost layer. Then we can easily reexpress Eq. (2) in terms of coordinates of the atoms at the step edges and the atoms of the terraces.

$$I = | \sum_{m=1}^{T} \sum_{n=1}^{N_T} f_n e^{i\Delta k \cdot (n-1)\vec{r} + \vec{R}_m} |^2 , \tag{3}$$

where the symbols are defined in Fig. 12. Next we assume that the form factors of all terrace atoms are identical and that only the form factors of the edge atoms are different

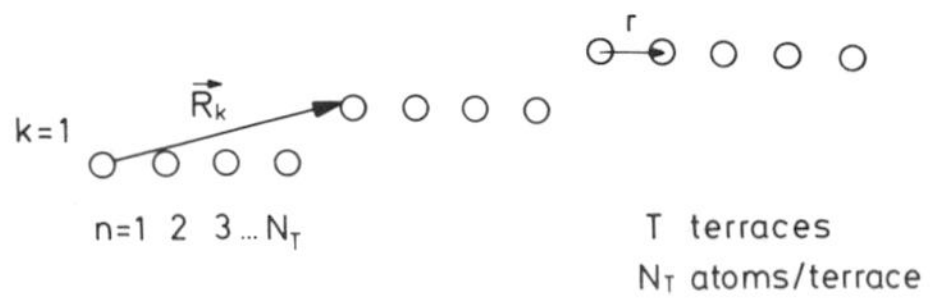

k=1

$\vec{R}_k$

n=1 2 3...N_T

$\vec{r}$

T terraces
N_T atoms/terrace

Fig. 12 Coordinates used for calculating diffraction from a regularly stepped surface.

88

$$f_n = f_{terr} \qquad n \geq 2 , \tag{4a}$$

$$f_1 = f_{terr} + f_{edge}(\Delta k) . \tag{4b}$$

Substituting into Eq. (3) we get

$$I = |\sum_{m=1}^{T} e^{i\Delta k \cdot \vec{R}_m} (f_{edge} + \sum_{n=1}^{N_T} f_{terr} e^{i\Delta k \cdot (n-1)\vec{r}})|^2 \tag{5}$$

$$= |S_{edge}(E)|^2 \{|f_{edge}|^2 + 2 \text{ Re } f_{edge} \cdot f_{terr} S_{terr}(T)$$

$$+ |f_{terr} S_{terr}(T)|^2\} . \tag{6}$$

where $S_{edge}(E)$ is the structure factor of the edges and $S_{terr}(T)$ is the structure factor of the individual terraces. E and T denote the corresponding reciprocal lattices.

In principle Eq. (6) applies equally to LEED and HAS. However, in applying Eq. (6) to LEED, we must account for the large form factors of the individual atoms. The change in the form factors at the edges is very small and results only from the small changes resulting from the different atomic environment at the step edge. If we neglect this we get

$$I_{LEED} = f_{terr}^2 S_{edge}^2(E) \cdot S_{terr}^2(T) . \tag{7}$$

For HAS we have just the opposite situation. As illustrated in Fig. 4 the terrace form factor is negligibly small away from the specular condition. However, as shown in Fig. 8 the step edge form factor is large depending on whether the step is upward or downward with respect to the direction of the parallel momentum transfer ΔK. Neglecting the terrace form factor altogether we get

$$I_{HAS}(\Delta K > 0) = f_{down \atop step}^2 S_{edge}^2 (E) \tag{8a}$$

and

$$I_{HAS}(\Delta K < 0) = f_{up \atop step}^2 S_{edge}^2 (E) , \tag{8b}$$

where $f_{up \atop step}^2 = f_{down \atop step}^2$.

Thus HAS is sensitive almost exclusively to the step edges and there is only as small effect from scattering from the individual terraces. Moreover as indicated in Eqs. (8a) and (8b) HAS can discriminate between upward and downward directed steps. LEED on the other hand is only sensitive to the terrace atoms and virtually insensitive to the edges.

These differences are nicely illustrated by referring to the Ewald diagrams for a single one dimensional scattering from a simple periodically stepped surface. Figure 13a shows the terrace reciprocal lattice rods as vertical lines spaced at intervals of $2\pi/a$ in the ΔK_x direction (a is the spacing of atom rows in the terrace planes). The diagonal lines show the reciprocal-lattice rods from the step edge lattice as observed by LEED which are inclined at the angle of the vicinal cut. The points of intersection of the two systems of lattice rods are coincident with the 3D-Bragg conditions. In fact the lattice rods of all vicinal planes pass through these points. At the 3D-Bragg points the peak width is unaffected by disorder or roughness of the surface. In LEED the maximum sensitivity to surface disorder occurs at the terrace lattice anti-Bragg points, which lie on the terrace lattice rods halfway between the 3D-Bragg points. At these anti-Bragg conditions, regularly spaced step edges lead to a splitting or, in the case of a surface with disorder, to a broadening of the peaks.

Figure 13b shows the corresponding diagram appropriate to HAS. In this case the lattice rods are those of the step edge lattice and the terraces have virtually no effect on the scattering. Thus in a HAS diffraction experiment it is possible to observe scattering along the entire extent of the step edge lattice rods. The 3-D Bragg points have the same meaning as in LEED. The anti-Bragg points are now located on the step edge lattice rods half-way between the 3D-Bragg points. In the ΔK_x coordinates they now are situated at positions half-way between the terrace reciprocal-lattice rods. Thus disorder will lead to a maximum broadening of the HAS peaks at entirely different positions in reciprocal lattice space than in LEED.

In conclusion the differences in intensities of the LEED and HAS reciprocal-lattice rods shown in Fig. 13 bring out nicely the complementarity of the two techniques. To a first approximation electrons are diffracted from the atoms of the terraces and the steps are invisible while helium atoms are diffracted from the step edges, with the terrace atoms invisible.

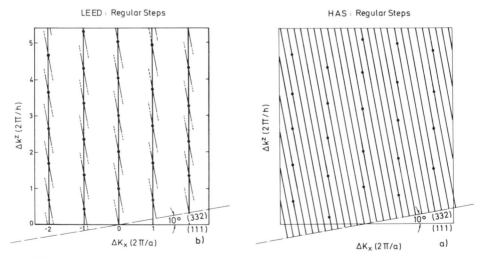

Fig. 13 Comparison of reciprocal-lattice rods for LEED and HAS from a regular stepped surface. In LEED the lattice rods of the terrace atoms predominate, whereas in HAS the lattice rods of the steps predominate.

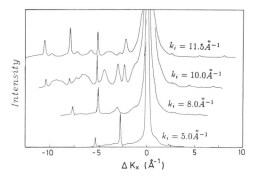

Fig. 14 Some typical angular distributions measured at four different
wave vectors in scattering from a nominal (332) vicinal surface
of aluminum.

This sensitivity of HAS to step edge lattices has recently been
demonstrated by a study of a (332) aluminum vicinal surface.[26] This
surface consists of (111) terraces with about 5.3 rows of atoms between
the step edges. The atoms at the step edges form densely packed rows.
In view of the close packing of the terrace and step edges this surface
was expected to be particularly stable.

Figure 14 shows a series of angular scans taken for four different
incident wave vectors.[26] The angular distributions are plotted with
respect to the momentum transfer along the terraces. A large number of
irregularly spaced peaks are found, which have the greatest intensity in
the region of $\Delta K < 0$. The interpretation of these structures is
facilitated with the aid of an Ewald construction similar to the one
shown in Fig. 13b. The results of an analysis of these and 30 additional
angular distributions measured for different incident wave vectors is
shown in Fig. 15. There the locations of all the maxima observed in the
angular scans have been plotted. It is important to realize that peaks
of greatly differing heights varying over three to four orders of
magnitude are included and that no distinction with respect to peak
intensity was made in order to keep the picture as simple as possible.
In the fixed angle arrangement $\theta_{SD} = 91.5°$ used in these experiments each
of the scans for a given value of incident wave vector corresponds to a
circle centered at the origin. This is illustrated by the two circles
which show the regions swept through by the largest and smallest wave
vectors used in the measurements of Fig. 14. An examination of the
entire set of data reveals two sets of parallel equispaced straight
lines. The angles of inclination and their spacing coincide with the
reciprocal-lattice rods of the (221) and (331) vicinal faces.
Surprisingly there is no clear evidence for the existence of the original
(332) vicinal surface. A partial explanation for the disappearance of
the (332) face will be presented later on.

Figure 15 also explains the reason why those peaks which lie close
to the (111) terrace lattice rods are so sharp. These peaks do, in fact,
all correspond rather closely to the 3D-Bragg points in reciprocal space
so that they are not affected by disorder. The other peaks half-way
between these positions are much broader in accord with their position
close to the step lattice anti-Bragg points. Thus the HAS angular

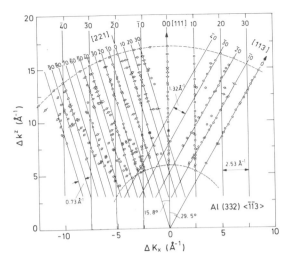

Fig. 15 HAS peak positions in reciprocal-lattice space: facetted
Al(332). Circles indicate the positions of all peak maxima
observed with angular scan measurements from a nominal Al(332)
crystal at temperatures of 150 K or below. An individual scan
describes an arc of a circle in reciprocal-lattice space. For
this scattering geometry, the scan with lowest incident energy
(9.6 meV) describes a circle of radius 6.08Å^{-1} and highest
incident energy (74.8 meV) the uppermost circle with radius
16.91Å^{-1}. The solid lines are the reciprocal-lattice rods from
(221), (11ī) and (113) facets of the surface. The uppermost
part of the figure displays the diffraction order of each
reciprocal-lattice rod. Short dashes, through circles,
illustrate the angular range corresponding to 1/3 FWHM of each
peak. Many peaks are very sharp showing little or no
broadening. This is indicated simply by the absence of dashes.
Note that the intensities of the peak vary over orders of
magnitude.

distributions provide very direct evidence for the existence of a high
concentration of (221) and (331) facets. From the very low signals in
the $\Delta K_x > 0$ part of the angular distributions we conclude that the
proportion of [331] facets is quite small. From the large intensity of
the specular peak at $\Delta K_x = 0$ it would appear that the surface also has a
high density of [111] faces. Thus the HAS experimental data indicate
that the surface has extensively restructured into a mixture of (111),
(221) and (331) facets, which when averaged over large distances, have an
average inclination equal to that of the original [332] vicinal surface.

TEMPERATURE INDUCED EFFECTS ON REGULARLY STEPPED SURFACES

With increasing temperature single-crystal surfaces are expected to
exhibit an increased disorder. This effect is particularly apparent on
periodically stepped surfaces where the disorder is largely localized in
the region of the step edges since the edge atoms are more loosely bound.
This phenomenon is called roughening[28] and has been recently extensively
studied by several techniques including HAS.[29]

In discussing the qualitative changes seen in the diffraction scans
with increasing surface temperature two effects must be accounted for:
(1) the increase of thermal vibrations at higher surface temperatures T_s

reduces the elastic scattering from the surface, $I = I_0 (\exp(-2W(T_s)))$, where the Debye-Waller factor, $W(T_s)$, is expected to be of the same order of magnitude for each diffraction peak. (2) Associated with the loss in elastic signal there is a corresponding increase in the phonon inelastic signal which appears at angles on both sides of the diffraction peaks.[10] This can be distinguished from the true elastic intensity using the time of flight (TOF) capabilities of the apparatus.

The importance of this TOF analysis at elevated temperatures is illustrated in Fig. 16.[26] The top panel shows an angular distribution measured without TOF analysis at the comparatively high crystal temperature of 606.4K. As compared with the angular distributions measured at room temperature (see Fig. 14) the structure is largely smeared out and the peak at $\Delta K_x = 0$ from (111) facets is extensively broadened. At the higher temperature the fraction of coherent elastically scattered intensity is significantly reduced by about a

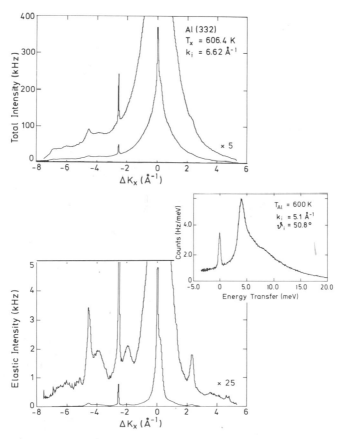

Fig. 16 Discrimination of elastic scattering for peak Form analysis. The top panel shows the total intensity as a function of scattering wavevector at $T_s = \sim 600$ K for a nominal (332) Al surface. The middle panel shows a typical time-of-flight spectrum with the elastic peak at energy transfer of zero meV. The bottom panel shows the angular distribution of only the elastic peak for the same temperature of 600 K.

factor 2.5 compared to T_S = 308 K. The middle panel shows a typical TOF spectrum for a surface temperature of 600 K. At this high temperature it is seen that the amount of inelastic scattering has increased significantly while the elastic peak has been diminished as a result of the Debye-Waller factor. By subtracting off the multiphonon background the true elastic intensity is obtained. Similar TOF measurements were made over a wide range of angles and from the elastic peak intensities the angular distribution shown at the bottom panel of Fig. 16 was obtained. The elastic angular distribution, although now a small proportion of the total HAS intensity, contains much more structure and is, in fact, very similar to the elastic distribution at the top which was taken at 300K. Thus the broadening and smearing out of the feature in the top panel of Fig. 16) is not due to any structural changes of the surface but simply due to an increase in the inelastic scattering in the angular distributions measured without TOF analysis. This is clear evidence that TOF energy resolution is essential in order to obtain structural information with quantitative accuracy. This has also been pointed out by several other groups.[30] Although aluminum has a particularly large phonon inelastic scattering cross section it is expected that similar effects will hold for other stepped surfaces for which roughening has been studied by HAS albeit without TOF resolution.

This TOF technique has been used to follow the full width half maximum, FWHM, of diffraction peaks in the region of $\Delta K = -2.5\text{Å}^{-1}$ as a function of crystal temperature.[26] The results of such a study for several of the peaks seen in the angular distribution is shown in Fig. 17. The one peak at $\Delta k^z = 8.86\text{Å}^{-1}$ (triangles) corresponds to a 3D-Bragg condition of the specular peak of the (221) facet and 10 peak of the (111) face. As expected from the properties of the 3D-Bragg points this peak is insensitive to the surface temperature. The constant width of this peak in reciprocal-lattice space of $\Delta K_x = 0.06\text{Å}^{-1}$ compares well with the angular resolution of the apparatus of $\Delta\theta = 0.35°$. This result illustrates once more that the sharpest peaks at the 3D-Bragg points are insensitive to virtually all types of surface disorder. In contrast the curves (b), circles, and (c), squares, are for two peaks that are roughly midway between 3D-Bragg conditions. Peak (b) is identified as the 1st order peak, and (c) is the 2nd order peak from the (221) facets. Both peaks have perpendicular momentum transfers (relative to the (111) coordinate frame) of about 13.0Å^{-1}, which for a (111) surface would be described as an anti-Bragg condition. These peaks show a measurable broadening over this temperature range and are hence sensitive to an increasing surface roughness. The two curves, Fig. 17(b) and (c), demonstrate that the roughness of (221) facets increases continuously in the range 300K to 600K. Thus the surface changes significantly over this temperature range indicating a great mobility of step edges. In this range of temperatures the disorder is apparently not frozen in. Measurements were not performed systematically at lower temperatures, but the available data suggest that the half-widths will always be greater than the angular resolution. This is consistent with the idea that the surface will have a certain amount of disorder even at lower temperatures.

Another interesting observation is the appearance of a new diffraction peak at higher temperatures close to 630K. Extensive TOF measurements similar to those presented in Fig. 16 make it possible to identify this peak as due a reciprocal-lattice rod of a (332) facet. The temperature behavior of this peak was extensively studied. This peak dropped off sharply at temperatures approaching 730K which is the temperature used in the initial preparation of the crystal. From this and other evidence we can postulate that the original (332) surface underwent considerable restructuring into the (111), (221) and (331)

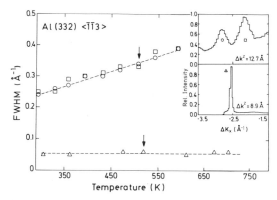

Fig. 17 The full widths at half maximum (FWHM) of diffraction peaks are
plotted against surface temperature. All elastic intensities
were determined from TOF spectra and fits were made to the data
assuming simple Gaussian peak shapes. a) Triangles: 3D-Bragg
condition, $\Delta k^z = 8.86$ Å^{-1} and $\Delta K_x = -2.25$ Å^{-1}; b) circles:
$\Delta k^z = 12.62$ Å^{-1}, $\Delta K_x = -2.88$ Å^{-1}; and c) squares: $\Delta k^z = 12.78$
Å^{-1} and $\Delta K_x = -2.22$ Å^{-1}. The inset shows two typical angular
scans at $\Delta k^z \simeq 12.7$ Å^{-1} (top) and at $\Delta k^z \simeq 8.9$ Å^{-1} (bottom) both
taken at surface temperatures of about 510 K as indicated by the
black arrows.

facets at temperatures above 550 K. During the usual cooling cycles
there appears to be insufficient time for sufficient mass transport to
enable the surface to return to its original structure.

STUDIES OF GROWTH AND ORDERING USING HELIUM ATOM SCATTERING

The standard method for monitoring the growth of thin films during
molecular beam epitaxy is reflection high-energy electron diffraction
(RHEED). The intensity of the specular peak is observed to oscillate
with deposition time. The extrema are associated with the completion of
individual monolayers. This experimental technique has the advantage
that it is easy to apply within a complicated molecular beam epitaxy
apparatus but the theory appears to be rather complex. Recently a group
in Madrid demonstrated that HAS can also be used for the same purpose.[31]
They found that two effects contribute to the oscillations which they
observed during growth of Cu monolayers on Cu(100). As in RHEED there is
a broadening of the specular peaks at the out-of-phase anti-Bragg
condition which results in a decrease in the specular intensity. This is
attributed to a loss of coherence due to the presence of many small
islands and flat spaces in between. In addition in HAS, however, there
is an additional attenuation of the specular peak due to the deflection
of the atoms by the step edges. Their concentration is greatest when the
number of islands or more precisely the total circumference of all
islands is greatest. This is the case at an intermediate stage of the
growth process between the completion of layers. This attenuation can
also occur for the in-phase Bragg condition. Figure 18 illustrates this
unique sensitivity of HAS on the concentration of steps during the layer
by layer growth.

The growth of Pb on Cu(111), which has been studied in detail by
LEED in the submonolayer range,[32] was recently studied by HAS in our
laboratory.[33] Over a large range of surface temperatures (140-300K) He
atom diffraction intensities also show the incommensurate overlayer

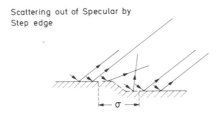

Scattering out of Specular by
Step edge

Attenuation by island edges

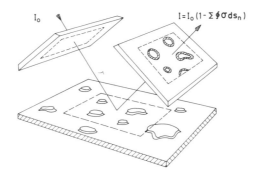

I_0

$I = I_0 (1 - \Sigma \oint \sigma ds_n)$

Fig. 18 Schematic diagram showing the mechanisms by which helium atoms
are attenuated in the intermediate stage of new layer-by-layer
growth of thin films.

growth at less than one monolayer, followed by the commensurate c(4x4)
phase as observed in LEED.[31] The experiments were performed with a high
resolution helium atom scattering apparatus, in which the specular beam
(E_i = 7-20 meV) is detected after scattering through a fixed angle of 90°
between incident and final beams. For these measurements the linear
angular resolution of the slit-like detector was reduce to as little as
0.2°. The specular intensity during growth of thicker Pb overlayers on
Cu(111) are presented in Fig. 19. They show clearly a single-layer
growth mechanism. The regularly spaced specular intensity oscillations
are attributed to the period of growth of a single monolayer.

A typical behavior of the specular beam intensity during the early
stages of the deposition of Pb on the Cu(111) surface is presented in
Fig. 20. This displays an unusually complex type of growth curve.
Periodicities corresponding to the single-layer growth and also twice
this periodicity are both clearly evident. The curve of Fig. 20 was
taken with a deposition rate of 0.4 ML/min and with a surface temperature
of 140 K. Very similar intensity patterns were observed also for other
deposition rates (0.2 ML/min to 0.5 ML/min.). Likewise the qualitative
features of the curves are also largely unaffected by changes in the
surface temperature in the range 60K to 200 K. At temperatures above
200K the size of oscillations has decreased significantly, as is to be
expected from a larger rate of diffusion of deposited adatoms at elevated
temperatures. Figure 20 compares the growth oscillation measured in the
anti-Bragg out-of-phase and in the Bragg in-phase conditions. The
amplitude of the oscillations is large in both cases with the greater
oscillation occurring for the in-phase condition.

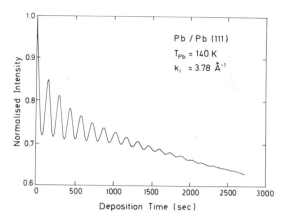

Fig. 19 Intensity oscillations of the specular helium atom beam
(k_i = 3.78Å^{-1}) during the deposition of Pb on a thick film of
previously deposited lead at a crystal temperature of T_s = 140K.

The observed oscillatory structure has been interpreted in terms of
a quantum size effect (QSE) which leads to an additional stabilizing
influence on layers of certain thicknesses. The QSE on the bulk and
surface properties of thin metal films was first studied theoretically
for a simple jellium model by Schulte.[34] His calculations revealed, with

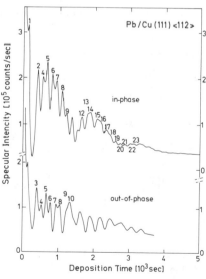

Fig. 20 Comparison of the growth oscillations of the helium specular
beam during growth of Pb on Cu(100) at T_s = 140K. The top
measurement was taken for the Bragg condition and the bottom
measurement was for the anti-Bragg condition. The numbers above
the maxima indicate the thickness of the corresponding Pb layer.

increasing layer thickness, oscillations in the electron density inside and outside the crystal. These oscillations were attributed to changes in the occupation of the quantized electronic energy levels. With increasing layer thickness the levels decrease in energy and more levels may fall below the Fermi energy ϵ_F. The onset of occupation of an additional level is expected to occur at a thickness D given by,

$$D = 0.5 \; \lambda_F \cdot n \quad n = 1,2,3,\ldots \; , \qquad (9)$$

where λ_F is the Fermi wavelength of the free electrons and n is an integer. This interpretation of the observed oscillations in Pb/Cu(111) is supported by a computer simulation which is able to fully reproduce the form of the oscillations and especially the attenuation between single and double-layer growth mechanisms.[33]

SUMMARY AND CONCLUSIONS

Helium atom scattering (HAS) has been demonstrated to have a remarkable sensitivity to all kinds of defects on smooth single-crystal-metal surfaces. As opposed to electrons which are scattered from the ion cores the helium atoms are scattered from the low electron densities prevailing at distances of about 3Å from the atoms of the surface. Because of the smoothing of the electron distributions at these large distances the single atoms in a terrace have only very small form factors whereas step edges and single adsorbates produce large disturbances. Thus the diffraction of helium atoms from single randomly distributed defects can be observed against a very small background from the smooth terraces. These scattering experiments from single adsorbates and step edges allow the determination of the size, shape and orientation of the defects. It has also been possible to monitor the adsorption of molecules at step edges.[35]

This remarkable sensitivity to step edges makes it possible to study periodically stepped surfaces. In particular it has been demonstrated that the individual facets of the multifacetted surfaces can be clearly identified from its reciprocal lattice rods even when several other facets are simultaneously present. This sensitivity to single facets makes it possible to study for the first time the kinetics of facetting induced by surface temperature changes.

It has also been shown that a time of flight (TOF) analysis of the scattered intensity is important for removing contributions from inelastic scattering from phonons. This is illustrated in a study of the peak shapes for scattering from a facetted surface. Some preliminary studies of roughening are presented based on only the elastic intensities. New technological developments will soon make it possible to carry out such TOF resolved measurements with much greater sensitivity.

Finally the sensitivity of HAS to step edges makes it an ideal technique for the in-situ study of the layer-by-layer growth of thin surface films. In one recent experiment the special sensitivity of HAS has made it possible to observe a new quantum size effect which leads to a remarkable stabilization of Pb layers on a Cu(111) substrate.

It is hoped that these advantages of HAS over the more established techniques of LEED and RHEED will soon find more widespread application in the study of complex surface morphology and kinetics.

Acknowledgement

We thank John Ellis for critically reading the manuscript.

REFERENCES

1. O. Stern, Naturwissenschaften 17, 391 (1929), see also T. H. Johnson, Phys. Rev. 31, 1122 (1928), 35, 1299 (1930).
2. C. J. Davisson and L. H. Germer, Phys. Rev. 30, 705 (1927).
3. H. E. Farnsworth, Phys. Rev. 40, 684 (1932).
4. A. Ellet, H. Olsen, and H. A. Zahl, Phys. Rev. 34, 493 (1929); I. Estermann, R. Frisch and O. Stern, Z. f. Physik 73, 384 (1931); Tykocinski - Tykocinev, J. Opt. Soc. Amer. 14, 423 (1927).
5. I. Estermann, R. Frisch and O. Stern, Z. f. Physik 73, 348 (1932).
6. E. W. Becker and K. Bier, Z. Naturforschung 9A, 975 (1954).
7. J. P. Toennies and W. Winkelmann, J. Chem. Phys. 66, 3965 (1977).
8. G. Lilienkamp and J. P. Toennies, J. Chem. Phys. 78, 5210 (1983).
9. T. Engel and K. H. Rieder in: Structural Studies of Surfaces (Springer Tracts in Modern Physics, vol. 31) Springer, Berlin, 1982 p. 55. G. Boato and P. Cantini, Adv. Electron. and Electron Phys. 60, 95 (1983). J. A. Barker and D. J. Auerbach, Surface Science Repts. 4, (1984).
10. J. P. Toennies in: Springer Series in Surface Science, (F. W. de Wette, ed.), Springer, Berlin 1988; p. 248, J. P. Toennies, J. Vac. Sci. Technol. A5, 440 (1987); J. Vac. Sci. Technol. A2, 1055 (1984).
11. V. Bortolani, A. Franchini, G. Santoro, J. P. Toennies, Ch. Wöll and G. Zhang, Phys. Rev. B., in press.
12. R. Smoluchowski, Phys. Rev. 60, 661 (1941).
13. B. Poelsema, R. L. Palmer and G. Comsa, Surface Sci. 136, 1 (1984).
14. J. P. Toennies in: Physical Chemistry, An Advanced Treatise, vol. VIA (W. Jost, ed.), Academic, New York 1974, p. 228-332.
15. L. K. Verheij, B. Poelsema, and G. Comsa, Surface Sci. 162, 858 (1985).
16. A. M. Lahee, J. R. Manson, J. P. Toennies and Ch. Wöll, J. Chem. Phys. 86, 7194 (1987).
17. A. M. Lahee, J. R. Manson, J. P. Toennies and Ch. Wöll, Phys. Rev. Lett. 57, 471 (1986), B. Poelsema, L. K. Verheij and G. Comsa, Phys. Rev. Lett. 53, 2500 (1984).
18. B. Poelsema, L. K. Verheij and G. Comsa, Phys. Rev. Lett. 49, 1731 (1982); B. Poelsema and G. Comsa, Faraday Discuss. Chem. Soc. 80, 24 (1989).
19. G. Drolshagen and R. Vollmer, J. Chem. Phys. 87, 4948 (1987).
20. B. J. Hinch, Phys. Rev. B38, 5260 (1988).
21. B. J. Hinch and J. P. Toennies, submitted to Phys. Rev.
22. J. D. McClure, J. Chem. Phys. 51, 16687 (1969); ibid, 52, 2712 (1970); ibid 57, 2810 (1972); J. Chem. Phys. 17, 2823 (1972).
23. B. J. Hinch, A. Lock, J. P. Toennies and G. Zhang, J. Vac. Sci. and Technol. B, in press.
24. J. Lapujoulade and Y. Lejay, Surface Sci. 69, 354 (1977), J. Lapujoulade, Y. Lejay and N. Papanicalaou, Surface Sci. 90, 133 (1979).
25. G. Comsa, G. Mechtersheimer, B. Poelsema and S. Tomoda, Surface Sci. 89, 123 (1979).
26. B. J. Hinch, A. Lock, H. H. Madden, J. P. Toennies and G. Witte, to be submitted to Phys. Rev. B.
27. M. Gester, A. Lock, J. P. Toennies and G. Witte, to be submitted to Surf. Sci.

28. W. K. Burton, N. Cabrera and F. C. Frank, Philos. Trans. R. For a review see H. van Beijeren and I. Nolden in: <u>Structures and Dynamics of Surfaces II</u>, W. Schommers and P. von Blanckenhagen eds. (Springer, Heidelberg 1987).

29. L. Lapujoulade, J. Perreau and A. Kara, Surface Sci. <u>129</u>, 59 (1983); E. H. Conrad, R. M. Aten, D. S. Kaufman, L. R. Allen, T. Engel, M. den Nijs and E. K. Riedel, Chem. Phys. <u>84</u>, 1015 (1986); ibid <u>85</u> 4856 (E) (1986); F. Fabre, D. Gorse, J. Lapujoulade and B. Salomon, Europhys. Lett. <u>3</u>, 737 (1987).

30. E. H. Conrad, L. R. Allen, D. L. Blanchard and T. Engel, Surf. Sci. <u>184</u>, 227 (1987), Surf. Sci. <u>198</u>, 207 (1988), P. Zeppenfeld, K. Kern, R. David and G. Comsa, Phys. Rev. Lett. <u>62</u>, 63 (1989).

31. L. J. Gomez, S. Bourgeal, J. Ibanez, and M. Salmeron, Phys. Rev. <u>B31</u>, 2551 (1985).

32. G. Meyer, M. Michailov and M. Henzler, Surface Sci. <u>202</u>, 125 (1988).

33. B. J. Hinch, C. Koziol, J. P. Toennies and G. Zhang, Europhys. Lett. <u>10</u>, 341 (1989).

34. K. F. Schulte, Surface Sci. <u>55</u>, 427 (1976).

35. R. Berndt, B. J. Hinch, J. P. Toennies and Ch. Wöll, J. Chem. Phys., in press.

GROWTH AND ORDERING OF Ag SUBMONOLAYERS ON Ge(111)

M. Henzler, H. Busch and G. Friese

Institut für Festkörperphysik, Universität Hannover
Appelstr. 2
3000 Hannover, Federal Republic of Germany

ABSTRACT

The evaluation of LEED spot profiles is extended by including the interference between scattering from clean and adsorbate covered surfaces. In this way the growth of incommensurate Ag islands on Ge(111) 2x8 is analyzed providing growth mode and scattering parameters including the phase shift. Also the faceting due to heating of a misoriented Ge(111) with submonolayers of Ag is studied in the same way. It is shown that quantitative analysis of growth and ordering is now possible for inhomogeneous surfaces as done so far with homogeneous surfaces.

INTRODUCTION

Growth and ordering may be studied with different techniques. Recently microscopy with atomic resolution has provided a lot of informations by direct imaging of growth features and changes of defects like steps, island shape point defects during changes by deposition or annealing. Techniques like STM or TEM (decoration) are presented in this volume,[1-4] other techniques like LEEM, REM or TEM (direct imaging or shadowing) are described in an earlier volume.[5-8] Those images provide direct and the best qualitative information on the kind and shape of a defect, quantitative information requires averaging over many images. Quantitative information is more easily and more accurately obtained by diffraction, which provides average values automatically. Those techniques like RHEED, X-ray scattering and LEED are included in the present and an earlier volume.[9-15]

So far the diffraction techniques have only be applied for homogeneous surfaces where the shapes of the normal spots (or of the extra spots in case of adsorbate induced superstructure) are described by a single scattering factor and a pair correlation function of the surface atoms. There are many cases, however, where an extension is needed. If a different material is deposited, two different scattering factors have to be used. With this extension a variety of new systems and new informations are available via a careful evaluation of a diffraction pattern. It is the goal of this paper to describe the kinematical approximation including two scatterers and to show with several examples its application to both commensurate and incommensurate monolayers on crystalline substrates.

Kinetics of Ordering and Growth at Surfaces
Edited by M. G. Lagally
Plenum Press, New York, 1990

In LEED (Low-Energy Electron Diffraction) the intensity of the diffracted electrons has to be described including multiple scattering. The integral intensity of the beam is used to determine the atomic position within the unit mesh of a periodic arrangement. There is additional information available in the spot profile depending on periodicity at the surface and deviations from it. It has been shown that a variety of defects like distribution of atomic steps, domain size and facets are quantitatively derived within the kinematic approximation.[11,13,16,17] The basic assumption is that the sample may be divided into units consisting of one surface atom (or one surface unit mesh) and all underlying atoms and that each unit has the identical scattering factor (depending only on incoming and outgoing wavevector k_o and k respectively). This assumption is definitely not applicable for He atom scattering or high-energy electron diffraction at grazing incidence since here, for example, the step edges have a drastically increased scattering factor. In Fig. 1 first for a homogeneous, stepped surface it is shown that the intensity is easily calculated from the individual (complex) scattering factor to provide a spot profile consisting of central spike and shoulder in this case. The validity of the approximation is experimentally checked by the integral intensity, which is constant during variation of the step density (e.g. during epitaxial deposition of half a monolayer) and by the ratio of peak to total intensity, which varies strictly periodically with the scattering vector $(k - k_o)$, where the period is given by any vector of the reciprocal net. The periodicity normal to the surface is used to derive the distribution of the surface atoms over the different layers.[18] The shoulder itself provides via its profile the pair correlation and therefore the average size and size distribution of terraces.

For an inhomogeneous surface two different scattering factors f_1 and f_2 are used (Fig. 2). For each scattering condition (k_o, k) a relation $f_2 = f_1 \cdot R \cdot \exp(i\phi)$ may be used. If the fraction θ of the surface is covered with material of scattering factor f_2, the scattered amplitude for specular reflection (so that all identical scatterers are in phase) is given by

$$A = (1 - \theta) f_1 + D f_2 = <f> ,\qquad(1)$$

which is the weighted average of f (including phase). The peak intensity is therefore

$$I_{peak} = A \cdot A* = |f_1|^2 |(1-\theta)+\theta \cdot R \cdot \exp(i,\phi)|^2 = |<f>|^2 .\qquad(2)$$

The total intensity (integrated over a full 2D-Brillouin zone) is given by

$$I_{total} = (1 - \theta) |f_1|^2 + D |f_2|^2 = <|f|^2> ,\qquad(3)$$

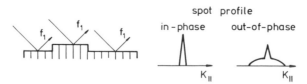

Fig. 1   Generation of a diffraction spot with shoulder due to steps. The surface is chemically homogeneous.

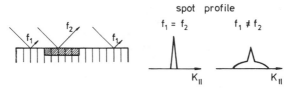

Fig. 2   Generation of a diffraction spot with shoulder due to
          inhomogeneities.   The surface is flat but with chemically
          different regions.

which is independent of phase. Then the integrated intensity of the
shoulder is given by

$$I_{shoulder} = I_{total} - I_{peak} = \theta \, (1-\theta) \, |\Delta f|^2 , \tag{4}$$

which is only observed for coverages different from zero or one and for
energies with different scattering factors for the two different
scatteres. To illustrate the dependance in Fig. 3 the variation of the
total and the peak intensity is shown. It should be noted that for $\phi = \pi$
a distinct zero is found, which enables an unambiguous determination of
the ratio $R = 1/\theta - 1$ and phase $\phi = \pi$. If R is very small or very high,
the intensity ratio is just given by coverage,

$$\frac{I_{peak}}{I_{total}} = \begin{Bmatrix} 1 - \theta & \text{for } R \ll 1 \\ \theta & \text{for } R \gg 1 \end{Bmatrix} .$$

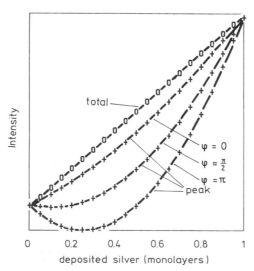

Fig. 3   Calculation of the total and peak intensity for a flat surface
          covered with an increasing amount of a different material
          (silver) within one level.   The ratio of the scattering
          amplitudes f_2/f_1 is 3. The curves for different phase shifts
          between f_1 and f_2 are shown.

Therefore the measurement of peak and total intensity provides a lot of information on both growth mode, coverage and relative scattering factor. As long as the domains of homogeneous structure are not too small, the evaluation will be quantitatively correct within the kinematic approximation, which can be checked by comparison of evaluations with different spots or energies. The shoulder again contains the lateral distribution, as discussed in detail in.[19]

A similar discussion is presented in Ref. 11 for a random distribution of the second type of scatterers. There a separation from thermal diffuse background is not possible in experiment. Here the domain or island structure provides shoulders which are well separated from background.

EXPERIMENTAL EXAMPLES

Due to the extension of the theory in the previous section, many inhomogeneous surfaces may be studied by spot profile analysis of a diffraction pattern. Figure 4 shows some schematic possibilities. For perfect epitaxy with the same lattice constant, but different scattering factor the theory applies easily (Fig. 4a). Even for incommensurate layers the spot profile of the 00-beam is described in the same way (Fig. 4b). If the new layer forms a superstructure, the profile of the extra spots is described with f_1 or $f_2 = 0$ (Fig. 4c). For reactive layers (Fig. 4d) the case of Fig. 4a, b or c may apply. Finally a quite different geometry should be included: facetting due to overlayers, which provides spots with a peculiar energy dependence of spot position (Fig. 4e). Their shape is described as in the case of the superstructure.

There are many examples in the literature where the spot profile of adsorbate induced superstructures has been used to evaluate the domain or island structure of the adsorbate. They include studies of domain growth kinetics[20,23] or studies of phase transitions.[21,27] It corresponds to $f_1=0$, since the portions, which are not covered or covered in a disordered way, do not contribute to the superstructure spot. If no superstructure is formed, a new domain may have a negligible scattering factor, as has been found for the Si/SiO_2 interface with patches, which were probably disordered due to suboxide formation.[22]

Examples of inhomogeneous surfaces

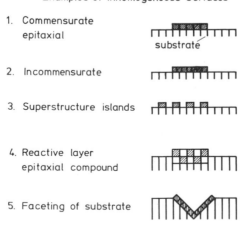

1. Commensurate
   epitaxial

   substrate

2. Incommensurate

3. Superstructure islands

4. Reactive layer
   epitaxial compound

5. Faceting of substrate

Fig. 4   Several possibilities for inhomogeneous surfaces.

Here for the first time an example is described, which takes into account two different finite scattering factors, so that also the relative phase is determined. The second example will describe faceting at a stepped Ge(111) surface due to interaction with submonolayers of Ag, which again makes use of the new possibilities.

INCOMMENSURATE Ag LAYERS ON Ge(111) 2x8

Submonolayers of Ag on Ge(111) have been studied to some detail.[23] If deposition or annealing temperature is higher than 100°C a 2x4 superstructure is found for a coverage of θ = 0.25 and a $\sqrt{3}$-structure for θ = 0.8. At intermediate coverages domains of the clean 2x8, the 2x4 and the $\sqrt{3}$-structure are seen, the size of which depends essentially on temperature and time of annealing.[23] If Ag is deposited at room temperature onto the clean Ge(111) with 2x8 structure, incommensurate islands are formed, which are well oriented with respect to the substrate: (111) || (111) and [110] || [110]. Only after heating to a temperature of 130°, when already part of the surface is converted to a commensurate superstructure, a Novaco-McTague rotation of 1.3° has been found. The incommensurate lattice constant of the Ag islands is seen as an additional first order Ag diffraction spot, which shows the lattice constant of the Ag(111) face. For the following discussion not the 10-beam is used, which contains only informations of the Ag islands. Rather the 00-beam is studied, which is the result of interference between Ag islands and the portions with bare Ge surface. In Fig. 5 the

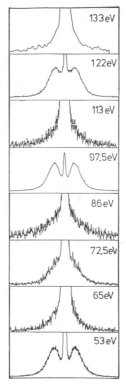

Fig. 5   Measured profiles of the 00-beam after deposition of 0.4 monolayers of silver onto Ge(111) 2x8 at room temperature.

profile of the 00-beam for different energies is shown after deposition of $\theta = 0.6$ with respect to atom density which corresponds to a covered area of about 35%. It is seen, that the profile consists of a central spike and a shoulder, the ratio of which varies drastically with energy due to variation of amplitude and phase of the relative scattering factor. For a quantitative analysis the profiles have been measured for coverages from zero up to a nominal full coverage (one monolayer of a Ag(111) plane). The total intensity (integration over full Brillouin zone) and the peak intensity (integration over central spike) is presented in Fig. 6 for an electron energy of 54 eV. The energy has been

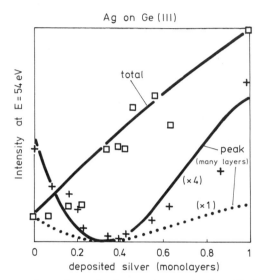

Ag on Ge (III)

Fig. 6   Measured total and peak intensity of the 00-beam at E = 54 eV after deposition of an increasing amount of silver onto Ge(111) 2x8. The curves are calculated (see Fig. 7).

selected for the zero in the peak intensity. For clarity reasons the intensity of the peak is expanded by a factor of four, it coincides with total intensity for the clean surface. Figure 6 also shows calculated curves, which do not completely correspond to Fig. 3. After deposition of a monolayer the profile still consists of peak and shoulder, so that growth in the second layer has started before the first one is completed. In the model therefore a probability of 0.4 has been used that atoms deposited on top of the first level islands stay there for second-layer growth. The probability increases to unity with completion of the first level. For a strict layer-by-layer growth a straight line is expected for total intensity as in Fig. 3. The model calculations including multilayer growth is shown in Fig. 7. The good fit in Fig. 6 (using the calculations of Fig. 7) provides a lot of information. The ratio of the scattering factors $R^2 = |f_2|^2/|f_1|^2 = 10$ for this energy. This ratio is obtained both from the increase of the total intensity and from the coverage for the zero of the peak intensity. The phase shift between f_2 and f_1 is $\phi=\pi$ for this energy. Due to multilayer growth the peak intensity is at monolayer deposition not the same as the total intensity; 80% of total intensity is found in the shoulder. A quantitative evaluation yields 19% of the surface area uncovered, 81% in the first level, 17% already in the second one and 2% in the third one. Therefore coverages and growth mode are derived out of a careful measurement of the spot profile. To show the variation of phase and ratio R the same measurements as in Fig. 6 have

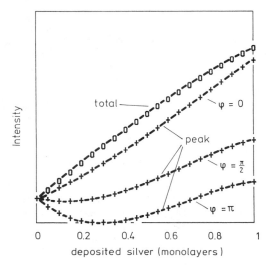

Fig. 7   Calculation of the total and peak intensities for the same
         conditions as in Fig. 3, except that a finite probability of
         second-layer growth has been included.

been repeated for an electron energy of $E = 86$ eV (Fig. 8). Using the
same growth parameters a different ratio $R^2 = 3$ had to be taken to fit
the variation of the total intensity. For the peak intensity only the
phase is left for fitting. The best fit is obtained with $\phi = 0.6 \cdot \pi$, so
that again ratio and phase are determined as given by the calculated
curves in Fig. 8.

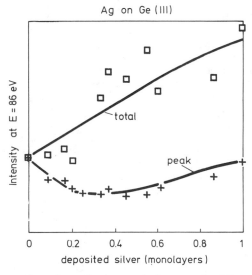

Fig. 8   Measured total and peak intensities as in Fig. 6 for an electron
         energy of $E = 86$ eV.

If the phase shift as measured between 50 and 150 eV is taken as due to the distance from the Ag atoms to the top Ge layer, a distance of 0.47 nm would be required, which is much larger than any reasonable value. There has to be an appreciable energy dependent phase shift between the clean and silver covered germanium corresponding to a geometrical shift of 0.1 to 0.2 nm.

These examples show, that here for the first time a phase difference and an amplitude ratio of scattering factors of two different materials has been derived experimentally.

FACETING OF STEPPED Ge(111) DUE TO Ag SUBMONOLAYERS

Former studies have shown that a vicinal surface of Ge(111) is stable in the clean state, it may, however, turn to a faceted face if heated together with adsorbed Ag in submonolayer quantities[24]. Since now LEED sytems with high resolution are available,[25] the faceting has been studied in more detail.

A germanium crystal has been cut and polished with 3° misorientation with respect to the (111) surface, so that a (19,17,17) surface has been obtained. After the usual cleaning with ion bombardment and heating in uhv a regular step array with the predicted terrace width of 18 atomic rows was obtained.

If 0.1 to 0.5 monolayers of Ag (measured with a quartz microbalance as number of silver atoms relative to germanium surface atoms in a (111) plane) have been deposited at room temperature and then heated to 400°C for 10 min, drastic changes could be observed after cooling to room temperature. A qualitative result is shown in Fig. 9 as a 2D plot of the diffraction pattern. The pattern of the Ge(111) $\sqrt{3}$-Ag is observed, which is reported for the Ge(111) surface after deposition and annealing of 0.8 monolayers of Ag.[23] There are additional spots, which do not belong to the (111) nor to the (19,17,17) face. To determine the orientation of the corresponding facet, the position of those additional spots has been determined relative to the spots of the (111) face for many different energies and plotted in K-space (Fig. 10). Here the component of the scattering vector parallel to the (111) surface is shown in multiples of a reciprocal lattice vector. 100% is therefore the position of the 10-beam of the (111) face. As ordinate the normal component of K is given in multiples of the normal reciprocal-lattice vector (111). The energy range covers the range between the reciprocal-lattice vectors (444) and (555). In Fig. 10 therefore the normal rods provide the position of the 00-beam and the 10-beam of the (111) face in that energy range. The additional spots of Fig. 9 are very nicely connected by a parallel set of inclined rods, which are easily identified as the (10,8,8) facet in this case.

It should be noticed that all possible lattice rods are seen in this case, whereas for the clean stepped surface only the facet rods close to rods of the (111) face are visible.[26] It is therefore needed that the steps are decorated by Ag atoms to provide an increased scattering factor of the step edges.

After evaluation of the measurement of coverages of 1/8, 2/8, 3/8, 4/8 and 5/8 of silver (each deposited onto the clean surface and then heated) the faceting may be described as shown in Fig. 11 and Table 1. The clean face shows just the (19, 17, 17) face, well ordered (as taken from half width of diffraction spots), only a few etch pits with flat (111) bottom are seen, as confirmed by SEM images. With 0.25 monolayers the surface splits up into facetted portions with (13, 11, 11)

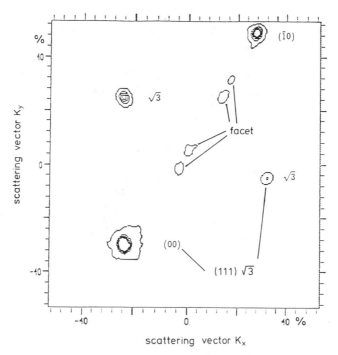

Fig. 9   Part of the diffraction pattern of a Ge(19,17,17) surface after
          deposition of 3/8 of a monolayer of Ag and heating.   The spots
          belong to the (111) √3 face and (10,8,8) facets.   Spots of the
          (19,17,17) are no longer detectable.

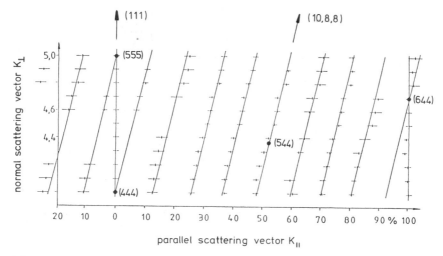

Fig. 10   Position of diffraction spots in K-space.   The parallel
          component of the scattering vector K is given as percentage of
          the (10) beam of the (111) surface.   The normal component is
          given in multiples of the (111)-vector.   The normal rods belong
          to the (111) face, the inclined rods to the (10,8,8) facet.

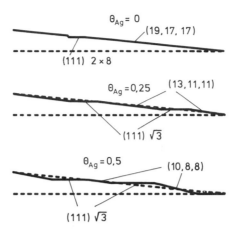

Fig. 11 Schematic presentation of the faceting of Ge(19,17,17) with different amounts of Ag.

Table 1:   Faceting of Ge (19, 17, 17)

Ag coverage	0	0.125	0.25	0.375	0.5
stepped portion					
orientation	(19,17,17)	m	(13,11,11)	m	(10,8,8)
coverage %	>95	i	75	i	50
ordered size (atomic distances)	>100	x	20	x	30
		t		t	
		u		u	
flat portion(111)		r		r	
structure	2x8	e	√3	e	√3
coverage %	<5		25		50
ordered size (atomic distances)	<100		80		80

orientation which are less ordered. 25% of the surface is very close to well ordered (111) faces. They show a very small inclination (regular step array with 120 atomic rows per terrace, which probably is given by the kinetics of facet formation). Nearly all silver is found on the (111) patches, only the amount needed for decoration is on the (13, 11, 11) facet.

The relative coverage has been estimated both from average inclination and relative intensities, which is a rough estimate. For half a monolayer the result is nearly the same, if the (10, 8, 8) facet is taken for 50% of the surface and the (111) with $\sqrt{3}$ superstructure (and broad terraces) for the rest. Again most of the silver is found on the (111) patches. For intermediate coverages a mixture of the adjacent structures and more disorder has been found and heavy disorder for more than half coverage. It should be noticed, that after desorption of silver and annealing always the structure of the clean surface could be restored.

The experiments show that the stability of a surface may be completely different with the presence of a fraction of an adsorbate monolayer, which causes the displacement of many monolayers of the substrate for approaching equilibrium. In the present case the driving force may be given by a highly favorable Ge(111) $\sqrt{3}$-Ag, which are formed until nearly all Ag atoms are incorporated. The facet represents the inclination needed for the average inclination of the starting surface. Further experiments with different annealing times should show, if the observed structure is kinetically determined or already close to thermodynamic equilibrium.[27]

CONCLUSION

The present paper should show that analysis of LEED spot profiles provides a lot of additional information, if inhomogeneous surfaces are studied with a high-resolution SPA-LEED system. A simple extension of the kinematic evaluation procedure by two different scattering factors yields quantitative information on kinetics and ordering in growth. It therefore is a necessary counterpart to the more qualitative microscopy information.

Acknowledgement

Helpful discussions with W. Moritz and M. Lagally are gratefully acknowledged. The studies have been supported by the Deutsche Forschungsgemeinschaft.

REFERENCES

1. M. G. Lagally, Y.-W. Mo, R. Kariotis, B. S. Swartzentruber, and M. B. Webb, this volume.
2. E. J. van Loenen, A. J. Hoeven, D. Dijkkamp, J. M. Lenssinck, H. Elswijk, J. Dielemann, this volume.
3. O. Jusko, U. Köhler, M. Henzler, this volume.
4. H. Bethge, this volume.
5. E. Bauer, W. Telieps in: Reflection High-Energy Electron Diffraction and Reflection Electron Imaging of Surfaces, P. K. Larsen and P. J. Dobson eds., NATO ASI Series, Vol. 188, Plenum New York 1988, p. 381.

6. J. M. Cowley in: <u>Reflection High-Energy Electron Diffraction and Reflection Electron Imaging of Surfaces</u>, P. K. Larsen and P. J. Dobson eds., NATO ASI Series, Vol. 188, Plenum New York 1988, p. 261.

7. K. Yagi, S. Ogawa, Y. Tanishiro in: <u>Reflection High-Energy Electron Diffraction and Reflection Electron Imaging of Surfaces</u>, P. K. Larsen and P. J. Dobson eds., NATO ASI Series, Vol. 188, Plenum New York 1988, p. 285.

8. M. Ichikawa, T. Doi in: <u>Reflection High-Energy Electron Diffraction and Reflection Electron Imaging of Surfaces</u>, P. K. Larsen and P. J. Dobson eds., NATO ASI Series, Vol. 188, Plenum New York 1988, p 343.

9. T. Sakamoto, K. Sakamoto, K. Miki, H. Okumura, S. Yoshida, and H. Tokumoto, this volume.

10. P.I. Cohen, G. S. Petrich, A. M. Dabiran and P. R. Pukite, this volume.

11. M. G. Lagally, D. E. Savage, M. C. Tringides in: <u>Reflection High-Energy Electron Diffraction and Reflection Electron Imaging of Surfaces</u>, P. K. Larsen and P. J. Dobson eds., NATO ASI Series, Vol. 188, Plenum New York 1988, p 139.

12. B. Bölger, P. K. Larsen, G. Meyer-Ehmsen in: <u>Reflection High-Energy Electron Diffraction and Reflection Electron Imaging of Surfaces</u>, P. K. Larsen and P. J. Dobson eds., NATO ASI Series, Vol. 188, Plenum New York 1988, p 201.

13. M. Henzler in: <u>Reflection High-Energy Electron Diffraction and Reflection Electron Imaging of Surfaces</u>, P. K. Larsen and P. J. Dobson eds., NATO ASI Series, Vol. 188, Plenum New York 1988, p. 193.

14. A. Lock, B. J. Hinch, and J. P. Toennies, this volume.

15. K. Kern and G. Comsa, this volume.

16. M. G. Lagally, Appl. Surf. Sci. <u>13</u>, 260 (1982).

17. M. Henzler in: <u>Structure of Surfaces II</u>, J. F. van der Veen, M. A. Van Hove eds., Springer Berlin 1988, p. 431.

18. R. Altsinger, H. Busch, M. Horn, M. Henzler, Surf. Sci. <u>200</u>, 235 (1988).

19. J. Wollschläger, J. Falta, M. Henzler, Appl. Phys. <u>A50</u>, 57 (1990).

20. O. G. Mouritsen, this volume.

21. H. Pfnür, P. Piercy, Phys. Rev. <u>B40</u>, 2515 (1989).

22. J. Wollschläger, M. Henzler, Phys, Rev. <u>B39</u>, 6052 (1989).

23. H. Busch, M. Henzler, Phys. Rev. <u>B</u>, in press.

24. E. Suliga, M. Henzler, J. Vac. Sci. Techn. <u>A1</u>, 1507 (1983).

25. U. Scheithauer, G. Meyer, M. Henzler, Surf. Sci. <u>178</u>, (1986) 441.

26. M. Henzler in: <u>Electron Spectroscopy for Surface Analysis</u>, H. Ibach ed., Springer Berlin (1977) p. 117.

27. R. J. Phaneuf, E. D. Williams, N. C. Bartelt, Phys. Rev. <u>B38</u>, 1984 (1988).

KINETICS OF STRAIN-INDUCED DOMAIN FORMATION AT SURFACES

M. B. Webb, F. K. Men, B. S. Swartzentruber, R. Kariotis, and M. G. Lagally

University of Wisconsin-Madison
Madison, WI 53706 - USA

ABSTRACT

By applying an external and variable strain to a Si(100) sample, it is possible to alter the relative populations of the 2x1 and 1x2 domains in a controlled and reversible way. This means one alters the configuration of monatomic steps which are the domain boundaries. This is conveniently observed by bending a thin bar of Si and observing either the superlattice LEED reflections or STM images. This effect is driven by the relaxation of the energy associated with long-range strain fields in the bulk which are due to the anisotropy of the intrinsic surface stress tensor of the two reconstructed domains. It is similar to the reduction of magnetic field energy by the configuration of magnetic domains. Here, we first briefly review the experimental observations for both nominally flat and vicinal surfaces and show that they are consistent with the theory of Alerhand et al. While the kinetics to produce or remove the unequal domain populations depend on the temperature, the steady state depends only on the strain and the vicinality but not on the temperature. The kinetics closely follow a simple relaxation $\Delta I(t) = \Delta I(\infty)\,(1 - e^{-t/\tau})$. $1/\tau$ is thermally activated with an activation energy of 2.2 ± 0.2 eV. One striking feature is that the time constants for establishing the asymmetric population upon applying the external strain are the same as those for removing the asymmetry after relieving the strain. Changing the populations requires moving steps and mass transport. It involves the same microscopic processes that are important in other phenomena like coarsening, step bunching, etc., and therefore understanding the kinetics in the present experiments may contribute to the understanding of a broader range of problems involving step motion.

INTRODUCTION

Both theoretical calculations and experiments with alloys and thin overlayers show that strain is an important ingredient of the physics of surface reconstruction and other properties.[1] In order to have strain as an independent variable, it is desirable to be able to apply an external, uniform, and continuously variable strain. The simplest scheme is to bend a cantilevered bar putting the surface in either uniaxial compression or tension. In such experiments we observed that it is

Kinetics of Ordering and Growth at Surfaces
Edited by M. G. Lagally
Plenum Press, New York, 1990

possible to reversibly manipulate the relative populations of the 2x1 and 1x2 domains on the Si(100) surface.[2] Under strain the domain populations become unequal with the domain compressed along the dimer bond direction being favored. A surprising result is that the kinetics for establishing the unequal populations upon applying strain are the same as for returning to equal populations upon relieving the strain. The domain boundaries are single-atom-height steps, so that changing the populations requires the motion of steps on the surface.

Alerhand, Vanderbilt, Meade, and Joannopoulos[3] developed a theory which suggest. l a mechanism for these observations. They suggested that for surfaces which reconstruct in domains of different orientation with different anisotropy of their intrinsic surface stress tensors, there are long range strain fields extending into the bulk. The energy associated with these strain fields can be minimized by the appropriate domain wall configuration. The theory gives all the qualitative behavior observed and quantitatively agrees with the experiments within their combined uncertainties.

In this paper we first briefly review the experiments including results on vicinal samples and STM observations. Then we briefly summarize the theoretical results. Finally we present and discuss detailed observations of the kinetics.

EXPERIMENTS

The Si(100) surface reconstructs by dimerizing surface atoms in adjacent rows giving a doubled periodicity. On crossing a single-atom-height step the orientation of the dimer bond is rotated by 90° so there are both 2x1 and 1x2 domains. These domains give distinct half-order LEED superlattice reflections whose integrated intensities are proportional to the domain populations.

The samples are bars (0.3 x 2 x 19 mm) cut from a Si(100) wafer. One end is clamped and the other is positioned between two Ta anvils which can be moved up or down with a high precision micrometer feedthrough to bend the bar. This produces a uniaxial compressional or tensile strain varying linearly along the bar. The strain at the surface is taken from the elasticity theory for a loaded cantileved bar. In the experiments discussed here, all strains were within the elastic limit as checked by seeing that the bar returned to its original position upon relieving the strain.[4] Strains of 0.3% are often achieved before breaking the sample.

We now list the experimental observations:

1. Figure 1 shows an example of the relative domain populations as a function of strain.[2] Here p is the fractional change in the population of a given domain. The sum of the superlattice intensities from the two domains remains constant. These data are for a nominally flat Si(100) surface with originally equal domain populations which was strained at an elevated temperature. The sense of the population asymmetry is that the domain compressed along the dimer bonds is favored. The results are very sensitive to surface preparation and contamination. Contamination or poor preparation inhibit the population asymmetry. The solid line is a fit to

$$p = (2/\pi) \tan^{-1} [\pi \epsilon^{ext}/2\epsilon(\ell)] , \qquad (1)$$

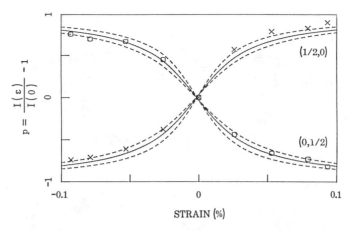

Fig. 1 Asymmetry of the domain populations as a function of strain. $I(\epsilon)$ is the intensity of the superlattice reflection, which is proportional to the area occupied by the corresponding domain. The ratio of majority and minority domain populations is $(1+p)/(1-p)$. The domain compressed along the dimer bond is favored. The solid line is a fit to the data with $\epsilon(\ell) = 0.048\%$. The dashed lines are for $\epsilon(\ell) = 0.048 \pm 0.01\%$.

with a value of $\epsilon(\ell) = 0.048\% \pm 0.01\%$. We discuss this below.

2. While the kinetics are temperature dependent, the steady state is temperature independent.[2] This suggests that the effect is due to a mechanical energy rather than a free energy. If the sample is cooled under strain and then unloaded, the asymmetry remains because of the slow kinetics at low temperature.

3. Measurements as a function of strain taken by moving along the bar with a fixed deflection of the end agree with those taken at a fixed position as a function of the deflection.[2] This shows that the effect depends on the strain and not on the strain gradient.

4. STM images taken before and after strain are shown in Fig. 2. The scans extend one micrometer in the horizontal direction. The nominally flat sample shown in Fig. 2a actually has an inadvertent miscut of 0.13° with monatomic steps running in approximately the [100] direction, or at 45° with respect to the dimer bonds. The average terrace width is 580Å and the two domain populations are equal to within 1%. After compressive strain along the [110] direction, Fig. 2b, the average terrace widths are 150 and 1040Å for the minority and majority domains respectively and the majority domain occupies 87% of the surface. The average of the sum of adjacent domain widths remains the same as for the unstrained sample within 2%. We shall see below that this is consistent with the striped phase and the local optimization procedure in the theory of Alerhand et al.[3]

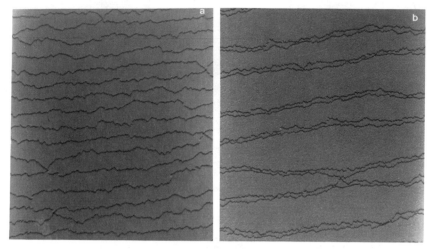

Fig. 2 STM images of a Si(100) surface with an inadvertent miscut of
0.13°. The nominal step direction is [100], i.e., 45° away from
the dimer row directions. The scans extend over one micrometer.
Left: Before strain. The average terrace width is 580Å. The
domain populations are equal to within 1%. Right: After straining
the sample in compression along the [110] direction the average
majority and minority terrace widths are 1040 and 150Å
respectively. For the majority domain the dimer bonds are
parallel to the applied compressive stress.

5. Figure 3 shows the LEED intensity results for strain experiments
on vicinal samples miscut 1° in the [110] direction. LEED
angular profiles of the specular beam show the expected
splitting as a function of electron energy. The different
symbols are for samples cut so that the steps are either
parallel (0's) or perpendicular (+'s) to the length of the bar
and thus to the direction of the strain. Thus for the two
samples opposite domains are favored under compression. The
populations in the unstrained samples are originally unequal and
for the plot the zero of the strain axis has been shifted to the
point of equal populations. Upon straining the widths of the
LEED angular profiles of the majority (minority) superlattice
reflections narrow (broaden) but again indicate that the sum of
the average terrace widths remains unchanged. The population
asymmetry for the vicinal sample is a more gradual function of
strain than for the flat sample. The solid lines in Fig. 3 are
again a fit to Eq. (1), but with $\epsilon(\ell) = 0.11\%$. The fact that
the data agree for both orientations of the steps relative to
the strain indicates that the effect is not a direct interaction
between the strain and the detailed step structure but rather
depends on the orientation of the dimer bonds relative to the
strain.

We now review the theoretical expectations of Alerhand et al.[3] They
consider surfaces which reconstruct with different orientational domains
having anisotropic intrinsic surface stress tensors. The surface
stresses produce long-range strain fields in the bulk, the energy in
which can be relaxed with the appropriate domain wall configuration. An

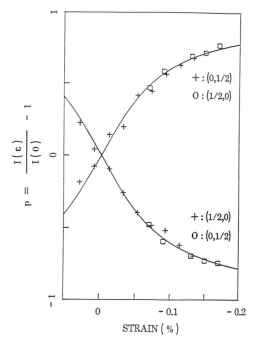

Fig. 3 Asymmetry of the domain populations for a 1° vicinal surface. The +'s (o's) are for samples cut so the steps are perpendicular (parallel) to the applied strain. The line is a fit with $\epsilon(\ell)=0.11\%$. For the fit the zero of the external strain has been shifted to the strain corresponding to equal populations.

external strain lifts the degeneracy and one domain grows at the expense of the other. They consider striped domains with alternating widths $\ell(1 + p)$ and $\ell(1 - p)$ where ℓ is the average domain width. Their expression for the total energy per unit area is

$$E = \epsilon^{ext}pF/2 + C_1/\ell - (C_2/\ell)\ln[(\ell/\pi a)\cos(\pi p/2)] , \qquad (2)$$

where a is a microscopic cut-off of the order of an atomic spacing, C_1 is the energy per unit length of a step apart from the long range elastic energy, $F = \sigma_{/\!/} - \sigma_{\perp}$ and the σ's are the components of the surface stress tensor parallel and perpendicular to the dimer bonds. The last term is the elastic relaxation energy, where

$$C_2 = F^2(1 - \nu)/2\pi\mu \qquad (3)$$

and where μ and ν are the bulk modulus and Poisson ratio. This energy is plotted vs. p for both $\epsilon^{ext} = 0$ and 0.1% and for $\ell = 80\text{Å}$ in Fig. 4. This energy, which is actually contained in the long-range strain fields, amounts to the order of 0.5 meV per surface atom and so is small compared to binding energies and potential barriers and should have negligible effect on the basic local atomic structure and processes on the surface.

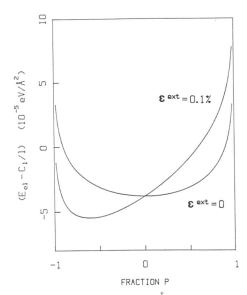

Fig. 4 The total elastic energy per unit areas as a function of p. The plots are for ℓ = 80Å and for ϵ^{ext} = 0 and 0.1%.

Alerhand et al.[3] minimize this energy both globally and locally and for our purpose we are concerned with the local minimization, where ℓ is fixed and p is allowed to vary corresponding to the experimental observations. The optimum p is given by Eq. (1)

$$p = (2/\pi) \tan^{-1} [\pi\epsilon^{ext}/2\epsilon(\ell)] ,\tag{4}$$

where

$$\epsilon(\ell) = - \pi F (1 - \nu)/4\ell\mu.\tag{5}$$

Using a semiempirical tight-binding theory for structural energies, Alerhand et al. find $\sigma_\perp$ = - 0.035 eV/Å^2 and $\sigma_{/\!/}$ = +0.035 eV/Å^2 so that F = 0.07 eV/Å^2. Other estimates vary by about a factor of two[3].

For our vicinal sample we know both ℓ and $\epsilon(\ell)$ so we can determine F. These data give F = 0.06 ± 0.02 eV/Å^2. Another measurement from STM images also gives F = 0.06 ± 0.02 eV/Å^2. This agreement, within combined uncertainties, between the experimental and theoretical values and the qualitative behavior leaves little doubt that the proposed mechanism is correct.

KINETICS

We now discuss the observed kinetics. Figure 5 shows an example of the time dependence of the LEED superlattice beam intensities. For the open symbols in Fig. 5, a nominally flat sample was heated to 500°C, and then strained to 0.04% and the intensity of the (0,1/2) beam was measured as a function of time. This is the majority domain and its intensity

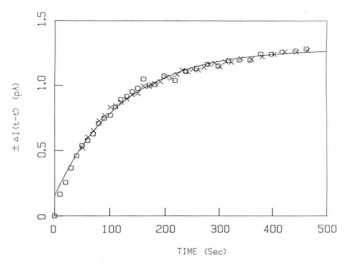

Fig. 5 The superlattice beam intensity as a function of time after applying (o's) and relieving (x's) the external compressive strain. The data after relieving the strain are plotted as -[I(t)-I(t')] and both t' and I(∞) were shifted for sake of comparison. The data for the o's and x's are for initial applied strains of -0.04% and -0.02% respectively. The solid line is a fit to a simple exponential relaxation with a time constant of 122s. The crystal temperature was 500°C.

increases. We plot [I(t) - I(t')] both because it takes several seconds to apply the strain and because after straining the diffracted beam has moved slightly so the Faraday collector must be moved. The data begin at t' some 10 to 20 seconds after beginning the strain. The crosses in Fig. 5 are the time dependence of the same beam after removing the strain. On unstraining, the intensity decreases. In order to compare the kinetics for straining and unstraining, we have here plotted - [I(t) - I(t')]. We have shifted the time axis because t' is different in the two cases. We have also shifted I(∞) to be the same for both curves. The data for the O's and X's are for strains of -0.04% and -0.02% respectively. The agreement between the two sets of data shows that the kinetics for straining and unstraining are the same and that they are independent of the initial strain. The solid line is a fit to a simple relaxation

$$[I(\infty) - I(t - t')]/[I(\infty) - I(t')] = e^{-(t - t')/\tau} , \qquad (6)$$

with τ = 122 sec.

There is a subtle but consistent deviation from this functional dependence that we ignore now but will discuss below.

Figure 6 shows a plot of $\ln(1/\tau)$ vs $1/T$ for a number of experiments on flat samples miscut by ∿0.5° toward [110]. All data points with the open symbol are from experiments unstraining the sample; all crosses are points from experiments straining the sample. Within the scatter of the data, the time constants are the same for straining and unstraining. The figure includes data for different samples and for initial strains

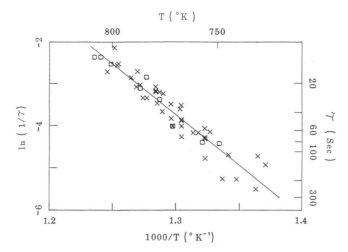

Fig. 6 Arrhenius plot of ln(1/τ) vs 1/T. All experiments were done on surfaces miscut $\sim$0.5° toward [110] with a range of initial applied strains between -0.02% and -0.1%. The slope gives an activation energy 2.2±0.2 eV.

varying between -0.02% and -0.1%. Again within the scatter, the time constant is independent of the initial strain. The slope in Fig. 6 gives an activation energy of 2.2 ± 0.2 eV and the intercept gives 1/τ $\approx$ 4x10^{12}s^{-1}.

Figure 7 is an example of the time dependence of the intensity for a 1° vicinal sample strained to 0.04% at T $\simeq$ 300°C. Here the time constant is very much shorter than for the flat sample. These short time constants are difficult for us to measure because of the time interval before we can take the first data, but in this example τ is roughly 6 seconds. It is clear that the time constant is a sensitive function of the average terrace width, but more experiments on a faster time scale and at lower temperatures are needed to characterize this dependence.

We now return to the fitting and point out that in all the curves like those in Fig. 5 the data at short times vary a bit faster than the fitted curve. This is more clear in Fig. 8. Here we show another example of the time dependence. The solid line is the best fit to a simple relaxation using all the data; the dashed line is the best fit using only the data in the first 30 seconds. We see such deviations in all the data, but the extent is slightly different from sample to sample and experiment to experiment. (We shall see below that one might expect a slightly different functional form for the time dependence but this slight difference is not sufficient to reconcile the data). A possible explanation is suggested by the data for the vicinal surface. As seen in the STM images, there is a distribution of terrace widths. Then, since the time constant is shorter for smaller ℓ, the disappearance of the contribution to the population asymmetry from narrow terraces will be faster than the average, making a steeper change at short time.

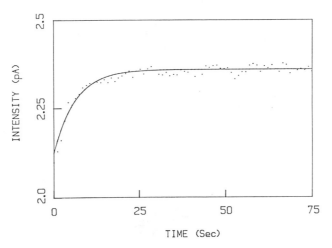

Fig. 7 The superlattice beam intensity as a function of time for a 1°
vicinal sample strained to -0.04% at about 300°C. The line
corresponds to a time constant of 6s.

We now consider what might be learned from the measured time
constant and its activation energy. Clearly the domain boundaries have
to move by the addition or deletion of atoms from the steps. Since the
STM shows that the surface remains in a striped phase, the transport has
to be across the terraces rather than, for example, some process
involving intersecting steps and transport along them. Also the energy in
the strain fields is very small compared to binding energies and so we
suppose that the atomic processes are essentially unchanged. We then
consider the steps fluctuating in position due to fluctuations in the net
flux of some equilibrium species diffusing on the surface, e.g. surface
adatoms, vacancies or divacancies. Then as the steps fluctuate they tend
to their lowest-energy configuration. The activation energy would then
be the sum of the formation and migration energies of the diffusing
species.

We attempt to make a simple one-dimensional model of such a process
where steps are assumed straight over a segment ξ along the step.
Consider a particular position on a step whose displacement from its
equilibrium position is y. The time rate of change of y is given by the
Langevin equation

$$dy/dt = -\gamma \delta H/\delta y + f(t) \ , \tag{7}$$

where γ is the kinetic coefficient, H is the energy per length ξ of step
as a function of displacement, and f(t) is a random function describing
the background fluctuations. Equation 2 gave the energy per unit area,
therefore

$$H = -C_2 \xi \ln[(\ell/\pi a)\cos(\pi y/2\ell)] \ , \tag{8}$$

where $y = p\ell$. The background fluctuations are assumed to be white and
Gaussian with their mean square displacement given by

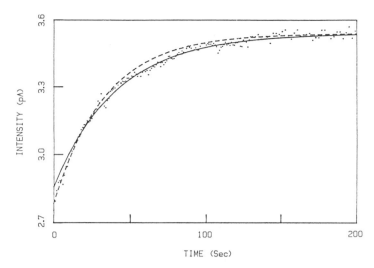

TIME (Sec)

Fig. 8 Deviation from a simple relaxation at short times. The solid line
is the best fit using the entire range of data with a time
constant of 41s. The dashed line is a fit using only those data
within the first 30s and has a time constant of 34s. The sample
was strained to -0.04% at 550°C.

$$<f(t)f(t')> = \sigma a^2 n_o D \delta (t - t') \, , \tag{9}$$

where n_o is the equilibrium concentration of the diffusing species, D is
the diffusion coefficient, and σ is the sticking coefficient of the
diffusing species on a step. The kinetic coefficient, γ, and the
amplitude of the background fluctuations are related by the
fluctuation-dissipation theorem

$$\gamma = \sigma a^2 n_o D/2kT \, , \tag{10}$$

where T is the temperature, and where we have used equipartition.
Substituting H into the Langevin equation gives

$$dy/dt = - (\gamma C_2 \xi \pi/2\ell) \tan(\pi y/2\ell) + f(t) \, . \tag{11}$$

The saddle point equation of this system can be solved exactly to give

$$y = (2\ell/\pi) \sin^{-1}[\sin(y_o \pi/2\ell) \, e^{-t/\tau}] \, , \tag{12}$$

where

$$1/\tau = \gamma C_2 \xi \pi^2/4\ell^2 \, . \tag{13}$$

For the times accessible in the experiments this time dependence of y is
indistinguishable from

$$y = y_o e^{-t/\tau} \, , \tag{14}$$

which we have used to fit the experimental data.   Substituting for γ we have

$$1/\tau = \pi^2 \sigma a^2 \xi n_o D C_2 / 8kT \ell^2 .\tag{15}$$

From this model and from the determination of F and thus C_2 discussed above, we can evaluate the product $\xi \sigma n_o D$. We take $\xi \approx \ell/4$ from STM observations.   Then

$$\sigma n_o D \approx 10^{15} e^{-2.2} \text{ eV}/kT .\tag{16}$$

There is considerable uncertainty in the experimental determination of both the activation energy and the prefactor. Also this simple model neglects step meandering and the distribution of terrace widths.   More kinetic experiments, particularly on a range of vicinal samples, and more theoretical work are clearly needed.

In spite of these uncertainties, we take the simplest approach by setting

$$D = a^2 \nu e^{-E_{migration}/kT}\tag{17}$$

and

$$n_o = a^{-2} e^{-E_{formation}/kT} .\tag{18}$$

This gives $\sigma \nu \approx 10^{15}$ sec^{-1} and the sum of the activation energies for the formation and migration of the diffusing species as 2.2eV.

SUMMARY

In summary, experiments using an externally applied uniform and variable strain allow the manipulation of the step configuration on the Si(100) surface. The effect is due to the relaxation of the small energy contained in long-range strain fields extending into the bulk due to the intrinsic surface stress tensor. All the qualitative behavior suggested by the theory of Alerhand et al.[3] is consistent with the observations and their parameters are at least in semiquantitative agreement with the experiments.   Since the energies are small and not localized at the surface, the strain should not affect the local binding and structure. Thus the experiments may provide a convenient and controlled way to study the basic atomic processes that are involved in, e.g., step configurations and migration, growth, and surface diffusion.   Here we have studied the kinetics of the step migration in response to the strain field and modeled it with a simple theory involving surface diffusion.

Acknowledgements

This work was supported by NSF Grant No. DMR87-20778 and in part by NSF Grant No. DMR 89-18927.   We thank Dr. P. Wagner of Wacker Chemitronic, Burghausen, Germany, for supplying the Si wafers used in the STM studies.

REFERENCES

1. See for example R. D. Meade and D. Vanderbilt, Phys. Rev. B40, 8905
     (1989); A. Ourmazd, D. W. Taylor, J. Bevk, B. A. Davidson, L. C.
     Feldman, and J. P. Mannaerts, Phys. Rev. Lett. 57, 1332 (1986) and
     reference therein.

2. F. K. Men, W. E. Packard, and M. B. Webb, Phys. Rev. Lett. <u>61</u>, 2469 (1988).

3. O. L. Alerhand, D. Vanderbilt, R. D. Meade, and J. D. Joannopoulos, Phys. Rev. Lett. <u>61</u>, 1973 (1988).

4. There should be a residual curvature after removing the strain because of the unequal populations of the two domains with different intrinsic surface stress tensors, but, with the relatively thick samples used here, this is negligible. Experiments have been done after plastic deformation and are reported in Ref. 2.

MOLECULAR KINETICS ON STEPS

H. Bethge

Institute of Solid State Physics and Electron Microscopy
Academy of Sciences of the GDR
Halle (Saale), German Democratic Republic

ABSTRACT

A summary of investigations of step structures, step motion and decomposition, and other kinetic processes at steps using the decoration technique is given. Diffusion along edges is considered. Heterogeneous thin-film growth under the influence of steps is discussed in terms of their impact on the well-known modes of epitaxial growth.

INTRODUCTION

I have been asked to give a survey of our former research on the morphology of surfaces and the kinetic processes on steps. As there are several extensive publications available,[1-3] in the present report I want to describe those investigations only briefly in two sections, with a third following informing about recent investigations showing the influence of steps on the 2-D overgrowth.

We use the electron microscope, and our special technique of imaging steps is the decoration effect, which will be described below. It was first introduced by Basset.[4] A small amount of Au is evaporated on a NaCl surface, suitably at an elevated temperature (150°C). As Fig. 1a shows the gold atoms form small 3-D nuclei with a preferred nucleation taking place at the steps. A carbon film also vapour-deposited covers the Au-particles and serves as a replica during the electron microscope investigation. In order to determine the step height and the level on either side of the step (higher or lower), it is necessary to discuss the planes beside the step by using the "twofold decoration".[6,7] The Au-decoration marks the position of the step, followed by a second evaporation using a low-melting metal yielding well-distinquishable particles, sometimes larger, which are always formed on the lower side beside the step (see Fig. 1b). It should also be mentioned that besides Au also Pt may be used causing smaller nuclei. Thus it is possible to detect step distances of even some 10 Å.

The decoration effect, however, implies certain conditions,[6-8] though it should be mentioned here that alkali halides and the surfaces of a number of minerals are easy to decorate. Silver halides require perfectly clean surfaces, while for Cu and Ag (dealt with in the last

Kinetics of Ordering and Growth at Surfaces
Edited by M. G. Lagally
Plenum Press, New York, 1990

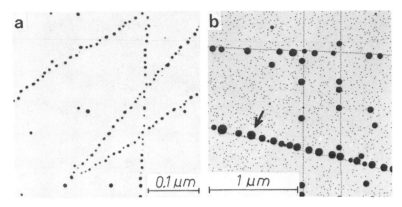

Fig. 1   Decoration of steps on (100) NaCl cleavage faces.   a)
Au-decoration of monatomic steps; b) "twofold" decoration (Au
and Bi), the arrow marks a higher step besides monatomic steps.

section) obviously the sufficient smoothness of the surface is the main
problem.

Recently, variants of the decoration technique have been reported
which enable the direct investigation of surfaces without using a
replica. Si and graphite were decorated with various ion crystals with
their step structures being imaged in the scanning microscope.[3] Here it
is the larger coefficient of the secondary emission of the decorating
material that yields the good contrast. Bauer et al.[9] have shown that in
a photoelectron microscope steps can be imaged on metal surfaces
decorated with a metal of lower work function.

ORIGIN OF STEPS

The step structure of a surface depends on the growth and the
further treatment of a crystal, particularly of its surface. The defect
structure of a crystal, especially the dislocations decisively influence
the surface morphology on an atomic scale.

Of all the crystal surfaces investigated with respect to the step
structure those of the NaCl have been studied most extensively. This is
due to the facts that the decoration technique is easy to employ and that
low-indexed surfaces are easy to prepare by cleavage. Here, the
characteristic step structures on (100) NaCl-faces will be presented as
they most generally represent the origin of steps.

If a NaCl crystal is cleaved in free atmosphere, followed by the
decoration in vacuum, step structures are revealed as typically shown
Fig. 2. The steps, more or less curved, have been designated as
"sinus-shaped cleavage pattern". This description is more than dubious
as the steps are not formed by cleavage. A cleavage face produced in
free atmosphere at once is covered with a water film with the surface
being dissolved. The NaCl crystal has a liquid-like surface. During
evacuation as a preparation step the water evaporates with the surface
recrystallizing. According to Fig. 2 the steps represent borders formed
due to edge tension, which consist of liquid-like monolayers just before

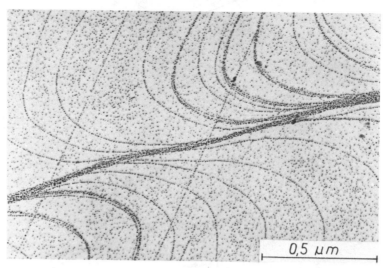

Fig. 2   Step pattern after dissolution of the NaCl surface by water vapor.

solidification.[10]   Real cleavage faces are obtained only if the samples are cleaved in absolute dry atmosphere or even better in vacuum.

As mentioned above, the dislocation structure of a crystal is decisive for the step structure of its surface. For discussing the connections Fig. 3 schematically shows the dislocation structure of a crystal.   Dislocations cannot freely end in a crystal; they form a three-dimensional network, they may end at the surface, or they form loops or rings.   Furthermore, a nearly perfect crystal consists of subgrains, i.e. of areas having a low misorientation.   This misorientation is compensated by low-angle boundaries, consisting of a two-dimensional network of dislocations.   The types and the arrangement of the dislocations along a low-angle boundary are determined by the extent and type of misorientation; the correlations are well-defined.

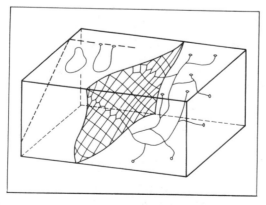

Fig. 3   Schematic drawing of a dislocation structure in a crystal.

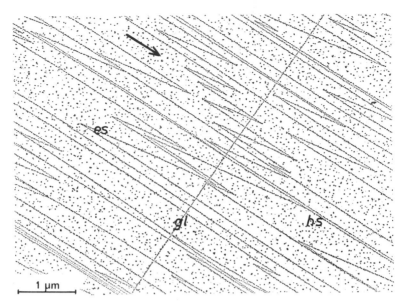

Fig. 4 Cleavage pattern of a cleavage structure on a (100) NaCl cleavage face. The arrow marks the cleavage direction.

If the Burgers vector of a dislocation emerging to the surface has a component normal to the surface, we simply speak of a screw dislocation, on the surface a step has formed starting from the point of emergence. As a rule this is a monatomic step. The point of emergence of an edge dislocation is indicated solely by a displacement of the atoms around the dislocation (see Fig. 5).

Figure 4 shows a real cleavage structure. The lightning-shaped cleavage steps are called "elementary cleavage structures". The steps forming a tip are monatomic (ms). If several ms meet they may form higher cleavage steps (hs). The slip-line sl is formed by a dislocation activated after cleavage. It was possible to explain the process producing the elementary cleavage structure,[11] but here it will be pointed out only briefly. In Fig. 4 the cleavage structure arises from the interaction of the crack with the dislocations within the crystal. The angle enclosed at the tip as well as the density of tips or steps, respectively, depend on the crack velocity. The latter has to be as high as possible, i.e., considerably higher than the dislocation mobility, in order to yield a cleavage face of step density as low as possible.

In Fig. 5 a subgrain boundary is cut by a cleavage crack. The boundary is of mixed type, but mainly of tilt character. Monatomic cleavage steps are produced starting from screw dislocations present in the boundary. Between the screw dislocations about 10 edge dislocations emerge to the surface, causing the tilt of the subgrains. The higher-magnified inset in the micrograph shows that, unlike the decoration along a step, Au-nuclei are arranged equidistantly. The distance corresponds to the tilt angle. The atomic displacement as

128

illustrated in Fig. 5 is the reason why the point of emergence of an edge dislocation may act as an active centre to the preferred Au-nucleation.

Particularly interesting are the step structures formed by thermal treatment. Decisive factors are the temperature and the time with the easiest conditions, however, for a treatment in vacuum.

The thermally activated processes are schematically drawn in Fig. 6. At temperatures above 300°C in step-free areas so-called "Lochkeime" arise (stages A→B→C) induced by vacancies always present in a surface. At the steps molecules dissociate from kinks, thus being able to migrate along the step. At temperatures high enough they may diffuse from the step to the surface, finally vaporizing. The individual stages correspond to the kinetic model that Kossel and Stranski[12] postulated for the growth or vaporization of a crystal.

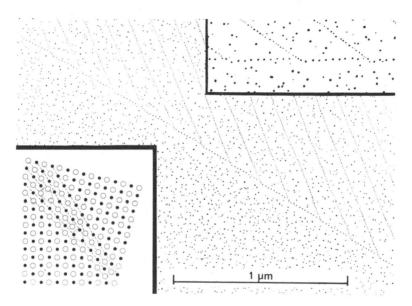

1 μm

Fig. 5   Cleavage pattern produced by a low-angle boundary.

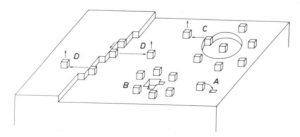

Fig. 6   The molecular concept of the thermal decomposition of crystal surfaces.

The two micrographs in Fig. 7 present surface structures at the beginning of the evaporation of a NaCl-crystal. At a temperature of 350°C 1 ML of the surface is evaporated in some 10 minutes, mostly by the formation and propagation of "Lochkeime". Step patterns on the surface as shown in Fig. 8 preferentially at elevated temperatures and at a higher evaporation rate originate. The step structure is determined by lamellar systems induced by dislocations emerging to the surface (see Fig. 3). The early stage of the formation of a spiral system, which according to Frank's concept[13] takes place at a step starting from a screw dislocation, is illustrated in Fig. 7b (see arrow). The different types of lamellar systems are determined by the Burgers vectors of the initial dislocations, which is explained in (1). Here it should be pointed out that all curved steps are monatomically high. The straight steps of the square spirals are double as high (lattice unit); always two curved steps coexist with a straight step (see arrow in Fig. 8). The distance of steps in a lamellar system depends on the temperature, but also on the elastic strains around the dislocation core. The latter cannot be determined as they additionally depend on the orientation with respect to the surface. Evaporation structures such as in Fig. 8 have the advantage that it is possible to define the step density per area sufficiently well.

This was a brief survey of step structures on NaCl-surfaces, which we know in detail. Far poorer is our knowledge of step structures on surfaces of metals and semiconductors, which, however, are far more interesting for topical problems of surface science than ionic crystals. The main problem of determining the step structure of a surface of these materials lies in the fact that, in general, the surfaces are produced by cutting a crystal. Slight deviations from the ideal face, mostly low-indexed, already yield a high step density. The way the steps are arranged depends on the special treatment of the crystal after cutting. The dislocation structure of single metal crystals is similar to that of NaCl-crystals. Individual dislocations as well as low-angle boundaries in interaction with steps primarily due to cutting yield a lightly defined step structure little investigated up to now.

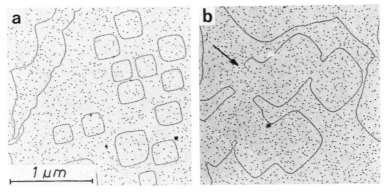

Fig. 7   Initial stages of decomposition by evaporation. a) 330°C, 30 min; b) 330°C, 45 min.

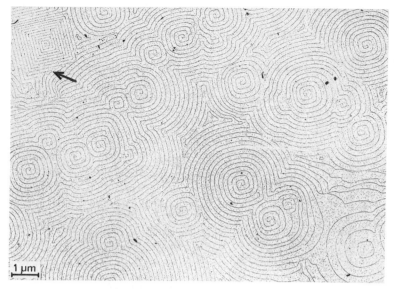

Fig. 8   Dislocation-induced   lamellar   step   pattern   after   intense
evaporation (400°C, 90 min.).

KINETICS ON STEPS

Molecular processes on steps as they are schematically drawn in Fig.
6 and well-known as the Kossel-Stranski Mechanism are decisive for the
growth of a crystal from the vapor phase as well as for the decomposition
during evaporation.   A more quantitative description requires the more
detailed knowledge of kinetic processes of the molecular quantities,
particularly the diffusion parameters.

As it is illustrated in Fig. 6 the initial process of the
decomposition at a step is the detachment of a molecule out of a kink
position and the diffusion along the step.   This edge diffusion has been
studied[14] by the so-called matched face technique.[15]

Figure 9 demonstrates the method and the observed effect of the edge
diffusion.   For the matched face technique the NaCl-crystal is cleaved
under ultra-high vacuum conditions with both the cleavage faces being
used for the investigations.   Corresponding structures on either side of
the surface have to be imaged.   The change in the step structure due to
thermal treatment, for instance, can be detected most precisely by
comparing it with the structure of a surface not previously treated.
Figure 9a shows the unaffected cleavage structure, whereas in Fig. 9b an
edge diffusion has taken place at a temperature of 275°C, which has
caused a change in the step configuration at the tip of the cleavage
steps.   The sketch in Fig. 9c shows that the initially pointed tip has
become club-shaped, with the sum of the transported molecules being zero.
The latter conclusion is drawn from a more detailed analysis of the
micrographs, revealing that the "movement" of the step is induced solely
by edge diffusion.   Obviously, the temperature of 275°C is too low to
enable the detachment of the molecules from the step onto the surface.

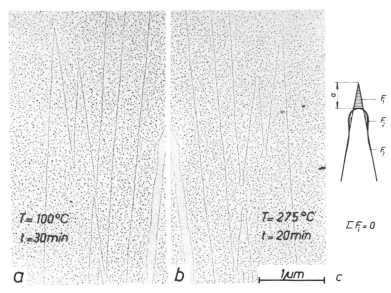

Fig. 9 Corresponding cleavage structures (matched-face technique) (a) without and (b) with thermally activated change of the tip configuration (c).

The driving force for the edge diffusion and the step movement can be derived from the gradient of the chemical potential which is determined by locally differing step curvatures. The following partial differential equation was developed under the assumption of isotropy of specific edge free energy and edge diffusion coefficients

$$\partial s/\partial t = G(T)\partial^2 k/\partial b^2, \qquad G(T) = D_r \rho n_r f_o^2/kT , \qquad (1)$$

where ∂s is the outward normal distance covered by the step element ∂b during the time increment ∂t, and ∂k is the curvature change of ∂b. $G(T)$ is a quantity containing the edge diffusion coefficient D_r, the specific edge free energy ρ, the number of molecules able to diffuse per unit of ledge length, and the square of the required area of a crystal unit f_o^2.

As this partial differential equation is not analytically solvable, it was integrated by means of finite difference. By fitting the numerically simulated changes to the experimentally observed ones the factor $G(T)$ could be determined for various temperatures.[14]

In Fig. 10 experimentally and numerically determined tip profiles are compared. These examples indicate that the apex angle of 2α essentially influences the developed structures. Cleavage tips with apex angle of $2\alpha > 1°$ form club-shaped steady-state profiles, whereas tips with nearly parallel steps ($2\alpha < 1°$) cord up completely thus producing isolated circle structures. The very good correspondence observed of the numerical results with the experimental ones justifies the fundamental conception of the structural change by isotropic edge diffusion. The temperature dependence of $G(T)$ determines the activation energy of the edge diffusion along the monatomic NaCl surface steps to be $E_r = 1.05$ eV. By means of conventional assumption it was possible to estimate the edge

132

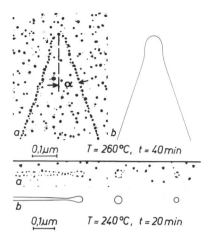

Fig. 10 Profiles observed and numerically calculated in dependence on the apex angle in the tip.

diffusion coefficient

$$D_r(T) = 1.02 \cdot 10^{-6} \exp(-1.05 \text{ eV}/kT) \text{ cm}^2/s.$$

Another aspect of the step kinetics is the interaction of a step with the molecules diffusing on the surface, yielding the velocity at which a step moves during the growth of a crystal from the vapor phase or during its decomposition while it evaporates. Special experiments enable the step velocity to be measured. It should be shown that it is possible to determine such fundamental parameters as the mean free diffusion path of a diffusing molecule.

At first the step velocity was measured in special growth experiments.[16] Two crystals were evaporated in vacuum to produce clean and defined surfaces. Then the two crystals were confronted in such a way that the distance between the surfaces was only 1 mm. The crystals were heated at different temperatures, e.g. 330 and 350°C. The supersaturation corresponding to this temperature difference results in a defined growth for the cooler crystal. For the observation of the step movement during a certain time we used the so-called double-decoration technique. Before both the crystals are confronted a first decoration is carried out to fix the starting position of the steps. When the growth is finished a second decoration fixes the attained position. Figure 11 shows an example. The broken line shows the starting position and the full line presents the position of the moving step. One can, of course, not exclude that the first decoration inhibits the step movement. But this effect can be avoided experimentally, if the step movement is determined from two experiments carried out at different growth times and the differences are made use of.

The step velocities measured were analyzed as a function of the distance of neighboring steps. In Fig. 12 the step velocity is shown in dependence on the distance for different supersaturations. At first the step velocity rises with the step distance and reaches at constant value

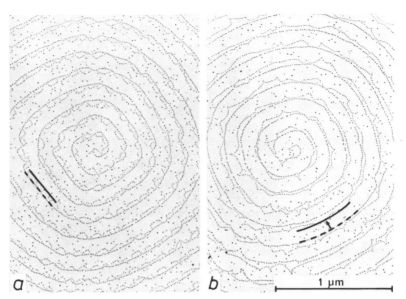

Fig. 11 Re-growth of evaporation spirals imaged by double decoration.
---- step front after evaporation, ———— growth front.
Conditions for the NaCl crystals used: 330/350°C = 340%
supersaturation. a) 8 min., b) 15 min.

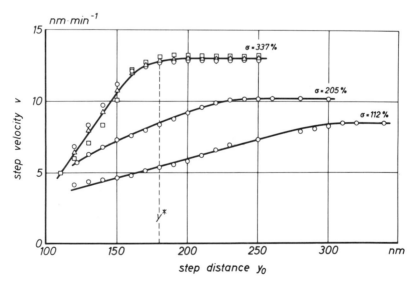

Fig. 12 Step velocity in dependence on the step distances for different
supersaturations.

y*. For a step distance larger than y* it is no longer possible for molecules impinging in between the steps to reach the step. The step velocity as a constant maximum value. We define a "trapping range" x_{tr} and use y* = $2x_{tr}$. The mean diffusion path can be evaluated by using the theory by Burton, Cabrera and Frank,[17] which implies a formula derived for the step velocity v. Assuming y* as the step distance for v_{max} reveals that the trapping range x_{tr} is about double the width of the mean free diffusion path x_s (cf. Table 1).

The double-decoration technique was also used to investigate the step velocity during evaporation though the matched face technique enables more precise measurements.

An example of those later investigations[18] is given in Fig. 13. The higher step (hs) serves as a fixed point to the exact correlation of both the micrographs. For separate steps (e.g. the step to the very left) the maximum step velocity v can be determined; and applying again the BCF-theory enables x_s to be evaluated, which is necessary for the computer simulation of the movement of special step arrangements (e.g. isolated step pairs or bunching processes of step systems).

The values of surface diffusion experimentally determined are listed in Table 1.

Table 1

Experimental Conditions and Techniques	Supersaturation or Surface Temperature	Trapping Range x_{tr}	Mean Diffusion Path $\overline{x}_S$
Growth (330°C) (double-decoration technique)	337% 112%	900Å 1500Å	430Å 720Å
Evaporation (double-decoration technique)	330°C	5200Å	2480Å
Evaporation, UHV (matched-face technique)	314°C 278°C 263°C		2500 ± 400Å 3600 ± 700Å 4500 ± 1000Å

As to the growth the values are lower relative to the experiments of vaporization. This is easy to understand as the high density of the diffusing molecules restricts x_s due to the dependence on the supersaturation at the same temperature. For the evaporation an interaction of diffusing molecules with each other can be excluded, but there is a strong dependence on the temperature.

More serious is the fact that the values measured of x_s are 1 to 2 orders of magnitude larger than those theoretically calculated. One reason for these discrepancies may certainly be that the diffusion kinetics based on parameters such as the Debye frequency and jump

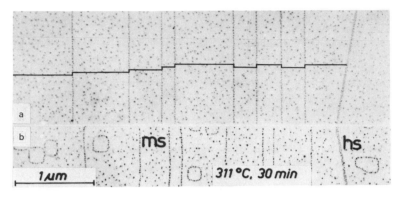

Fig. 13 Matched-face structures: a) original cleavage structure, b) counter surface evaporated (ms = monatomic step, hs = higher step).

distances of lattice units is oversimplified. Furthermore, the question arises whether molecules really diffuse as it is assumed a matter of course for the NaCl crystal. Though, however, mainly molecules are measured within the vapor, one can not exclude that it is only during the transition from the surface into the vapor phase that NaCl molecules are dissociated from diffusing complexes. For example, two molecules having a strong bond to each other and forming a diffusing particle might have an interaction energy with the surface lower than that of single molecules.

In conclusion one may say that the kinetics of surface diffusion still cannot satisfactorily be explained or understood, respectively. Certainly, its detailed theoretical consideration is difficult, but the problems as to the kinetic energy of a diffusing atom or molecule are significant as the following section also proves.

STEPS AND 2-D OVERGROWTH

In a final chapter thin film growth will be considered with the influence of steps being discussed. Some generally known concepts serve as an introduction while others to be newly defined will be the basis of a molecular-kinetic consideration revealing the significance of processes on steps.

Figure 14 shows the common modes of growth mechanisms. Modes 1 and 3 occur for systems of strong interaction; we speak of two-dimensional (2-D) growth. In mode 2 the interaction is weak, the layer growth begins with a three-dimensional (3-D) nucleation (according to the decoration effect) at steps on the surface. It may be assumed that as to the

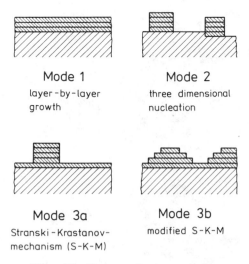

Mode 1
layer-by-layer
growth

Mode 2
three dimensional
nucleation

Mode 3a
Stranski-Krastanov-
mechanism (S-K-M)

Mode 3b
modified S-K-M

Fig. 14 Modes of overgrowth.

epitaxial orientation the three-dimensional nuclei along the step have a more perfect orientation than those on the smooth surface. Respective experimental results are not unambiguous. For the formation of epitaxial layers according to mode 2 the process of coalescence is of particular significance.

For the layer-by-layer growth (mode 1) Frank and van der Merwe[19] considered the fitting mechanisms of two lattices across the interface and accordingly introduced the interface dislocations. In former time in our group Woltersdorf[20] investigated different systems with the formation of interface dislocations being studied in dependence on the layer thickness. There was a good agreement between experiment and precise calculations.

Considering the influence of steps the so-called Stranski-Krastanov-Mechanism (mode 3) is of particular interest. First, there is a two-dimensional (2-D) growth on the substrate followed by a three-dimensional (3-D) growth after one or two monolayers have formed, as shown in Fig. 14. Usually, the Stranski-Krastanov-Mechanism (S-K-M) is described as mode 3a. Regarding the primary 2-D growth as essential to S-K-M we may designate mode 3b as the modified Stranski-Krastanov-Mechanism. We think that the formation of a monolayer is followed by the growth of terraced hillocks, which is evidenced by convincing experimental results as exemplified, e.g., by Fig. 15. The hatched area stands for a monolayer from which additional layers of the hillock start, which due to different contrast are easy to count.

The different modes 1-3a briefly described above have theoretically been discussed in numerous works. A detailed excellent summary is given by Kern.[21] Following the concepts first developed by Bauer[22] the binding energy between substrate A and deposited material B is highly significant. According to Dupre's relation elucidated in Ref. 21, for the specific free interfacial energy $\sigma*$ one can write

$$\sigma* = \sigma_A + \sigma_B - \beta , \qquad (2)$$

Fig. 15 Overgrowth of Au on (111) Ag, demonstrating the modified Stranski-Krastanov-Mechanism, imaged in transmission by extinction contrast.

with σ_A and σ_B being the free surface energies for the substrate and the deposited material, respectively, and β being the adhesion energy. The ratio of β to the σ-values determines which modes are governed by what growth process. The sufficiently precise measuring of the above-mentioned parameters is the main problem. Nevertheless, it was possible to explain reasonably well a number of systems investigated.

As to the temperature dependence one can say that formally it is contained in the more detailed thermodynamic concepts. But the inaccuracy of the surface energies is even more true of the temperature dependence, as the special formation of thin film often most sensitively depends on the temperature, which is exemplified by growing Au on Ag. We have investigated this system in detail, and to us it seems that a more molecular-kinetic consideration very well describes the results, particularly the temperature dependence. Here the interaction of the diffusing atoms of the material deposited with the steps is treated.

The experimental conditions are demonstrated in Figs. 16 and 17. Starting from molten drops spherical Ag crystals are produced of well-defined surfaces after electrolytical deposition.[23] The surfaces are almost perfect having a low step density. For imaging the steps the decoration technique may be employed as formerly described for NaCl, though requiring particular care to remove the replica. Figure 16 shows the atomic step of a growth spiral on a (111) face. Perfectly clean surfaces are attained by evaporating a certain surface layer by a heat treatment at higher temperatures under ultra-high vacuum conditions. Subsequent evaporation at lower temperatures finally yields a clean surface of low step density. It was on such surfaces that the first stages of a 2-D growth of Au on Ag were investigated, as Fig. 17 reveals.[24] In Fig. 17d Auger spectroscopy was used to prove the growth of approximately half a monolayer; but, of course, also LEED was used to study the system Au/Ag.

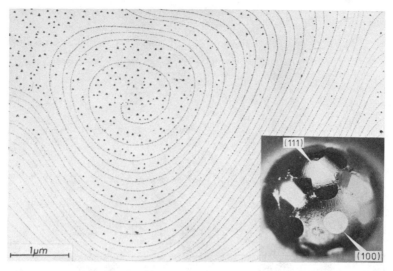

Fig. 16 Monatomic spiral step on a (111) Ag face electrochemically grown
on a spherical Ag crystal.

It is well-known that the first growth stage begins with a preferred
nucleation at the steps, as shown by Figs. 17a, b and c. But the
question arises whether in the initial nucleation at a step diffusing Au
atoms on its upper side participate in the same way as on its lower side.
Here the step barrier energy has to be taken into account and to be
studied. In the micrograph of Fig. 18 the density of nuclei was
determined on the surface with respect to the distance from the steps,
with the distribution being demonstrated in a histogram.[25] The
measurements as a function of the temperature significantly show that
below a certain temperature there is a far more intense nucleation on the
surface above a step close to it. Here, there was a higher
supersaturation as part of the Au atoms diffusing on the higher surface
cannot overcome the step barrier. It should be mentioned that this step
effect can be observed more directly in the field ion microscope as shown
by Ehrlich.[26]

These former results show that kinetic processes on steps may
influence the growth of thin films. Figure 19 is to demonstrate this
influence more generally, solely considering rather simple
molecular-kinetic concepts. Figure 19a shows the potential distribution
generally described of a surface having a step. We define a step barrier
energy, denoting it E_{SB}. A distinction has to be drawn (Fig. 19b)
between the heterogeneous step, i.e. the step of one monolayer of
substance deposited on the substrate, and the homogeneous steps of a
growing layer. Because of the interaction with the interface the latter
may additionally depend on the layer thickness, marked by indices
(drawing on the right of Fig. 19b). Furthermore, we introduce a
diffusion energy E_D (Fig. 19c), which, however, is more difficult to
define. It denotes the migration energy of a diffusing adatom reaching
the step. The ratio of E_D to E_{SB} determines the probability of
overcoming the step.

With respect to S-K-M (Fig. 19d), for the epitaxial growth of one
first monolayer E_D should be larger than E_{SB}^{ht} and also than $E_{SB}^{ho'}$ if
two monolayers have formed before a 3-D growth starts. For the modified
S-K-M it follows that

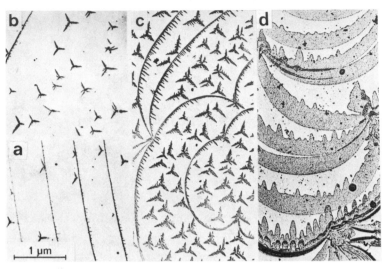

Fig. 17 Nucleation and growth of Au on Ag(111) with increasing
deposition.

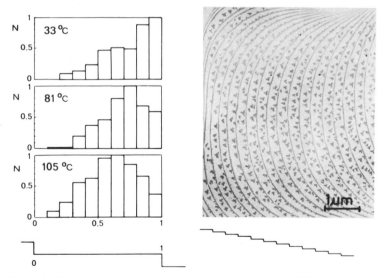

Fig. 18 Step-influenced density of Au-nuclei on Ag(111) as a function of
temperature.

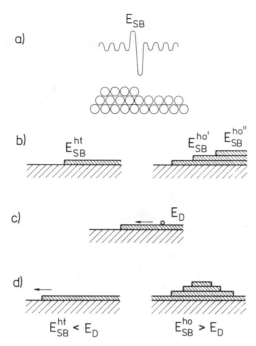

Fig. 19 Illustration of the different step barrier energies (a,b), the diffusion energy (c), and respective relations (d) to explain the modified Stranski-Krastanov mechanism.

$$E_{SB}^{ho} > E_D, \tag{3}$$

and for the layer-by-layer growth

$$E_{SB}^{ho^n} < E_D. \tag{4}$$

As to the temperature dependence, it surely is very low for the step barrier energy E_{SB} relative to the diffusion energy E_D.

Figure 20 demonstrates that over the range of some 10°C E_D may change to such an extent that it also causes the epitaxial growth modes to change. In a number of experiments[27] the temperature of the substrate and the thickness of the Au layers were varied at constant deposition rates of about 1.5 ML/min.

For the epitaxial growth at room temperature Fig. 20a shows that first a homogeneous monolayer is formed. For a larger mean layer thickness (Fig. 20b) small three-dimensional Au-crystallites are formed of a size of 200-300Å. This is a good example of S-K-M as illustrated in Fig. 14 as mode 3a.

The most remarkable fact within the Au/Ag system is that over a relatively small temperature range the terraced hillock growth (modified S-K-M in Fig. 14) turns over into a perfect layer-by-layer growth which may be explained by the above-mentioned assumptions, i.e., that at about 150°C the diffusion energy E_D will be larger than E_{Sb}^{ho}.

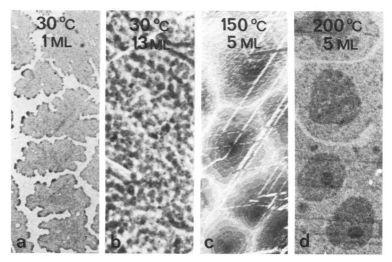

Fig. 20 Growth modes for different temperatures and mean layer
thicknesses, Au on Ag(111).

Summarizing the investigations on the system Au/Ag one can say that
the experimental conditions are extraordinarily favorable. The step
structure of the substrate crystal is easy to detect by means of the
decoration effect. Extremely thin films may be prepared from the
deposited gold to be then examined in the electron microscope. The
conditions for the extinction contrast are favorable and atomic lamellar
structures yield a good image contrast.

The above discussion is a first attempt to include the interaction
with steps into the kinetics of the 2-D growth. Some simple
molecular-kinetic assumptions allow a qualitative interpretation to be
made of the results obtained from experiments on the system Au/Ag for the
different epitaxial growth modes. For a profound theoretical treatment
the dependence on the supersaturation (i.e. on the evaporation rate)
should also experimentally be investigated. Furthermore, detailed
knowledge is necessary of the parameters of surface diffusion, including
the theoretical interpetation of the latter. Besides, the step barrier
energy should also further be evaluated.

In addition to the above-described step effects, for the S-K-M, in
particular, we have to consider structure anomalies of the overlayers.
The question arises whether the first monolayer of a deposited substance
A is "reconstructed" relative to the surface structure of a bulk crystal
A. This seems to hold true of some of the systems as well as of the
first monolayer of Au on (111) Ag-faces.[28]

In conclusion of this chapter one can say that the 2-D overgrowth of
thin films is very complex. Well-defined model systems have to be
investigated in more detail. The combination of methods, enabling only
the integral investigation of a surface, as e.g. LEED, with imaging
techniques seems to be necessary. A respective excellent method is LEEM
(where LEED is combined with the imaging by backscattered electrons),
which was developed by Bauer.[29] Likewise promising is the first imaging
of surfaces by photoelectrons (PEEM,[30]).

High-Resolution Scanning Tunneling Microscopy allows the imaging down to atomic dimensions. But there is the problem of classifying the atomic details according to the results obtained from measurements of more macroscopic scale. For the future somewhat more distant, the direct combination of PEEM with the energy analysis of the imaging photoelectrons (PES) will be the most useful method of solving open questions. Besides the sensitive imaging of the surface it would also be possible to measure the electronic state which is inevitable to the theory of the 2-D overgrowth with the interface included.

Acknowledgement

The experimental investigations presented above are part of the studies carried out by the group working on surface physics and thin films in the Institute of Electron Microscopy and Solid State Physics of the Academy of Sciences of the GDR. For many years I have had a good co-operation with this group. I am particularly grateful to Professor Krohn, and to Drs. Keller, Höche, Klaua and Meinel.

REFERENCES

1. H. Bethge, Phys. Stat. Sol. 2 (3), Part I 3, Part II 775 (1962).
2. M. Krohn and H. Bethge, in: Crystal Growth and Materials, E. Kaldis and H. J. Scheel eds. (Amsterdam, 1977), 142.
3. H. Bethge, in: Interfacial Aspects of Phase Transformations, B. Mutaftschiev ed. (Dordrecht, 1982), 669.
4. G. A. Bassett, Phil. Mag. 3, 1042 (1958).
5. H. Bethge and K. W. Keller, Optik, Stuttg. 23, 462 (1965/66).
6. M. Krohn, G. Gerth and H. Stenzel, Phys. Stat. Sol. (a) 55, 375 (1979).
7. M. Krohn and H. Stenzel, in: Electron Microscopy in Solid State Physics, H. Bethge and J. Heydenreich eds. 202 (Amsterdam, 1987).
8. M. Krohn, Vacuum 37, 67 (1987).
9. M. Mundschau, E. Bauer and W. Swiech, Surface Science 203, 412 (1988).
10. H. Bethge and M. Krohn, in: Adsorption et Croissance Cristalline (Paris, 1965), 24.
11. H. Bethge, in: Proc. Conf. Physical Basis of Yield and Fracture (Oxford, 1966), 17.
12. W. Kossel, in: Leipziger Vorträge: Quantentheorie und Chemie (Leipzig, 1928), 1.
13. F. C. Frank, Disc. Faraday Soc. 5, 48 (1949).
14. H. Höche, J. Crystal Growth 33, 255 (1976).
15. H. Höche and H. Bethge, J. Crystal Growth 33, 246 (1976).
16. H. Bethge, K. W. Keller and E. Ziegler, J. Crystal Growth 3, 184 (1968).
17. W. K. Burton, N. Cabrera and F. C. Frank, Proc. Roy. Soc. (London) A243 358 (1951).
18. H. Hoche and H. Bethge, J. Crystal Growth 52, 27 (1981).
19. F. C. Frank and J. H. van Der Merwe, Proc. Roy. Soc. (London) A198, 205 (1949).
20. J. Woltersdorf, Thin Solid Films 85, 241 (1981).
21. R. Kern, G. Le Lay and J. J. Metois, in: Current Topics in Materials Science, E. Kaldis ed. (Amsterdam, 1979), Chapter 3.
22. E. Bauer, Z. Kristallographie 110, 372 (1958).
23. H. Bethge and M. Klaua, Ann. Physik 17, 177 (1966).
24. M. Klaua and H. Bethge, Ultramicroscopy 17, 73 (1985).
25. M. Klaua, Rost Kristallov 11, 65 (1975).
26. G. Ehrlich and F. G. Hudda, J. Chem. Phys. 44, 1039 (1966).

27. K. Meinel, M. Klaua and H. Bethge, J. Crystal Growth <u>89</u>, 447 (1988).
28. K. Takayanagi, Y. Tanishiro, K. Yagi, K. Kobayashi and G. Honjo, Surface Science <u>205</u>, 637 (1988).
29. W. Telieps and E. Bauer, Ultramicroscopy <u>17</u>, 57 (1985).
30. H. Bethge, G. Gerth and D. Matern, in: <u>Proc. 10th Intern. Congr. in Electron Microscopy</u>, (Hamburg, 1982) vol. I, 69.

MICROSCOPIC ASPECTS OF THE INITIAL STAGES OF EPITAXIAL GROWTH:

A SCANNING TUNNELING MICROSCOPY STUDY OF Si ON Si(001)

M. G. Lagally, Y.-W. Mo, R. Kariotis,
B. S. Swartzentruber, and M. B. Webb

University of Wisconsin-Madison
Madison, WI 53706 USA

ABSTRACT

Scanning tunneling microscopy offers the opportunity to investigate growth and ordering processes at the atomic level and hence to identify and distinguish different kinetic and energetic mechanisms that may be operative. A review of the fundamental mechanisms of growth is given. Scanning tunneling microscopy measurements of the ordering and growth of Si on Si(001) are used to illustrate these mechanisms. Diffusion coefficients, growth anisotropy and lateral accommodation coefficients, equilibrium island shapes and free energies, edge desorption energies, and diffusional anisotropy are discussed.

INTRODUCTION

Carefully engineered growth of crystals through atomistic processes like molecular beam epitaxy (MBE), chemical vapor deposition (CVD), or variants on these has formed the basis of a large and still expanding technology. Although many exotic devices have been developed, frequently by empiricism, much is still not understood, in particular about the kinetics of growth and about the interactions between atoms that control the kinetics and thermodynamics of growth. The lack of such understanding is becoming more evident with the attempts to develop atomistic-scale structures, such as ones based on vicinal substrates or on a particular distribution of defects in surfaces. Because of the microscopic view that it affords, scanning tunneling microscopy provides an excellent opportunity to investigate in a quantitative manner atomistic processes during growth. In this article, we briefly review basic aspects of the kinetics of growth at the atomic level and illustrate these with examples taken from STM measurements of Si on Si(001).

The theoretical framework for investigating crystal growth has been in hand for a long time, and has, in some cases, been developed to a considerable degree.[1-5] Consider the simplest case: the homoepitaxial growth of material on itself (A-on-A). Atoms arrive from the vapor phase at a surface. To remain on the surface ("accommodate") they must give up their heat of condensation through one or a series of inelastic collisions. Once bound in the holding potential of the surface, an atom

Kinetics of Ordering and Growth at Surfaces
Edited by M. G. Lagally
Plenum Press, New York, 1990

will diffuse until it finds another of its kind, either at the edge of a terrace or another freely diffusing atom. In the latter case, a critical nucleus may form, which will grow with the addition of a third atom. On a surface of a material with strong bonding, two atoms may actually form already a stable nucleus. Nucleation is thus the first stage, followed subsequently by growth, and eventually by "coarsening" (or "ripening"). The growth and coarsening processes, and the influence of kinetic and thermodynamic factors on them, are now discussed in greater detail.

In any growth process, arriving atoms "like each other", i.e., they desire thermodynamically to form densely packed islands. Depending on the relative interfacial free energies[4] of the adsorbate atoms and substrate, these islands may be three-dimensional (3D) clusters (Volmer-Weber growth), two-dimensional (2D) islands (Frank-van-der Merwe growth), or an intermediate situation (Stranski-Krastanov growth) in which an initial 2D layer is followed by 3D clusters. Clearly a homoepitaxial system desires to form 2D islands. The thermodynamics of all three cases can be described by a simple phase diagram illustrating phase coexistence (i.e., first-order phase transformations). The two phases are the islands and the "2-D vapor", which consists of those atoms (monomers) that are trapped in the holding potential of the surface but are not attached to islands. They represent the lateral vapor pressure of the islands at the temperature of the system. This vapor pressure is determined, as usual, through a Clausius-Clapeyron relation with the barrier being the effective lateral desorption energy of an atom from an island. (No equilibrium with the 3D vapor is assumed to exist). Figure 1 shows the relevant phase diagram. Below some transition temperature related to the cohesive energy of the island structure, 2D "solid" and 2D "vapor" coexist. The equilibrium concentration of each at a particular surface coverage and temperature is given by the phase boundaries at that temperature. Consider now a growth process. If the shutter from a deposition source is opened (and subsequently closed again) at sufficiently low temperature (below T_c), the coverage rapidly changes to some value inside the two-phase coexistence region through an increase in monomers, producing the equivalent of a random disordered phase quenched to a thermodynamic condition at which it is unstable ("chemical-potential" quench). The quench is shown by the horizontal arrow in Fig. 1. The disordered phase represents a condition of supersaturation of monomers, which then desire to attach themselves to existing islands or form new ones by nucleation. The process of island growth continues until the supersaturation is eliminated and islands are again in local 2D equilibrium with their vapor. As long as the supersaturation is maintained, all islands grow in an attempt to reduce the supersaturation (LeChatelier's Principle). MBE, or any other growth process, is simply a continuing series of chemical-potential quenches to maintain the supersaturation.

The most important microscopic quantity involved in the formation of stable nuclei and their growth is the diffusion coefficient for monomers on the surface at the surface temperature, consisting of an activation energy, E_{migr}, and a preexponential factor, D_0, that involves the usual quantities of attempt frequency, geometric factors, and an entropy term. The transfer of the heat of condensation of the arriving atom to the lattice may influence the measurement of E_{migr} over a particular (low) temperature range (see below). Nucleation and growth are assumed to be unactivated. The growth rate of an island can then be simply written in terms of a flux to the island and a flux from the island:

$$r(T) \alpha \nu_o e^{-E_{migr}/kT} \left[1-Ae^{-E_{form}/kT} \right] . \tag{1}$$

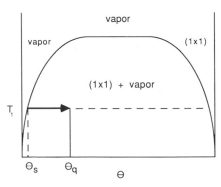

Fig. 1   Phase diagram of a lattice gas system in which two phases may
coexist.   At any temperature T_1 and coverages within the
boundaries of the coexistence region, a solid phase (2D islands,
here labeled (1x1)) coexists with a vapor at the appropriate
concentration (the lateral vapor pressure), $\theta_s(T_1)$.   The
horizontal arrow indicates a chemical potential quench to a
supersaturated state θ_q that will then decay by diffusion and
island growth to the equilibrium state.

The flux from the island involves the creation of a diffusable species,
with an activation energy, E_{form}, which represents the effective lateral
desorption ("evaporation") energy of an atom from the island.   The
constant A is the preexponential factor for desorption (or more generally
the ratio of desorption prefactor to accomodation coefficient).   The
attempt frequency, ν_o, is explicitly shown.   In a simple bond-breaking
model (applicable reasonably well to metals), one expects the activation
energy for self-diffusion, E_{migr}, to be of the order of 10% of the
cohesive energy of the solid, and $E_{form} > E_{migr}$, because more bonds need
to be broken to create a diffusable species.   It is not certain how these
numbers might differ for semiconductors, but it is possible that E_{migr}
may actually be quite low.[6]   This conclusion arises from the concept that
the minimum-energy path for the motion of a diffusing species does not
involve the breaking and subsequent reforming of bonds, but rather
involves some form of continuous charge transfer in going from the
initial to the final site. In Volmer-Weber (i.e., B-on-A) growth, E_{form}
may be less than E_{migr}.[7]   In the homoepitaxial system considered here, we
expect, for any reasonable degree of supersaturation (in MBE,
supersaturations are typically very large) that the flux away from
islands is small.   It is then possible to check the value of E_{migr}
independent of E_{form} by investigating the increase in density or size of
islands.   It is also possible to determine at least limits on E_{form} with
the appropriate experiments.   These are described later.

When the flux of atoms arriving from the source is turned off,
islands will grow until the mean supersaturation is eliminated.   The net
flux to and from each island reaches zero as the island establishes its
local lateral equilibrium vapor pressure.   Because of fluctuations in
their initial formation and growth, there will be a size distribution of
islands.   The free energy of an island determines its local vapor

pressure; smaller islands will have a greater boundary free energy (more unsaturated edge bonds relative to the island perimeter), therefore be less stable, and hence have a larger lateral vapor pressure. The difference in concentrations due to differing vapor pressures at large and small islands leads to a further ordering mechanism, "coarsening" or "ripening", which is driven by the difference in boundary free energy of islands of different sizes. Figure 2 shows this process schematically. The growth (decay) rate of the large (small) islands can be written (for an assumed bimodal size distribution) as

$$r(T) \; \alpha \; \nu_o e^{-E_{migr}/kT} \left[e^{-E_{form}(1)/kT} - e^{-E_{form}(2)/kT} \right], \qquad (2)$$

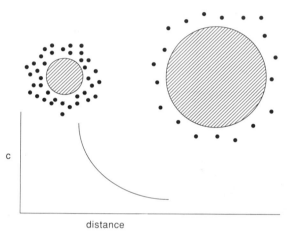

c

distance

Fig. 2   Schematic diagram of the coarsening process. Small islands have a larger local vapor pressure than large islands, giving rise to a concentration gradient that drives diffusion. As the islands become similar in size, the gradient disappears and coarsening stops.

where $E_{form}(1)$ and $E_{form}(2)$ represent the desorption energies of atoms from the small and large islands respectively. The preexponential factors for desorption from large and small islands have been assumed to be the same. Equation 2 can be written as

$$r(T) \; \alpha \; \nu_o e^{-E_{migr}/kT} \; e^{-E_f(1)/kT} \left\{ 1 - e^{-[E_f(2)-E_f(1)]/kT} \right\} . \qquad (3)$$

One can expand the last exponential:

$$\left\{ 1 - e^{-[E_f(2)-E_f(1)]/kT} \right\} = \left\{ 1 - 1 - \Delta E_f \right\} , \qquad (4)$$

where $\Delta E_f = E_f(2) - E_f(1)$, and hence

$$r(T) \; \alpha \; \Delta E_f \; \nu_o \; e^{-E_{migr}/kT} \; e^{-E_f(1)/kT} . \qquad (5)$$

148

The growth rate during coarsening thus always involves both a formation and a migration barrier. Therefore a pure migrational activation energy can never be obtained when desorption from islands or edges is involved. Most macroscopic measurements of diffusion (e.g., spreading of a sharp distribution) average over such effects and thus ought to generally produce an activation energy that does not represent just the activation energy for monomer self-diffusion.

The rate of coarsening also contains the term ΔE_f, which is related to the difference in island sizes. As this difference gets small, the driving force for coarsening goes to zero, and a distribution of islands all approximately the same size results (Ostwald ripening). Theoretical determinations of the asymptotic time dependence for the coarsening process[8] for a 2D system give $t^{1/3}$. In the limit of high dilution, migration controls the rate of growth of islands, because of the large path required between islands.[7]

So far we have ignored discussion of the shape of islands, and have implicitly assumed (e.g., in Fig. 2) that they were round both during growth and at equilibrium. Islands will be round at equilibrium at any temperature if there is no anisotropy in the free energy. The free energy is

$$F = E - TS, \tag{6}$$

where E is the energy (i.e., representative of the atomic interactions in the system and any anisotropies in them) and S is the entropy. The effect of entropy is to diminish any anisotropies that exist in the mechanical energy, so that at higher temperatures generally the equilibrium shape will be smoother and less anisotropic than at 0K. In systems that form distinct anisotropic structures, like the dimer rows in Si(001), it is likely that interactions are anisotropic and hence that the equilibrium island shape may be anisotropic as well, at least at sufficiently low temperatures.

In addition to an anisotropy in the equilibrium shape, there may be a growth anisotropy, given by kinetic parameters, such as the ease of accommodation of an arriving atom at an island edge with a particular orientation. The more anisotropic this accommodation is, the more anisotropic will be the growth shape. The problem can be thought of as a probability of sticking. There are several mechanisms that may be responsible for growth shapes. The most straightforward is energy accommodation of the arriving atom, in analogy with gas accommodation at surfaces. In order for an arriving monomer to stick, its energy must be transferred through inelastic collisions to the edge before it reflects and escapes the holding potential of the edge. There are numerous examples of surface-orientation-dependent gas sticking coefficients. How large an effect this can be for adsorption of atoms at different island edges is not known. Other possibilities for anisotropic accommodation are discussed below when STM data for Si(001) are considered. If the accommodation coefficient is not equal to one, Eq. (1) must be modified to include this effect. This can be accomplished by modifying the constant A, as indicated below Eq. (1).

A growth structure can clearly depend on kinetic parameters such as deposition rate (degree of supersaturation), temperature, and total coverage. At any instant, there will be competition between minimization of free energy to seek the equilibrium shape (for example by edge diffusion) and establishment of the growth shape through kinetic limitations. If the deposition rate is, e.g., very low relative to the rate at which an island can reach its equilibrium shape, the latter will be observed. Conversely, if the deposition rate is high, a shape more related to kinetic factors will be observed.

If the flux is turned off after a growth shape is established, annealing to allow the system to coarsen also results in an island shape change. Both edge diffusion and migration between islands can affect the shape; only the latter affects island area. Differences in growth shapes and equilibrium shapes can be investigated using STM to study the coarsening process. The shapes obtained after long-time annealing represent the equilibrium shapes at that temperature; any difference between these and the initial shapes must be ascribed to kinetic factors.

No surface is truly flat. Edges of terraces separated by atomic steps can be thought of in the same context as above, i.e., as boundaries of very large islands. The structure of these edges is determined by the same rules as described above. An edge may have an equilibrium configuration or it may have a growth or evaporation shape. Anisotropy in the interaction energies may make edges with one orientation different from those with another. Anisotropy in kinetic factors can also cause different types of edges to appear different, e.g., to have different degrees of roughness or to grow (i.e., displace laterally) at different rates during deposition. In all cases, the behavior of edges and the dependence of edge roughness on temperature must be consistent with the equilibrium and kinetic behavior of islands that form during deposition.

The rest of this paper addresses in a quantitative (or at least semiquantitative) manner the issues described qualitatively above, using as an example the growth of Si on Si(001). Initial measurements for Ge growth on Si(001) are also reported.

Si(001) SURFACES

The growth of Si on Si(001) offers the opportunity to observe and explore all the processes mentioned above. The surface forms a (2x1) reconstruction, consisting of rows of dimerized surface atoms. Two reconstructions are possible, with dimer rows running along orthogonal <110> directions. Terraces separated by a single atomic step will have opposite reconstructions. In general, in diffraction experiments, both (2x1) and (1x2) domains are observed, a consequence of the fact that it is difficult, if not impossible, to prepare a surface that is flat (containing no steps) over an area equal to the size of the best incident probe ($\gtrsim 10^5 \text{Å}^2$). Theoretical considerations[9] of inherent strain in Si(001) predict that even a surface oriented perfectly toward [001] will break up into "up" and "down" domains, thus showing both reconstructions. There is no experimental evidence for this prediction, however, up to domain sizes > 2000Å. Terraces this large, representing the best-aligned surfaces that have so far been available, have been observed in the STM.[10]

Because of the two reconstructions it is possible to create steps that have fundamentally different properties. These properties are most clearly illustrated for surfaces miscut so that the surface normal points away from (001) toward a <110> direction. For such vicinal surfaces, the dimer rows run alternately perpendicular and parallel to the terrace edges. As a convention, we shall call a (2x1) domain one in which the dimer rows run normal to the edges and (1x2) one in which they run parallel. The most visual evidence that these edges are energetically different[11] comes from STM observations of edge roughness, as shown in Fig. 3. These observations will be discussed in greater detail below. For miscuts toward <110>, as the vicinality is increased beyond about 2°,

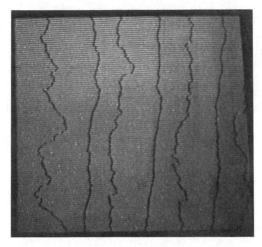

Fig. 3   STM micrograph of edge roughness in Si(001) cut toward <110>.
Edges are alternately "rough" (S_B steps) and "smooth" (S_A
steps), indicating an anisotropy in the edge energies.
Horizontal scale: ∿7,000Å.

double-atomic-height steps form. The nature of this transformation is
not fully understood,[12] nor is the precise angle known or to what degree
kinetic limitations play a role in determining the step configurations
that are typically observed.[13,14] Finally, externally applied stress
affects the step configuration,[15,16] causing one terrace to shrink at the
expense of the other. The sign of the strain determines which
reconstruction predominates.[15]

MBE of Si on Si(001) can be performed at temperatures of ∿500°C with
high-quality results. At room temperature and typical MBE deposition
rates, several RHEED intensity oscillation cycles are observed,
indicating that ordering is taking place, but that the surface quickly
gets rough. Sakamoto[17] has reported a number of interesting results in
Si MBE. It is evident that the growth kinetics in this system are quite
complex.

EXPERIMENTAL

Measurements were made in a vacuum chamber containing a STM,
low-energy electron diffraction (LEED) optics, and a Si evaporator.
Pressures are routinely in the 10^{-11} Torr range. The STM consists of a
quadrant tube scanner mounted on a walker that allows accessing various
parts of the sample. Samples are transferred from a multiple-sample
"parking lot" to the STM via a manipulator. The Si surface is cleaned by
a heating procedure described elsewhere.[14] It leaves the surface clean
with a minimum number of dimer vacancies (<3% of surface atoms) and no
other defects. Nominally flat surfaces generally had a vicinality of
∿0.1°, giving mean terrace sizes of ∿500Å. Recently samples oriented
within ≤0.03° of [001] have been used.

Depositions are performed at various substrate temperatures by
evaporation from another Si wafer. The substrate temperature is measured
using optical and IR fine-focus pyrometers that have been calibrated
against a thermocouple for a Si sample mounted in a constant-temperature
furnace. Samples are subsequently quenched to room temperature and
transferred to the STM. Depositions were typically made to coverages of

151

a fraction of a monolayer (ML), but in some cases to as much as 10 ML.

In all of the discussions, it is assumed that the Si vapor falling on the surface consists of monomers. Mass spectrometric data for Si evaporation from polycrystalline Si "rocks" show about 10% multimers in the flux.[18] It is possible that evaporation from particular surface orientations produces preferentially dimers. We are unaware that this is known. We have generally used a (111) wafer as an evaporation source.

DETERMINATION OF MIGRATION AND FORMATION ENERGIES FOR Si ON Si(001)

In the introduction we discussed the roles in growth and coarsening of E_{migr} and E_{form}, respectively the activation energies for migration and for lateral desorption of a diffusable Si entity (here assumed to be a Si monomer), from an edge. One can separate these and (at least in principle) determine them quantitatively using STM.

A lower limit to E_{form} can be obtained simply from STM observations of possible fluctuations in the structure of terrace or island edges. Atoms desorbing from an edge and readsorbing somewhere else on the edge change the edge structure over time. If no changes occur in the structure, one can put a lower limit on E_{form}. The rate of lateral desorption from an edge is given again by

$$r(T) \; \alpha \; \nu_o e^{-E_{form}/kT} . \tag{7}$$

Terrace edges in Si(001) of both kinds shown in Fig. 3 are stable even locally over a period of at least hours at room temperature. Using $\nu_o = 10^{12}$/sec gives a limit for $E_{form} > 0.8$ eV/atom (compare to the cohesive energy of Si, $E_{coh} = 4.64$ eV/atom). Any value smaller than this would result in observable differences in edge structure over a period of one hour. Its magnitude suggests that it is quite difficult to remove an atom from either type of edge. The desorption barrier is not the same quantity as the relative energy of a step, as calculated, e.g., by Chadi.[11a] Chadi calculates the difference in energy between a flat surface and one containing a step, per unit length of edge. This number may be small while the desorption energy is large. The value of $\nu_o = 10^{12}$/sec is the standard estimate of attempt frequency. It may be smaller, but in order to reduce E_{form} significantly, its value would have to be unreasonably small.

One further conclusion of the stability of edges is that migration along the edge is also negligible at room temperature. One expects that a lower barrier would in general need to be overcome for edge diffusion than for lateral desorption. Hence the above limit more properly refers to edge diffusion, with desorption energies probably even larger. It should be noted that the above result gives no conclusion about the nature of the species that desorb or diffuse along the edge when the kinetics do become significant. It is possible that this unit is a dimer at least in some situations.

A large value of E_{form} confirms our previous assumption that the influence of the lateral desorption term is negligible in growth of Si on Si(001) even at relatively low supersaturations and high temperatures. One can therefore determine E_{migr} in an experiment that measures the surface self-diffusion coefficient. A physical picture of this procedure can be obtained by considering a bare terrace bounded by steps, one "up" and one "down", to which one atom is added from the 3D gas phase. This atom will diffuse on the surface until it finds the step. The parameter important here is the diffusion coefficient of the monomer and its activation energy. If more than one atom arrives, they may meet each other before finding the step. The chance then exists to form an island, which will subsequently grow as other monomers strike it. If the edges of the terrace are far away, essentially all the monomers will form islands. Initially, almost all monomers stick to each other and form new islands, leading to an early rate of increase of the number of islands proportional to t^3. As the number of stable islands increases, this rate decreases because monomers will have a probability of finding existing islands instead of other monomers. The surface diffusion coefficient determines how large an area a monomer can interrogate in a given time, and thus determines the probability of a monomer finding an existing island before a new monomer is deposited in its vicinity (for a given deposition rate R). Therefore the surface diffusion coefficient ultimately determines the number density of stable islands after deposition to a certain dose with a given rate R. Thus simply by counting the number of stable nuclei that have formed in a known substrate area at a particular substrate temperature, one can estimate the diffusion coefficient at that temperature.[5,19] Thus the island density will reflect the deposition rate as well as the substrate temperature (changing D). The activation energy for diffusion can be extracted from measurements at different temperatures.

Figure 4 shows STM micrographs of the distribution of Si islands after deposition to four doses at a fixed rate and a substrate temperature of 475K. Similar measurements have been made down to room temperature.[19] From them several conclusions are possible: 1) Islands form at room temperature, implying diffusion is occurring at room temperature. The islands are of monolayer height, and the stable nucleus appears to be one dimer. 2) The density of islands increases with increasing dose, initially rather rapidly and then more slowly. Islands also grow. 3) Islands have anisotropic shapes; this anisotropy is more pronounced at higher growth temperatures. 4) For fixed dose there are fewer islands for deposition at higher temperatures than at lower. 5) Although defects occur in the substrate, there is not a pronounced decoration effect, i.e., defects do not play the decisive role in determining the density of islands that form. 6) Once the flux is off, the island distribution is invariant at room temperature; a measurable rate of island coarsening takes place only for annealing temperatures above about 520K. This fact is important because it implies that the images in Fig. 4 represent effectively the island configuration at the end of deposition. The last observation confirms the result that E_{form} is large.

Figure 5 shows a plot of the stable-nucleus density as a function of dose at otherwise identical conditions but at two substrate temperatures. As expected from the above qualitative argument, the higher diffusion coefficient at higher temperatures should produce a lower stable-nucleus density. Fits to these data with a rate equation model of surface diffusion[20] that takes into account island nucleation, loss of monomers to edges, and adsorption of monomers at existing islands are shown in Fig. 5. From these fits we extract diffusion coefficients of roughly 1×10^{-12} cm^2/sec and 2×10^{-12} cm^2/sec respectively at 300 and 475K. The model gives the rate of change with dose (time) of the number of monomers (diffusable species) $n(x,t)$ in terms of monomer gain and loss terms:

$$\frac{\partial n(x,t)}{\partial t} = R + D\nabla^2 n(x,t) - Dn^2(x,t) - D^2 n(x,t) \int_0^t dt' n(x,t')^2, \qquad (8)$$

where R is the deposition rate and D is the surface self-diffusion coefficient. The integral over time of $Dn^2(x,t)$ represents the total number density of islands. Edges bounding a terrace of finite size and

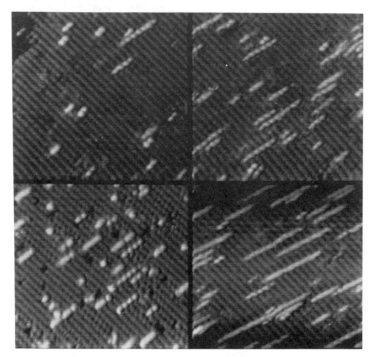

Fig. 4   STM micrographs of the distribution of Si islands after
deposition to four doses at a fixed rate and substrate
temperature T = 475K.   Islands show a considerable shape
anisotropy.   Each panel is 230Åx230Å.   Left to right, top: 0.05,
0.1 ML; bottom: 0.15, 0.2 ML.

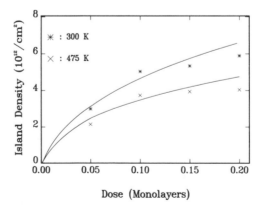

Dose (Monolayers)

Fig. 5   Determination of diffusion coefficients of Si in Si(001).   The
number density of stable nuclei is plotted vs dose at two
temperatures.   The fit is from Eq. (8).

154

acting as sinks can be introduced in the boundary conditions. For sufficiently large terraces and low temperatures (short diffusion lengths) the loss to edges is negligible. Island coalescence is not considered here. The model is applicable only at low doses.

The values of D at the two different temperatures are not greatly different. In view of the large uncertainties, these differences may not be significant. Nevertheless, if we are permitted to assume that they are, and that two points determine a line, we can extract an (at this stage very rough) activation energy for diffusion. The value is 0.04 eV/atom. This is a very small value based on usual concepts of activation energy for surface diffusion. A simple calculation illustrates this. Using the conventional value for the attempt frequency $\nu_o = 10^{12}$/sec, a value of $D \approx 10^{-12}$ cm^2/sec requires an activation of energy of the order of $E_{migr} = 0.5$ eV if thermal activation is limiting the rate of diffusion. Alternatively, $E_{migr} \sim 0.04$ eV requires $\nu_o \sim 10^2$-10^3. If this small experimental value for E_{migr} proves to be correct, it suggests one of at least two possibilities: 1) The activation energy for diffusion of Si on Si is a small fraction of the energy required to break bonds and reform them,[6] i.e., there is a minimum-energy path, as for cleavage[6b] that is much lower than one might expect from a simple bond-breaking argument. This possibility does not address the problem of the very small prefactor that would then be required. 2) Diffusion in this temperature range is not determined by thermal activation but rather by the heat of condensation and accommodation of the arriving Si species. This possibility can be pictured as follows. To accommodate and come to rest, an incoming atom has to lose its heat of condensation. The atom will have a diffusional range that depends on how well this energy is transferred to the lattice. If the heat of condensation is controlling the diffusion, there should be no temperature dependence to the diffusion coefficient. Eventually at higher temperatures the thermal activation must begin to dominate and one would expect to see a knee in the Arrhenius relation between lnD and 1/T.

A second way to determine diffusion coefficients on surfaces rests on the same principle of an interrogated area, but uses the edges of terraces as sinks, in competition with stable nuclei. The consequence should be a denuded zone near terrace edges, i.e., a mean capture distance associated with an edge in which the nucleated-island density is much smaller. If the edge is assumed to be a perfect sink, a simple analysis[21] gives directly the diffusion coefficient. An STM micrograph demonstrating the effect is shown in Fig. 6. The micrograph suggests several interesting extensions and also complications. If the edge is not a perfect sink, (see also below), the diffusion coefficient is underestimated. Different types of edges may have different accommodation coefficients and hence show different denuded zones. The accommodation of atoms arriving at the edge from the upper terrace and from the lower terrace may be different, giving different denuded-zone widths for the same diffusion coefficient. Finally, the diffusion coefficient may itself be anisotropic, producing different denuded-zone widths in different directions even if the accommodation is the same. The different aspects of the problem can be sorted out by the appropriate experiments,[21] as discussed briefly below.

ANISOTROPIC ISLAND SHAPES: GROWTH OR EQUILIBRIUM STRUCTURES

In the introduction we differentiated between growth shapes, dependent on kinetic factors, and equilibrium shapes, controlled by the free energy of the structure. Figure 4 showed a considerable anisotropy

Fig. 6   STM micrograph of Si growth on nearly perfectly oriented (≲0.03°
          misoriented) Si(001) at 540K, showing denuded zones free of
          islands near certain terrace edges. This sample was strained so
          that the majority domain terraces become very large, while the
          minority domains shrink to < 300Å and to near zero in some
          spots. Five levels are shown. The denuded zone implies that
          migration downwards over the rough edge is highly likely and
          thus does not see a large barrier. The lack of a significant
          denuded zone at the smooth ("up") edge (lower right) further
          confirms that atoms do not stick very well on this edge. Dose:
          ∿0.1 ML. Scale 1x1μm.

in island shape, observed as well by others.[22,23] Coarsening experiments
can differentiate between growth and equilibrium shapes, as the islands
approach this equilibrium configuration during annealing. Figures 7 and
8 show examples of the consequence of annealing.[24] In all cases we have
found a shape change toward lesser anisotropy. Generally we observe,
after annealing, aspect ratios of about 2 and certainly less than 3.
Even anneals at T > 800K produce about the same anisotropy. We can not
conclude from Figs. 7 and 8 that the shapes shown there are at
equilibrium. Because of the limited terrace sizes in the substrates to
which we have access there is continued loss of atoms to edges and
further annealing will eventually lead to terraces free of islands.
Therefore true equilibrium can not be reached if any information on
islands is desired; however, all of our measurements suggest that true
equilibrium would be in the direction of rounder islands. This
conclusion affects shapes from higher-temperature anneals more than those
from lower temperature anneals. We conclude that the boundary
free-energy anisotropy in Si(001) in orthogonal directions is ∿2:1 for
T≈600K, a temperature at which coarsening and boundary migration are
sufficiently rapid to allow us to observe shape and size changes, and
does not change drastically at higher temperatures. Because we do not
observe significant kinetics below 500K in laboratory times, we can draw

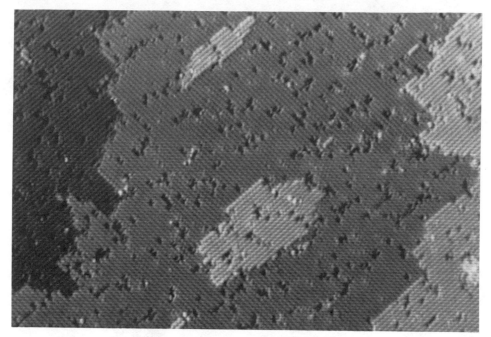

Fig. 7 STM micrographs of the distribution, shape, and size of Si
        islands after annealing at ∿625K. The initial dose is about 0.2
        ML.   By annealing, some adatoms merge into substrate steps.
        Scale 1000x700Å.

no concrete conclusions about the equilibrium anisotropy at room
temperature, except to suggest its direction: it will be slightly larger
than at higher temperatures. Entropy tends to equalize the boundary free
energies in different directions. The mechanical energy contribution to
the boundary free energy is positive and can be thought of as the
interaction energy stored in the unsaturated bonds at the edge. Assume
now that bonds are strong in one direction and weak in the orthogonal
one. In the direction in which the broken bonds are weak, the mechanical
energy is a small positive number; the corresponding boundary free energy
in that direction will be smaller than in the orthogonal one, in which
the mechanical energy is high. It should be noted that it is expensive
to create a kink on the low-energy edges (high kink excitation energy),
because that kink will have a segment of the high-energy step.
Conversely, kinks are easily created on the high-energy edges, because it
costs very little mechanical energy to create them. This is why the
high-energy edges of islands or terraces are "rougher". A schematic
boundary free-energy anisotropy plot[25] is shown in Fig. 9. As the
temperature is increased, entropy reduces the free energy in all
directions, but by different amounts. This occurs because entropy is
itself a function of the temperature and the strength of the interactions
and hence different in different directions. The entropy is given by

$$S = k\left[\ln Z - \beta \frac{\partial \ln Z}{\partial \beta}\right] ,$$ (9)

where

$$Z = \sum_n e^{-\beta E_n} , \quad \beta = 1/kT$$

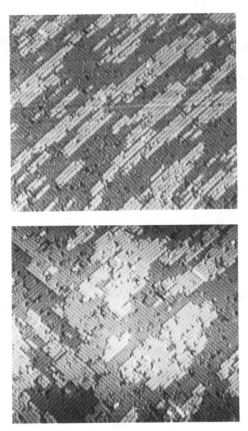

Fig. 8   STM micrographs of Si islands on Si(001). Top: grown at 575K at
         1/20 ML/sec, coverage ∿0.5 ML; bottom: after annealing at 575K
         for 10 min.   Scale is 500Å x 500Å.

and E_n is the mechanical energy of a configuration and the sum is over
all configurations. The free energy of edges on which the kink
excitation energy is small will be reduced more by an increase in
temperature than that of the stable edges, on which the kink excitation
energy is large. Consequently islands become rounder at higher
temperatures. This argument implies that the anisotropy in the
mechanical energy (equal to the free energy at 0K, or the unsaturated
bond strengths) will be larger than the ∿2:1 ratio of free energies at
500K determined in the experiments. We have made calculations,[20] based
on a 2:1 free-energy ratio at 500K, the apparent weak temperature
dependence of this ratio, and estimates of the correlation lengths of the
roughness of S_A and S_B steps that suggest kink excitation energies of
∿0.08 eV and 0.2 eV/atom for the S_A and S_B steps in Si(001) respectively.
Because the smallest observable kinks in both directions consist of two
dimers, the corresponding mechanical energies are half of these values,
or 0.04 eV and 0.10 eV/atom. These should be compared with the values of
0.01 eV and 0.15 eV/atom determined by Chadi.[11a] Figure 10 shows
quantitatively what we stated qualitatively above, that the free energy
of the high-energy (S_b) edge is much more dependent on temperature than

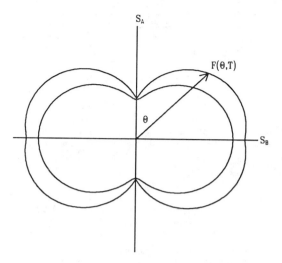

Fig. 9   Schematic polar plot of the boundary free energy of 2D islands
with anisotropic interactions, given in text, at two
temperatures. At higher temperature, entropy reduces the free
energy, but by different amounts in different directions. The
equilibrium island shape is obtained in the standard manner[25].

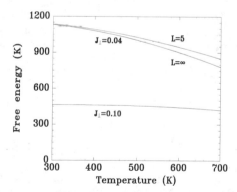

Fig. 10. Calculation of the temperature dependence of the free energy for
S_A and S_B steps in Si. Values of 0.04 and 0.1 eV/atom are used
for the edge energies. Excitation energies used in the
calculation are twice this large because we observe chiefly
double-dimer kinks. Also shown is the free energy for the S_B
edge for an island constrained to a width of five dimer rows.

159

the low-energy (S_A) edge. To the extent that the two free-energy curves converge, the island shape anisotropy should be temperature dependent. We do not observe a significant effect of this sort, although we can only make measurements over a limited temperature range. The free-energy curves can be flattened by shifting them both upward in mechanical energy, making it more costly to create kinks. For small islands, there is another factor that needs to be considered, namely that, for a limited line length, fewer configurations are possible and the partition function in Eq. (9) should be summed over only a finite range. Hence the entropy contribution should be less important for small islands. A calculation for the S_B edge of an island with a width of five dimer rows is also shown in Fig. 10. It can be seen that the dependence of the free energy on temperature is somewhat weaker. For smaller islands, the effect of entropy is even weaker; for a single line of atoms, the free energy reflects just the mechanical energies. The islands on which we have measured anisotropy range from four or five dimer rows wide up to 50 or more. In all, the anisotropy is of the order of 2 or 3. This result suggests that the mechanical energies must be high to begin with, so that entropy is not extremely important in controlling the shapes. The values of .04 eV/atom and 0.1 eV/atom that we suggest above for the S_A and S_B edge energies are the consequence of this reasoning.

The observed anisotropy during growth is as large as 15-25 at temperatures at which the equilibrium shape anisotropy is ∿2:1. This is shown in Fig. 11 and also in Fig. 4. Hence a kinetic limitation must be present. In the limit of high deposition rates and temperatures low compared to coarsening temperatures, the kinetic process dominates and the island shape anisotropy reflects essentially the accommodation coefficient anisotropy at the sides and ends of dimer rows. We conclude that this anisotropy may be as much as a factor of 10 or more.[24] We have already mentioned in the introduction the most generally quoted cause for an accommodation coefficient differing from one, namely the ease of transfer of the arriving atom's energy to the stable structure. Such transfer probabilities may differ for different edges, much as they do at surfaces with different orientations. It may be difficult to justify a factor of ten in this manner, however. A more likely explanation is that an arriving monomer can bond more easily and directly on the end of a dimer row than at a side. Attachment at a side would be equivalent to the incipient formation of a new row, and probably requires a dimer, not just a monomer. A monomer then would have a very short residence time on the side (very low binding energy). On the end the picture is different. Here the substrate bond configuration is such that it should be easy for an atom to attach. In fact, the bonding arrangement of Si atoms in the substrate requires that there are two types of bonding sites for an arriving atom at the end of an existing dimer row, one right at the end, and the other displaced by one row lattice constant. We refer to the latter as the "diluted-dimer" structure.[19]. We have observed that the latter frequently occurs during the early stages of growth,[19] suggesting that it is a slightly more favorable site for initial binding than the "close-packed-row site". Figure 11 shows examples of the diluted-dimer structures. If, as mentioned earlier, the diffusing species is a dimer and not a monomer, the above arguments on accommodation transfer straightforwardly. It would be energetically more favorable as well for a dimer to attach at the end of a dimer row than to attempt to form a new row.

To summarize, equilibrium island shapes in Si at T∿500K have a maximum aspect ratio of <3:1, implying that the boundary free energies in orthogonal directions (normal and parallel to dimer rows) differ by no more than this factor. Because entropy reduces the island shape

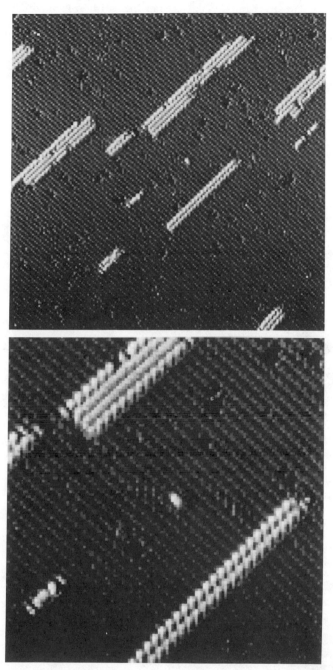

Fig. 11 STM micrographs of growth structures of Si on Si(001) at high
resolution; top 600Å x 600Å, bottom: 250Å x 250Å. Islands are
quite anisotropic. At the ends of many rows, one can observe
the "diluted-dimer" structure, in which a dimer is missing
between the end of the completed row and the final adsorbed
dimer. This is particularly evident in the leftmost two islands
in the lower micrograph. A single dimer is also observable in
the center of this micrograph.

anisotropy, the interaction energies can be more anisotropic; preliminary calculations suggest an anisotropy of ∿2.5:1 in kink excitation energies and values of ∿0.04 eV and 0.1 eV/atom for the edge energies of type S_A and S_B steps respectively. Because the equilibrium shapes are only mildly anisotropic most of the shape anisotropy observed during growth must be a kinetic phenomenon. We have suggested accommodation coefficient models to explain the anisotropy, with an anisotropy in this coefficient of ≳10:1.

Similar results are observed for submonolayer Ge deposited on Si(001).[21] Equilibrium islands with a (2x1) structure (dimer rows) form. Growth and equilibrium shapes are shown in Fig. 12. The mechanisms for

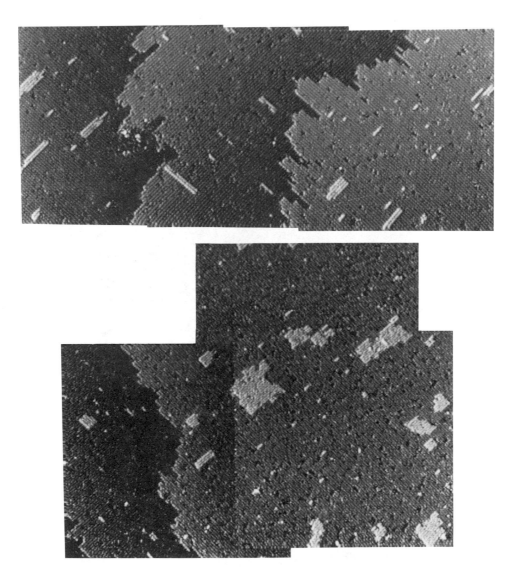

Fig. 12 STM micrographs of 2D Ge islands on Si(001) top: deposited at 500K, ∿0.05 ML, R = 1 ML/40 min., scale: 1500x600Å. Islands are oblong as for Si bottom: annealed at 575K for 10 min., scale: 1400x1150Å. Islands are nearly equiaxed.

Si and Ge growth and ordering appear to be quite similar. For growth at sufficiently high temperature, Ge accomodates at steps in the Si substrate, creating bands of Ge that follow the Si edges. This conclusion is obvious from topographic pictures, although we have not directly observed electronic difference across the junction area. By comparing Si and Ge growth at similar coverages, rates, and substrate temperatures, one can, in principle, obtain a comparison of their migration and formation energies and the relative degree of anisotropy in their accommodation at different edge types during growth.

TERRACE EDGES: EQUILIBRIUM STRUCTURES AND GROWTH SHAPES

Terraces are simply large islands. The behavior of terrace edges, in particular their roughness and the behavior during growth, must follow the same thermodynamic and kinetic laws as smaller islands, with the assumption, already discussed, that size effects are absent. As described above, for the proper vicinal miscut (toward <110>) the two types of edges correspond to island edges with dimer rows parallel and perpendicular to the edge. The dimer row orientation for which the "down" step is rough is also the one for which the island edge is ragged. Edge roughness must be explained on the same basis and be consistent with islands shapes. At any temperature, the edge free energies guide the degree of roughness. To agree with the island shape anisotropy they can differ by no more than a factor of $\sim$2 at 600K, which must suffice to explain the difference in edge roughness of the two types of edges in Si(001) cut toward <110>. (This statement assumes no island size effects in the free energy). In a direction that produces a rough edge, the total "line length" of strong bonds can be maintained while inserting kinks of the other edge type that cost only a small mechanical energy. By doing so a reduction in free energy is achieved through the larger increase in entropy resulting from using a larger region of phase space (more possible configurations of edge atoms). Interaction energy ratios can be higher than 2:1, as described, because the entropy reduces this anisotropy. At any finite temperature, the free energy determines the configuration, however. We are attempting to determine the correlation lengths of edge roughness directly from STM micrographs. For the terrace edges we do not know as well as we do for the islands the temperature for which the configuration shown in the micrographs is the equilibrium one. Clearly it is not the equilibrium configuration for any temperature below $\sim$500 K, at which we first observe sufficient kinetics for coarsening to occur. It is possible that the cooling rate causes some "freezeout" to occur already at higher temperatures. We estimate that the configuration shown, e.g., in Fig. 3, represents equilibrium at a temperature between 500 and 600 K. Experiments are in progress using LEED to measure the edge roughness directly at various temperatures by analyzing the diffracted-beam profile in the appropriate direction.[26] Calculations have been performed to show that diffracted-beam profiles can be quantitatively related to the correlation function of edge roughness and that from the latter, edge energies can be deduced.[27] With such information, it will be possible to extract the entropy contribution to edge roughness and the absolute ratio of interaction energies at the two steps, and to establish consistency with the equilibrium island shapes.

Edge roughness and atomic configuration are affected by other factors, notably by strain due to externally applied stress and by degree of vicinality. In both cases, one can think of a constraint on the roughness due to the constraint of terrace size. Strain causes one type of terrace to grow at the expense of the other.[15] Increased vicinality makes all of them small. Not all configurations are possible when terraces are small, causing terrace edges eventually to become straighter. Figure 13 shows a micrograph for a 4° miscut surface. The edges are quite straight. A direct comparison with Fig. 3 is not possible, of course, because a 4° vicinal surface has double-height steps. However, samples stressed in such a manner that the (1x2) terrace is small do appear to give less local roughness in the S_B terrace edges than do unstrained samples. The kinetics of step motion in a strained surface is discussed elsewhere in this volume.[28]

During deposition, edges may develop growth shapes in the same manner as islands. In particular, for vicinal surfaces cut toward <110>, edges that have dimer rows normal to them grow at a faster rate than do those with dimer rows parallel, catching up to the latter to form double-height steps after deposition of $\sim 1/2$ ML. This phenomenon has been observed both by RHEED[17] and by STM.[23] The boundary free-energy differences between the two edges at typical growth temperatures can not differ by more than a factor of ~ 2. Any effect beyond this must again be ascribed to anisotropic accommodation at the different edges. We believe that such anisotropic accommodation is one of the fundamental aspects of growth in materials like Si, Ge, and possibly also GaAs.

DIFFUSIONAL ANISOTROPY AND GROWTH

We have shown that anisotropies in boundary free energy, in energy, and in accommodation occur in Si-on-Si(001) ordering and growth and suggest that they can occur in general in materials that have anisotropic surface structure. Additionally in Si the intrinsic surface stress is anisotropic.[9,15,28] It would not be surprising if the surface diffusion coefficient were also anisotropic. Such an anisotropy is difficult to verify because of the existence of some of the other effects described above.

The simplest way to check diffusional anisotropy would be to search for differences in denuded zones (such as those shown in Fig. 6) on terraces with edges aligned parallel and orthogonal to the expected direction of rapid diffusion. In the direction of rapid diffusion, a larger denuded zone should be observed. In Si(001), one might expect physically the direction of rapid diffusion to be along dimer rows, and the slow direction across dimer rows (although the opposite has also been suggested). For this form of diffusional anisotropy, a miscut toward <110> assures the correct geometry, producing alternate terraces that look identical with respect to diffusion behavior. For Si(001) miscut by more than 2°, all terraces have dimer rows running perpendicular to the edges, and it becomes impossible to study diffusional anisotropy. In other materials, such as GaAs(001), which forms double-layer steps, it is necessary to produce vicinal surfaces with different terminations, which then have the crystallographic axes in one case parallel and in the other perpendicular to the steps, with all terraces of a given termination being identical. In any case a determination of the diffusion coefficient from the denuded zone gives too low a value for D if one assumes perfect sticking, but the accommodation coefficient is actually less than one.

If the edges are equivalent in their ability to capture the diffusing species, then a larger denuded zone will form near those edges for which the rapid-diffusion direction is normal to the edges. If, additionally, the capture probability of a "down" step is the same as that of an "up" step, the denuded zone on each terrace will be symmetric (although different on alternating terraces in vicinal Si(001)). Neither of these conditions is assured. We have already demonstrated that the

Fig. 13 STM micrograph of vicinal Si(001) cut 4° toward <110>. The surface has double-height steps. Scale: ∿2000Å along each diagonal. Compare with Fig. 3.

accommodation coefficient at S_B and S_A edges is different. Hence the denuded zones can not be symmetric on any terrace for Si(001) cut toward <110> because each terrace is bounded by an S_A and an S_B edge. We can, however, definitively show that the capture at an S_B "down" step is as great as an S_B "up" step,[21] implying that there is no (or at most a very small) barrier to mass transport "downstairs". Mass transport "downstairs" is an essential ingredient for epitaxial growth, because lower levels must be perferentially filled to obtain layer-by-layer growth. It is frequently assumed in MBE modeling, however, that the downward flow is small and that most atoms deposited on a terrace migrate to and adsorb on the "up" edge.

The information that accommodation is similar, (if not the same) on an S_B edge for atoms arriving from above and below the edge can be used to determine the anisotropy in diffusion coefficient, because on the "up" terrace the dimer rows are normal to the edge, while on the "down" terrace they are parallel. From observations of this sort, we determine qualitatively a faster diffusion (near room temperature) along dimer rows. The same measurement is in principle possible at S_A edges, except that the accommodation at these edges is so small that at any reasonable temperatures and coverages the denuded zones are too small to make a clear differentration between them. An absolute value of D can not be obtained in this manner unless one assumes the accommodation coefficient equal to one. This is probably a good assumption for S_B steps over a wide temperature range. One may also be able to separate the diffusion coefficient if it and the accommodation coefficient have different temperature dependences.[29] One may expect that the accommodation coefficient, if it is an electronic effect as we suggest, has only a weak temperature dependence. Then from the temperature dependence of the denuded zones, the activation energy and preexponential factor of D can be obtained. Such measurements are in progress.

If diffusional anisotropy exists, can it affect island shapes during growth? We have addressed this question using Monte Carlo calculations and have concluded that it can not.[24] This conclusion can also be reached on physical grounds, by considering the availability and arrival of atoms at an island in a 2D lattice. Hence the growth shapes of islands and the observation of double-height layer formation during Si epitaxy[17,23] are likely not a consequence of any possible anisotropy in diffusion.

SUMMARY AND CONCLUSIONS

We have provided a brief overview of mechanisms operative in the ordering and growth of two-dimensional layers and have illustrated those with STM measurements for the submonolayer growth of Si and Ge on Si(001). Several classical concepts in the thermodynamics and kinetics of growth can be observed in these 2D systems, and in some cases quantified. The concepts considered include self-diffusion on surfaces, thermodynamics of island shapes and sizes, and kinetics of growth and coarsening, including accommodation.

In the diffusion measurements, we distinguish in the activation energy for diffusion between a migration term and an energy for formation of a diffusable species (an energy for desorption from an edge). We show that the latter must be large (≥ 0.8eV) for Si on Si(001). We demonstrate the appropriate measurements and theoretical evaluation to determine a surface self-diffusion coefficient based purely on migration. We obtain a value of $\sim 10^{-12}$ cm^2/sec at room temperature for Si on Si(001), sufficient to cause ordering to occur. One might offhand conclude that this result would predict that MBE of Si on Si(001) is possible at room temperature. In fact, several RHEED oscillations, indicating islanding for several layers before the surface becomes uniformly rough, are observed.[30] However, good growth at room temperature is not possible because other processes, including migration over edges and desorption from existing islands are part of the total epitaxial growth process. Desorption from edges appears, at least to be a slow process.

In experiments in which Si is deposited onto the surface for T<500°K, our measurements suggest that the effective activation energy for migration is very small. We speculate that we may be observing a heat-of-condensation effect, in which the diffusion is no longer purely controlled by thermal activation through the substrate, but rather by the efficiency with which the heat of condensation of the incoming species is given to the lattice.

We have shown, through coarsening experiments, that the free-energy anisotropy of Si islands on Si(001) can be no more than ~ 2:1 at T>500K. Fitting island shapes to a model calculation of the boundary free energy suggests values for the step energies of 0.04 and 0.10 eV/atom respectively for S_A and S_B steps.

We observe, during growth, an island shape anisotropy more like 15:1 or greater. This large anisotropy must be a kinetic effect, i.e., an accommodation coefficient anisotropy. We suggest a qualitative model that invokes the difference in ability of monomers to bind at the ends and sides of existing dimer chains. We suggest that these probabilities of accommodation differs by a factor of ten. We do not know the absolute values of accommodation at either edge.

Any time accommodation less than one exists at an edge, a measurement of the diffusion coefficient that depends on a denuded zone or the absence of islands in the terrace (e.g., RHEED intensity oscillation damping at high growth temperatures in GaAs) will give an answer that is too low. A factor of ten decrease in accommodation will lead to a factor of 10 increase in the measured value of D. In different terminations of vicinal GaAs(001) surfaces, the terrace edges are different,[26] and it is quite possible that the accommodation coefficient is as well. An accommodation coefficient whose magnitude depends on the nature of the edge would give different measured values for D on the two types of terraces, even if it actually were the same.[26c] Anisotropic accommodation can affect our measurement of diffusion just as they can the RHEED measurements above if accommodation is less than one at the ends of dimer chains. The ideal situation for the model is that the accommodation coefficient equals 1 at the ends of dimer chains (S_B steps) and 0 on the sides (S_A steps). This result is approximately correct, since there is an observed factor of >10 difference in its value on the sides and ends. We expect our determination of D to be, therefore, only weakly affected by the anisotropy in the accommodation coefficient, and to be relatively accurate.

Acknowledgements

This research has been supported by the U.S. Office of Naval Research and in part by the U.S. National Science Foundation, Grant No. DMR 86-15089. We thank Dr. P. Wagner, Wacker Chemitronic, Burghausen, Germany for supplying us with high-quality, well oriented Si wafers for part of this work.

REFERENCES

1. Epitaxial Growth, J. D. Matthews ed., Academic, New York (1975).
2. B. Lewis and J. C. Anderson, Nucleation and Growth of Thin Films, Academic, New York, (1978).
3. W. K. Burton, N. Cabrera, and F. C. Frank, Phil. Trans. Roy. Soc. (London) A243, 299 (1951).
4. E. Bauer, Z. f. Kristall. 110, 372 (1958).
5. J. A. Venables, G. D. T. Spiller, and M. Hanbücken, Rept. Prog. Phys. 47, 399 (1984).
6. a) P. J. Feibelman, private communication.
    b) J. E. Northrup and M. L. Cohen, Phys. Rev. Letters 49, 1349 (1982).
7. M. Zinke-Allmang, L. C. Feldman, and S. Nakahara, this volume.
8. a) I. M. Lifshitz and V. V. Slyozov, J. Phys. Chem. Sol. 19, 35 (1961).
    b) D. A. Huse, Phys. Rev. B34, 7845 (1986).
9. O. L. Alerhand, D. Vanderbilt, R. Meade, and J. Joannopoulos, Phys. Rev. Letters 61, 1973 (1988); V. I. Marchenko, Pis'ma Zh. Eksp. Teor. Fiz. 33, 381 (1988) [JETP Lett. 33, 381 (1988)].
10. Y.-W. Mo and B. S. Swartzentruber, unpublished.
11. a) J. D. Chadi, Phys. Rev. Letters 59, 1691 (1987).
    b) D. E. Aspnes and J. Ihm, Phys. Rev. Letters 57, 3054 (1986).
12. O. L. Alerhand, A. N. Berker, J. D. Joannopoulos, D. Vanderbilt, R. Hamers, and J. E. Demuth, Phys. Rev. Letters 64, 2409 (1990).
13. C. E. Aumann, D. E. Savage, R. Kariotis, and M. G. Lagally, J. Vac. Sci. Technol. A6, 1963 (1988).

14. B. S. Swartzentruber, Y.-W. Mo, M. B. Webb, and M. G. Lagally, J. Vac. Sci. Technol. $\underline{A7}$, 2901 (1989).
15. F. K. Men, W. F. Packard, and M. B. Webb, Phys. Rev. Letters $\underline{61}$, 2469 (1988).
16. B. S. Swartzentruber, Y.-W. Mo, M. B. Webb, and M. G. Lagally, J. Vac. Sci. Technol. $\underline{A8}$, 210 (1990).
17. T. Sakamoto, K. Sakamoto, K. Miki, H. Okumura, S. Yoshida, and H. Tokumoto, this volume.
18. Handbook of Thin-Film Technology, L. I. Maissel and R. Glang eds., McGraw-Hill, New York, NY (1970).
19. Y.-W. Mo, R. Kariotis, B. S. Swartzentruber, M. B. Webb, and M. G. Lagally, J. Vac. Sci. Technol. $\underline{A8}$, 201 (1990).
20. R. Kariotis, Y.-W. Mo, B. S. Swartzentruber, M. B. Webb, and M. G. Lagally, in preparation.
21. Y.-W. Mo and M. G. Lagally, in preparation.
22. R. Hamers, U. K. Köhler, and J. E. Demuth, Ultramicroscopy $\underline{31}$, 10 (1989).
23. E. J. van Loenen, H. B. Elswijk, A. J. Hoeven, D. Dijkkamp, J. M. Lenssinck, and J. Dieleman, this volume; A. J. Hoeven, J. M. Lenssinck, D. Dijkkamp, E. J. van Loenen, and J. Dieleman, Phys. Rev. Letters. $\underline{63}$, 1830 (1989).
24. Y.-W. Mo, B. S. Swartzentruber, R. Kariotis, M. B. Webb, and M. G. Lagally, Phys. Rev. Letters $\underline{63}$, 2393 (1989).
25. C. Herring, Phys. Rev. $\underline{82}$, 87 (1951); in Structure and Properties of Solid Surfaces, R. Gomer and C. S. Smith eds., U. Chicago Press, Chicago (1953).
26. a) D. Saloner, J. A. Martin, M. C. Tringides, D. E. Savage, C. E. Aumann, and M. G. Lagally, J. Appl. Phys. $\underline{61}$, 2884 (1987); b) M. G. Lagally, D. E. Savage, and M. C. Tringides, in Reflection High-Energy Electron Diffraction and Reflection Electron Imaging of Surfaces, P. K. Larsen and P. J. Dobson eds., Plenum, New York (1989); c) P. I. Cohen, G. S. Petrich, A. M. Dabiran, and P. R. Pukite, this volume.
27. R. Kariotis, B. S. Swartzentruber, and M. G. Lagally, J. Appl. Phys. $\underline{67}$, 2848 (1990).
28. M. B. Webb, F. K. Men, B. S. Swartzentruber, R. Kariotis, and M. G. Lagally, this volume.
29. S. Stoyanov, Europhys. Lett. $\underline{11}$, 361 (1990).
30. P. K. Larsen, personal communication.

SURFACE KINETICS AND GROWTH MORPHOLOGY:

THEORETICAL MODELS AND PHYSICAL REALIZATIONS

Boyan Mutaftschiev

Centre National de la Recherche Scientifique
Laboratoire Maurice Letort
54600 Villers les Nancy, France

ABSTRACT

In the first part of the paper we discuss the shape and phase shift of the reemitted beam intensity when a stepped crystal face is exposed, in vacuum, to periodic square shaped pulses of a molecular beam. Two mathematical solutions are proposed: the periodic steady state solution which allows the determination of mean stay time and scaled interstep distance on the surface from measurements of phase shift and amplitude attenuation of the reemitted beam first harmonic, and a solution valid for long incident beam pulses, according to which the same parameters can be determined from the shape of the reemitted beam intensity.

In the second part are given some experimental results concerning the condensation of submonolayers of platine on Pt(111). A theoretical model analogous to that of the first part but taking into account the irreversible adsorption in this case, allows to interpret the different roughness of the Pt-surface during growth in the temperature range of 130-400 K, as due to the competition between propagation of growth fronts and two-dimensional nucleation on the flat regions between them.

The third part of the paper deals with the changes in the electronic structure of Pt-thin films on Ni(111) after annealing. It is shown that, while in the most cases the work function of the films has an intermediate value between those of Pt and Ni, due to the formation of surface alloy, in the cases when no alloy formation is visible (by LEED), the work function can be very different. This result is explained by the elastic stress of the Pt-overlayers.

INTRODUCTION

The increasing requirements of modern microelectronic devices call upon interfaces between thin layers with well defined morphology on the atomic scale. It cannot result but from a perfectly controlled surface kinetics during the growth of every individual atomic layer, e.g., by, molecular beam epitaxy (cf, for example, the chapter by P. M. Petroff: "Direct epitaxial growth of quantum structures with two and three-dimensional carrier confinements" in this volume). The relationship between surface growth kinetics and the resulting surface morphology is, therefore, presently, at least of such a great importance

Kinetics of Ordering and Growth at Surfaces
Edited by M. G. Lagally
Plenum Press, New York, 1990

for semiconductor physics as was the relationship between transport phenomena, or heat conduction, and the perfection of single crystals pulled from the melt, during the early seventies.[1-3]

Surface kinetics depend on some physical parameters, the experimental determination of which under growth conditions is achieved in relatively few cases, e.g. during MBE-growth of GaAs,[4-6] of pure and mixed alkali halides[7-10] and of metals on insulators.[11,12] The pulsed beam technique, in which the time dependent intensities of a chopped incident and a scattered molecular beams are compared, offers both the qualitative insight of the elementary mechanisms involved, and the possibility of quantitative fit was suitable theoretical models. One should remind that the latter approach is largely used in chemical kinetics on surfaces[13,14] and in electrochemical systems,[15,16] where the pulsed molecular beam is replaced by an alternative voltage applied to the electrolytic cell.

In the first part of this paper we discuss the mathematics of the transient surface kinetics in a relatively simple but fairly representative model: that of a singular face containing quasi-periodic growth lamellae of monatomic height. We will show two different manners of treating the problem and the two associated experimental techniques resulting in the determination of mean residence times and scaled mean diffusion lengths, i.e. of the activation energies of desorption and surface diffusion.

Since pulsed-beam experiments are performed in reversible, near to equilibrium, conditions, the impact of surface kinetics on surface morphology, on the atomic scale, is hard to check in situ. Microscopic techniques require "frozen" surface states and the link with surface morphology during growth (and related growth conditions) cannot be but indirect.[17,18]

In the second part of the paper we will show an example of an experimental check of some theoretical predictions for a very simple case of surface kinetics: that of the steady propagation of growth fronts due to the incorporation of irreversibly adsorbed molecular species, in competition with "non-classical" two-dimensional nucleation on the terraces between two successive growth fronts. The comparison between theory and practice, still limited to the orders of magnitude of the kinetics concerned, stresses the effort which should be made to get precise, reliable information on microscopic surface morphology under well known growth conditions.

THE PULSED BEAM TECHNIQUE AS APPLIED TO THE BCF GROWTH MODEL

In 1950 Burton, Cabrera and Frank (BCF)[19] solved the problem of the steady state kinetics of propagation of periodic growth lamellae on the surface of a singular (flat) crystal face. The BCF model assumed a dynamic equilibrium (or infinite exchange rate) between molecules in the vapor and those adsorbed on the flat regions of the surface at an infinite distance from any step. Therefore, the thermodynamic supersaturation,

$$\sigma = \ln \frac{p}{p_0} \approx \frac{p}{p_0} - 1, \qquad (1)$$

in the vapor phase (where p is the actual pressure of the supersaturated vapor, and p_0 is the saturated vapor pressure at the same temperature) turns out to be identical with the one,

$$\sigma \approx \frac{n(\infty)}{n_o} - 1,$$ (2)

reigning on the surface ($n(\infty)$ is the adlayer concentration far from any step, and n_o the concentration of the adlayer in equilibrium with the saturated vapor).

Steps are atomically rough even at temperatures near the absolute zero point.[20] For this reason, there are no significant hindrances for incorporation of molecules coming from their immediate neighborhood. As a consequence, the adlayer concentration at a step is always the equilibrium one, n_o.

A model of the stepped surface, together with the involved elementary processes, are sketched in Fig. 1. For sake of the mathematical treatment, steps are assumed to be straight and equidistant, although this is only approximately true. The origin of the coordinate normal to the step is taken in the midpoint of the terrace of width λ between two successive steps.

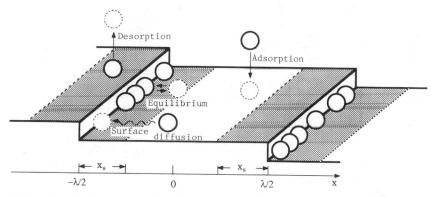

Fig. 1 Model of the stepped surface used in the calculations. The first row of incorporated atoms is represented by spheres, like the gas and adsorbed atoms. Shadowed areas are the "active bands" bordering the steps.

When the surface is exposed to a vapor with the supersaturation σ, a flux of admolecules heads towards the steps through surface diffusion, the deficit of matter on the terraces being made up by impingement of vapor molecules. The deficit is zero at an infinite distance from the steps, where the impinging flux j_i is equal to the desorbing flux, $n\uparrow_\infty$

$$j_i = n\uparrow_\infty = \frac{n(\infty)}{\tau} = \frac{n_o}{\tau}(\sigma+1) ,$$ (3)

(τ is the mean residence time of an admolecule on the surface), the adsorption probability being set equal to unity.

171

The BCF model excludes the possibility that the adatom concentration on the terraces between steps becomes enough high to give rise to a two dimensional nucleation. In this sense, it is coherent with the pulsed-beam method which can be applied only in cases of near-to-equilibrium growth, (high desorption fluxes and low surface concentration of adatoms), typical for high surface temperature.

The other extreme case, treated in Section II of this paper, that of irreversible adsorption, is clearly not suitable for pulsed beam studies. It allows the surface concentration to reach high values and the nucleation to be become overwhelming if the surface temperature is low enough.

1.   Short Periodic Pulses; the Phase Shift Method[21]

The incident intensity during pulsed molecular beam experiments can be written with the help of a Fourier series, as follows:

$$j_i = \frac{n_o}{\tau} (\sigma+1)g(t) = \sum_{k=-\infty}^{+\infty} c_k \exp(ik\omega t), \tag{4}$$

where $g(t)$ is the gating function of the chopper.

Surface diffusion towards linear steps is a unidimensional problem. The mass conservation law at any point of the surface with coordinate x and adlayer concentration $n(x,t)$ is given by

$$D \frac{d^2n}{dx^2} - \frac{dn}{dt} - \frac{n}{\tau} + \frac{n_o}{\tau} (\sigma+1)g(t) = 0. \tag{5}$$

The first term accounts for the flux due to surface diffusion with diffusion coefficient D, the third is the desorption flux and the fourth is the impinging flux j_i from equation (4). As mentioned, the boundary conditions is

$$n\left(\pm \frac{\lambda}{2}\right) = n_o \quad \text{at any time } t. \tag{6}$$

The general solution of Eq. (5) is

$$n(x,t) = \sum_{k=-\infty}^{+\infty} a_k(x) \exp(ik\omega t) . \tag{7}$$

The Fourier coefficients $a_k(x)$ are time independent and have the values

$$a_k(x) = \left[a_k\left(\pm\frac{\lambda}{2}\right) - \frac{c_k}{q_k^2 D}\right] \frac{\cosh(q_k x)}{\cosh(q_k \lambda/2)} + \frac{c_k}{q_k^2 D} , \tag{8}$$

$$q_k = \frac{(1+ik\omega\tau)^{1/2}}{x_s} ,$$

where x_s is the mean diffusion length of an admolecule on the surface.

A straightforward solution of Eqs. (7) and (8) is obtained for the steady state, or for the end of very long square-shaped pulses, whose gating function g(t) is either 1 (growth) or 0 (evaporation). Setting k = 0, we have

$$n(x) = a_o(x) \; ; \; n\left(\pm\frac{\lambda}{2}\right) = a_o\left(\pm\frac{\lambda}{2}\right) \; \text{and} \; c_o = \begin{bmatrix} (n_o/\tau)(\sigma+1) \\ 0 \end{bmatrix}$$

which yields

$$n(x) = n_o(\sigma+1) - n_o\sigma \; \frac{\cosh(x/x_s)}{\cosh(\lambda/2x_s)} \tag{9}$$

for the growth period (a result already obtained by BCF), and

$$n(x) = n_o \; \frac{\cosh(x/x_s)}{\cosh(\lambda/2x_s)} \; , \tag{10}$$

for the evaporation period.

The periodic solution (k ≠ 0) of Eq. (7) is obtained from the boundary condition a_k (± λ/2) = 0. One has

$$n(x,t) = \sum_{k=-\infty}^{+\infty} \frac{c_k}{q_k^2 D} \left[1 - \frac{\cosh(q_k x)}{\cosh(q_k \lambda/2)} \right] \exp(ik\omega t) \; . \tag{11}$$

The knowledge of the exact form of the gating function, i.e. of the Fourier coefficients c_k allows, thus, in principle the calculation of n(x,t). However, more important is the experimentally measurable intensity of the reemitted flux from unit surface area, j_r, obtained by integration of the desorption frequency n(x,t)/τ over the terrace width λ (and unit length of the step):

$$j_r = \frac{1}{\lambda} \int_{-\lambda/2}^{+1/2} \frac{n(x,t)}{\tau} \; dx \; . \tag{12}$$

The ratio $j_{1,r} / j_{1,i}$ of the first harmonics of the reemitted and of the incident beam intensities, called "reaction product vector" can be measured directly by lock-in detection, and gives access to the amplitude attenuation A_1 and the phase shift ϕ_1, related by the expression:

$$\frac{j_{1,r}}{j_{1,i}} = A_1 \exp(-i\phi_1) \; ,$$

where, according to Eqs. (12) and (7),

$$\tan \phi_1 = \omega\tau - (1+\omega^2\tau^2)$$

$$\times \frac{V\sinh2U - U\sin2V}{(U^2+V^2)(\cosh 2U+\cos 2V) - (U-\omega\tau V)\sinh2U - (V+\omega\tau U)\sin2V} \tag{13a}$$

and

$$A_1 = \left[\frac{2U\sinh 2U - \cosh 2U + 2V\sin 2V + \cos 2V}{(1+\omega^2\tau^2)(U^2+V^2)(\cosh 2U+\cos 2V)} \right]^{1/2} . \tag{13b}$$

Here are U and V the real and imaginary parts of $q_1\lambda/2$ respectively and have the values

$$U = \frac{\lambda}{2x_s} \frac{1}{\sqrt{2}} \left[(1+\omega^2\tau^2)^{1/2} + 1 \right]^{1/2}$$

and

$$V = \frac{\lambda}{2x_s} \frac{1}{\sqrt{2}} \left[(1+\omega^2\tau^2)^{1/2} - 1 \right]^{1/2} .$$

When the terraces are of infinite width $(\lambda/2x \to \infty)$, the mean residence time τ is directly measured from the phase shift:

$$\tau = \frac{\tan\phi_1}{\omega} .$$

For smaller interstep distances, an effective mean residence time can be measured, whose scaled value,

$$\frac{\tau_{eff}}{\tau} = \frac{\tan\phi_1}{\omega\tau} ,$$

is lower, the smaller is the scaled terrace width $\lambda/2x_s$, and the higher the frequency of the chopper, as shown in Fig. 2.

We will close this section by the statement that phase shift measurements can lead to the determination of both mean residence time τ on the flat regions of the surface, and scaled mean diffusion length $\lambda/2x_s$. A nearly perfect surface and low frequency are needed for the determination of τ. Once this parameter is known, measurements at different frequencies on imperfect surfaces ($\lambda/2x_s$ of the order of 1-10) allows the determination of $\lambda/2x_s$. If by other independent methods one has access to the average interstep distance λ, the absolute value of x_s and hence the coefficient of surface diffusion and their temperature dependence can be obtained.

The above method was successfully applied to the slightly different case of foreign adsorption of metallic (Cd) vapor on insulator (NaCl) substrates.[11,12] More simple models and mathematical solutions have been used in other experimental studies.[4-10]

2.   Long Square-Shaped Incident Pulses; the Reemitted Pulse Shape Method[21]

The exact shapes of the reemitted pulse intensities versus time, can be easily treated mathematically in the case of long square shaped incident pulses, i.e. when the gating function is alternatively equal to 1 and 0. The notion "long pulses" implies that the boundary conditions on the surface at the beginning of each new pulse are identical with those corresponding to the steady state reached at the end of the

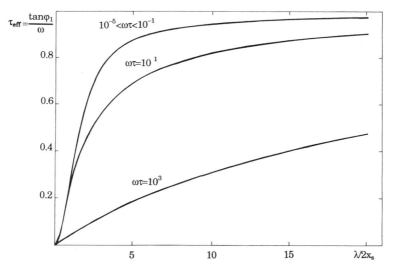

Fig. 2 Scaled effective mean residence time as a function of scaled
terrace width for different values of the product $\omega\tau$, calculated
with the help of Eq. (13a).

preceding one. Although more restrictive than the method treated in the
preceding section, the present method has the advantage of giving a clear
physical description of the elementary processes of
adsorption-desorption, surface diffusion, and incorporation taking place
on different parts of the surface.

It is well known that the surface of a perfect singular face (no
kinks!), exposed alternatively to a constant vapor flux j_i (cf Eq. (3)),
and to the vacuum, is "charged" with, and "discharged" of admolecules,
according to the law of charging a capacitor. The reemitted flux from
unit surface is, thus, equal to

$$j_g = \frac{n_o}{\tau} (\sigma+1) [1-\exp(-t/\tau)] ,$$ (14a)

during the growth period, and to

$$j_e = \frac{n_o}{\tau} (\sigma+1) \exp(-t/\tau) ,$$ (14b)

during the evaporation period, provided that the initial surface
concentrations were the steady state ones of the preceding period, i.e.,
zero in the first case and $n_o(\sigma+1)$ in the second case.

The steady state intensities reemitted by a singular face
containing periodic steps at a distance λ are

175

$$j_{g,s} = \frac{n_o}{\tau}(\sigma+1)\left[1 - \frac{2x_s}{\lambda}\tanh\left(\frac{\lambda}{2x_s}\right)\right] + \frac{n_o}{\tau}\frac{2x_s}{\lambda}\tanh\left(\frac{\lambda}{2x_s}\right), \qquad (15a)$$

and

$$j_{e,s} = \frac{n_o}{\tau}\frac{2x_s}{\lambda}\tanh\left(\frac{\lambda}{2x_s}\right), \qquad (15b)$$

respectively. For $\lambda \gg 2x_s$, those equations simplify to

$$j_{g,s} = \frac{n_o}{\tau}(\sigma+1)\left[1 - \frac{2x_s}{\lambda}\right] + \frac{n_o}{\tau}\frac{2x_s}{\lambda}, \qquad (16a)$$

and

$$j_{e,s} = \frac{n_o}{\tau}\frac{2x_s}{\lambda}. \qquad (16b)$$

The physical meaning of Eqs. (16a) and (16b) is enlightening. While a perfect crystal face is reemitting in the steady state the entire incident beam (cf Eq. (14a) for $t \gg \tau$), only a fraction $1 - 2x_s/\lambda$ of the stepped surface area behaves in that way (first term in Eq. (16a)). Two bands of width x_s situated on both sides of every growth front accept a fraction of the impinging flux equal to $(n_o/\tau)(\sigma+1)(2x_s/\lambda)$, and reemit in the same time a flux $(n_o/\tau)(2x_s/\lambda)$ (cf the second term of Eq. (16a)) independently of the supersaturation. The same flux is reemitted during the evaporation period when the impinging flux is nil. The described behavior of the "active bands" with relative width $2x_s/\lambda$ is exactly that of a rough kinked face which, in contrast to the singular flat face incorporates and evaporates molecules without specific hindrances, like the surface of a liquid.

The combined effect of adsorption-desorption and surface diffusion thus results in the formal division of the stepped surface into active bands supplied directly from the vapor as if they were rough faces, and a passive fraction which totally reemits the impinging flux. The exact equations (15a) and (15b) show simply that when the active bands from both sides of a terrace overlap, their "degree of coverage", $2x_s/\lambda$, must be weighted by a factor $\tanh(\lambda/2x_s)$. When $\lambda \ll 2x_s$, the entire surface behaves as that of a rough kinked face.

The reemitted flux in the transient period of growth is

$$j_g = \frac{n_o}{\tau}(\sigma+1)\left[1-\exp\left(-\frac{t}{\tau}\right)\right] - \frac{n_o}{\tau}(\sigma+1)\frac{2\tilde{x}}{\lambda}\tanh\left(\frac{\lambda}{2x_s}\right)$$

$$+ \frac{n_o}{\tau}\frac{2x_s}{\lambda}\tanh\left(\frac{\lambda}{2x_s}\right), \qquad (17a)$$

and that in the transient period of evaporation

$$j_e = \frac{n_o}{\tau}(\sigma+1)\exp\left[-\frac{t}{\tau}\right] - \frac{n_o}{\tau}(\sigma+1)\left(\frac{2x_S}{\lambda} - \frac{2\tilde{x}}{\lambda}\right)\tanh\left(\frac{\lambda}{2x_S}\right)$$

$$+ \frac{n_o}{\tau}\frac{2x_S}{\lambda}\tanh\left(\frac{\lambda}{2x_S}\right) , \tag{17b}$$

where x is now a time dependent effective width given by the equation

$$\tilde{x} = x_S\left\{\mathrm{erf}\left[\sqrt{\frac{t}{\tau}}\right] - 2\sqrt{\frac{t}{\pi\tau}}\exp\left(-\frac{t}{\tau}\right)\right.$$

$$+ \sum_{k=0}^{\infty}(-1)^k\left[4\sqrt{\frac{t}{\pi\tau}}\exp\left(-\frac{t}{\tau}\right)\exp\left(-\frac{(k+1)^2(\lambda/2x_S)^2}{t/\tau}\right)\right.$$

$$- 4(k+1)\frac{\lambda}{2x_S}\exp\left(-\frac{t}{\tau}\right)\mathrm{erfc}\frac{(k+1)(\lambda/2x_S)}{\sqrt{t/\tau}}$$

$$+ \exp\left[2(k+1)\frac{\lambda}{2x_S}\right]\mathrm{erfc}\left(\frac{(k+1)(\lambda/2x_S)}{\sqrt{t/\tau}} + \sqrt{\frac{t}{\tau}}\right)$$

$$\left.\left. - \exp\left[-2(k+1)\frac{\lambda}{2x_S}\right]\mathrm{erfc}\left(\frac{(k+1)(\lambda/2x_S)}{\sqrt{t/\tau}} - \sqrt{\frac{t}{\tau}}\right)\right]\right\}/\tanh\left(\frac{\lambda}{2xs}\right) , \tag{18}$$

the value of which varies from zero, for $t = 0$, to x_S for $t \to \infty$.

In Eq. (17a) one can distinguish:

- the "charging" of the passive part of the surface (first term);

- the time independent evaporation from the active bands with constant width x_S (third term);

- the incorporation of molecules in active bands with time dependent width x, nil at $t = 0$ and equal to x_S at $t \gg \tau$ (second term).

From Eqs. (3), (15a) and (17a) for the respective intensities of the incident, reemitted steady state, and reemitted transient fluxes during a growth period, one has:

$$J_g = \frac{j_{s,g} - j_g}{j_i} = \exp(-t/\tau) - (1-\Psi)\frac{2x_S}{\lambda}\tanh\left(\frac{\lambda}{2x_S}\right) ,$$

where $\Psi = x/x_S$ from Eq. (18). An identical expression, with opposite signs is obtained for the evaporation period ($J_e = -J_g$). The development of Eq. (19) to the linear term with respect to t, taking into account that $(d\Psi/dt)t=0=0$, yields

177

$$J_g \approx - \frac{t}{\tau} + 1 - \frac{2x_s}{\lambda} \tanh \left(\frac{\lambda}{2x_s} \right). \qquad (20)$$

Figure 3 shows the reduced fluxes J_g and J_e during an entire growth-evaporation period of length $T = 10\tau$, calculated from Eq. (19) for different values of the scaled terrace width $\lambda/2x_s$. One can notice that:

- the slope of all curves at $t = 0$ and $t = T/2$ are identical, as expected from Eq. (20),

- the absolute values of the function at $t = 0$ and $t = T/2$ decrease with decreasing scaled width $\lambda/2x_s$, as $1 - (2x_s/\lambda) \tanh(\lambda/2x_s)$. It is sensibly different from that of the perfect surface (dashed line) for terrace widths λ as large as $16x_s$.

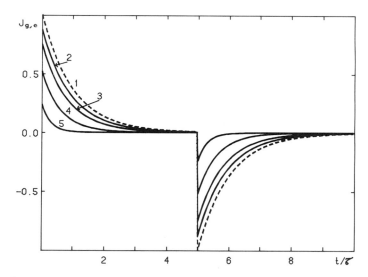

Fig. 3 Scaled reemitted flux density during an entire growth-evaporation period of length 10τ. The different curves correspond to different values of the scaled terrace width $\lambda/2x_s$ as follows: ∞ (curve 1), 8 (curve 2), 4 (curve 3), 2 (curve 4), 1 (curve 5).

THE GROWTH KINETICS OF THE Pt(111) SURFACE DURING VACUUM DEPOSITION

1.   Experimental Result[22]

The UHV deposition of platinum on Pt(111) has been studied in situ by different techniques, namely:

- Auger electron spectroscopy (AES) and low energy electron diffraction (LEED) which give information on coverage and structure of the deposited layers;

- ultraviolet photoelectron spectroscopy (UPS) and photoemission of adsorbed xenon (PAX), used respectively for the determination of the overall work function of the surface and as local probe of the work function on distinct xenon adsorption sites. In the latter case it has been shown[23,24] that the binding energy of the Xe 5p levels differs from xenon atoms adsorbed on terrace sites and on sites along steps by as much as 1 eV, independently of the step orientation.

The work function of the initial Pt(111) surface, after prolonged heating at 1200 K under 10^{-7} mbar of oxygen followed by several cycles of argon ion bombardment and annealing at 1000 K, had the average value of 5.85 ± 0.05 eV. In Fig. 4 is plotted the work function variation versus platinum coverage θ_{Pt} at three surface temperatures. The different qualitative behavior of the Pt-submonolayers in the three cases is evident. At 130 K the maximum variation of the work function is already reached at a coverage $\theta_{Pt} = 0.12$ and it levels off for higher coverages. At 250 K the decrease of the work function is continuous and fits roughly with a $\theta_{Pt}^{1/2}$ law. At 400 K no appreciable change of the work function with Pt-coverage could be observed.

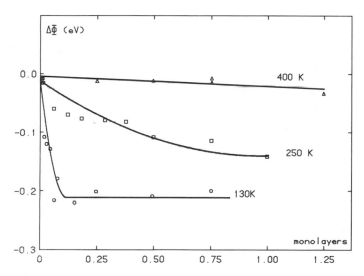

Fig. 4 Work function of the Pt(111) surface versus platinum coverage for three surface temperatures.

Figure 5 shows two series of UPS He I difference spectra in the energy range of the Xe 5p levels corresponding to Xe adsorption with different degrees of coverage on a well annealed Pt(111) surface and on a Pt(111) surface coverage by a 0.12 monolayer of platinum at 130 K. The surface temperature during Xe adsorption has been 95 K in both cases.

On annealed Pt(111) the binding energy of the Xe $5p_{1/2}$ level (referred to the Fermi level of the platinum) is 6.03 eV at low xenon coverages, and increases to 6.10 eV at the monolayer coverage.

The Pt coverage of 0.12 corresponding to the spectra of Fig. 5b has been chosen with respect to the maximum change of the work function observed at this surface temperature of (cf Fig. 4). At xenon pressures lower than several 10^{-8} mbar, adsorption is irreversible and a 7.0 eV

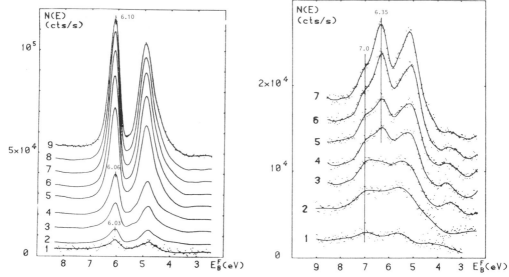

Fig. 5 UPS He I difference spectra in the Xe 5p levels energy range at different degrees of coverage of the surface by adsorbed xenon and constant surface temperature, 95K, left (a): on a well annealed Pt(111) surface: the equilibrium Xe pressure is varied from 5×10^{-8} mbar (curve 1) to 5×10^{-6} mbar (curve 9); right (b): on a Pt(111) surface with 0.12 monolayer of deposited platinum: curves 1, 2 and 3 correspond to xenon exposures of 1, 2 and 3.5 L respectively, curves 4 to 7 are obtained under equilibrium conditions at xenon pressures from 4×10^{-8} mbar to 5×10^{-6} mbar.

peak appears on the spectra of Fig. 5b. In the pressure range 4×10^{-8} to 5×10^{-6} mbar, spectra are obtained in equilibrium conditions and a 6.35 eV peak becomes predominant.

The total Xe coverage can be determined from the area under the Xe 5p levels. The resulting adsorption isotherms, degree of coverage versus equilibrium pressure, relative to the series of spectra of Fig. 5, are shown in Fig. 6. The Pt surface coveraged by 0.12 monolayer of deposited platinum is seen to be more active with respect to xenon adsorption than the flat (111)-surface.

The Xe 5p levels of Fig. 5b have been decomposed into two sets of peaks using asymmetric Gaussian lineshapes. The areas under the Xe $5p^{1/2}$ peaks at 7.0 and 6.35 eV are plotted, together with the work function variation due to Xe adsorption, in Fig. 7 as a function of Xe coverage. One notices the relation that exists between occupation of the more active adsorption sites, corresponding to the peak at 7.0 eV and the decrease of the work function. The deduced dipole moment of the Xe atoms in this range of coverage is as high as one Debye. The adsorption in the sites corresponding to the peak at 6.35 eV does not result in a further

180

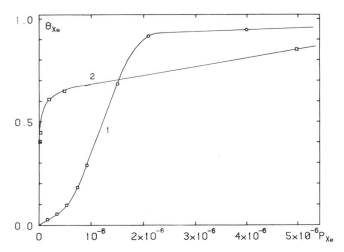

Fig. 6 Adsorption isotherms of xenon at 95 K on a well annealed Pt(111)
surface (curve 1), and on a surface with 0.12 monolayer of
deposited platinum (curve 2), obtained by integration of the
spectra under the Xe 5p levels of Fig. 5.

change of the work function, although their degree of coverage continues
to increase up to the completion of the monolayer.

2. A Tentative Interpretation[25]

The above experimental results give precious indication with
respect to the microscopic morphology of the growing surface and its
relation to the growth conditions, in this case, the surface temperature.

Platinum has an enthalpy of sublimation at 300 K of
134.5 cal/g·atom or 5.85 eV per atom. The adsorption energy of an
isolated Pt atom on the Pt(111) face is measured to be 3.66 eV.[26]
Accordingly, the calculated equilibrium concentration n_o of the isolated
adatoms on the surface[19] in the considered temperature interval, 130-400
K, is extremely low. The impinging flux j_i during vapor deposition, and
the net flux of matter remaining on the surface, are thus the same: the
adsorption is completely irreversible.

Platinum atoms on the Pt(111) surface start to be mobile just above
the temperature of the liquid nitrogen, 77 K. The corresponding
activation energy for surface diffusion is, therefore, at most
$\epsilon = 0.25$ eV, and the diffusion constant can be expressed by[26]

$$D = \frac{k_B T \delta^2}{2h} \exp\left(-\frac{\epsilon}{k_B T}\right) = 2.6 \times 10^{-6} \, T \exp\left(-\frac{2900}{T}\right) ,$$

where k_B and h are the respectively the Boltzmann and the Planck
constants, T is the absolute temperature and δ is the smallest hopping
distance. δ^2 is roughly equal to the area a of an adsorption site. On
the Pt(111), $a = 6.53 \times 10^{-16}$ cm^2.

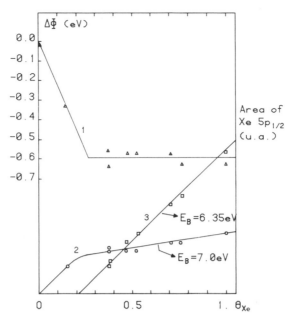

Fig. 7 Work function variation (curve 1) and areas under the Xe $5p_{1/2}$ peaks at 7.0 eV (curve 2) and 6.35 eV (curve 3) of the spectra of Fig. 5b, versus total xenon coverage.

Large mobility and lack of desorption in the considered temperature range lead the Pt adatoms on the Pt(111) surface to behave as a highly supersaturated but still very diluted two-dimensional lattice gas, diffusing toward the growing steps, and undergoing simultaneously a clustering in the most supersaturated regions of the surface. Setting, as before, the average distance between neighboring steps equal to λ, the mean time τ_d during which the most distant adatoms reach a step is calculated from the Einstein formula:

$$\frac{\lambda}{2} = \sqrt{D\tau_d} \ .$$
(21)

In our experiments λ was estimated to be about 250 Å, due to the possible misorientation ($\approx 0.5°$) of the substrate surface from the (111) direction, caused by polishing.

The second and the third columns of Table I give the values of the coefficient of surface diffusion and the time for an atom to reach a step at the three substrate temperatures T of our experiments.

Due to the irreversibility of the adsorption and to the negligible surface concentration at equilibrium, n_o, the differential equation for the balance of matter at any time t after the shutter opening is a simplified version of the Eq. (5):

$$D \frac{d^2n}{dx^2} - \frac{dn}{dt} + j_i = 0,$$
(22)

Table I

Coefficient of surface diffusion D; mean time τ_d for surface diffusion of an atom over the half-width of a terrace; its ratio, τ_d/τ_s, with respects to the time τ_s between the passage of two successive growth fronts through a given point of the surface; maximum adatom concentration n_{max}, and total number of binary collisions, N_λ, on a section of the terrace with length λ during the time τ_s.

T(K)	D $(cm^2 s^{-1})$	$\tau_d(s)$	τ_d/τ_s	$n_{max}(cm^{-2})$	N_λ
130	6.98×10^{-14}	2.24×10^1	8.07×10^{-2}	6.73×10^{13}	3.20×10^6
250	6.00×10^{-9}	2.60×10^{-4}	1.02×10^{-6}	7.81×10^8	3.72×10^1
400	7.44×10^{-7}	2.10×10^{-6}	8.23×10^{-9}	6.30×10^6	3.12×10^{-1}

with the boundary conditions

$$n = n_o (\approx 0) \quad \text{for all } x\text{'s and } t = 0, \tag{23a}$$

$$n = n_o (\approx 0) \quad \text{for } x = \pm\lambda/2 \text{ and } t > 0. \tag{23b}$$

The solution of Eq. (22) shows that steady state is reached at a time nearly equal to τ_d. The steady-state solution yields the simple equation

$$n(x) = \frac{j_i}{2D} \left(\frac{\lambda^2}{4} - x^2 \right), \tag{24}$$

where the equilibrium concentration n_o has been neglected. The highest concentration is reached, of course, at the point $x = 0$:

$$n_{max} = \frac{j_i \lambda^2}{2D \; 4} = \frac{j_i}{2} \tau_d, \tag{25}$$

where τ_d has been introduced by virtue of Eq. (21).

The flux of atoms, j_s, incorporated in unit length of the step, the rate of advancement of the step, V_s, and the time period τ_s between the passage of two successive steps through a given point of the surface, are respectively:

$$j_s = -2D \left(\frac{dn}{dx} \right)_{x=\lambda/2} = j_i \lambda, \tag{26a}$$

$$V_s = j_s a = j_i \lambda a, \tag{26b}$$

$$\tau_s = \frac{\lambda}{V_s} = \frac{1}{j_i a}. \tag{26c}$$

τ_s is seen to be independent of the distance between steps. This becomes clear if one considers that the passage through the same point of the surface of two successive steps is equivalent to the deposition of one monatomic layer and hence depends on the impinging flux j_i only. The impinging flux in our experiments being[21] about 6×10^{12} atoms cm^{-2} s^{-1}, $\tau_s = 254$ s.

The fourth column of Table I gives the ratio between the mean time for surface diffusion over the half-width of the terrace between two steps (approximately equal to the lag of steady growth) and the time necessary for the coverage of the same by a new condensed layer at the three substrate temperatures, as above. The extremely small values of this ratio for substrate temperatures of 250 and 400 K justify:

- the diffusion model used here, where steps have been assimilated to a train of immobile sinks;

- the consideration of steady state only, the transient period being a negligible fraction of the time necessary to coat a terrace by a new layer.

At a surface temperature of 130 K the transient period is about 8% of the time necessary to condense a new layer on the terrace. The rate of propagation of steps, calculated by the assumption of steady state, must be in this case overestimated.

The fifth column of Table I give the maximum concentration of adatoms on a terrace (for x = 0 and steady state). The important values of the latter, compared to the negligible values of the equilibrium concentration, show that if, inside the adsorption layer away from the steps, a condensation takes place, it must follow a "non-classical" mechanism, i.e., that every isolated adatom is potentially a critical nucleus, and every doublet is an overcritical cluster able to grow irreversibly.[27] In this case the nucleation kinetics is governed by the rate of binary collisions.

The frequency of binary collisions between hopping adatoms on the surface of the substrate is given by[28]

$$\dot{Z} = \frac{\alpha a n^2}{\tau_{ss}} , \tag{27}$$

where α is the maximum number of first neighbors of an adsorbed atom in the plane of the surface (two-dimensional coordination number) and τ_{ss} is the mean stay time of an atom in an adsorption site,

$$\tau_{ss} = \frac{h}{k_B T} \exp \left(\frac{\epsilon}{k_B T} \right) = \frac{a}{2D} . \tag{28}$$

The adatom concentration profile along the surface containing immobile steps, Eq. (24), can be transformed, when steady state is achieved, into a time dependence of the same concentration in one point of the surface, crossed periodically by advancing equidistant steps.

The collision frequency, $\dot{Z}$, at all points having coordinate $x = [(2j+1)/2]\lambda$, $j = 0,1,2,\ldots$ is thus, according to Eqs. (24), (26c), (27) and (28),

$$\dot{Z} = \frac{\alpha j_i}{2D} \lambda^4 \left(\frac{t}{\tau_s} \right)^2 \left(1 - \frac{t}{\tau_s} \right)^2 . \tag{29}$$

184

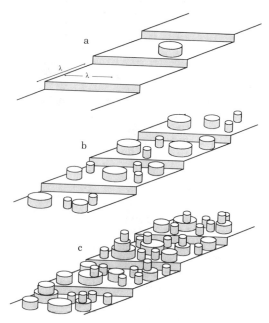

Fig. 8 Artist's view of the surface morphology of a stepped Pt(111) during Pt deposition at: (a) 400 K, (b) 250 K, and (c) 130 K.

The question, whether or not a terrace remains smooth in the time period τ_s between two successive passages of steps can be answered by estimating the total number N_λ of binary collisions on an area λ^2, i.e. on a length comparable to the terrace width (Fig. 8). This is achieved by integrating Eq. (29) twice, over τ_s for the time, and over λ for the distance normal to the steps. The numerical results are given in the sixth column of Table I and can be understood as follows:

- at the surface temperature of 400 K, there is hardly a formation of a single cluster throughout the supersaturated layer of adatoms on the terrace before it is swept by the transit of the next propagating step (Fig. 8a);
- at 250 K, the number of formed two-dimensional islands is such that the total length of their ledges during growth overwhelms the (constant) length of the bordering steps (Fig. 8b);
- at 130 K, the number of binary collisions and hence the number of clusters on the terrace is so large that the surface gets rapidly a time independent high roughness (Fig. 8c).

The three qualitatively different surface morphologies, deduced above, fit quite well with the experimental results for the overall run of the work function of the platinum surface at the three indicated surface temperatures, as shown in Fig. 4. The maximum roughness of this surface is likely to result in a work function decrease of about 200 meV. This roughness is almost immediately reached when condensation takes place at the temperature of 130 K (Fig. 8c). When nucleation is predominant but not excessive, as it looks to be at 250 K (Fig. 8b), the maximum surface roughness could also be reached after deposition of one or more monolayers, by simultaneous deposition on many atomic levels. Meanwhile, the total ledge length should be proportional to $\theta_{Pt}^{1/2}$. Finally, the independence of the work function of the Pt-coverage can't be understood but as corresponding to the situation of Fig. 8a.

The more detailed microscopic information obtained from the PAX results is also the more difficult to interpret. The Xe atoms adsorbed in the most active sites, probably along the steps of small two-dimensional islands shown in Fig. 8c, are irreversibly bound and strongly polarized. Their maximum degree of coverage is about 0.25, which is compatible with the presumed surface roughness. Whether the next type of sites, characterized by the peak of 6.35 eV, is due to adsorption on areas among islands, whose ledges are already screened by the first adsorbed Xe atoms, or to adsorption on the top of those islands, is not yet clear. The statement that the originally smooth Pt(111)-surface does not recover its smoothness after the deposition of a complete monolayer of platinum is confirmed, however, by additional annealing experiments which show that the spectrum of a xenon monolayer on a Pt-surface covered by 0.12 monolayer of platinum start to show the Xe $5p_{1/2}$ peak at 6.03 eV, characteristic for the flat (111)-face, only after one minute annealing at 473 K.

VACUUM DEPOSITION AND THERMAL EVOLUTION OF Pt-FILMS ON Ni(111)[29]

In order to stress the potentiality of the experimental methods described in the previous section for the in situ studies of the electronic properties of metal surfaces during vacuum deposition, we will present some recent results for Pt-films deposited on Ni(111) under conditions similar to those for the deposition on Pt(111).

Figure 9 gives on the one hand the UPS HeI spectra and the work

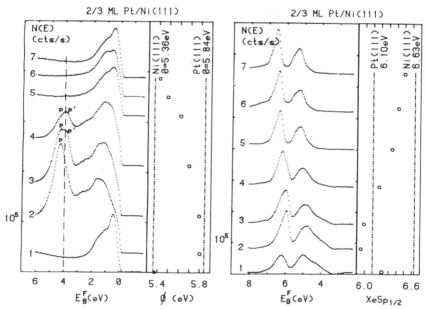

Fig. 9 UPS and work function measurement for Pt/Ni(111). Left (a): UPS spectra and work function variation of: (1) Ni(111), (2) 2/3 monolayer of Pt/Ni(111) as deposited, (3) +30s at 300°C, (4) +30s at 350°C, (5) +30s at 400°C, (6) +30s at 500°C, and (7) 1 min at 600°C; Right (b): UPS difference spectra and evolution of the binding energy of the Xe $5p_{1/2}$ peak of one monolayer of xenon adsorbed on a 2/3 monolayer of Pt/Ni(111): (1) as deposited, (2) +1 min at 200°C, (3) +30s at 300°C, (4) +30s at 350°C, (5) +30s at 400°C, (6) +30s at 450°C, and (7) +1 min et 600°C.

function variation of the Ni-surface after deposition of 2/3 monolayer of Pt at 160 K (a), and on the other hand, the spectra and the position of the Xe $5p_{1/2}$ peak of the same surface after adsorption of one monolayer of Xe at 95 K (b). The different curves in each plot correspond to annealing periods, of different time and temperature, preceding the spectra recording. It is quite obvious that after a long, high-temperature annealing, the two types of spectra of Figs. 9a and b and the two measured parameters (work function and energy of the Xe 5p peak) must recover their values corresponding to the pure Ni-surface, what is indeed the case. The most striking feature is, however, the evolution of the binding energy of the Xe $5p_{1/2}$ peak. After "annealing" at room temperature of this low temperature deposited Pt-film, and up to 200°C, the Xe $5p_{1/2}$ peak is shifted to binding energies by 200 meV lower than that for the (111) face of the pure platinum. The electronic properties of the surface, therefore, far from being intermediate between those of the two pure constituents, as could be expected for a regular alloy, are probably due to the formation of ordered, highly stressed platinum overlayers, as shown by subsequent LEED studies.[30]

Acknowledgements

Some parts of this paper are based on unpublished works of the author in collaboration with C. R. Henry, J. J. Ehrhardt, J. A. Barnard and M. Alnot. Other parts are published results of J. J. Ehrhardt, J. A. Barnard and M. Alnot. The author is indebted to all these persons for precious help.

REFERENCES

1. J. C. Brice, Growth of Crystals from the Melt, North Holland, Amsterdam, 1965.
2. D. T. J. Hurle, in: Crystal Growth: an Introduction, P. Hartman, ed., North Holland, Amsterdam, 1973, pp. 210-247.
3. B. Mutaftschiev, in: Dislocations in Solids, vol. 5, F. R. N. Nabarro, ed., North Holland, Amsterdam, 1980, pp. 57-126.
4. J. R. Arthur, Surf. Sci. 43, 449 (1974).
5. J. H. Neave, B. A. Joyce, P. J. Dobson and N. Norton, Appl. Phys. A31, 1 (1983).
6. A. Y. Cho, in: Molecular Beam Epitaxy and Heterostructures, L. J. Chang and K. Ploog, eds, NATO Adv. Stud. Inst. Ser., vol. E 87, Nyhoff, Dordrecht, 1985, pp. 191-226.
7. H. Dabringhaus and H. J. Meyer, J. Crystal Growth 16, 17, 31 (1973).
8. B. J. Stein and H. J. Meyer, J. Crystal Growth 49, 696 (1980).
9. H. Dabringhaus and H. J. Meyer, J. Crystal Growth 61, 85, 95 (1983).
10. H. Dabringhaus and H. J. Meyer, Surf. Sci. in press.
11. C. R. Henry, PhD Thesis, University of Aix-Marseille III, Marseille, 1983.
12. C. R. Henry, C. Chapon and B. Mutaftschiev, Surf. Sci. 163, 409 (1985).
13. S. T. Ceyer and G. A. Somorjai, Ann. Rev. Phys. Chem. 28, 477 (1977).
14. M. P. D'Evelyn and R. J. Madix, Surf. Sci. Rep. 3, 412 (1984).
15. V. Bostanov, J. Crystal Growth 42, 194 (1977).
16. E. Budevski, in: Interfacial Aspects of Phase Transformations, B. Mutaftschiev ed., NATO Adv. Stud. Inst. Ser., vol. C, 87, Reidel, Dordrecht, 1982, pp. 559-604.
17. H. Bethge, K. W. Keller and E. Ziegler, J. Crystal Growth 3/4, 194 (1968).
18. H. J. Meyer, H. Dabringhaus, A. Maas and B. J. Stein, J. Crystal Growth 30, 225 (1975).

19. W. K. Burton, N. Cabrera and F. C. Frank, Phil. Trans. Roy. Soc. (London) A 243, 299 (1951).
20. W. K. Burton and N. Cabrera, Disc. Farad. Soc. 5, 38, 40 (1949).
21. C. R. Henry and B. Mutaftschiev, to be published.
22. M. Alnot, J. J.Ehrhardt and J. A. Barnard, Surf. Sci. 208, 285 (1989).
23. K. Wandelt, J. Vac. Sci. Technol. A2, 802 (1984).
24. K. Wandelt, in: Thin Metal Films and Gas Chemisorption, P. Wissmann ed., Elsevier, Amsterdam, 1987, pp. 280-371.
25. M. Alnot, J. A. Barnard, J. J. Ehrhardt and B. Mutaftschiev, J. Crystal Growth, in press.
26. D. W. Basset and P. R. Webber, Surf. Sci. 70, 520 (1978).
27. G. Zinsmeister, Vacuum 16, 529 (1966); Thin Solids Films 2, 497 (1968).
28. C. Chapon and B. Mutaftschiev, Z. Phys. Chem. (Neue Folge) 77, 93 (1972).
29. J. A. Barnard, J. J. Ehrhardt, H. Azzouzi and M. Alnot, Surf. Sci. 211/212, 740 (1989).
30. J. A. Barnard and J. J. Ehrhardt, to be published.

STRUCTURE AND ORDERING OF METAL OVERLAYERS ON Si(111) AND

Ge(111) SURFACES

R. Feidenhans'l
Risø National Laboratory
DK-4000 Roskilde, Denmark

F. Grey
Max Planck Institute for Solid State Research
Heisenbergstrasse 1
D-7000 Stuttgart 80, Federal Republic of Germany

M. Nielsen
Risø National Laboratory
DK-4000 Roskilde, Denmark

R. L. Johnson
II. Institute for Experimental Physics
Hamburg University
D-2000 Hamburg 50, Federal Republic of Germany

ABSTRACT

We have performed x-ray diffraction studies of two examples of metal overlayers on semiconductor surfaces in the monolayer coverage regime: (1) Pb on Si(111) and Ge(111) and (2) Au on Si(111).

LEED patterns show that Pb has two different $\sqrt{3}$ structures on both Si(111) and Ge(111) as a function of coverage. The low coverage structure is the same on both surfaces with one Pb atom per unit cell. The high coverage phase is a slightly distorted Pb (111) overlayer for Ge(111). For Si(111) the structure is an incommensurate Pb(111) overlayer, as revealed by the x-ray diffraction data. We have also made studies of the high temperature 1x1 phase of Pb/Ge(111) where we observe a halo of diffuse scattering about the origin due to a molten Pb overlayer.

Au on Si(111) is known to have a 5x1 structure saturating at about 0.4 ML, a $\sqrt{3}$ structure at 1 ML and a 6x6 at 1.5 ML. We have performed x-ray diffraction studies of all three structures and will discuss the structural analysis and the complications arising from disorder.

INTRODUCTION

Ultrathin metal overlayers on semiconductor surfaces may serve as model systems for the metal/semiconductor interface formation. The phase diagrams in the monolayer coverage regimes can be quite complex[1] and are important for the growth of thicker films. This has recently been demonstrated in the case of the growth of Pb on Si(111).[2] At 95 K the growth proceeds in a layer by layer fashion as revealed by oscillations

Kinetics of Ordering and Growth at Surfaces
Edited by M. G. Lagally
Plenum Press, New York, 1990

in the reflection high-energy electron diffraction (RHEED) intensities. The oscillation pattern for the first few layers of Pb depends strongly upon the initial surfaces which were either the clean Si(111)7x7, Au/Si(111)√3x√3R30° or Au/Si(111)6x6. At higher temperatures Pb grows on Si(111) and Ge(111) in a Stranski-Kastranov mode - i.e. completion of a two-dimensional overlayer before the development of three-dimensional islands. Au, on the other hand, develops a gold-silicide.[1]

It therefore seems worthwhile to investigate the atomic structure of these systems in the monolayer coverage regime. X-ray diffraction is very well suited for such a task. It provides direct insight into the atomic structures and can give accurate in formation about the long-range order (for a review, see Ref. 3, 4). We report here on structural investigations on Pb/Ge(111), Pb/Si(111) and Au/Si(111). These systems have been studied in some detail by other techniques, but as will be shown later, X-ray diffraction can easily distinguish between proposed models even in such complex cases as Au/Si(111). We will also show how disorder in the form of random occupancies or liquid overlayers can be studied.

EXPERIMENTAL

The Si(111) and Ge(111) samples were cut to within 0.1° and etched and polished at the Max-Planck-Institute, Stuttgart to give a mirror-like surface. The samples were cleaned and characterized at the FLIPPER II photoemission station at HASYLAB (Hamborg Synchrotron Radiation Laboratory).

The Si(111) samples were cleaned by annealing to 900°C and slowly cooling down to produce a sharp 7x7 Low Energy Electron diffraction (LEED) pattern. The Ge(111) samples were cleaned by cycles of 500 eV Ar[+] ion sputtering, annealing at 700°C and slowly cooling down until a sharp c(2x8) LEED pattern emerged.

Angle-integrated photoemission spectra were taken to check for impurities. Au and Pb were deposited from BN effusion cells with rates around 0.2-0.3 ML/min. After preparation, the samples were transferred to a portable UHV-chamber for X-ray diffraction. The X-ray diffraction measurements were performed at the 32-pole wiggler beamline W1.

The X-ray chamber was mounted on a vertical scattering diffractometer. The X-rays were monochromated by Si(111) crystals and focussed on to the sample by a gold-coated toroidal mirror. The X-ray wavelengths used for the studies described here were around 1.3 Å. The samples were aligned such that the X-rays had a fixed glancing angle of incidence to the surface. The angle of incidence was usually set to the critical angle for total reflection, typically around 0.2°-0.3°. A position sensitive detector (PSD) subtending 4° perpendicular to the surface and 0.8° in the surface plane was used to measure the scattered X-rays. Integrated intensities of the reflections were collected by sample rotation, ω scans. A more detailed description of the apparatus and the experimental procedures is given in Ref. 3.

SURFACE X-RAY DIFFRACTION

As surface X-ray diffraction recently has been reviewed in detail (Ref. 3,4) we will briefly review the essential features needed for the further discussion.

All the information of the atomic geometry inside the unit cell is contained in the structure factor

$$F(\underline{q}) = \sum_j f_j e^{-M_j} e^{i\underline{q}\cdot\underline{r}_j} \quad , \qquad (1)$$

where f_j are the form-factors for the individual atoms, e^{-M_j} the Debye-Waller factors, $\underline{r}_j$ the position of the atoms inside the unit cell and $\underline{q}$ the momentum transfer. Unfortunately, only the structure factor intensities $|F(\underline{q})|^2$ can be measured. They are derived from the integrated intensities which are given by[3]

$$I_{hk}(q_z) = \frac{I_o}{\Omega} \frac{A\lambda^2 P}{a \; sin(2\theta)} \frac{e^4}{m^2 c^4} |F_{hk}(q_z)|^2 \frac{\lambda}{2\pi} \Delta q_z \cdot |T_i|^2 |T_f|^2 \quad , \qquad (2)$$

where (h,k) are the indices for the surface reflection, q_z is the momentum transfer normal to the surface, A is the total area of the surface participating in the diffraction, λ is the wavelength, P the polarization, which for vertical scattering at a synchrotron is $P = 1$, Ω is the angular speed in the ω-scan, a is the area of the surface unit cell, 2θ is the in-plane scattering angle, and Δq_z is the resolution along the Bragg rod normal to the surface.

Refraction effects are contained in transmission coefficients T_i and T_f for the incoming and diffracted beam, and finally I_o is the incoming photon flux. I_o is not precisely known but is around 10^{11} photons/sec/mm^2. Often both the incoming and the diffracted beam have glancing angles to the surface, then $q_z = 0$ and Δq_z is given by[3,4]

$$\frac{\lambda}{2\pi} \Delta q_z = \frac{L}{R} \quad , \qquad (3)$$

where L is the length of the detector normal to the surface and R the distance from the detector to the sample. Apart from I_o, all factors in equation (2) are known and the structure factor intensities $|F_{hk}|2$ can be derived on a relative scale. By measuring the intensities at $q_z \approx 0$ one gets only the projection of the structure on to the surface plane. To get the full structure, out-of-plane data are necessary. We use the short notation F_{hk} for F_{hk} ($q_z = 0$). In Fig. 1 a typical glancing incidence/ glancing exit geometry is shown.

Because the phases of the structure factors are unknown, one cannot directly derive the electron density from the experimental data. However, very useful information can nevertheless be obtained from the 2-dimensional pair-correlation or Patterson function

$$P(x,y) = \sum_{hk} |F_{hk}|^2 cos(2\pi(hx+ky)) \quad . \qquad (4)$$

A contour map of $P(x,y)$ gives a map of the interatomic vectors within the unit cell. This can contain information enough to make a trial model and can also reject wrong models. Refinement of the model to the experimental data is often done with a χ^2 test

$$\chi^2 = \frac{1}{N-p} \sum_{hk} \left(\frac{|F_{hk}^{exp}|^2 - |F_{hk}^{model}|^2}{\sigma_{hk}} \right)^2 \quad , \qquad (5)$$

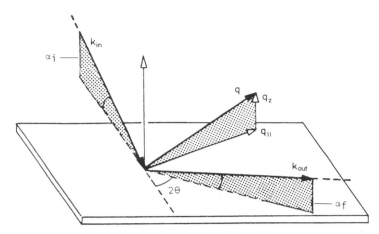

Fig. 1   A typical glancing incidence/glancing exit geometry. The x-ray
         beam impinges onto the surface at an angle α_i. After
         diffraction through an angle 2θ the x-rays exit the surface at
         an angle α_f. Both α_i and α_f are around 0.3° for in-plane
         studies.

where N is the number of measured structure factor intensities, p the
number of free parameters and σ_{hk} the uncertainties on $|F_{hk}^{exp}|^2$. When
$X^2 \approx 1$ the model is considered to be satisfactory.

Pb ON Ge(111)

        The Pb/Ge(111) system has been studied extensively by a variety of
methods. It was first studied by Reflecting High-Energy Electron
Diffraction (RHEED) by Ichikawa[6,7] and by LEED, Auger Electron
Spectroscopy (AES) and Scanning Electron Microscopy (SEM) by Le Lay and
Metois.[8-10] In the monolayer regime two stable $\sqrt{3}\times\sqrt{3}$ R30° structures (in
short $\sqrt{3}$) exist at room temperature, but the two structures have
different saturation coverages. The low coverage structure is labelled α,
the high-coverage structure β. The atomic geometries of the α and β
phases were revealed by surface x-ray diffraction[11,12] and are shown in
projection in Fig. 2. The α-phase consists of one Pb atom per $\sqrt{3}$ unit
cell situated in the T_4 site above the second layer of Ge-atoms. The Pb
atom induces relaxations in the first few Ge layers below the surface.
The relaxations can be explained as elastic distortions.[13] The structure
has been confirmed by LEED[14] and theoretical calculations.[15] The β-phase
saturates at the coverage θ = 1/3 ML (1 ML is identified as one Pb atom
per one Ge atom at the surface). The β-phase is a distorted (111) Pb
overlayer. Note that a (111) Pb layer rotated 30° with respect to the
Ge-lattice only needs a 1% compression in order to be in registry with
the substrate (lattice constants: a_o (Ge) = 5.65754 Å, a_o (Pb) = 4.9505 Å
at room temperature). This would, however, be a 1/2 $\sqrt{3}$ structure. The
atoms at the positions labelled B in Fig. 2 are placed in highly
asymmetrical sites and displace towards the sites labelled A. The
registry of the Pb overlayer relative to the Ge lattice has been
determined by x-ray diffraction using the interference between scattering
from the bulk crystal truncation rods from the bulk and the scattering
from Pb overlayer. For details see Ref. 3 and 11. The Pb atoms labelled A
are situated in the three-fold hollow sites above the fourth layer of Ge
atoms, called H_3 sites, with the atoms at B displaced towards the nearest

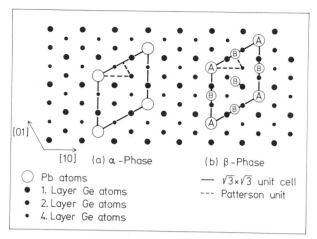

Fig. 2   The $\sqrt{3}$ structure of Pb/Ge(111) in (a) the α-phase, Pb coverage
1/3 ML and (b) the β-phase, Pb coverage 4/3 ML. In the β-phase
the Pb atoms are located in T_4 sites. In the β-phase the atoms
labelled A are located in H_3 sites. The atoms labelled B are
displaced towards the first layer of Ge atoms, giving a distance
between atom A and B of 3.11 ± 0.02Å.

first-layer Ge atoms. This model has been confirmed by a dynamical
LEED study,[14] where the vertical displacements were also derived. It was
found that the atoms at B are buckled out of the surface to accommodate
for the 1% compression. The presence of two inequivalent Pb sites has
also been proposed on the basis of photoemission measurements.

Note that the β-phase is not simply obtained by "filling in" sites
in the α-phase, but requires an additional shift of registry. Therefore,
the α- and β-phase are immiscible and phase-separate at intermediate
coverages. The α- and β-phases do not scatter coherently and the relative
amount of each can be determined using the lever rule:[17]

$$|F_{hk}^{obs}|^2 = \left(\frac{4}{3} - \theta\right)|F_{hk}^{\alpha}|^2 + \left(\theta - \frac{1}{3}\right)|F_{hk}^{\beta}|^2 . \tag{6}$$

Pb desorbs at a significant rate from the surface at temperatures higher
than 350°C. Hence it is possible in a controlled manner to lower the
coverage and investigate the phase diagram using Eq. (6).[17] The result is
shown in Fig. 3. The diagram is similar to the diagram sketched by
Ichikawa,[6] but with more detail. An interesting feature is the sharp
decrease of the transition temperature of the $\sqrt{3} \rightleftharpoons 1\times1$ transition which
goes from 330°C at θ = 1.33 ML to 180°C at θ=1.25 ML. A similar feature
was observed by RHEED.[6]

The 1x1 phase has been the subject of some controversy. On the basis
of RHEED measurements, it has been proposed to be an isotropic
two-dimensional liquid,[7] whereas x-ray standing waves measurements favour
a model with Pb atoms locally in the same sites as in the β-phase, but
without any long-range order.[18] Surface x-ray diffraction can elucidate
on this controversy.[19]

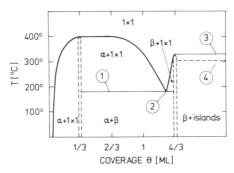

Fig. 3   Sketch of the phase diagram of Pb/Ge(111). Vertical dashed
lines bound single phase regions (suggested only). The number
features are: (1) A triple line at 180°C, where the β-phase
melts. (2) An "eutectic point" at $\theta \approx 1.25$ ML. (3) Beyond 4/3
ML the β-phase melts at 330°C. (4) The Pb island melting
temperature is $T_m = 302$°C for the coverages (<10ML) we have
studied. This is 25°C below the bulk Pb melting temperature.

We have observed diffuse scattering in a halo around the origin,
similar to scattering observed from 3D liquids.[20] There is one important
difference, however, the radius of the halo. The scattered intensity from
an isotropic 3D liquid is given by[20]

$$I(q) = 1 + \rho_o \int_o^\infty \frac{4\pi r}{q} [g(r) - 1]\sin(qr)dr \; , \tag{7}$$

where $g(r)$ is the pair-correlation function. For a 2D liquid the volume
integration element is different and

$$I(q) = 1 + \rho_o \int_o^\infty 2\pi r[g(r) - 1] J_o(qr)dr \; , \tag{8}$$

where J_o is the Bessel function of zero order and ρ_o the density.[20] The
difference between Eqs. (7) and (8) can be illustrated by a model
pair-correlation function $g(r) = \delta(r-a)+1$, which corresponds to a
diatomic ideal gas. Equations (7) and (8) then give

$$I(q) = \begin{cases} 1 + \dfrac{\sin(qa)}{qa} & , \; 3D \\ \\ 1 + J_o(qa) & , \; 2D \; . \end{cases} \tag{9}$$

With $a = 3.5$ Å, the Pb-Pb bond length, the scattering from a 3D liquid
peaks at 2.21 Å^{-1}, while the scattering from a 2D liquid peaks at
$q = 2.01$Å, a difference of 9%.

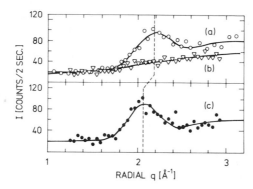

Fig. 4   (a) Radial scan 10° to (1,1) through liquid halo, at θ = 4ML and
T = 310°C; 3D phase liquid, 2D phase solid.   (b) Same scan at
T = 230°C; 3D and 2D phases solid.   (c) Same scan at = 223°C and
θ = 1.25 ML; 3D phase absent, 2D phase liquid.   Dashes vertical
line indicates fitted midpoint of halo.

We can observe 3D-scattering by depositing more than 1 ML Pb on the
surface. Then Pb grows as 3D islands in a Stranski-Krastanov mode. In
Fig. 4 we show radial scans through the halo for both the 2D overlayer
and the 3D islands. The scans are along non-symmetry directions, where no
Bragg peaks from either the bulk or the overlayer are observed. Scan (a)
is a scan through the halo from 3D liquid islands on a surface with
nominal 4ML Pb coverage and the scattering peaks at 2.18 ± 0.02 Å^{-1}. Scan
(b) is a scan below the melting temperature where only background
scattering is seen. In scan (c) the coverage is reduced to 1.25 ML and
all the islands have disappeared. At T = 223°C we again observed a halo
of scattering, however, the peak has shifted to 2.05 ± 0.02 Å^{-1}. This 6%
shift inwards agrees well with the expected effect of reduced dimension
discussed above. The effect could be used to study surface melting on a
Pb surface.[21] When only one layer is molten, the peak in the halo would
be at 2.05 Å^{-1}. As more and more layers melt when the melting temperature
is approached from below, the peak should gradually move towards 2.18Å^{-1}.
Unfortunately, we are not able to do this in this experiment, because Pb
does not wet the Ge(111) surface.

We have also measured the intensity of the halo as a function of
azimuthal angle and seen an anisotropy.[17,19] We conclude that 1x1 phase
is a 2D liquid weakly modulated by the underlaying solid.

Pb ON Si(111)

How large is the effect of the underlying substrate on the Pb
overlayer? This can be studied by changing the substrate to Si(111). The
lattice mismatch is then changed from 1% to 5.3%. We will discuss the
effect induced on the atomic structure on two differently prepared
surfaces (1) Pb deposited on Si (111) at RT and (2) Pb deposited on Si
(111) at 250°C.

The Pb/Si(111) system was first studied by Estrup and Morrison. When Pb was deposited onto the Si(111) 7x7 surface at RT, they observed changes in the intensities of the 7x7 LEED pattern and with increasing Pb coverage extra spots gradually became visible in a hexagonal arrangement around the Si integer-order reflections. The "extra spots" finally dominate the diffraction patterns. After annealing to 300°C, the structure changes irreversibly into a $\sqrt{3}$ structure. Annealing to 400°C, and desorbing Pb from the surface, the structure changes into a different $\sqrt{3}$. In the low density structure all third-order reflections are present, whereas in the high density structure only reflections corresponding to a 1/2 $\sqrt{3}$ structure could be observed, very similar to Ge(111). The Pb/Si(111) system has been studied by LEED and low energy ion scattering spectroscopy (ISS) by Saitoh et al.[23] and by LEED and AES by Le Lay et al.[24] We will label the low-density and high-density structures α and β[25], respectively, as for Pb/Ge(111).

We start by discussing the metastable Pb/Si(111) 7x7 structure formed by deposition at RT. Saitoh et al. attributed the 7x7 LEED pattern as arising from double diffraction between an incommensurate Pb(111) overlayer and the Si(111) substrate. We also observe the 7x7 pattern with x-ray diffraction.[26] Since x-rays are scattered weakly at the surface,[3] multiple scattering does not occur and the double diffraction model can be ruled out. The strongest peak is the (8/7,0) with h = 1.1427±0.0002 (8/7 = 1.1429). Since Pb (Z = 82) scatters x-rays much more efficiently than Si(Z = 14), we are nearly only sensitive to Pb and must conclude that the Pb overlayer is not incommensurate, but has a genuine 7x7 periodicity. The strongest seventh-order reflections are always found in the hexagon around the Si integer-order reflections. An example is shown in Fig. 5. The widths of the reflections corresponds to a spatial coherence length on the surface of about 700 Å. We collected

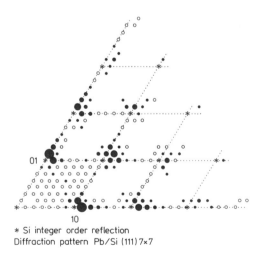

* Si integer order reflection
Diffraction pattern Pb/Si (111) 7×7

Fig. 5   In-plane diffraction pattern for Pb/Si(111) 7x7. The areas for the filled circles are proportional to the structure factor intensities. The open circles are measured reflections with zero intensity. Note that the strongest peaks are all on the high-q side of the hexagons around the integer-order Si-reflections.

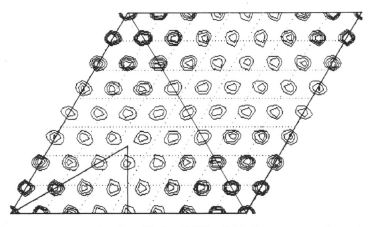

Fig. 6 Contour map of the Pb/Si(111) 7x7 Patterson function. Only
positive contours are shown. The solid right angle triangle is
the irreducible unit.

145 fractional order integrated intensities from which we derived 97
non-symmetry equivalent structure factor intensities. A contour map of
the Fourier inversion of those, the Patterson function in Eq. (6), is
shown in Fig. 6. It clearly suggests an 8x8 mesh of atoms on a 7x7 unit
cell. This would give a Pb-bondlength of 3.36 Å, substantially shorter
than the bulk bondlength of 3.50 Å. An unrelaxed 8x8 mesh is a 1x1
structure with a lattice parameter of 7/8 of the Si (1,0) vector. The
reflections in reciprocal space will appear at (8/7,0) and multiples
thereof. This model has X^2 = 23. Relaxing the 8x8 mesh is necessary and a
least-squares analysis show that the Pb atoms in the center adopts a
lattice constant closer to the bulk value, while the atoms at the corners
disappear and those nearest to the corner have a very low occupancy of
0.25±0.10. This gives X^2 = 8. The reason why the Pb overlayer adopts a
7x7 periodicity at all, is presumably because that the underlying Si
retains the stacking fault of the 7x7 Dimer-Adatam-Stacking fault (DAS)
structure.[27] The stacking fault in half the unit cell does not disappear
at RT where the deposition occurs. Hence, the Pb overlayer must be
modulated by the domain walls (the dimers) between the regular and the
faulted parts of the Si surface. A strong modulation must be expected
where the dimers row meet, at the large corner holes in the 7x7
DAS-structure.[27] This probably is where there is low occupancy of Pb.
Including an unrelaxed DAS-structure (without adatoms) below the Pb
overlayer, gives a slight improvement of the fit, X^2 = 7. Although Pb
scatters much more than Si, the underlaying Si 7x7 structure is important
at the weak reflections where the scattering from the Pb is nearly
cancelled. However, our limited data-set prevents a full analysis
including all relaxations of the Si-substrate. Further refinement of the
model can be expected when more and better data become available.

    Interestingly, the stacking fault can be removed by a very mild heat
treatment (200°C), where the Pb/Si(111)7x7 structure transforms
irreversibly into the $\sqrt{3}$-β phase. In contrast, on the clean surface the
Si(111)7x7≈1x1 transition occurs at 850°C.

197

The $\sqrt{3}$-β phase is quite different from the Pb/Ge(111) β-phase; the larger lattice mismatch has made the buckled, compressed overlayer unfavorable and instead the Pb overlayer becomes incommensurate. This is demonstrated in Fig. 7, where a radial scan (full line) through the expected (2/3, 2/3) position is shown. The peak intensity is shifted to (0.6515, 0.6515) ± (0.0005, 0.0005). The transverse and radial widths of the peak corresponds to a spatial coherence of only 130 Å, compared to the 700 Å coherence length of the 7x7 structure. This incommensurability has not previously been detected by LEED.[22-24] If the scans are performed along direction 2 (see Fig. 7), the reflection clearly has a shoulder on the high momentum transfer. A two-Gaussian fit to this line shape gives a second peak at 0.690±0.005. Transverse scans through the satellite peak and its symmetry equivalent revealed that they are along the high-symmetry directions relative to the incommensurate peak.

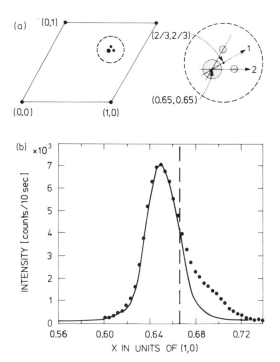

Fig. 7   (a) Sketch of the reciprocal space for the incommensurate Pb/Si(111) surface.   The expanded region around the (2/3,2/3) position shows the fundamental Pb peak of the incommensurate layer at (0.65,0.65).   The expected peaks, due to domain-wall formations are indicated. (b) Solid line: measured radial scan through the fundamental Pb reflection at (0.65,0.65), scan (1) in (a).   The vertical dashed line indicates the commensurate (2/3,2/3) position.   Points: measured scan in the (1,0) direction, scan 2 in (a).   A shoulder can be seen near the expected satellite position.

The origin of the satellite peaks is found in the interaction between the Si substrate and the Pb overlayer, which modulates the Pb overlayer by producing domain walls. A prominent example of such a modulation is the physisorption of inert gases on graphite surfaces.[29] In the weak interaction regime, it can be shown that the characteristic reciprocal lattice vector of the superstructure is $3 \cdot (2/3 - 0.6515) = 0.0455$.[29] This would give a satellite at $0.6515 + 0.0455 = 0.6970$, close to the observed position. Unfortunately, the spatial coherence length of the Pb overlayer is too small to resolve the satellite peaks more clearly, the coherence length is only 130 Å while the size of the superstructure domains is 85 Å. Therefore, coherence only exists over about two domains.

Finally, we note that the Pb/Si(111) $\sqrt{3}$-α phase is similar to the Pb/Ge(111) $\sqrt{3}$-α with one Pb atom per unit cell.[26]

Au ON Si(111)

Au/Si(111) is very different from Pb/Ge(111) and Pb/Si(111), probably because Au diffuses into Si and forms a gold-silicide layer[1], whereas Pb does not intermix with either Ge(111) or Si(111). We will discuss recent x-ray diffraction measurements on the superstructures that occur in the monolayer regime. The different structures as a function of coverage are shown in Fig. 8.[1,31,32] A 5x1 structure is formed at about 0.4 ML Au, a $\sqrt{3}$ structure is formed at 1.0 ML and a 6x6 structure at 1.5 ML. In between these coverages there are disordered structures with diffuse LEED patterns and/or mixed phases.[32] In order to clarify this complicated pattern we have started a structural investigation by x-ray diffraction to detail the atomic geometry of each structure. We will report some preliminary results.

As for Pb, Au (Z = 79) scatters x-rays much more than Si(Z = 14) so x-ray diffraction is mostly sensitive to the topology of the Au atoms and only marginally sensitive to the Si atoms. We will concentrate on the in-plane structure, i.e. the structure projected onto the surface plane.

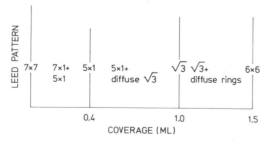

Fig. 8   Diagram of the LEED pattern for Au/Si(111) as a function of Au coverage.   After Refs. 1, 31 and 37, although there is not complete agreement on the saturation coverages.

a.   The 5x1 Structure

The 5x1 structure was first seen by Bishop and Riviere.[33] The 5x1 LEED pattern is accompanied by diffuse streaks running through the half-order position as clearly observed by Lipson and Singer.[34] A model with two Au atoms in equivalent high symmetry sites has been suggested on basis of low-energy ion scattering measurements.[32,35] The model is also favoured by x-ray standing waves measurements if the Au atoms are placed in a site embedded in the substrate bridging two Si atoms in the lower half of the (111) double layer.[36] The structure of the Au atoms is shown in Fig. 9c.

The 5x1 structure has three equivalent domains, related by a 120° rotation. We were not able to detect differences in the integrated intensities from equivalent reflections from each domain, so the data were reduced to a set from one domain. In total a set of 52 inequivalent, in-plane fractional-order structure intensities was obtained. The width of the fifth-order diffraction peaks corresponded to a coherence length of about 400 Å.

A contour map of the corresponding Patterson function is shown in Fig. 9a. The full 5x1 unit all is shown and contains two irreducible units, which are related by inversion symmetry. The map contains important pieces of information: (1) The map has four strong non-origin peaks. These must correspond to Au-Au interatomic vectors, since Si is hardly visible due to its low atomic number. Hence, at least four Au-atoms must be present in the 5x1 unit cell. But if the 5x1 unit cell contains four Au-atoms, the coverage would be 0.8 ML, which contradicts previous coverage determinations.[31,32] (2) none of the interatomic vectors are close to vectors between atoms residing on lattice sites. This means that the Au-atoms are situated at low-symmetry sites, in agreement with the x-ray standing waves measurements.[36] A model reproducing the main features is shown in Fig. 9b including only the four atoms shown as full circles. The structure is least-square refined by allowing the atoms to move along the bulk mirror line in the (2,1) direction. With three atomic displacements and one isotropic Debye-Waller factor for all four atoms, the best fit gives X^2 = 18. Note that the atoms cannot all be in high-symmetry sites, although we do not have any information about the registry. The Patterson map can give a third hint; all peaks in the map are elongated in the (0,1) direction. This suggest anisotropic thermal vibrations or static displacements breaking the symmetry. A dramatic improvement can be obtained by including an anisotropic Debye-Waller factor. The best fit has X^2 = 8.1 with a Debye-Waller factor $B_{//}$ along the (2,1) direction of $B_{//}$ = 0.4±0.4 $Å^2$ and $B_{\perp}$ = 12.9±2.8 $Å^2$ perpendicular to it. The latter corresponds to a one-dimensional r.m.s. amplitude of 0.4Å! A thermal, isotropic B-factor has been determined by LEED[37] to be $\sim$ 0.62$Å^2$. The reason for the large $B_{\perp}$ can therefore not be thermal vibrations but must be static disorder. An error map, a map of the electron density difference between the real structure and the model structure assuming the phases of the model structure to be correct, suggest that three more atoms can be included. Assuming that they are Au and fitting the occupancy we arrive at the model listed in Table I with X^2 = 3.8. The final refinement must also include reconstruction or relaxation in the underlaying Si-substrate.

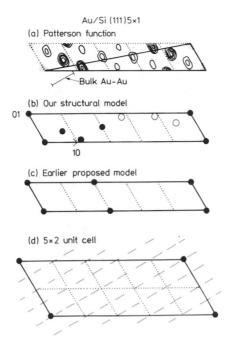

Fig. 9 Au/Si(111) 5x1: (a) A contour map of the Patterson function.
The origin rises 10 contour levels. The triangle is the
irreducible unit. (b) Our model for the Au-structure in the
projected 5x1 unit cell. The atoms indicated with full circles
have full occupancy, the atoms with open circles have partial
occupancy (see Table I). (c) The model for the Au-atoms as
proposed in the Ref. 40. (d) A 5x2 unit cell. Note the 5x2
unit cell breaks the mirror line (dashed) from the Si bulk,
whereas the 5x1 unit cell does not.

Table I

Atomic Coordinates and Debye-Waller Factors of the Au Atoms Within
the 5x1 Cell. $X^2 = 3.9$

x	y	occ	$B_\parallel$	$B\perp$
0.000	0.000	1.00	1.06	10.6
1.212	0.106	1.00	1.06	10.6
0.800	0.400	1.00	1.06	10.6
2.072	0.536	1.00	1.06	1.06
2.796	0.898	0.44	1.06	1.06
4.386	0.693	0.12	1.06	1.06
3.860	0.930	0.30	1.06	1.06

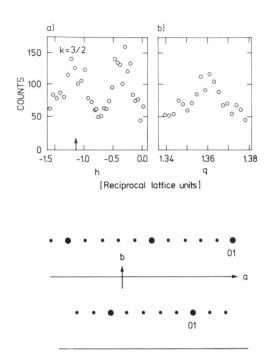

Fig. 10 Top part: (a) Scan along the streak at k = 3/2. (b) Scan in the
direction normal to the streak crossing at (-1.15, 1.50).
Bottom part: Sketch of the reciprocal lattice. Big circles are
Si integer-order reflections, small circles fifth-order
reflections. The diffuse streaks are indicated showing the
scans corresponding to (a) and (b).

We will now return to the diffuse streaks in the 5x1 LEED pattern,[34]
because they can give answers to some of the puzzles arising from the
above 5x1 structural analysis. We have also observed the streaks by x-ray
diffraction as shown in Fig. 10. The scan marked a is a scan along the
diffuse streak in the (1,0) direction keeping k fixed at k = 3/2. Scan b
is in the (1,2) direction which is perpendicular to the streak. The
streak is sharp in that direction, the width of the peak corresponds to
the same coherence length as derived from the width of the 5x1
diffraction spot. This implies disorder in the 5x1 structure. The
streaks have also been observed by RHEED.[38] A model for the disorder was
given by Lipson and Singer.[34] They proposed that the structure is not 5x1
but a random mixture of 5x2 and c(10x2) unit cells. An example of that
is shown in Fig. 11. By only measuring reflections corresponding to a 5x1
structure, the two 5x1 subunits in the 5x2 cell are projected into one
5x1 unit cell. So if the structure is 5x2 with two Au atoms in each 5x1

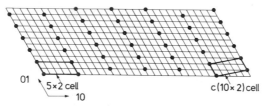

Fig. 11 Random disorder between rows of 5x2 and c(10x2) cells. Note the
         c(10x2) cells arise by shifting one row of 5x2 cells by the
         vector (0,1).

subunit, it can show up in the Patterson map as a four atom 5x1 model.
Four atoms in a 5x2 cell gives a coverage of 0.4 ML, in agreement with
previous coverage determinations.[31,32] The disorder can also explain the
large static displacements normal to the bulk mirror lines. The mirror
lines along the (2,1) direction are also mirror lines in the 5x1
structure, but not in a 5x2 or c(10x2) structure, as shown in Fig. 9d.
Therefore, the mirror symmetry must be broken which gives rise to the
large B-factors in the 5x1 analysis. The 5x2 structure has then two
domains, one of each mirror symmetry (plus the three 120° rotations of
each).

    A full structural analysis is clearly now very complicated, because
we have (1) partial occupancies, (2) a broken symmetry, and (3) disorder
between two possible units cells. The intensity variation along the
streaks will reflect the atomic structure in the 5x2 cell, but we have
not yet been able to derive a satisfactory model, bearing in mind the
complications mentioned above.

b,   The √3x√3R30° Structure

    For the √3 structure two different models have recently been
proposed, a trimer model[39] and a honeycomb model[40] see Fig. 12. Both
models are based on low energy ion scattering measurements. The trimer

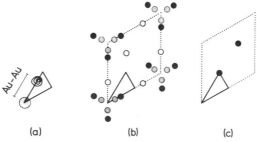

                  (a)   (b)   (c)

Fig. 12 (a) A contour map of the Patterson function for the
         Au/Si(111)√3 x √3R30° structure. The origin rises 10 contour
         levels.   (b) The trimer model derived from the x-ray diffraction
         data.   Full circles are Au-atoms with full occupancy, shaded
         circles are partially occupied Au-atoms and open circles Si
         atoms.   (c) The honeycomb model proposed by Huang and Williams.

model has a saturation coverage of 1.0 ML, the honeycomb model 0.67ML. We have measured 49 fractional-order, in-plane integrated intensities from which we deduced 16 inequivalent structure factor intensities. The spatial coherence length was around 450 Å. A contour map of the Patterson function is shown in Fig. 12a. There is only one non-origin peak in the irreducible unit. This must then be a Au-Au interatomic vector. The distance corresponds very well to a bulk Au-Au bondlength as indicated. The vector can clearly not be reproduced by the honeycomb model, which hence must be excluded. However, the vector is contained in the trimer model. A least-square analysis with only the trimer atoms gives $X^2 = 27$. The scattering from the relaxed Si layers is important for reflections where the scattering from the Au-trimer is weak. We can improve the fit by including a relaxed Si-layer plus partial occupancy of Au in the inner of the trimers, see Fig. 12. This gives $X^2 = 5.0$. The atomic coordinates are given in Table II. Au-atoms with partial occupancy could arise from the nucleation of the 6x6 structure, which may happen before the $\sqrt{3}$ structure is completed. The atoms with partial occupancy could also be Si atoms with full occupancy since $Z_{Si}/Z_{Au} = 14/79 = 0.18$, close to the measured 0.21 ± 0.04 Au-occupancy, but the Si-Si bondlength clearly becomes too short since the Si atoms for symmetry reasons must lie in the same plane.

Table II

Atomic Coordinates and Debye-Waller Factors Within the $\sqrt{3}$
Cell. $X^2 = 5.0$

Atom	x	y	occ	B
Au	0.497	0.249	1.00	3.5
Au	0.243	0.121	0.21	3.5
Si	1.289	0.645	1.00	0.5

We cannot draw any conclusions about the out-of-plane structure with only in-plane data, but can suggest two possibilities. The Au-trimer could be sitting on top of a truncated Si-bulk with a slightly relaxed first layer. Another possibility would be that the outermost layer consists of the Au-trimer layer plus half a bilayer of Si, the latter slightly relaxed from ideal 1x1 symmetry. Rodscans i.e. out-of-plane diffraction data are needed to distinguish between the two cases.

c.   The 6x6 Structure

For the 6x6 structure we have measured 139 in-plane, fractional-order structure factor intensities. A contour map of the Patterson function is shown in Fig. 13. Two immediate conclusions can be drawn. (1) The peak marked 1 is the same as the peak in the map for the $\sqrt{3}$ structure. The peak marked 2 is the vector between two $\sqrt{3}$ unit cells. Therefore, the 6x6 structure must at least partly be similar to the $\sqrt{3}$ structure. This has also been proposed by scanning tunneling microscopy (STM).[43] (2) The distance between the $\sqrt{3}$ subunits is slightly longer in the 6x6 structure than in the $\sqrt{3}$ structure. This is seen by the peak marked 2 which is shifted from the expected position. We have not yet

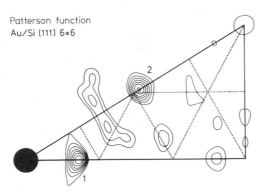

Patterson function
Au/Si (111) 6×6

Fig. 13 Patterson map for Au/Si(111)6x6. The origin rises 10 contour
levels.

been able to deduce the detailed atomic geometry of the 6x6 structure,
but can conclude that the Au-trimer from the √3 structure is an important
part of it.

CONCLUSION

In conclusion, we have studied the structure of two quite different
metals on semiconductor systems.

1) Pb on Si(111) and Ge(111). This is a system where the structural
    analysis is straightforward. On both surfaces, Pb has a low (α) and
    a high coverage (β) phase. The α phase is the same for Si(111) and
    Ge(111) with one Pb atom per unit cell. The β-phase is in both cases
    a (111) overlayer rotated 30° with respect to the underlying
    substrate. For Ge(111), where the lattice mismatch is 1.0%, the Pb
    overlayer is commensurate, and slightly distorted to take up the
    strain. At higher temperatures the Pb overlayer melts and is weakly
    modulated by the substrate. For Si(111) the lattice mismatch is
    5.3% which is too large to be accommodated by a distortion. Instead,
    the Pb overlayer becomes incommensurate and is modulated by the
    substrate as seen as satellites close to the main peaks. If the
    β-phase is created by deposition on Si(111) at RT, the (111)
    overlayer adapts the 7x7 periodicity of the underlying Si-surface.

(2) Au on Si(111). This system is very different from Pb on Si(111).
    Both the 5x1 and √3 structures contains disorder or atoms with
    partial occupancy. The 5x1 structure is a random mixture of rows of
    5x2 and c(10x2) unit cells. A structural model for the projected 5x1
    subunit was derived. The ideal √3 structure would consist of an
    Au-trimer and a slightly relaxed Si layer. However, apparently the
    6x6 structure starts before the √3 structure is fully saturated. The
    6x6 structure is locally very similar to the √3 structure, as seen
    in the contour maps of the Patterson function.

We think that the two above mentioned studies demonstrate the
potential of x-ray diffraction for studying the ordering and structure of
overlayers on surfaces, even in quite complex cases.

Acknowledgement

   We acknowledge useful discussions with Jan Skov Petersen, who also
participated in the Pb/Ge(111) experiments, and with Jakob Bohr, who
participated in some of the Au/Si(111) experiments. We also want to thank
Detlev Degenhardt for participating in some of the Au/Si(111)
measurements, and to thank the staff at HASYLAB for their help. This work
was supported by the German Federal Ministry for Science and Technology
under contract number 05390CAB, the Max Planck Society and the Danish
National Research Council.

REFERENCES

1. G. Le Lay, Surf. Sci. 132, 169 (1983).
2. M. Jalochowski and E. Bauer, J. Appl. Phys. 63, 4501 (1988).
3. R. Feidenhans'l, Surf. Sci. Rep. 10, 105 (1988).
4. I. K. Robinson, in: Handbook on Synchrotron Radiation, Vol. 3, D. E.
    Moncton and G. S. Brown eds. (North-Holland, Amsterdam, to be
    published).
5. P. R. Bevington, Data Reduction and Error Analysis for the Physical
    Sciences (McGraw-Hill, New York) 1968.
6. T. Ichikawa, Solid State Comm. 46, 827 (1983).
7. T. Ichikawa, Solid State Comm. 49, 59 (1984).
8. J. J. Metois and G. Le Lay, Surf. Sci. 133, 422 (1983).
9. G. Le Lay and J. J. Metois, Appl. Surf. Sci. 17, 131 (1983).
10. G. Le Lay and J. J. Metois, J. Phys. Colloq. C5, 427 (1984).
11. R. Feidenhans'l, J. S. Pedersen, M. Nielsen, F. Grey and R. L.
    Johnson, Surf. Sci. 178, 927 (1986).
12. J. S. Pedersen, R. Feidenhans'l, M. Nielsen, K. Kjr, F. Grey and R.
    L. Johnson, Surf. Sci. 189/190, 1047 (1987).
13. J. S. Pedersen, Surf. Sci. 210, 238 (1989).
14. H. Huang, C. M. Wei, H. Li, B. P. Tonner and S. Y. Tong, Phys. Rev.
    Lett 62, 559 (1989).
15. J. N. Carter, V. M. Dwyer and B. W. Holland, Sol. St. Comm. 67, 643
    (1988).
16. B. P. Tonner, H. Li, M. J. Robrecht, M. Onellion and J. L. Erskine,
    Phys. Rev. B36, 989 (1987).
17. F. Grey. Ph.D. Thesis, Copenhagen University (1988). Available as
    Report no. Risø-M-2737 on request to the Risø Library.
18. B. N. Dev, F. Grey, R. L. Johnson and G. Materlik, Europhys. Lett. 6,
    311 (1988).
19. F. Grey, R. Feidenhans'l, J. S. Pedersen, M. Nielsen and R. L.
    Johnson, to be published.
20. A. Guinier, X-ray Diffraction, W. H. Freeman and Company, (San
    Francisco and London), 1963.
21. P. H. Fuoss, L. J. Norton and S. Brennan, Phys. Rev. Lett. 60, 2046
    (1988).
22. P. J. Estrup and J. Morrison, Surf. Sci. 2, 465 (1964).
23. M. Saitoh, K. Oura, K. Asono, F. Shoji and T. Hanawa, Surf. Sci. 154,
    394 (1985).
24. G. Le Lay, J. Peretti, M. Hanbucken and W. S. Yang, Surf. Sci. 204,
    57 (1988).
25. Note that this is reversed to the labelling by Saitoh et al.
26. F. Grey, R. Feidenhans'l. M. Nielsen and R. L. Johnson, Journal de
    Phys. Coll. C7, 181 (1989) and to be published.
27. K. Takayanagi, Y. Tanishiro, S. Takakashi and M. Takakashi, Surf.
    Sci. 164, 367 (1985).
28. J. J. Lander, Surf. Sci. 1, 125 (1964).
29. M. Nielsen, J. Als-Nielsen and J. P. McTague, in: Ordering in Two
    Dimensions, S. K. Sinha ed., North Holland (1980).

30. R. Feidenhans'l, F. Grey, J. Bohr, M. Nielsen and R. L. Johnson. Journal de Physique Coll. C7, 175 (1989).
31. K. Higashiyama, S. Kono and T. Sagawa, Jpn. J. Appl. Phys. 25, L117 (1986).
32. Y. Yabunchi, F. Shoji, K. Oura and T. Hanawa, Surf. Sci. 131, L412 (1983).
33. H. E. Bishop and J. C. Riviere, Brit. J. Appl. Phys. (J. Phys. D) 2, 1635 (1969).
34. H. Lipson and K. E. Singer, J. Phys. C7, 12 (1974).
35. J. H. Huang and R. S. Williams, Surf. Sci. 204, 445 (1988).
36. L. E. Berman, B. W. Battermann and J. M. Blakely, Phys. Rev. B38, 5397 (1988).
37. B. A. Nesterenko and A. D. Borodkin, Sov. Phys. Sol. Sta. 12, 1621 (1971).
38. M. Ichikawa, T. Doi and K. Hayakawa, Surf. Sci. 159, 133 (1985).
39. K. Oura, M. Katayama, F. Shoji and T. Hanawa, Phys. Rev. Lett. 55, 1486 (1985).
40. J. H. Huang and R. S. Williams, Phys. Rev. B38, 4022 (1988).
41. I. K. Robinson, Phys. Rev. B33, 3830 (1986).
42. S. R. Andrews and R. A. Cowley, J. Phys. C18, 6247 (1985).
43. F. Salvan, H. Fuchs, A. Baratoff and G. Binnig, Surf. Sci. 162, 634 (1985).

DEVELOPMENT OF STRUCTURAL AND ELECTRONIC PROPERTIES IN THE

GROWTH OF METALS ON SEMICONDUCTORS

G. Le Lay* and M. Abraham*

CRMC2-CNRS, Campus de Luminy, Case 913
13288 Marseille Cedex 9, France
*Also UFR Sciences de la Matière
Université de Provence,
Marseille, France

K. Hricovini and J.E. Bonnet
L.U.R.E., Bât. 209 D - 91405 Orsay Cedex, France

ABSTRACT

We review recent work on the determination of the geometrical structures of several two-dimensional phases appearing in the early stages of the formation at and beyond room-temperature, of prototypical metal-semiconductor interfaces, namely Ag/Si(111), Pb/Ge(111) and Pb/Si(111). In parallel we present new data obtained by synchrotron radiation photoemission experiments on the same systems. Comparison of the structural results with the development of the electronic properties sheds light on numerous crucial points like the nature of the reversible 2D phase transitions that occur at 250-300°C for Pb on Ge(111) and Si(111).

INTRODUCTION

In this work, we contribute to the ongoing discussion on the structural and electronic properties of "prototypical" metal-semiconductor (Me/Sc) interfaces during the first stages of their formation, that is for submono- and monolayer metal films. We define as "prototypical" Me/Sc couples which are widely recognized as non-reactive and thus form abrupt junctions without intermixing. As a matter of fact, very few couples obey this definition (most Me/Sc systems present at least a certain degree of reactivity), but their study is of basic interest because such "simple" systems may provide a good medium in which to investigate coverage-dependent structural arrangements, bonding characteristics, and growth modes along with the development of the electronic properties. Of special interest is the identification of the mechanisms that determine the Schottky barrier (SB) height, to serve as a guideline for parallel or future research on other more "realistic" and often more complex Me/Sc couples.

The SB problem has been treated by theorists since the 1930's and since the 1970's Synchrotron-Radiation (SR) studies of Me/Sc interfaces, that is the initial formation of SBs, have been one of the most active

Kinetics of Ordering and Growth at Surfaces
Edited by M. G. Lagally
Plenum Press, New York, 1990

areas of solid-state science. Despite the large advances made in this field, an unified description of the many phenomena observed for different kinds of Sc interfaces is not yet available. One of the probable reasons is that for most, if not all, Me/Sc systems, a detailed picture of the adsorption and growth behaviour is still lacking and, as a consequence, a clear understanding of the mechanisms responsible for the building-up of the barrier.

Typically, silver overlayers on silicon (111) surfaces have attracted a great deal of attention especially because of the mysterious Ag-induced Si (111)-($\sqrt{3}$x$\sqrt{3}$)R(30°) reconstruction (in short AgR3).

Recently, lead monolayers on silicon or germanium have also been extensively studied, since both Pb/Si (111) and Pb/Ge (111) display a perfectly reversible two-dimensional (2D) phase transition, unique, to our knowledge, in the strong chemisorption regime of Me/Sc systems and attributed, in case of Ge, to a 2D melting transition.[1]

Indeed, a comprehensive review of those "prototypical" Me/Sc couples is beyond the scope of this presentation. We will thus summarize a few recent results, mostly derived from SR studies (valence band and core-level studies, surface X-ray diffraction (XRD) experiments), in conjunction with other surface sensitive techniques like scanning tunneling microscopy (STM) and X-ray photoelectron diffraction (XPD) with the aim of helping solve some open questions.

Ag ON Si (111)

1. The AgR3 Structure

All conceivable kinds of investigation techniques have been used to characterize this "model" interface. Despite all these efforts, the geometry of the AgR3 reconstruction obtained after annealing and observed as a beautiful honeycomb structure in real space by STM[2,3] (Fig. 1) is still controversial. The key point is the nominal saturation coverage of this structure. Is it one monolayer (ML) (in substrate units, i.e., 3 Ag atoms per $\sqrt{3}$ mesh) or two-thirds of a ML? This nominal coverage should be in principle directly obtained from quartz crystal monitor measurements

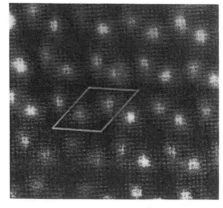

Fig. 1   STM image of the Si(111) $\sqrt{3}$x$\sqrt{3}$ R(30°) Ag structure (from Ref. 3).

of the impinging silver flux in conjunction with coverage determination on the silicon substrate surface using techniques such as Auger electron spectroscopy (AES) or X-ray photoelectron spectrocopy (XPS). However, until now, no consensus has been reached, although most recent works find the onset of three-dimensional overgrowth of silver nuclei (break point followed by a plateau in intensity versus coverage plots according to the well-established Stranski-Krastanov (SK) growth mode) at 1 ML. The reasons for the disagreement on the atomic arrangement of the Ag and Si atoms, even between different groups using the same analytical methods: [STM (IBM team at San José[2] and IBM team at Yorktown Heights),[3] XRD (FOM group at Daresbury[4] and Japanese team at the Photon Factory,[5] XPD (Japanese team[6] and Hawaiian team)[7]], despite the similarities of the collected data are, in fact, two-fold. The nominal coverage θ_S may be an input parameter in the analysis of the data (e.g. θ_S = 2/3 → Ag honeycomb[2], θ_S = 1 → Si honeycomb[3], in the interpretation of identical STM images) and/or may be indirectly determined in standard type analysis of the scattering data using χ^2 values for XRD or LEED type R-factors for XPD. Indeed, in the last cases there is no proof that the best found structure is actually the true structure.

The only method to obtain a high degree of certainty of a proposed model (as for example the DAS model[8] for the clean Si/(111)7x7 surface) is the support or at least the absence of severe rebuttal from a large variety of different experimental investigations and theoretical considerations.

2.   The Honeycomb Model

In the context of the AgR3 structure, most new results (including ion scattering spectroscopy)[9] support the two-domain missing-top-layer Ag honeycomb model of Bullock et al.[7] (Fig. 2) (in short honeycomb-MTL, 2 domain). Indeed, this opinion is at variance with the conclusions drawn by Fan et al.[10] discussing LEED results for both ($\sqrt{3}\times\sqrt{3}$)R(30°) Ag on Si and a ($\sqrt{3}\times\sqrt{3}$)R(30°) structure observed on a clean Si (111) surface; from the similarity of the LEED I(V) curves, they claim that the AgR3 one is in fact a true silicon structure, the silver atoms being randomly dispersed below the surface forming a diffuse interface.

Nevertheless, I think, as Bullock et al.[7] do, that the "vacancy model" they propose for this structure provides strong support for the honeycomb-MTL, 2 domain model, as their proposed structure is precisely domain 1 in Fig. 2 if the Ag atoms are replaced by a honeycomb of Si adatoms on the second Si layer. This has further consequences since the assertion by LEED specialists that "phases which are structurally identical but contain different scatterers produce different LEED spectra"[11] might well be balanced by the counter assertion that similar structures should yield nearly identical I(V) curves.

However, two major objections against a honeycomb model may be put forward. First, as mentioned above, the measured completion of the 2D Ag-adlayer is rather at 1 ML (or often, depending on preparation conditions between 2/3 and 1 ML[12] which is thus consistent with either a 2/3 or a 1 ML coverage and does not help much in clarifying the question). Second, honeycomb models with this coverage (2/3 ML) contain an odd number of valence electrons per $\sqrt{3}$ unit cell and demand a metallic surface as underlined by Van Loenen et al.,[3] a point further confirmed in a recent theoretical calculation of the surface density of states by Chan and Ho.[13] On the contrary, the AgR3 surface is found to be semiconducting both by photoemission[14] and scanning tunneling spectroscopy[3] experiments. To solve these dilemmas, two explanations have been proposed by Kono et al.[15,16] On the one hand, a 1/3 ML dilute Ag adatom layer might ride on

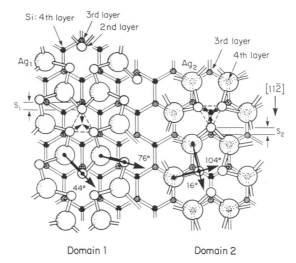

Fig. 2 Two-domain missing-top-layer Ag honeycomb model of the Si(111) $\sqrt{3}x\sqrt{3}$ R(30°) Ag structure (from Ref. 7). The structure is characterized by the vertical heights z_1 and z_2 of the Ag atoms relative to the top Si layer (the second layer of the full Si(111) surface) and by the contraction parameters S_1 and S_2 of the Si trimers in the top layer. The long arrows in the bottom half of the figure point to the two sets of nearest-neighbor-Si forward scattering peaks that produce the 4-peak structure seen at low polar angle values in azimuthal scans of Ag $3d_{5/2}$ intensity (see Fig. 2 of Ref. 7).

the 2/3 ML honeycomb AgR3 layer being so loosely bound to it that these adatoms move from the tip of the STM. On the other hand, observing a dramatic shift of the bulk Si 2p component for the AgR3 surface in SR core-level spectroscopy experiments, these authors attribute this shift to an inherent negatively charged surface (by one electron per $\sqrt{3}$ unit cell to make the number of electrons even) making it semiconducting.

The weak points with these explanations are that i) no direct evidence of any weakly bound adatom layer has been found up to now and ii) that the dramatic shift of the bulk Si 2p component observed by Kono et al.[16] has been severely questioned by others: the mode of preparation of the Si n-type sample to produce a clean 7x7 surface (direct-current heating at $\sim$ 1150°C) used by these authors is likely to give rise to a p-type overdoped selvedge.[17]

3. Band Bending Changes and Schottky Barrier Height

We have checked this crucial point with new SR-soft X-ray photoemission experiments at Hasylab.[18] Our Si (111) samples were cleaned in situ by mild sequences of Ar-ion sputtering and annealings. The results displayed in Fig. 3 point only to a slight shift ($\sim$ 0.18 eV) towards lower binding energy upon room temperature adsorption of 2 Å of Ag on Si (111) 7x7 and no further change after further annealing to obtain the AgR3 structure. Thus our work corroborates an early measurement by F. Houzay[19] and also the more recent ones obtained at RT by Wachs et al.[20] Only small, if any, band bending changes occur upon Ag adsorption. As the Fermi level for the 7x7 surface is at 0.65 eV above the valence band maximum (VBM), this small shift leads to a n-type

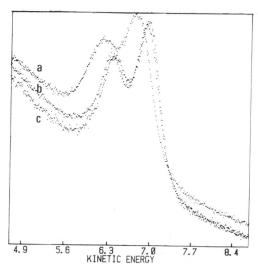

Fig. 3   Si 2p core levels recorded at hν = 110 eV (bulk sensitive mode)
for a) clean Si(111) 7x7, b) 2Å Ag deposited at RT onto Si(111)
7x7, c) Si(111) √3x√3 R(30°) Ag.

Schottky barrier height of 0.65 eV in rather poor agreement with the
experimental value 0.79 eV measured on diodes.[21]

4.   Si 2p Line Shape Analysis

Interestingly, as deduced from the decomposition of the Si 2p core
lines, both for the RT Ag adlayer on top of Si (111)7x7,[20] Fig. 4, and
the AgR3 structure,[10] Fig. 5, a single component of surface Si 2p levels
(Si component shifted by 0.36 eV (RT adsorption) or 0.24 eV (AgR3) to
higher binding energies relative to the bulk line) persists: in the
submonolayer range, Ag causes removal of the second component
(S_2 component shifted to lower binding energies by ~0.72 eV,[16] 0.76 eV[20]
relative to the bulk line) seen on the clean Si (111)7x7 spectra. We
believe this implies a single type of Si atoms at the AgR3 surface thus
ruling out Si honeycomb models on top of Ag trimers,[3,4] as on the RT Ag
adlayer. It is also worth mentioning, as underlined by Kono et al.,[16]
that the 0.26 ML spectra shown in Fig. 5 are fitted literally by simple
superposition of the spectra for the 7x7 and the AgR3 surfaces. This
means that upon Ag deposition onto the heated substrate the AgR3
structure grows in the form of 2D islands that develop on the surface by
Ag migration over the surface forming domains of AgR3 layer leaving the
rest in a Si 7x7 surface. In earlier work,[22] we studied the energetics of
the Ag/Si (111) system by isothermal desorption spectroscopy, and reached
the same conclusion from a model interpretation of the zero-order
kinetics.   As far as the RT Ag adlayer is concerned, it is quite
remarkable that it also develops in flat triangular 2D islands (now in a
close-packed Ag (111) structure) located preferentially at low coverages,
on the faulted halves of the 7x7 unit cells, as shown by the beautiful
STM work of Tosh and Neddermeyer[23] (see Fig. 6). This confirms that the
Si lattice below the adatom layer is not severly disrupted by Ag
adsorption.

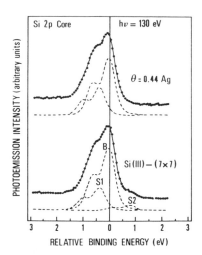

Fig. 4   Surface sensitive ($h\nu$ = 130 eV) Si 2p core level spectra from clean and RT Ag-covered Si(111) surfaces.   The solid curves are fit to the data with bulk (B) and surface-derived (S_1,S_2) components.   Binding energies are referred to the bulk component of the clean Si $2p_{3/2}$ line (after Ref. 20).

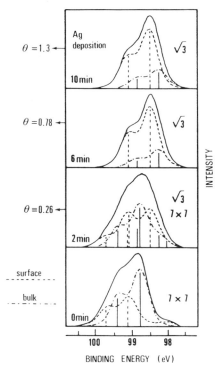

Fig. 5   Surface sensitive ($h\nu$ = 130 eV) Si 2p core level spectra of the growing Si(111) $\sqrt{3}x\sqrt{3}$ R(30°) Ag structure (from Ref. 16).

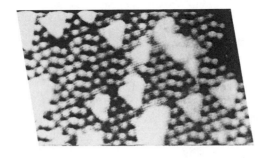

Fig. 6   STM image of 1/3 ML Ag adsorbed at RT on Si(111) 7x7 (from Ref. 23).

Pb ON Ge (111)

1.   2D Surface Phases

Upon condensation at RT, Pb which grows on Ge (111) according to the layer-plus-islands mode (SK growth mode), displays two different surface structures (α and β) with the same R3 unit mesh but markedly different intensities in their LEED spots. The α phase (Fig. 7) is completed at 1/3 ML and the Pb atoms occupy the so-called T_4 sites (three-fold eclipsed sites) as established by surface XRD experiments,[25] which are the preferential (lowest energy) chemisorption sites according to pseudo-potential total energy minimization calculations when substrate relaxations are taken into account[24]. The completion of the α phase is taken to be 4/3 ML[25]* and is essentially a 1% compressed, Pb (111) close-packed layer rotated by 30° with respect to the underlying Ge (111) plane**. Basically, the Pb atoms occupy two different and new sites: 1/3 ML occupies the three-fold hollow (H_3) sites and 1 ML is between the top (T_1) and three-fold eclipsed (T_4) site. Let us mention that the rotated Pb (111) layer was proposed some years before by Ichikawa[27] who noticed, by RHEED, the occurrence of a perfectly reversible 2D phase transition $PbR3_\beta \rightleftharpoons Pb$ 1x1. Based on the azimuthal isotropy of the RHEED beams above the transition temperature, Ichikawa argued for a 2D melting transition. This point was a matter of controversy[28] but recent surface XRD experiments[29] have shown the existence of an isotropic halo of scattering in the 1x1 phase in agreement with the earlier RHEED observations. Above the transition, the Pb monolayer becomes a 2D liquid, weakly modulated by the substrate.

Intuitively, one would expect a signature of those structural changes in the electronic properties of the system. We have performed a detailed in situ spectroscopic ellipsometry study which, as expected, reveals distinct characteristics of each phase (α, β, 1x1),[30] as well as

*Even though others found rather 1 ML: here we meet the same problem that we discussed above for the AgR3 phase.
**However an intriguing fact, discussed recently by Huang et al.,[26] is that after completion of the 2D β phase three-dimensional Pb islands grow in parallel (unrotated) orientation according to the following epitaxial relationship: (111) Pb // (111) Ge[27] - <110> Pb // <110> Ge.

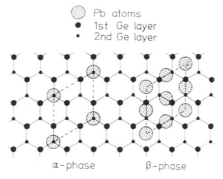

α-phase   β-phase

Fig. 7   Geometrical models of the α and β phases of Pb on Ge(111) as
          derived from surface XRD (from Ref. 25).

SR valence band and core-level photoemission experiments[31] which are
summarized in the next paragraph.

2.   New SR Photoemission Results

      As shown in Fig. 8, a very strong surface state at 0.8 eV below the
Fermi level is seen in the angle-resolved VB photoemission spectra of

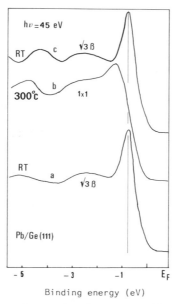

Binding energy (eV)

Fig. 8   Normal-emission valence band spectra (hν = 45 eV) of Pb/Ge(111):
          (a) β phase at RT; (b) 1x1 phase at 300°C (same sample as in
          (a)); (c) β phase again after cooling back to RT (same sample as
          in (a) and (b)).

216

the β phase at RT; it is completely washed out at 300°C, past the phase transition β → 1x1 at 250°C and fully restored upon cooling from the 1x1 to the β phase again. In addition, no change is noticed in the shape of the substrate Ge 3d core levels, measured in a very surface sensitive mode with hν = 60 eV, while a ∿ 5% increase in the FWHM of the overlayer lead 5d core level peaks is revealed upon going from the β to the 1x1 phase.

3.   Interpretation

All these findings support the description given above: i) the disappearance of the surface state in the VB spectra testifies to a disordering of the Pb adlayer, ii) the increase in full width at half maximum of the Pb 5d core levels is expected in the case of a 2D melting transition because in the molten adlayer the Pb atoms will have slightly different local arrangements with respect to the ordered β phase, iii) the absence of change in the FWHM of the substrate Ge 3d core lines surely reflects the compactness of the Pb adlayer.

Pb ON Si (111) AND Pb ON Si (100)

In this section, we concentrate on a few critical points of these two adsorbate systems. A complete discussion can be found in Refs. 9 and 32.

1.   Pb/Si (111): Surface Structures

First note the strong similarity between the Pb/Si (111) and Pb/Ge (111) surfaces, i.e., same unreactivity and SK growth mechanism, similar R3 LEED patterns, same parallel orientation of the 3D crystallites beyond the completion of the 2D adlayer, similar reversible 2D phase transition. This similarity is deceptive, however, as we discuss below.

Here we focus on the last 2D phases that coexist with the 3D crystallites and discuss the most recent results. First, upon RT condensation, a metastable phase results which has been shown to be a commensurate Si (111) 7x7-Pb structure by surface XRD.[33] This phase is essentially a close-packed Pb layer (4% compressed relative to a bulk Pb (111) layer) in parallel orientation relative to the Si (111) substrate with a 8x8 mesh of Pb atoms in the 7x7 unit cell. The silicon 7x7 structure underneath is not severely altered as compared to the clean reconstruction given by the DAS model[8]: retaining the stacking faulted Si bilayer and Si dimers (but no Si adatoms) led to a slight improvement in the least-squares structure factor refinement.

Upon heating at ∿ 200°C, or deposition at high temperature, this metastable 7x7 phase transforms irreversibly into a stable phase which displays a R3 structure in LEED[34] or RHEED[35] but according to the more precise surface XRD measurements,[33] is an incommensurate structure denoted Pb/Si(111)R30°i. This incommensurate phase is also a 2D Pb close-packed structure but with its [1,0] direction rotated by 30° relative to Si [1,0]. Indeed, this last structure resembles the β phase of Pb/Ge (111) discussed in the previous section. Nevertheless if on Ge (111) the dense 2D layer can adopt a commensurate R3 structure because of little misfit (1% compression relative to a bulk Pb (111) layer), this is no longer the case on Si (111): because of the smaller lattice parameter of silicon, the misfit is larger and a commensurate R3 structure would require à 5% compression which appears too much; instead the Pb adlayer adopts an incommensurate structure with a smaller misfit (2.3 % compression relative to a bulk Pb (111) plane).

This incommensurate phase, just as the β phase, also undergoes a perfectly reversible 2D phase transition to a 1x1 structure: Pb/Si (111) R 30°i ⇌ 1x1 which again bears strong similarity with the Pb/Ge (111) β ⇌ 1x1 one and thus could be thought to be also a 2D melting transition.

2.   Adsorption/Desorption Kinetics Experiments

It should be mentioned that this overall picture is at variance with two experiments performed separately in Marseille: Quentel et al.[36] used in-situ ellipsometry in conjunction with quartz microbalance coverage measurements, while Le Lay et al.[37] used, in a different vacuum vessel, simultaneously Auger electron spectroscopy (AES) and isothermal desorption spectroscopy (ITDS), as well as in-situ LEED and ex situ scanning electron microscopy (SEM) observations together with quartz microbalance coverage determinations. Both groups found that the 2D Pb adlayer is completed at 1 ML (instead of $\sim$ 4/3 of a monolayer: 8x8 = 64/7x7 = 49 $\sim$ 1.31 ML at RT), which once more raises questions about the discrepancy found between some direct measurements of the saturation coverage of a given 2D phase upon adsorption kinetics experiments which rely on quartz crystal impinging flux calibration and the indirect determination from least-square refinements in structure factors or R-factors analysis in standard X-ray or LEED intensity experiments. But what is much more striking, notwithstanding the absolute coverage determination, with totally different experiments: in situ ellipsometry, an optical method; LEED, an electron scattering method and ITDS, a kinetic method, the two groups found that the Pb adlayer forms in three steps. As a matter of fact, three different surface phases are successively identified by their different optical response after equal evaporation times in the ellipsometric data[36] (see Fig. 9), while three

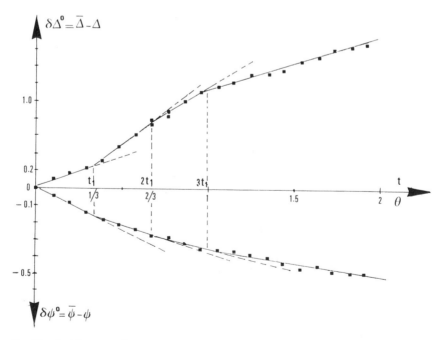

Fig. 9   Variations of the ellipsometric signals $\delta\Delta$ and $\delta\Psi$ upon lead deposition onto Si(111) 7x7 (from Ref. 36).

different superstructures with the same R3 unit cell but markedly different intensities in their LEED patterns are successively observed. Moreover the isothermal desorption process occurs also in three kinetic regimes.[37] Indeed, three successive phases respectively completed at 1/3, 2/3 and 1 ML as proposed by the Marseille groups straightforwardly explain these results. How to reconcile this with the determination of a ∿ 4/3 ML completion of the adlayer, as described before, remains an open question.

3. Photoelectron Spectroscopy Results

The study of the evolution of the electronic properties of the interface in the course of its formation may shed some light on some obscure points and help in clarifying the situation. In the following, we summarize some interesting results for the saturated 2D adlayers derived from photoelectron spectroscopy experiments.

i) The Pb 5d core level spectra show practically no change for the incommensurate phase and the 1x1 phase at high temperature beyond the temperature of the phase transition and are slightly narrower than for RT condensation.

ii) The Si 2p core levels (recorded in a surface sensitive mode at $h\nu$=130eV) of the Pb/Si (111) 7x7 phase at RT and of the Pb/Si(111)R30°i phase after annealing have nearly identical shapes as shown in Fig. 10, but for a slight decrease in FWHM after annealing. No change is noticeable upon passing the temperature of the transition for the 1x1 phase. However, these lines are noticeably narrower than the clean Si (111) 7x7 spectrum.

iii) The Si 2p core levels, taken in a bulk sensitive mode at $h\nu$ = 108eV, are shifted by ∿ 0.4 eV towards higher kinetic energies (KE) both on n

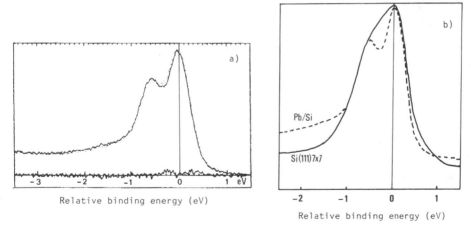

Fig. 10 Surface sensitive Si 2p core level ($h\nu$ = 130 eV) spectra of (a) the metastable Pb/Si(111) 7x7 phase and of the Pb/Si(111)R30°i phase and of their difference; (b) the Pb/Si(111)R30°i phase and the clean Si(111) 7x7 surface.

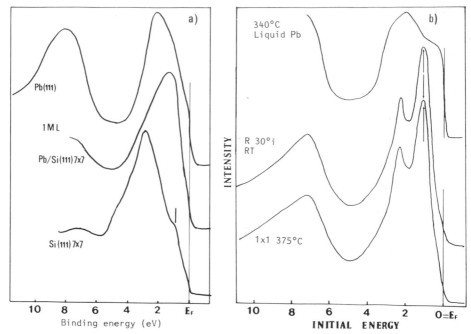

Fig. 11 (a) Valence band spectra (hν = 45 eV) of the full monolayer of
Pb adsorbed and measured at RT on Si(111): Pb/Si(111) 7x7
structure. For comparison the spectra of the clean Si(111) 7x7
surface with markers at the intrinsic surface states and of bulk
solid Pb(111) are also displayed. (b) Valence band spectra
(hν = 48 eV) of the full monolayer of Pb adsorbed on Si(111)
measured at RT (Pb/Si(111)R(30°)i structure) and at 375°C (1x1
structure). The spectrum of liquid Pb is shown for comparison.

and p type samples. After annealing, an additional ∿ 0.1 eV shift also
towards higher KE is noticed.

iv) At any temperature, the intrinsic surface states of the initial clean
7x7 surface are rapidly washed out in the submonolayer range. At
completion of the adlayer upon RT condensation , i.e., the Pb/Si (111)
7x7 phase, the spectrum presents a broad peak (maximum at 1.2 eV below
the Fermi level E_F). After annealing, i.e., the incommensurate phase, a
new extrinsic and very strong surface state (non dispersing with $k_\perp$) at ∿
0.9 eV below E_F is prominent. It persists, practically unchanged at high
temperature, i.e. the 1x1 phase, as shown in Fig. 11.

4.   Interpretation

This last result, the persistence of the sharp surface state at HT
together with the absence of broadening of the Pb 5d lines expected in
case of a melting transition because of the different local environment
of the Pb atoms with respect to the substrate, is taken as a clear
indication that the Pb/Si (111) [R30°i ⇌ 1x1] phase transition, at
variance with the Pb/Ge (111) [β √3 ⇌ 1x1] one, is not a melting
transition ; the nature of this transition remains a question for future
research. However, we suggest that the disappearance of the extra spots
in the LEED "√3x√3" pattern, corresponding to the R30°i phase, might be
related to a varying degree of incommensurability due to the different
expansion coefficients of the Pb adlayer and of the Si (111) substrate.

The slight narrowing of the Pb 5d and Si 2p core levels after annealing, passing from the metastable Pb/Si (111) 7x7 phase to the incommensurate one, probably reflects different local environments in each structure and a relaxation of the strain in the dense Pb adlayer after the 30° rotation. Much more striking is the development of the very sharp extrinsic surface state at $\sim$ 0.9 eV below E_F in the VB spectra. This prominent feature is to be related to the similar strong surface state at 0.8 eV below E_F mentioned above for the Pb/Ge (111) β phase (which is, we recall, a commensurate 30° rotated Pb dense layer on Ge (111) corresponding to a 1 % compression of the Pb layer relative to a bulk Pb (111) layer). We attribute this Pb induced surface state on Si (111) to a change of the underlying silicon from the 7x7 structure to a more bulk like termination as the close-packed Pb adlayer rotates by 30°. This is supported by the associated slight decrease in FWHM of the Si 2p lines. This point, and others, will be discussed in a forthcoming paper where we will present a detailed line shape deconvolution analysis of the core lines for every surface phase of both Pb/Ge (111) and Pb/Si (111) systems.[40]

5.   Schottky Barrier Heights

The last point we would like to comment on is the large shift towards higher KE of the Si 2p core lines completed at saturation of the Pb adlayer (by $\sim$0.4 eV at RT or 0.5 eV after annealing) as compared to clean Si (111) 7x7 which is the same for both p-type and n-type substrates. These shifts are associated to band bending changes which are interesting to compare with the corresponding ones in case of Ag/Si (111). It is well established that for the 7x7 clean surface the Fermi level is pinned at 0.65 eV above the valence band maximum.[41] For a p-type sample, the same 0.4 eV shift leads to a Schottky barrier (SB)$\phi_{B,p}=(E_F-E_V)_s$ $\sim$0.25 eV from which one infers a n-type SB of $\phi_{B,n}=E_g-\phi_{B,p}$ =0.87 eV, in agreement with our independent measurements on n-type sample yielding Pb $\phi_{B,n}$7x7=$(E_g-E_F)_s$=1.12-(0.65-0.4)=0.87 eV. This $\sim$0.9 eV n-type SB for the RT Pb/Si (111) 7x7 phase is much larger than the one we determined for the corresponding RT Ag/Si(111)7x7: Ag $\phi_{B,n}$7x7 $\sim$ 0.65 (see earlier), despite the fact that upon RT adsorption for both metals the underlying silicon substrate which retains the 7x7 reconstruction under the Ag and Pb adlayers, appears only weakly perturbed. Besides it is not simply related to the difference in work function of the two metals (3.8 eV for Pb (111),[42] 4.48 eV for Ag(111)[43]): in the framework of the induced density of interface states model of Flores et al.,[44] the final position of the Fermi level should be at the vicinity of the charge neutrality level ϕ_0 according to the relation

$$E_F - \phi_0 \simeq S\ (x + E_g - \phi_0 - \phi_M)\ ,\tag{1}$$

where x and E_g are the electron affinity and the gap of the semiconductor, ϕ_M the work function of the metal and S a parameter which has been recently correlated by Mönch[45] with the electronic dielectric constant of the semiconductor and which is $\sim$ 0.1 for Si. The charge neutrality level of several semiconductors was calculated recently by Tersoff;[46] it is at 0.36 eV above the VBM for Si. With $x + E_g$ $\sim$ 5.33 eV[43] and the other values indicated, one calculates Ag $\phi_{B,n}$7x7 $\sim$ 0.69 eV and Pb $\phi_{B,n}$7x7 $\sim$ 0.64 eV. For silver, the agreement with our measured value is remarkable; on the other hand, the discrepancy for lead is important Moreover, in both cases, the comparison with the barrier heights deduced from electrical measurements on thick diodes: 0.79 eV for Ag,[21] 0.7 eV for Pb,[47] is poor!

Finally, we have carried out a comparative study of the Pb/Si(100)[32,48,49] one. We just briefly indicate, in the context of the Schottky barrier discussion, that we have measured on n-type Si (100) 2x1 substrates in a bulk sensitive mode at $h\nu$ = 108 eV, a $\sim$ 0.2 eV shift of the Si 2p lines towards higher kinetic energies after completion of the 2D Pb adlayer at RT (which still displays a 2x1 LEED pattern). For a n-type Si (100) 2x1 surface, the Fermi level lies at 0.45 eV above the VBM;[41] we derive a SB height $Pb\phi_{B,n}2x1 \sim 0.87$ eV identical to that found on Si(111) 7x7: $Pb \phi_{B,n}7x7 \sim 0.87$ eV.

6. Exciting New Results

We would not like to end this paper without citing the beautiful new results obtained by Jolochowski and Bauer[50] for condensation of Ag and Pb (as well as Au) onto Si (111) at low temperature where the growth is extremely laminar instead of following the layer-plus-islands mode. In Pb films, in addition to RHEED and classical resistivity oscillations with 1 ML periodicity caused by the periodic reproduction of the surface microstructure, quantum size effect (QSE) oscillations have been seen in the electrical resistivity. For Ag layers, only RHEED oscillations are reported by these authors but we note that QSE oscillations in the density of states of the sp band spectra of the silver thin films deposited onto Si (111) 7x7 at RT, where the growth is still fairly smooth, were noted by Wachs et al.[51] a few years ago, during UV photoemission experiments at normal emission.

CONCLUSION

We have shown, on a few "prototypical systems", Ag/Si (111), Pb/Ge (111), Pb/Si (111), the richness and also the complexity of these simple metal-semiconductor interfaces, when viewed at the first stages of their formation. Advances in the description of the different 2D superstructures and of the phase transitions were recently obtained thanks to the use of different new powerful experimental techniques for structure determination, like STM, surface X-ray diffraction, or photoelectron diffraction. In parallel to the structural analysis, the study of the development of the electronic properties of these systems, mostly using synchrotron radiation photoemission experiments, is extremely fertile.

Despite this, still several questions which we have outlined some of general far-reaching consequences like the Schottky barrier problem, or the discrepancy between the measured saturation coverages and the one determined from structural analysis, are still open. This leaves space for future research !...

Acknowledgements

One of us, G. Le Lay, wishes to thank the Risø team: F. Grey, R. Feidenhans'l, M. Nielsen as well as R. Johnson from Hasylab for their invitation to perform during May 1989 a cooperative work on a related system: Ag/Ge (111) both with XRD and photoemission experiments, in a fruitful, exciting and friendly atmosphere at Hasylab. The new Si 2p core level measurements for the Ag/Si (111) system mentioned in this paper were a by-product of this fecund collaboration.

REFERENCES

1. T. Ichikawa, Solid State Commun. 46, 827 (1983); 49, 59 (1984).
2. R. J. Wilson and S. Chiang, Phys. Rev. Lett. 58, 369 (1987).
3. E. J. Van Loenen, J. E. Demuth, R. M. Tromp and R. J. Hamers, Phys. Rev. Lett. 58, 373 (1987).
4. E. Vlieg, A. W. Denier Van der Gon, J. F. Van der Veen, J. E. Macdonald and C. Norris, Surface Sci. 209, 100 (1989).
5. T. Takahashi, S. Nakatani, N. Okamoto, T. Ishikawa and S. Kikuta, Jpn. J. Appl. Phys. 27, L 753 (1988).
6. S. Kono, T. Abukawa, N. Kakamura and K. Anno, Jpn. J. Appl. Phys. 28, 302 (1989).
7. E. L. Bullock, G. S. Herman, M. Yamada, D. J. Friedman and C. S. Fadley, Phys. Rev. Lett., submitted.
8. K. Takayanagi, Y. Tanishiro, M. Takahashi and S. Takahashi, J. Vac. Sci. Technol. A3, 1502 (1985).
9. For detailed works and extensive discussions, see the "Proceedings of the 2nd Int. Conf. on the Formation of Semiconductor Interfaces" (ICFSI-2), Osaka, Nov. 1988, to appear in Appl. Surf. Sci., and G. Le Lay in Les Interfaces et la Liaison Chimique, Eds de Physique, Paris 91 (1988).
10. W. C. Fan, A. Ignatiev, H. Huang and S. Y. Tong, Phys. Rev. Lett. 62, 1516 (1989).
11. W. S. Yang, S. C. Wu and F. Jona, Surface Sci. 169, 383 (1986).
12. G. Le Lay, A. Chauvet, M. Manneville and R. Kern, Appl. Surf. Sci. 9, 190 (1981).
13. C. T. Chan and K.M. Ho, Surface Sci., to be published.
14. T. Yokotsuka, S. Kono, S. Suzuki and T. Sagawa, Surface Sci. 127, 35 (1983).
15. S. Kono, K. Higashiyama and T. Sagawa, Surface Sci. 165, 21 (1986).
16. S. Kono, K. Higashiyama, T. Kinoshita, T. Miyahara, H. Kato, H. Ohsawa, Y. Enta, F. Maeda and Y. Yaegashi, Phys. Rev. Lett. 58, 1555 (1987).
17. J. P. Lacharme, S. Bensalah and C. A. Sebenne, to be published.
18. G. Le Lay, F. Grey and R. Johnson, unpublished results.
19. F. Houzay, Thesis Univ. Paris-Sud, Orsay, 1983.
20. A. L. Wachs, T. Miller, A. P. Shapiro and T. C. Chiang, Phys. Rev. B35, 5517 (1987).
21. E. H. Rhoderick, Metal Semiconductor Contacts, Oxford University Press, 1978.
22. G. Le Lay, M. Manneville and R. Kern, Surface Sci. 72, 409 (1978).
23. S. Tosch and H. Neddermeyer, Phys. Rev. Lett. 61, 349 (1988).
24. J. E. Northrup, Phys. Rev. Lett. 53, 683 (1984).
25. R. Feidenhans'l, J. S. Pedersen, M. Nielsen, F. Grey and R. L. Johnson, Surface Sci. 178, 92 (1986).
26. H. Huang, C. M. Wei, H. Li and B. P. Tonner, Phys. Rev. Lett. 62, 559 (1989).
27. J. J. Métois and G. Le Lay, Surface Sci. 133, 422 (1983); G. Le Lay and J. J. Métois, J. de Physique C5 427 (1984).
28. T. Ichikawa, Solid State Commun. 46, 827 (1983); 49 59 (1984).
29. F. Grey, Thesis Risø National Laboratory, Roskilde DK (1988).
30. M. Abraham, G. Le Lay and J. Hila, to be published.
31. K. Hricovini, G. Le Lay, M. Abraham and J. E. Bonnet, to be published.
32. G. Le Lay, K. Hricovini and J. E. Bonnet, Phys. Rev. B 39, 3927 (1989).
33. F. Grey, R. Feidenhans'l, M. Nielsen and R. L. Johnson, Proceedings of the 1st Int. Conf., Marseille, France, May 1989; to appear in J. de Physique, France.

34. P. J. Estrup and J. J. Morrison, Surface Sci. $\underline{2}$, 465 (1964); M. Saitoh, K. Oura, K. Asano, F. Shoji and T. Hanawa, Surface Sci. $\underline{154}$, 394 (1985).
35. H. Yabuchi, S. Baba and A. Kinbara, Appl. Surface Sci. $\underline{33/34}$, 75 (1988).
36. G. Quentel, M. Gauch and A. Degiovanni, Surface Sci. $\underline{193}$, 212 (1988).
37. G. Le Lay, J. Peretti, M. Hanbücken and W. S. Yang, Surface Sci. $\underline{204}$, 57 (1988).
38. F. Feidenhans'l, J. S. Pedersen, J. Bohr, M. Nielsen, F. Grey and R. L. Johnson, Phys. Rev. $\underline{B38}$, 9715 (1988).
39. K. Hricovini, G. Le Lay, M. Abraham and J. E. Bonnet, Phys. Rev. B, under press.
40. K. Hricovini, G. Le Lay, M. Abraham and J. E. Bonnet, in preparation.
41. F. J. Himpsel, F. R. Mc Feely, J. F. Morar, A. Taleb-Ibrahimi and J. A. Yarmoff, Lectures Notes of the CVIII Course Photoemission and Absorption Spectroscopy of Solids and Interfaces with Synchrotron Radiation, International School of Physics "Enrico Fermi", Varenna, Italy, July 1988.
42. K. Horn, B. Reihl, A. Zartner, D. E. Eastman, K. Hermann and J. Noffke, Phys. Rev. $\underline{B30}$, 1711 (1984).
43. D. Bolmont, Ping Chen and C. A. Sebenne, J. Phys. $\underline{C14}$, 3313 (1981).
44. F. Flores and C. Tejedor, J. Phys. $\underline{C20}$, 145 (1987).
45. W. Mönch in Festkörper Probleme (Advances in Solid State Physics) vol. XXVI, P. Grosse ed. $\underline{67}$ (1986).
46. J. Tersoff, Phys. Rev. Lett. $\underline{32}$, 465 (1984).
47. J. C. Freeouf, Surface Sci. $\underline{132}$, 233 (1983).
48. G. Le Lay, K. Hricovini, J. E. Bonnet and M. Abraham, to be published.
49. M. Sauvage, R. Pinchaux, G. Le Lay et al., to be published.
50. M. Jalochowski and E. Bauer, Surface Sci. $\underline{213}$, 556 (1989).
51. A. C. Wachs, A. P. Shapiro, T. C. Hsieh and T. C. Chiang, Phys. Rev. $\underline{B33}$, 1460 (1986).

FROM THERMODYNAMICS TO QUANTUM WIRES:

A REVIEW OF REFLECTION HIGH-ENERGY ELECTRON DIFFRACTION

P. I. Cohen, G. S. Petrich, A. M. Dabiran and P. R. Pukite

Department of Electrical Engineering
University of Minnesota
Minneapolis, MN 55455 - USA

ABSTRACT

The built-in staircase of steps on a vicinal surface has been used by Petroff and coworkers as a template for the growth of quantum wires. The limits of this technique depend upon the degree to which the steps are ordered before and during growth and on the extent of step meandering. Here we review those aspects of reflection high energy electron diffraction (RHEED) that allow characterization of these staircase steps during growth by molecular beam epitaxy. First we describe Henzler's methods of measuring the surface step distribution. We use Lagally's criterion to give an operational measurement of the transfer width of a RHEED instrument. We show that in the direction of the incident beam, coherent diffraction over distances as large as 8000 Å is detected. Simple one-dimensional models of terrace length disorder are considered to show the sensitivity of the measured pattern. The appropriate correlation functions of surface disorder are described. By directing the beam perpendicular and parallel to the staircase, we also examine the kink density. Differences that depend upon the two types of step termination possible on a zincblende structure are observed. Recent work of Bartelt and Einstein on scaling is described. The main difficulties are that to apply equilibrium arguments, sufficient temperature and flux stability is required to balance growth and sublimation. A second difficulty is that trace impurities can pin steps during growth.

INTRODUCTION

Quantum wire arrays[1,2] have been successfully fabricated by growing AlAs and GaAs on vicinal GaAs(100) substrates. They offer the opportunity to create novel device structures by virtue of their unique carrier confinement, transport, and electronic transitions and are significantly different from conventional quantum wells. The details and operation of these structures are described in more detail in these proceedings in the next paper by Petroff et al. Briefly, GaAs(100) substrates are misoriented by about 2° toward the [011] direction to create a staircase of steps over the surface. Then alternating submonolayers of GaAs and AlAs are deposited. By proper choice of growth conditions, step flow growth can be obtained allowing GaAs and AlAs to

Kinetics of Ordering and Growth at Surfaces
Edited by M. G. Lagally
Plenum Press, New York, 1990

cover separate portions of a terrace. Petroff will describe experiments to determine which species attach at the step edges. If the alternating sequence is continued, the individual sub-monolayers are enlarged by adding fractional layers of GaAs and AlAs on top of the like material already present, producing an array of parallel quantum wires. How well this procedure works depends on the regularity of the staircase formed by the slight misorientation and by the extent of the control over the growth kinetics that gives the separation of GaAs and AlAs. Both disorder in the terrace lengths and step-edge meandering are important. Our purpose here is to review the application of reflection high-energy electron diffraction (RHEED) to characterize these two types of disorder in order to fabricate optimal structures.

A second goal is to understand the thermodynamic aspects of disorder and growth of these structures. We want to measure the energetics of kink formation, the surface diffusion of adatoms that give step flow growth, and the extent to which the differing components of the wires phase separate. These two goals lie at opposite ends of the spectrum between basic research and applied technology. Thus the bulk of this paper is a review of those aspects of RHEED that bridge the large gap in between thermodynamics and quantum wires. Though the analysis will emphasize RHEED, many aspects are directly transferrable to low-energy electron diffraction (LEED).

In the next section we will describe the sensitivity of RHEED to ordered step arrays. The key issues to be addressed will be the distance on a surface over which diffraction is coherent and how different forms of terrace length disorder affect the characteristic diffraction pattern. In the third section the importance of the detector will be reviewed in terms of surface correlations that are measurable. The calculation of correlation functions needed for the determination of step meandering will be developed. In the fifth and sixth sections measurements from vicinal GaAs surfaces will be presented. Both the meandering of steps and terrace length disorder will be discussed. Finally recent predictions of scaling of step disorder in noninteracting step arrays during equilibrium will be considered.

THE SENSITIVITY OF RHEED

Diffraction From a Staircase

For the kind of quantum wire structures being proposed and investigated[1] and for typical films grown by MBE, misorientations of GaAs(100) of between 1 mrad (0.057°) and 2° are useful. For surfaces misoriented toward the <011> direction, these correspond to staircases with mean terrace lengths between 2800 and 80Å, or 700 to 20 atomic rows. Especially for the higher misorientations, growth conditions can easily be found that give terrace lengths comparable to Ga diffusion distances.[3]

Two kinds of stepped surfaces can be prepared in the III-V's due to the structure of the bulk zincblende lattice. The reason is the same as that which produces two types of (111) planes and comes about because the [111] and [$\bar{1}\bar{1}\bar{1}$] directions are inequivalent. In preparing a zincblende crystal with a {111} surface, one can terminate with one dangling bond per atom or three dangling bonds per atom, depending upon where a hypothetical surface plane is placed. The former is energetically favored. Thus the termination of a {111} is forced by bulk crystallography and not by the details of the growth. With the [111] direction defined from a Ga to As atom, a Ga terminated surface is (111)A and an As terminated surface is ($\bar{1}\bar{1}\bar{1}$) or (111)B. One should note that because of the partial ionic character of the bonding, the {111} is not the cleavage plane in zincblende so that this argument only applies to surface termination once the {111} is prepared. It also says nothing about the modification of the energetics due to surface reconstruction. Because of the same bulk lattice, a (100) surface can be misoriented toward either the [011] or [01$\bar{1}$] direction to give surfaces with (100) terraces and either (111)A or (111)B risers. In this case the (100) terrace termination is determined by the presence of sufficient excess As. These two types of steps are illustrated in Fig. 1. Which type is obtained is readily determined from the reconstruction observed in the diffraction pattern. One expects that these two kinds of stepped surfaces could have different reactivities, different kink energetics, different ledge reconstructions, different adatom incorporation rates - hence these are entirely different surfaces.

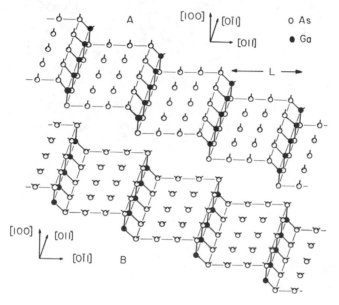

Fig. 1 The two types of steps that arise on misoriented GaAs(100) surfaces as a result of the bulk zincblende structure.

The characteristic diffraction pattern for these surfaces is shown in Fig. 2 for a 7 mrad GaAs surface. There are a number of features evident in this figure. The streaks perpendicular to the shadow edge correspond to the GaAs(100) 2 x 4 reconstruction with the incident beam in the [001] direction (half way between the two-fold and four-fold patterns). The broad light regions are spurious reflections from filaments and should be ignored. Below the shadow edge one can see the portion of the primary beam that misses the sample. The specular beam should ordinarily appear along the central or (00) streak, equidistant from the shadow edge. But here the glancing angle, ϑ_i, was chosen such that the diffraction from adjacent terraces is $(n+1/2)\lambda$ out-of-phase. This means that at the exact position of the specular reflection the diffracted intensity is a minimum (it is not zero because of disorder) and instead the peaks at higher and lower angles are observed where the path length difference from adjacent terraces is either λ or 2λ. The

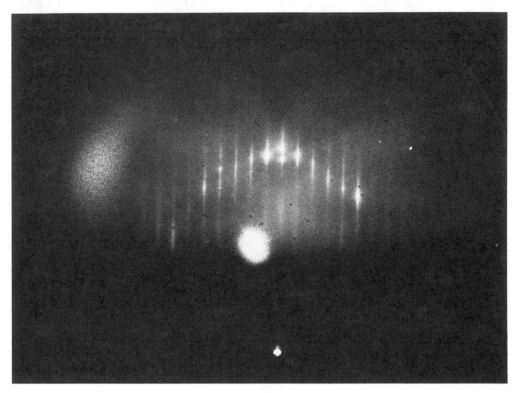

Fig. 2 Diffraction pattern from a GaAs(100) surface misoriented by 7
        mrad in the [001] direction. For the figure the beam is
        directed down the staircase. The surface has a 2x4
        reconstruction.

separation of these two components depends on the mean terrace length and
the scattering geometry.[4]

To calculate the diffracted intensity vs final glancing angle along
the streak, ϑ_f, one uses a theory that includes multiple scattering
within terraces, but not between terraces.[5] Further we treat the problem
of a one-dimensional staircase of rows of atoms; this is a special case
whose appropriateness is discussed in Sec. 3. Within these approximations
the diffracted intensity is the Fourier transform of the pair correlation
function of the surface,[6] i.e.,

$$I(S_x,S_y) = \int C(x,z)\, e^{i(S_x x + S_z d)}\, dx\, dz \; , \tag{1}$$

where S_x and S_z are the components of the momentum transfer $S = k_f - k_i$
and for specificity, the beam is directed down the staircase in the $\hat{x}$
direction. Here k_i and k_f are the incident and final wavevectors with
magnitude $2\pi/\lambda$. The $\hat{z}$ direction is normal to a low-index terrace.
$C(x,z)$ is a one-dimensional correlation function which is the probability
of finding a surface scatterer at the origin and at (x,z). To calculate
$C(x,z)$ assume that the terrace lengths L_i are distributed according to a
probability distribution $P(L_i)$. The specific ordering of the terraces is
assumed to be random, but consistent with the distribution function.

Then the correlation function is a sum over the probabilities of all possible distributions of terrace lengths that will fit between an origin and (x,z). For example, each term looks like

$$P_i(L_0)P(L_1)P(L_2) \ldots P(L_{n-1})P_f(L_n) , \tag{2}$$

where all of the $P(L_i)$ are identical except for the first and last since the starting and ending point could occur in the middle of a terrace. A procedure for determining P_i and P_f in terms of the $P(L)$ is discussed further in Ref. [7]. A simple distribution of terraces that gives a result similar to measurement uses $P(L) = \alpha \exp{-\alpha(L-L_d)}$ for $L_d \leq L \leq \infty$ and zero otherwise. Here L_d is the assumed minimum terrace length. Then the diffracted intensity at a glancing angle $\vartheta_i = (n+1/2)\pi/(kd)$, where n is an integer and d is a step height, is[7]

$$I(S_x, S_z = \pi/d) = \frac{4/\langle L \rangle}{2\alpha^2 + 2\alpha^2 \cos 2\pi\xi + (2\pi/L_d)^2 \xi^2 - (4\pi\alpha/L_d)\xi \sin 2\pi\xi} , \tag{3}$$

where $\xi = S_x L_d/2\pi$. Note that this glancing angle is measured with respect to a low-index plane and corresponds to an out-of-phase condition. This particular terrace length distribution has a mean terrace length $\langle L \rangle = L_d + 1/\alpha$ and an rms deviation of $1/\alpha$. Because of the cutoff at L_d, the diffracted intensity given by Eq. (3) can exhibit two peaks as in the photograph of Fig. 2. For $1/\alpha$ small compared to $\langle L \rangle$, the separation of the peaks ΔS_x is equal to $2\pi/\langle L \rangle$ and the FWHM of the component peaks is approximately

$$\frac{\delta S_x}{\Delta S_x} = 1.5 \left(\frac{\sigma}{\langle L \rangle} \right)^2 , \tag{4}$$

where δS_x is the width of the component peaks and σ is the rms deviation of the terrace lengths. This result agrees well with data measured by RHEED from GaAs surfaces prepared by MBE.[7]

An important point is that for small L_d the result that $\Delta S_x = 2\pi/\langle L \rangle$ does not hold and, in fact, the splitting can disappear. This is illustrated in Fig. 3 where, using Eq. (3), profiles for $\langle L \rangle = 200\text{Å}$ and several values of L_d are shown. A large value of L_d gives a sharply peaked terrace length distribution and well resolved peaks. But for low L_d, the peaks are only poorly resolved and the separation does not correspond to the mean terrace length. For other distributions,[7] this result holds as well. For a geometric distribution there is no splitting.[8] However, there still may be a characteristic change in the position of the specular beam and a way to determine the disorder.[8]

Instrument Response

When there is a sharply peaked distribution function $P(L)$, split peaks that are measurable can be used to determine the distance over the surface from which there is coherent diffraction. For example, Fig. 4 is a barely resolved peak from a vicinal surface of GaAs that is misoriented by 0.5 mrad (as determined with x-ray diffraction). Since at least two terraces must contribute to the diffraction, the distance over which there is coherent scattering from this surface is on the order of 1 μm. Note that this is the instrument response in the direction of the incident beam; the large magnitude results from the low glancing angle of incidence. Thus, operationally we define the coherently scattering distance[9,10] (not the coherence length of an electron which has to do with wave packets[11]) to be given by twice the step height divided by the smallest resolvable misorientation.

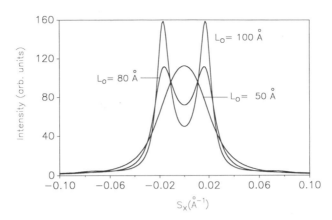

Fig. 3   A calculation of the diffraction profile for which the
distribution of steps is $P(L) = \alpha\exp[-\alpha(L-L_d)]$. The incident
beam is assumed to be directed down the staircase at a glancing
angle of incidence to correspond to an out-of-phase condition.
For the calculation [Eq. (3)] the mean terrace length is always
200Å. The values of L_d used were 50 (unresolved peaks), 80, and
100Å.

This coherently scattering distance follows from classical optics
and is easily seen to be due primarily to the range of angles that are in
the incident beam and that are admitted by the detector. For a flat
surface suppose a distance L gives a peak of FWHM $\delta\vartheta$ for an glancing
incident angle ϑ_i. Then as in single slit diffraction, the first minimum
will occur at $\vartheta_i + \delta\vartheta/2$ where there is destructive interference between
pairs of points in the slit. The width of the beam is then

$$\delta\vartheta = \frac{2\pi}{k \sin \vartheta_i L} .$$ (5)

If the range of angles in the incident beam is $\delta\vartheta_o$, then coherent
scattering over distances larger than

$$L - \frac{2\pi}{k \, \delta\vartheta_o \sin \vartheta_i}$$ (6)

cannot be detected. For RHEED, when $\vartheta_i = 1°$, $\delta\vartheta_o = 0.5$ mrad, and $k = 51.2$ Å^{-1} then $L = 10,000$Å, similar to the measured expectation. Parallel
to the direction of the incident beam the factor of $\sin \vartheta_i$ is not present
and the distance is much smaller.

MEASURABLE CORRELATIONS

For a general surface the diffraction in the single scattering
approximation is the Fourier transform of a two-dimensional correlation
function, $C(x,y)$, which describes the correlated probabilities of top
level scatterers separated by (x,y). The difficulty is that these are
nearly impossible to calculate even for Ising like models since steps in
one direction affect the other.[12,13] A simplification is achieved if a

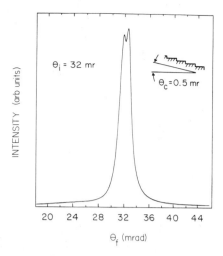

Fig. 4   A measurement on a GaAs(100) surface showing that even for a
surface misoriented by as little as 0.5 mrad, interference from
several steps is obtained.   This gives an operational definition
of the distance over which coherent scattering is obtained to be
of the order of 8000Å.

slit detector is used, reducing the problem to one dimension.

Starting with the general case

$$I(S_x,S_y) = \int C(x,y)e^{i(S_xx+S_yy)} \, dxdy \,,$$ (7)

where the integral extends over the surface of interest.   By using a slit
detector to integrate over a diffracted beam one obtains

$$\int I(S_x,S_y)dS_x = \int C(0,y) \, e^{iS_yy} \, dy \,.$$ (8)

Especially in RHEED where the instrument response is so asymmetric, this
happens to some extent in any case.   As a caveat, for this analysis to be
correct, the scattering factor cannot change too rapidly with angle.

For a vicinal surface with the beam directed down the staircase of
steps, the appropriate one-dimensional correlation function was described
in the previous section.   But with the beam directed parallel to the step
edges, a different formulation is needed.   Figure 5 shows the notation
for a surface with rough edges and with each step assumed identical.   We
divide a meandering step into a sequence of columns of length $\ell(x)$, where
x is the distance parallel to a step edge which is assumed very long.   We
assume that there are no isolated islands on the terraces.   With the beam
parallel to a step edge the diffracted beam is split into two components,
for reasons similar to the case described above, but now they are at the
same final glancing angle, ϑ_f, and separated in azimuthal angle.   The
separation of the two components is illustrated by the Ewald construction
of Fig. 6.   This wire-frame, perspective view shows the reciprocal
lattice composed of the intersection (product) of the reciprocal lattices
of the two gratings that form a staircase of steps.   The wide cylinder of

231

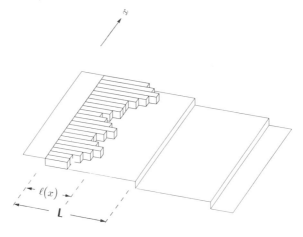

Fig. 5   Geometry and notation used for the calculation of the correlation function in Eq. (11).

diameter $2\pi/L$ corresponds to reciprocal lattice of a single terrace (L is the mean terrace length) and the sharp rods, with diameter 2π divided by the staircase length, corresponds to the grating of step edges and is at an angle to the low-index terrace given by the mean misorientation.   With the beam pointing down the staircase the locus of final k_f that conserve energy intersect the reciprocal lattice shown in two points.   These correspond to the split components of Fig. 2 and Eq. (3).   But with the beam directed parallel to the step edge, the Ewald sphere intersects in two places separated by $2\pi/L$.   The angle between them is $\delta\vartheta_f = 2\pi/kL$ and is small;   for misorientations slightly larger than 1°, however, it is resolvable.   Whether the components are resolvable, of course, depends on the order of the surface staircase.

To simplify the evaluation (or interpretation) of the correlation function, we evaluate the diffracted intensity at $(S_x, \pi/L, \pi/d)$, where L is the mean terrace length and d is the step height.   This is a point profile motivated by the slit integrated profile first calculated by Kariotis.[14]   At this value of $\underline{S}$, the diffracted amplitude is

$$A = \sum_{n=0,1} \int f_n(x,y)e^{iS_x x}\, e^{i\pi y/L}\, e^{i\pi n}\; dxdy , \qquad (9)$$

where $f_n = 1$ if there is a top layer scatterer at (x,y) and zero otherwise.   The column of length $\ell(x)$ consists of scatterers on level $n = 0$ and the underlying terrace has scatterers on level $n = 1$.   Using these values for f_n, the integral is evaluated to give

$$A(S_x) = \int dx\; e^{iS_x x}\, g(x) \qquad (10)$$

where

$$g(x) = \frac{2L}{i\pi}\, e^{i\pi\ell(x)/L} . \qquad (11)$$

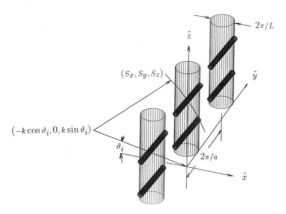

Fig. 6   Perspective view of the Ewald construction that shows the RHEED
         pattern for a misoriented surface. When the incident beam is
         directed down the staircase, the specular beam is split into two
         beams that are at the same final azimuth. When the incident
         beam is directed parallel to the step edges, the specular beam
         is split into two components at the same final glancing angle
         but separated in azimuth. For misorientations less than about
         1° this is difficult to resolve.

The intensity is the square modulus of $A(S_x)$ and is the autocorrelation
of $g(x)$. Since $\ell(x)$ is in an exponential, this is not quite a
length-length correlation function. However taking the length-length
correlation function, $P(u)$, from Kariotis's partition function
calculation,[14] the diffracted intensity is

$$I(S_x, \pi/L, \pi/d) - \int du\ e^{iS_x u} \int dx\ P(u)\ e^{i[\ell(x)-\ell(x+u)]} . \qquad (12)$$

To evaluate $P(u)$, Kariotis has developed a relation based on a
Hamiltonian describing kink and step formation.[14]

STEP BUNCHING

     The growth of $Al_xGa_{1-x}As$ shows a very clear case of step train
disordering. As a function of the Al mole fraction and substrate
temperature, the step train order as described by RHEED shows a strong
variation with time during growth. Both the Schwoebel effect[15] or step
pinning by impurities[16,17] could be involved. Figure 7 shows the RHEED
profile during growth of $Al_xGa_{1-x}As$ with $x = 0.25$ on a GaAs(100)
substrate that was misoriented by 2° toward the (111)A. For this growth
the ratio of column V to column III flux was about 2, the growth rate was
1 μm/hr, and the substrate was held at 675°C. This diffraction profile
was measured with the incident beam pointing down the staircase of steps
on the surface. First one should note that though the two components
forming the split specular streak were of nearly equal intensity and
width at 580°C, at this relatively high temperature where one can grow
smooth AlGaAs, the relative intensity of the two peaks are very
different. This is a reversible change with temperature[18] and could be
due to a reconstruction change from a 2 x 4 to the 1 x 1 that exists at

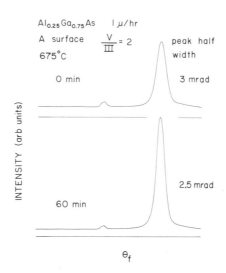

Fig. 7   The diffracted intensity along the specular streak during the growth of AlGaAs on GaAs(100) misoriented by 2° to expose Ga terminated steps. The width of the two components is related to the terrace length disorder. At this mole fraction of Al and substrate temperature, step bunching is not evident.

these high temperatures. Second, upon growth there is not much of a change, and for this case after an hour there might be a slight ordering. By contrast, for growth on a GaAs(100) surface misoriented toward the (111)B, at very nearly the same conditions, the staircase disordered. This is seen in Fig. 8 where after starting with split components that were sharp, growth for 40 min left a surface with only a weakly resolved, characteristic step diffraction pattern. This is reversible in the sense that if GaAs is grown on top of this surface, the original ordered step train is developed.

The results[16] for a variety of temperatures are shown in Fig. 9 for both A and B step terminations. For the data one sample of each surface was used; after each growth a buffer of GaAs was grown and the surface annealed in an As_4 flux to obtain a sharp diffraction pattern. Differences in the initial peak width reflect this procedure. The main point is that the amount of disordering is reduced if the substrate temperature is raised or if the A step termination is used. Further, there is some disordering even on the surface with Ga terminated steps. Finally, at some temperatures the disordering begins immediately, while at others there is a slow initial delay before the disordering begins in earnest. For both surfaces and at all temperatures, the disordering is more rapid as the Al mole fraction is increased.

Gilmer[19] has developed a simple rate equation model to describe the step bunching process in terms of one parameter. This parameter reflects the asymmetry in the ease at which an adatom can cross a step from different directions. Later the model was reworked by Tokura et al.[20] to obtain a relation for the time dependence of the variance in the terrace length from the mean. The basic mechanism is illustrated in Fig. 10. Here a staircase of steps is shown with terrace length T_n for the nth step. The growth rate is one monolayer in τ seconds and one assumes that

234

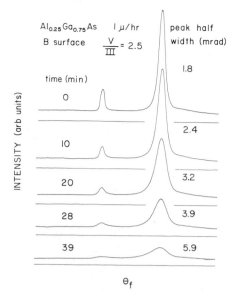

Fig. 8   The diffracted intensity along the specular streak during the
          same conditions as in Fig. 7 but for steps misoriented in the
          [01$\bar{1}$] direction to expose As terminated steps.   During growth of
          AlGaAs the peaks broaden, corresponding to step bunching.

there is only step flow. The parameter η describes the asymmetry in
attachment to the neighboring steps. If adatoms striking the nth terrace
could attach at the (n-1)th terrace as easily as the (n+1)th terrace then
$\eta = 0$. If, however, there is an asymmetric barrier for crossing a step
from the right over the left, then as shown there might be an $(1+\eta)/2$
enlargment of the (n-1)th terrace and an $(1-\eta)/2$ enlargment of the nth
terrace. There are four such terms which when summed, yield a rate
equation describing change in the length of the nth terrace, T_n, viz.

$$\frac{dT_n}{dt} = \frac{1 + \eta}{2\tau} (T_{n+1} - T_n) + \frac{1 - \eta}{2\tau} (T_n - T_{n-1}) . \tag{13}$$

To solve this,[20] expand it in a finite series and assume periodic
boundary conditions. Set

$$T_n = \sum_q T_q e^{iqn}$$

$$\frac{Nq}{2\pi} = 0, 1, \ldots, N-1$$

to obtain for the variance of the step terrace length distribution

$$\overline{\Delta T_n^2} = \sum_{q \neq 0} |T_q(0)|^2 \exp (-4\eta \frac{t}{\tau} \sin^2 q/2) . \tag{14}$$

235

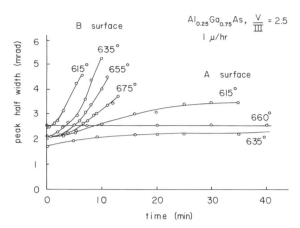

Fig. 9   A collection of data like that shown in Figs. 7 and 8 to show the importance of substrate misorientation and substrate temperature. The FWHM of the individual components is related to the terrace length disorder.

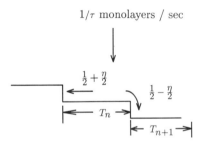

Fig. 10 Definition of the asymmetry parameter η that gives either step ordering or step bunching.

Depending upon the initial distribution of terrace lengths and on the asymmetry parameter, one can determine the subsequent behavior. If this asymmetry parameter η is positive then the staircase orders; but if it is negative, the fluctuations diverge. For ordering one requires that there is a feedback mechanism by which small steps grow faster than larger steps. One should note that the initial distribution is important since if there is no initial disorder, then there is no reason for the growth to prefer one step over another so that no change will take place. If however there is an initial difference between steps, then an asymmetry in attachment at up or down steps can take over. To illustrate this, assume that initially the T_n = T ± d, with a random distribution. Then the Fourier components are identical and a calculation of Eq. (14) for several values of η is shown in Fig. 11. Only slight values of asymmetry are needed to achieve modest agreement with the data. For some experimental curves there is an initial delay that cannot be fit with this assumed random initial terrace length distribution. Finally we should note that impurities could also be important[16,17]. They could adsorb randomly, pinning some steps at the expense[20] of others, producing terrace length disorder.

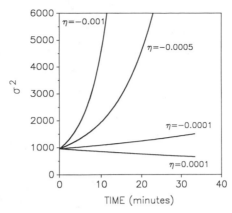

Fig. 11 Terrace length roughening or ordering calculation based on the
work of Gilmer et al.[19] and Tokura et al.[20] to show the
importance of the initial distribution of terrace lengths. The
small values of the asymmetry parameter give modest agreement
with experiment. The long delay before disorder initiation in
some of the data cannot be fitted with the assumed random
initial distribution of terrace lengths.

STEP MEANDERING

Though we do not know the detailed structure of the Ga terminated or
As terminated steps, it does appear from simple analysis of the
diffraction data that Ga terminated step edges are much straighter than
the As terminated ones. The picture of the GaAs(100) surface that is
deduced from the data is schematically illustrated in Fig. 12. This
result is similar to the difference in kink density of the two types of
steps that can form on Si(100).[21,22] The diffraction patterns from
surfaces with these two types of step termination are strikingly
different. If two samples are placed side by side on the holder, and the
beam switched from one to the other even the eye can distinguish the
degrees of order. Note that the A and B misorientations can be prepared
by taking a wafer that is polished on both sides, cleaving it into two,
and then mounting both but with one turned upside down. Examination of
the 2 x 4 reconstruction and determination of the staircase direction
from RHEED[23] or from x-ray diffraction quickly gives the step
termination.

To determine the terrace length order, the incident beam is directed
down the staircase of steps and the diffracted beam intensity is measured
along the length of the streaks. This was described in Sec. 2. The
results for the two types of step termination are shown in Fig. 13. For
these data two GaAs wafers were mounted side by side on the sample holder
to minimize differences in sample history, incident flux, and
temperature. A GaAs buffer was grown by MBE and the measurements were
made under an As_4 flux and with the surface in a 2 x 4 reconstruction.
Note that in the top curve of the figure, the cut off peak is an

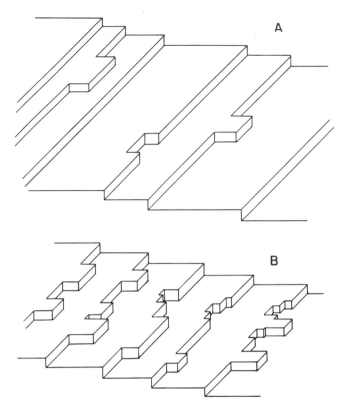

Fig. 12 Schematic diagram showing the order (disorder) in the two types
of stepped surfaces that can be created on GaAs(100). The data
to be presented in the next few figures show that Ga-terminated
steps are straight while there is severe terrace length
disorder. For As-terminated steps, the step edges meander
though the mean terrace lengths are better ordered. This model
holds for GaAs(100) at 600°C in an As_4 flux.

artifact. The main result is that the components of the peaks for the Ga
terminated surface are much broader than those for the surface with As
terminated steps. This broadening is removed if the temperature is raised
to cause the higher temperature 1 x 1 reconstruction and can be recovered
reversibly. Based on the analysis of Sec. 2 we estimate that the rms
deviation in terrace lengths is twice as high for the Ga terminated step
structure.

To examine the meandering the electron beam is directed parallel to
the steps. Once again the difference between the two step terminations
is striking. The data are shown in Fig. 14 and are from the same samples
as in Fig. 13. The top curve corresponds to the As terminated surface
(the beam is in the [011] direction parallel to the steps) and the bottom
curve corresponds to the Ga terminated step surface. For these data a
slit detector was used to integrate over one of the split components. At
this point these measurements have not been evaluated over the range of
scattering geometries, terrace lengths, and substrate temperatures to
make quantitative statements about observed trends. Nonetheless, the

238

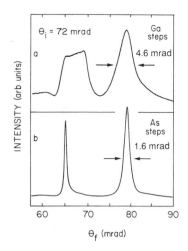

Fig. 13 Intensity profile along the specular beam for GaAs(100) misoriented by 2°. The staircase with Ga terminated steps has broad components and measurable terrace length disorder. The flat top of the left peak is an artifact. The staircase with As terminated steps exhibits much less terrace length disorder. There is an As_4 flux, but no Ga flux (i.e. no growth).

results are reproducible for a few misorientations and many sample preparations, indicating that the Ga terminated steps meander less than those that are As terminated. One issue that complicates this interpretation[24] is whether islands on the terraces contribute to the measurement. At this point we cannot conclusively answer this except to say that adatoms are mobile on flat terraces at these temperatures and should have sufficient time to reach the step edges. Intensity oscillations are not observed at modest growth rates, indicating that step flow is obtained. What is needed are a comprehensive set of data and analysis by the methods described in Secs. 2 and 3.

SCALING

The diffraction measurements should be expressed in terms of the natural length scales. Then those features which depend on real differences between surfaces could be compared; for example, step-step interactions on surfaces with different misorientations should be clarified if the natural scales are the same. Bartelt and Einstein[25,26] have recently determined the length scales for vicinal surfaces and calculated how the diffracted intensity should scale with terrace length and temperature. Their analysis assumes noninteracting steps and equilibrium conditions. For a vicinal surface of mean terrace length L, the natural length in the direction of the staircase is this terrace length, while in the orthogonal direction the natural length is the distance between step-step collisions. This is calculated for a one-dimensional random walk in which the hopping probabilities are temperature dependent. When applied to the diffracted intensity, the

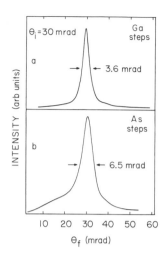

Fig. 14 Intensity profile of one of the split components with the incident beam directed parallel to the step edges. A slit detector is used. The Ga terminated steps meander less.

prediction is that for two surfaces with terrace lengths L_A and L_B in the $\hat{x}$ direction, the diffracted intensities of the two surfaces are related according to

$$I_A(S_x, S_y) = I_B \left[S_x \frac{L_B}{L_A}, \; S_y \left(\frac{L_B}{L_A} \right)^2 \right]$$ (15)

and the intensities measured at two temperatures according to

$$I(S_x, S_y, T) = I \left[S_x, S_y \frac{b^2(T)}{b^2(T_o)}, \; T_o \right],$$ (16)

where

$$b^2(T) = e^{-\Gamma/kT}/(1 + 2e^{-\Gamma/kT})$$ (17)

and Γ is the activation energy to form a kink.

There are several difficulties in applying these relations. First one must attain equilibrium conditions. For GaAs this situation might be approached by making measurements during slow growth at moderately high temperatures. To illustrate differences in measurements we show in Fig. 15 the FWHM of the diffracted beam directed parallel to step edges of a GaAs(100) surface misoriented by 1° toward the [001]. This surface probably behaves like one misoriented toward the [011]. Similar data for a 2° surface has been presented by the Santa Barbara group.[27] In Fig. 15 the growth rate is 1 layer per 11 secs. and the temperature was first increased and then decreased. Over this narrow temperature range, it is reversible. But the effect of growth conditions is seen in Fig. 16 where the Ga flux is removed, and all other conditions remain constant. Here the substrate temperature is increased and unlike the data in Fig. 15,

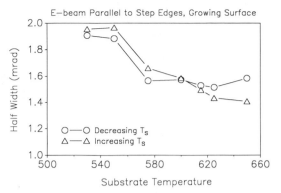

Fig. 15 The half width of slit-integrated diffracted beams for a GaAs(100) surface misoriented by 1° toward [001]. The substrate temperature is indicated on the abscissa, the growth rate is about 1 layer in 11 sec. The results are shown for both increasing and decreasing temperature as indicated on the abscissa. Similar results were reported by Chalmers [27].

the FWHM increases also. It increases until at modest temperature the 2 x 4 reconstruction changes to a 1 x 1 and the beam sharpens. At this temperature evaporation is observed[28] and probably the surface mobility is enhanced. Upon cooling, the phase changes again occur, with the half widths of the profiles following the same trends. We should also note that the profiles of the beams giving the data in Figs. 15 and 16 are very different. The presence of the two phase transitions indicated in Fig. 16 severely restricts the temperature range over which scaling might operate. Especially for these misoriented samples, finite size effects cause the phase transition to extend over a region perhaps 30°C wide. The data shown were taken at an out-of-phase condition. To test scaling, we need a comprehensive set of data at many glancing angles of incidence.

SUMMARY

We have described those features of RHEED which promise to aid current approaches to fabricate quantum wire structures on vicinal surfaces. These RHEED techniques were developed over an extended period and in large measure have been adapted from LEED. Since the work was not aimed at quantum wire structures until quite recently, much of the analyses described have not been thoroughly tested. For example, part of the analyses of the meandering of steps was developed at this workshop and the scaling arguments are still only available in preprint form. Neither has been examined systematically over a wide range of conditions and scattering geometries. However, we have shown that RHEED is sensitive to the kind of questions crucial to the optimal development of these device structures. For example, we showed that it is sensitive to the statistics of terrace length order over nearly a micrometer and gives quantitative measurements of kink generation. Further it now appears

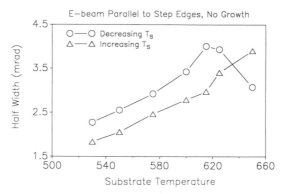

E−beam Parallel to Step Edges, No Growth

Fig. 16 Results similar to Fig. 15 but without a Ga flux. The results are very different possibly showing the role of equilibrium vs kinetics in the step meandering. A major worry in applying the theories of Bartelt and Einstein for this system is that the reconstructions change at 520°C and at 620°C. For these surfaces the phase transitions are not sharp due to finite-size effects. In addition, we do not know to what extent islands form on the terraces.

possible to start with a Hamiltonian that describes step and kink energetics and then calculate the correlation functions that could be compared to the diffraction analysis. These results are applicable to a range of materials so that near ideal growth parameters of quantum wire structures should be practicable. Obviously, whether or not this fundamental surface science technique can successfully impact these novel device structures must be tested by actual fabrication, ex situ electron microscopy and electrical characterization.

There are still clear obstacles to be overcome. First, the role of impurities in pinning steps has yet to be fully understood. We don't know whether the step merely bends around firmly anchored impurities or whether large step perimeters are immobilized. Alternatively, impurities may change the kink energetics of the step. Second, we don't understand the interactions between steps that correlate their motion. In principle strain fields are long range and should affect the ordering of steps in a vicinal staircase. But the distribution of steps on a vicinal surface during the growth of a strained layer has to our knowledge not yet been studied. Third, steps are surely reconstructed, but the atomic structure of steps is not yet resolvable. This is probably crucial for a microscopic understanding of the Schwoebel effect. Fourth, we do not know how slight deviations of a misorientation from a principle direction will affect the resulting staircase. If diffusion parallel to the step is sufficiently rapid, then presumably the equivalent of a equilibrium habit of the kinked surface will be obtained. Fifth, it is still difficult to decouple measurements parallel to step edges and down the staircase. For example, a rough step edge or isolated islands on a terrace will affect both. For all of these issues, the hope is that statistical information provided by RHEED can be combined with imaging data to eliminate uncertainties.

Acknowledgement

This research was partially supported by the U.S. National Science Foundation, Grant No. DMR 86-15207.

REFERENCES

1. P. M. Petroff, A. C. Gossard, and W. Wiegmann, Appl. Phys. Lett. 45, 620 (1984); M. Tsuchiya, J. M. Gaines, R. H. Yan, R. J. Simes, P. O. Holtz, L. A. Coldren, and P. M. Petroff, J. Vac. Sci. Technol. B7, 315 (1989).
2. T. Fukui and H. Saito, Appl. Phys. Lett. 50, 824 (1987).
3. J. M. Van Hove, P. R. Pukite, and P. I. Cohen, J. Vac. Sci. Technol. B3, 563 (1985).
4. P. R. Pukite, J. M. Van Hove, and P. I. Cohen, J. Vac. Sci. Technol. B2, 243 (1984).
5. M. Henzler, Advances in Solid State Physics 19, 193 (1979).
6. M. B. Webb and M. G. Lagally, Solid State Physics 28, 301 (1973).
7. P. R. Pukite, C. S. Lent, and P. I. Cohen, Surf. Sci. 161, 39 (1985).
8. T.-M. Lu and M. G. Lagally, Surf. Sci. 120, 47 (1982).
9. T.-M. Lu and M. G. Lagally, Surf. Sci. 99, 695 (1980).
10. J. M. Van Hove, P. R. Pukite, P. I. Cohen, and C. S. Lent, J. Vac. Sci. Technol. A1, 609 (1983)
11. G. Comsa, Surf. Sci. 81, 57 (1979).
12. C. S. Lent and P. I. Cohen, Surf. Sci. 139, 121 (1984).
13. J. M. Pimbley and T.-M. Lu, Surf. Sci. 159, 169 (1985).
14. R. Kariotis, B. S. Swartzentruber, and M. G. Lagally, J. Appl. Phys. 67, 2848 (1990).
15. R. L. Schwoebel and E. J. Shipsey, J. Appl. Phys. 37, 3682 (1967).
16. D. Saluja, P. R. Pukite, S. Batra, and P. I. Cohen, J. Vac. Sci. Technol. B5, 710 (1987).
17. D. C. Radulescu, G. W. Wicks, W. J. Schaff, A. R. Calawa, L. F. Eastman, J. Appl. Phys. 63, 5115 (1988).
18. G. S. Petrich, A. M. Dabiran, and P. I. Cohen, to be submitted.
19. P. Bennema and G. H. Gilmer, in: Crystal Growth: An Introduction, P. Hartman ed. (North Holland, 1973) ch. 10.
20. Y. Tokura, H. Saito, and T. Fukui, J. Cryst. Growth 94, 46 (1989).
21. N. Inoue, Y. Tanishiro, and K. Yagi, Jpn. J. Appl. Phys. pt. 2, 26, L293 (1987).
22. M. G. Lagally, Y.-W. Mo, R. Kariotis, B. S. Swartzentruber, and M. B. Webb, this volume.
23. P. R. Pukite, J. M. Van Hove, and P. I. Cohen, Appl. Phys. Lett. 44, 456 (1984).
24. P. M. Petroff, private communication.
25. N. Bartelt and T. L. Einstein, submitted to Surf. Sci., 1989.
26. M. Fisher, J. Stat. Phys. 34, 667 (1984).
27. S. A. Chalmers, A. C. Gossard, P. M. Petroff, J. M. Gaines, and H. Kroemer, J. Vac. Sci. Technol. B7, 1357 (1989).
28. J. M. Van Hove and P. I. Cohen, Appl. Phys. Lett. 47, 726 (1985).

DIRECT EPITAXIAL GROWTH OF QUANTUM STRUCTURES WITH TWO AND

THREE-DIMENSIONAL CARRIER CONFINEMENT

P. M. Petroff

Materials Department and Electrical and Computer
Engineering Department
University of California
Santa Barbara, CA 93106 USA

ABSTRACT

We present the principles of "directed epitaxy" methods which are used for the growth of novel superlattice types in the GaAs-AlGaAs system. The "tilted superlattices" (TSL) are produced by using the periodic step array of a vicinal surface as a template. The role of phase stability, growth kinetics, step orientation and periodicity on the TSL perfection are discussed. The "columnar superlattice" (CSL) also uses a periodic step faceting as a growth template; their growth requires additionally a self organization of the epitaxy. The CSL could be used for the direct growth of quantum box superlattices. The TSL have been used to produce quantum wire superlattices. Their unique optical properties are a direct consequence of two dimensional carrier confinement.

INTRODUCTION

Quantum structures which exhibit two and three-dimensional carrier confinement are traditionally produced by high resolution lithography or ion beam processing of a quantum well or two dimensional electron gas structure. Both techniques usually introduce a depletion layer with a controllable or fixed width and thus a quantum wire (QW) or quantum box (QB) structure is obtained. The quantum structures dimensions with this processing approach are usually in the range of 500Å to 1000Å and only isolated QWs and QBs can be made since the typical dimensions for the depletion layer are larger than 400Å in GaAs. In spite of its limited resolution and the drawbacks of a defect induced depletion layer, this approach leads to interesting structures which exhibit quasi-one-dimensional properties or to mesoscopic structures.[1-3] There is however no hope of producing true superlattices of coupled QWs and QBs with this type of processing. A more direct approach is eminently desirable since novel and attractive electronic properties have been predicted for QW and QB superlattices. For example, ultrahigh mobilities in coupled QB superlattices have been predicted at room temperature in the GaAs-GaAlAs system.[4]

Kinetics of Ordering and Growth at Surfaces
Edited by M. G. Lagally
Plenum Press, New York, 1990

A novel epitaxial approach can be used to directly grow QW superlattices. With this "directed epitaxy", applications of time and spatial constraints promote a controlled positioning of atoms on the semiconductor substrate. This approach relies heavily on the growth kinetics during epitaxy and assumes that near equilibrium growth conditions can be achieved. The tilted superlattice (TSL) growth[5] in which fractional submonolayers of two compound semiconductors are alternatively deposited on a vicinal (100) surface, is an example of this approach.

1. Principle

On a clean semiconductor surface, spatial constraints to atomic motion and atom incorporation are introduced by the presence of unreconstructed bonds. In regular epitaxy, these bonds provide a 2 dimensional template on which an ordered nucleus is formed (Fig. 1A). If a layer growth mode has been established, the nucleation will be initiated preferentially at sites along step edges. Thus, if step edges are positioned in an ordered array, as is the case with a vicinal surface, one may expect a quasi-ordered nucleation of layers (Fig. 1B). If along these step edges a regular array of nucleation sites is introduced by some means, preferential nucleation is obtained at these sites (Fig. 1C). We make use of these effects by defining for two III-V compounds e.g. GaAs and AlAs, the epitaxial growth conditions which will directly produce multiple carrier confinement.

Quantum wells, quantum wires and quantum boxes may be obtained directly by making use of two compound semiconductors during epitaxy. For example, by growing a thin layer of TSL between two wider band gap epitaxial layers, a quantum wire superlattice may be directly grown. The growth of these structures and their novel electronic properties has been detailed in the case of MBE[6,7] and MOCVD[8] deposition. A quantum box superlattice may also be obtained if we make use of the same deposition mode as that of the TSL on a vicinal substrate with nucleation site selectivity of the type shown in Fig. 1C. In Fig. 2, we have illustrated the idealized deposition sequence which could lead to the formation of a columnar superlattice (CSL).

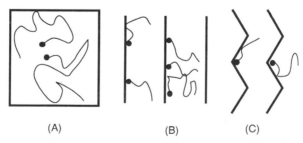

(A)   (B)   (C)

Fig. 1 Schematic of 3 cases of epitaxy. (A) Singular surface with random layer nucleation, (B) vicinal surface with periodic step array (thick lines), and (C) vicinal surface with periodic kink array along step edges. Atom trajectories are shown schematically by thin lines.

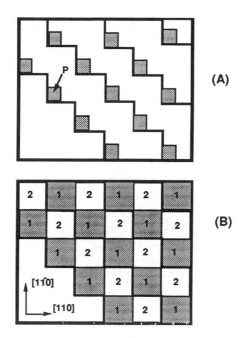

Fig. 2 Schematic of the epitaxial growth sequences leading to the
formation of a columnar superlattice. (A) On a vicinal surface,
with periodic kink sites at step edges (thick lines), a fraction
of monolayer (P) of AlAs is deposited. The kinetics favor AlAs
bonds over GaAs-AlAs bonds. (B) A half monolayer of AlAs(1) and
GaAs(2) have been deposited. A repetition of this deposition
cycle will yield a columnar superlattice.

The formation of small AlAs islands at concave kink sites along the
step edges is favored by the higher number of bonds available. The island
shape will remain as shown in Fig. 2A if the formation of AlAs-AlAs
bonds is favored over that of GaAs-AlAs. In reality, the surface and
interfacial energy and diffusion kinetics will produce islands with a
shape differing from that shown in Fig. 2. From previous experiments on
the equilibrium shapes of GaAs islands,[9] we expect their edges to be
parallel to <100>, <110> and <210> directions. The quantum box
superlattice is directly grown by sandwiching a thin columnar
superlattice layer between two wider band gap AlAs layers.

RESULTS

In this section, we examine some of the parameters that should
directly influence the tilted superlattice (TSL) and columnar
superlattice (CSL) perfection.

1. <u>Phase Stability and Long Range Ordering</u>

At the two extreme end of the TSL or CSL formation we find the
perfectly ordered and completely disordered structures. The final result
of the growth depends both on the stability of the ordered versus that of

the disordered state and the growth kinetics which are present during deposition.

If the ordered state is the lowest energy configuration, we expect that the TSL or CSL deposition should take place relatively easily. Such is the case of the AlAs-GaAs system where we have observed a self organization of the Al and Ga epitaxy on the GaAs vicinal surface.[10]

This process occurs when Al and Ga are deposited by migration enhanced epitaxy[11-13] (MEE). When the number of Al and Ga atoms deposited per cycle is equal to the number of surface lattice sites and if no As flux is incident on the surface, a coherent TSL is formed and observed by Transmission Electron Microscopy (TEM).[10] The deposition sequence illustrated in Fig. 3 which illustrates this process, requires that exchange interactions between Al and Ga (Processes 1,2 and 3 in Fig. 3A) are taking place prior to the As cycle deposition. Figure 3B shows the self organization of Al and Ga from the random incident flux of Al and Ga corresponding to a monolayer of surface atoms. We have shown directly by TEM that in the coherent TSL, the AlAs is located at the step risers while the GaAs is at the step edges. We have also shown using growth interruption that during the MEE deposition, the Al is already located at the step edges in the sequence corresponding to Fig. 3B. The large difference in formation enthalpy between AlAs (29kcal/mole) and GaAs (23 kcal/mole) may be responsible for the preferential formation of AlAs at the step edges. Since the binding energy of a surface atom arriving at a kink site is proportional to the formation enthalpy of the compound that is formed, we expect the Al atoms to have a greater sticking probability at step edges than the Ga atoms.

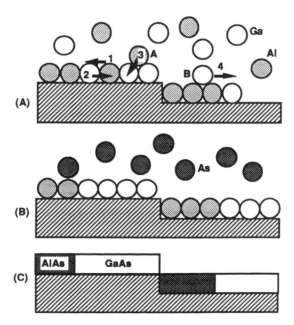

Fig. 3 Schematic of a self organized epitaxial system at various formation stages. A) deposition of the group III elements and B) presence of the self organized epitaxial layer prior to the group V deposition shown in C). A repetition of (A) to (C) will produce a coherent tilted superlattice[10].

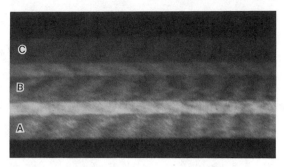

Fig. 4 Dark field TEM micrograph of 3 coherent TSL layers.   Each layer is
       200Å thick.   Layer A is deposited on top of a GaAs layer while
       layers B and C are deposited on top of a TSL layer.

       In Fig. 4 we show a TEM micrograph of 3 coherent TSL layers grown on
top of a GaAs layer and two TSL layers.

       The driving force for the exchange reactions 1,2 and 3 in Fig. 3A is
believed to be associated with the existence of a thermodynamically
favorable phase separation between the Al and Ga on an As saturated
surface. This type of phase separation between AlAs and GaAs which was
previously observed during MBE deposition of AlGaAs on vicinal {110}
surfaces[14] is similar to the present one and probably of the same origin.
Unlike a number of long range ordering reactions reported for a large
number of III-V compounds,[15-17] the AlGaAs phase separation is not strain
stabilized since the lattice mismatch at the growth temperature is in the
10^{-4} range. The scale of the ordering is controlled by the misorientation
angle of the vicinal surface.

       We expect a similar self organized epitaxy in III-V compounds e.g.
AlInAs, AlGaSb, which are metastable under the growth condition (MBE or
MOCVD) and for which there exist a large difference in the formation
enthalpy of the binary compounds forming the alloy.

2.   Surface Orientation and Perfection

       The effects of surface orientation and perfection on the TSL or CSL
are expected to be important. The layer growth regime required for the
TSL or CSL deposition is a function of the interfacial energy and
nucleation kinetics. Thus the bonding arrangements and nature of the
surface and the presence of impurities will dominate the layer growth
formation. Optimal MBE growth conditions on a {111} or {110} surface are
different from those required for a {100} surface.

       The orientational stability of the surface may be drastically
affected by the presence of impurities. For example the introduction of
small amounts of Carbon causes the Silicon {100} and {112} surfaces to
facet.[18]   The step orientation on Si {111} vicinal surfaces have also
been associated with a temperature dependent faceting.[19]. We should
expect similar effects for a GaAs or AlGaAs surface deposited by MBE or
MOCVD. This macroscopic faceting has deleterious effects on the TSL
perfection since a small change in the vicinal angle will result in a
large change of the TSL tilt angle.

       The step uniformity is critical in establishing the TSL perfection.
The growth of a well ordered step array is promoted by deposition of a
thick buffer layer (<1000Å) on the vicinal surface. The step periodicity

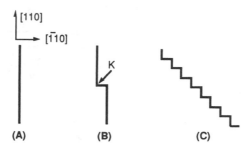

Fig. 5 Schematic of 3 step orientations on a GaAs(001) surface. A) [110]
As terminated step edge. B) near [110] step edge with a kink site
(K) and C) [100] oriented step edge.

can be followed by reflection high-energy electron diffraction
(RHEED).[20]. A step equalization process is found as the buffer layer
thickness increases.[20] A possible explanation involving the atomic
confinement of atomic motion to the step ledge on which the atom
initially lands has been proposed.[21] The ability of establishing the
growth conditions of coherent step motion during the TSL deposition using
RHEED is essential for their perfection.[6,7]

The nature of bonds and the kink site density at step edges are
expected to control the abruptness of the AlAs-GaAs interface in a TSL.
For example, the initiation of a layer growth is expected to differ
markedly at Ga or As terminated <110> steps on the {001} vicinal
surfaces. In principle the [110] oriented steps with the As bond normal
to the edge (Fig. 5A) should not initiate any layer growth. On the other
hand, a small misorientation from the [110] direction may cause these
steps to act as preferential nucleation centers for a layer growth mode
since there will now be some Ga sites available at the kink (Fig. 5B).
The reported differences in the TSL perfection for vicinal surfaces with
<110> Ga and As terminated steps[6] may well be associated with the easier
nucleation probability for the Ga terminated step edges. The <100> step
edge in Fig. 5C is shown with a high kink density and comprises equal
number of As and Ga sites. Presently, there is no data for GaAs growth on
this type of vicinal surface, but a high kink density is expected to
favor abrupt interfaces for the TSL structure. The above discussions
neglects bond reconstruction and step edge reorientation directed by
minimization of the surface energy. For example, the faceting of GaAs
islands deposited by MBE on (001) Ge surfaces occurs preferentially on
{110}, {112}, {100}, and {111} planes, indicating that step edges with
<110>, <100>, and <210> orientations are more stable.[9] This suggests that
the CSL growth should be attempted on (001) vicinal surfaces misoriented
to produce the above step orientations.

3.   Arsenic Pressure and Deposition Temperature Effects

The substrate temperature controls the growth kinetics and atomic
diffusion on the surface. It is now well established that atomic
diffusion is increased when deposition is carried out in the MEE mode.
The coherent TSL[10] cannot be grown if the As flux is incident on the
surface during the Al and Ga flux deposition. Under MEE conditions, the
coherent TSL is observed in the temperature range $550° < T_s < 650°C$. At

250

lower substrate temperature, either the nucleation kinetics at sites other than steps is dominant or the phase stability of the AlGaAs is more important. The TSL grown under regular MBE conditions is highly disordered.[6] All these results point out the role of As on the surface in drastically reducing atomic mobility and preventing quasi equilibrium growth conditions.

4.   Quantum Wire Superlattices and Tilted Superlattice Based Devices

Quantum wire superlattices are produced by deposition of a thin (<400Å) TSL layer between two wider band gap GaAlAs layers.[6-7,22] The quantum wire superlattices have shown novel optical properties associated with the two dimensional carrier confinement produced in the structure. When excited by polarized light, the quantum wire superlattice show a strong in-plane anisotropy in the ratio of the electron-light hole exciton peak luminescence (I_{lelh}) to electron-heavy hole exciton peak luminescence (I_{ehh}).[22] This behavior is due to anisotropic matrix elements of two dimensional quantum confined systems and is explained by a first order theory incorporating the optical selection rules.

A quantum wire laser based on the TSL growth has been fabricated and showed lasing characteristics corresponding to the QW state energy.[23]

The TSL has been used to produce a novel electron wave interference device.[24] This FET structure shows drain current oscillations due to electron wave interferences with the TSL at 4.2°K and large bias currents.

Quantum box superlattices in principle could be fabricated directly by growing a columnar superlattice between two wider band gap AlGaAs layers. Quantum box superlattices hold great promises for producing high mobility FET devices working at room temperature.[4]

CONCLUSIONS

The use of directed epitaxy for producing Tilted superlattices and columnar superlattices has been presented. We have discussed the role of phase stability, growth kinetics, step orientation and regularity on the TSL perfection.   The understanding of the growth process for these structures is still in its infancy and much remains to be done to obtain ultra sharp interfaces and periodic QW superlattices. Nevertheless, quantum wire superlattices and devices have demonstrated novel properties which have been briefly discussed. The exciting properties predicted for QW and QB superlattices warrant further research in the field of directed epitaxy.

Acknowledgements

This paper presents many results obtained by M. Tsuchiya, J. Gaines and S. Chalmers. The author thanks them for their enthusiastic and daring collaboration. Fruitful discussions with A. Gossard, H. Kroemer and L. Coldren are also acknowledged. This research was carried out under the QUEST program: an NSF sponsored Science and Technology Center and under an AFOSR grant.

REFERENCES

1. T. Hiramoto, K. Hirakawa and T. Ikoma, J. Vac. Sci. Technol. B6, 1014, (1988).

2. B. J. vanWees, H. VanHouten, C. W. Beenakker, J. G. Williamson, L. P. Kouwenhoven, D. van der Marel, and C. T. Foxon, Phys. Rev. Lett. 60, 848 (1988).
3. M. A. Reed, J. N. Randall, R. J. Aggarwal, R. J. Matyi, T. M. Moore, and A. E. Wetsel, Phys. Rev. Lett. 60, 535 (1988).
4. H. Sakaki, Jap. J. Appl. Phys. 28, L 314 (1989).
5. P. M. Petroff, A. C. Gossard, and W. Wiegmann, Appl. Phys. Lett. 45, 620 (1984).
6. J. M. Gaines, P. M. Petroff, H. Kroemer, R. J. Simes, R. S. Geels, and J. H. English, J. Vac. Sci. Tech. B6, 1378 (1988).
7. P. M. Petroff, J. M. Gaines, M. Tsuchiya, R. Simes, L. A. Coldren, H. Kroemer, J. H. English, and A. C. Gossard, J. Cryst. Growth 95, 260 (1989).
8. T. Fukui, and H. Saito, J. Vac. Sci. Technol. B6, 1373 (1988).
9. P. M. Petroff, J. Vac. Sci. Technol. B4, 874 (1986).
10. M. Tsuchiya, P. M. Petroff, and L. A. Coldren, App. Phys. Lett. 54, 1690 (1989).
11. Y. Horikoshi, M. Kawashima, and H. Yamaguchi, Jpn. J. Appl. Phys. 27, 169 (1988).
12. Y. Horikoshi and M. Kawashima, J. Cryst Growth 95, 17 (1989).
13. H. Yamaguchi, and Y. Horikoshi, Jap. J. Appl. Phys. 28, 352 (1989).
14. P. M. Petroff, F. Reinhardt, A. Y. Cho, R. Logan, and A. Savage, Phys. Rev. Lett. 48, 179 (1982).
15. M. Srivastava, G. P. Martins, and A. Zunger, Phys. Rev. B31, 2561 (1985).
16. T. S. Kuan, W. I. Wang, and E. L. Wilkie, Appl. Phys. Lett. 51, 51 (1987).
17. H. R. Jen, M. J. Jou, Y. T. Cherng, and G. B. Stringfellow, J. Cryst. Growth 85, 175 (1987).
18. E. Suliga and M. Henzler, J. Vac. Sci. Technol. A1, 1507 (1983).
19. R. J. Phaneuf and E. D. Williams, Surf. Sci. 195, 330 (1988) and Phys. Rev. Lett. 58, 2563 (1987).
20. S. Chalmers, A. C. Gossard, P. M. Petroff, J. Gaines, and H. Kroemer, J. Vac. Sci. Technol. B7, 1357 (1989).
21. Y. Tokura, H. Saito and T. Fukui, J. Cryst. Growth 94, 46 (1989).
22. M. Tsuchiya, J. Gaines, R. H. Yan, R. J. Simes, P. O. Holtz, L. A. Coldren, and P. M. Petroff, Phys. Rev. Lett. 62, 466 (1989).
23. M. Tsuchiya, L. A. Coldren, and P. M. Petroff, 100C 1989 Conference Proceedings, Kobe Japan p 104 (1989).
24. K. Tsubaki, Y. Tokura, T. Fukui, H. Saito, N. Susa, Electron. Lett. 25, 728 (1989).

KINETICS IN MOLECULAR BEAM EPITAXY - MODULATED BEAM STUDIES

C. T. Foxon*, E. M. Gibson+, J. Zhang+, and B. A. Joyce+

*Philips Research Laboratories
Redhill, Surrey, RH1 5HA England
+Semiconductor Materials Interdisciplinary
Research Centre
The Blackett Laboratory, Imperial College
London SW7 2BZ, England

ABSTRACT

This article reviews the present state of our knowledge of the growth of III-V compounds gained from modulated beam mass spectrometry measurements. We have discussed in particular the data relating to the loss of volatile group III elements at high temperatures and have shown that there is much confusion in the literature in this area. New data for the loss of Ga at high temperatures both during Langmuir evaporation and during growth with both As_2 and As_4 has shown that the As species used is not a key factor in determining the loss of Ga during growth and that the presence of Al also does not influence the Ga loss rate during growth.

INTRODUCTION

Molecular Beam Epitaxy (MBE) began as a study of the growth kinetics of GaAs from molecular beams of Ga and As_2 using modulated molecular beam mass-spectrometry (MBMS) measurements.[1] Our fundamental knowledge of the MBE process has come from the extensive use of the UHV techniques, particularly MBMS[2] and Reflection High Energy Electron Diffraction (RHEED),[3] in combination with theoretical studies of growth using thermodynamics,[4] Monte-Carlo techniques[5] and electron occupancy of surface bonds.[6]

The purpose of this paper is to review our current understanding of MBE from MBMS measurements and to provide data recently obtained at elevated growth temperatures. We will also discuss how this data agrees with the results obtained using the alternative methods outlined above.

EXPERIMENTAL TECHNIQUE

The chemistry of the growth process in MBE has been studied in some considerable detail using modulated beam mass spectrometry. Part of the flux desorbing from the surface is detected mass spectrometrically usually using a quadrupole mass spectrometer. The fundamental reason for using modulated beam techniques is to distinguish signals in the mass spectrometer arising from species desorbing from the surface from those produced by background gases in the UHV environment. This is particularly important for volatile species such as As_4 or P_4 which have a significant vapor pressure at ambient temperature.

The simplest method employed consists of studying the change in the intensity of the signal in the mass spectrometer produced by a step function change in the incident or desorbed beams. For this purpose the shutter in front of the molecular beam source may be used. The major problem with this technique is that there is often a change in the background pressure of As_4 in the MBE equipment produced by opening or closing either the group III (Al, Ga or In) or group V (P, As or Sb) shutter. A mass spectrometer unless very carefully arranged will not easily be able to distinguish the resulting time dependent change in desorption rate from the signal produced by the change in background pressure. This technique is therefore limited in general to the study of relatively low vapor pressure species such as the group III elements.

A much better method is to modulate either the adsorbed or the desorbed molecular beam at a frequency high enough to prevent any significant changing background effects. The critical frequency will depend upon the pumping speed in relation to the volume of the system and thus may depend on the species being studied. The critical test is to remove the sample and see if at the frequency used the time dependent signal disappears.

The resulting time dependent signal can be measured and analyzed in a variety of ways. The simplest method is to use phase sensitive detection. The details of this method have been reviewed by Jones et al[7] and Schwartz and Madix.[8] To obtain details of the surface chemistry it is essential to measure the change in amplitude and phase of the signal as a function of frequency and to fit the data to a model of the process. It is essential also to allow for the change in amplitude and phase produced by molecular flight times of the adsorbed and desorbed species and the ion flight time within the detector.

An alternative and somewhat more efficient method is to collect the information using a signal averager (if ion counting techniques are used this will simply be a multichannel scaler) and use Fourier transform techniques to analyze the data.[2] The effects of a finite flight time with a spread of velocities for both the adsorbed and desorbed beams together with the time delays introduced in the detector can all be deconvolved using this method. A timing reference for the modulation is provided by a small lamp and photodetector mounted on the beam modulator.

Inherent in the use of Fourier transform techniques is the assumption that the system behaves as a time invariant linear system. Non linear chemical processes will give rise to frequency components in the signal which are not present in the modulated beam. This can be conveniently recognized by using an excitation which is symmetrical and thus contains only odd harmonics; the presence of even harmonics in the signal indicates a non-linear response. This can be reduced by using a larger DC flux and a small AC perturbation. The signal then represents the derivative of the desorption rate with respect to the adsorption rate, integrating yields the required data.

From modulated beam measurements four quantities can be determined:

1)   the thermal accommodation coefficient of adsorbed molecules
2)   the surface lifetime and binding energies of desorbed species

3)   sticking coefficients
4)   the order of chemical reactions

Thus a complete model of the various reaction kinetic processes giving rise to growth by MBE can be obtained.

GROWTH OF BINARY COMPOUNDS

In MBE the group III elements Al, Ga and In are always supplied as the monomer by evaporation from the liquid. Over most of the temperature range used in MBE the group III elements have a unity sticking coefficient and therefore the growth rate and alloy composition are simply determined by relative supply rates to the surface. At high temperatures however, as Arthur first showed for Ga,[9] the group III element with the higher vapor pressure will have a finite lifetime on the surface and therefore a non unity sticking coefficient. This leads to a reduction in growth rate and a change in alloy composition (this point will be discussed in more detail below in relation to the growth of alloy films).

In MBE the group V elements P, As or Sb can be supplied either as the tetramer by sublimation from the solid or as the dimer by evaporating from a suitable III-V compound (As can be obtained from InAs, GaAs etc). The dimer can also be obtained by dissociation of the tetramer using a two-zone "cracker" furnace with the front end operating at about 900C.

The interactions of Ga and As_2 on a GaAs surface are shown in Fig. 1. According to this model the As_2 molecules are first adsorbed into a mobile, weakly bound precursor state in which the lifetime is less than 10^{-5} s. Growth takes place by As_2 being dissociatively chemisorbed on single Ga adatoms[9-11]. The sticking coefficient of As is therefore proportional to the Ga supply reaching the surface and any excess As is lost by re-evaporation. It follows that stoichiometric GaAs is always obtained providing an excess As flux is supplied. A sticking coefficient of unity is obtained for As_2 on a Ga terminated surface. At low temperatures, As_2 molecules can associate on the surface to form As_4 which subsequently desorbs. This is the only direct evidence for the surface migration of As_2. The other binary compounds formed from dimers probably behave in a very similar manner, but little direct MBMS data has been published to confirm this.

For most layers grown by MBE, As_4, from elemental As has been used to avoid impurities associated with dopants in GaAs or contamination from the hot zone of the cracker. The chemical reactions involved with the tetramer are more complex, as shown in Fig. 2. Pairs of As_4 molecules interact on adjacent Ga sites with the excess As atoms being desorbed.[12] The main experimental observations that led to this model are that the maximum sticking coefficient of As_4 is 0.5 even when a large excess of Ga is supplied to the surface and in addition there is a second order dependence of the As_4 desorption rate on the adsorption rate at low As coverages.

The behaviour with As_4 is in complete contrast to that of As_2 where a unity sticking coefficient can be achieved on a Ga rich surface. The lifetime of the tetramer is also much greater than the dimer on the GaAs surface. This in turn implies that the there is a much higher population density in the mobile precursor state for equivalent fluxes of As. The different behaviour of the two As species may be expected to influence the properties of films grown under otherwise identical conditions.[13] This has been confirmed by a number of observations; studies of the

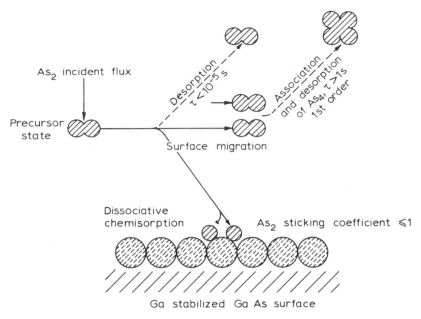

Fig. 1 Model of the growth of GaAs from molecular beams of Ga and As_2.

electronic structure of the GaAs surfaces grown with As_2 and As_4,[14] the lower concentration of deep levels in films grown with the dimer[15] and the improved minority carrier properties of AlGaAs-GaAs double heterostructures obtained with As_2.[16] In this last study both the GaAs and the AlGaAs-GaAs interfaces were improved when the dimer was used. At high substrate temperatures it is possible that As_4 may be decomposed to form As_2 on the surface but there is no direct evidence for this suggestion. The fact that in a two zone cracker furnace a temperature of 900C is required for complete dissociation suggests that this will not occur on the GaAs surface at 700C unless there is some catalytic process involving Ga taking place. In general films grown with the dimer are better than those grown with the tetramer especially at low substrate temperatures.

GROWTH OF ALLOY (MIXED BINARY COMPOUND) FILMS

There are two distinctly different situations in the growth of alloy films namely those in which different group III elements are used such as In_x, $Ga_{1-x}As$ or Al_x, $Ga_{1-x}As$ and mixed group V elements such as $GaAs_xP_{1-x}$ or $GaAs_xSb_{1-x}$. For simplicity we will refer to these as InGaAs, AlGaAs, GaAsP and GaAsSb respectively.

1.   Alloys With Mixed Group V Elements

The situation for mixed group V element alloys is quite simple when growth occurs at temperatures where there is no substantial loss of material from the surface. Under such conditions the element with the lower vapor pressure is incorporated preferentially. Thus, for GaAsP[17] or InAsP[17] the sticking coefficient of As is much greater than P and does not depend upon whether the dimer or tetramer is used. For GaSbAs[18] or

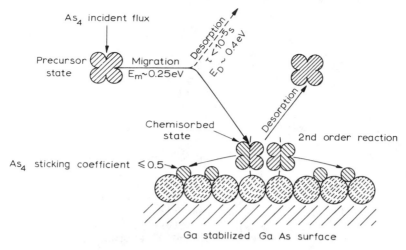

Fig. 2 Model of the growth of GaAs from molecular beams of Ga and As$_4$.

other antimony containing species Sb has the higher sticking coefficient. At low temperatures therefore the composition of the alloy can be easily controlled by limiting the amount of the preferentially incorporated element and providing an excess of the more volatile species. For example, if the As flux, J_{As}, reaching the GaAs$_y$P$_{1-y}$ surface is small compared to the Ga flux, J_{Ga}, and an excess P flux is supplied, the As fraction y will be given by

$$y = 2 \times J_{As}/J_{Ga} \quad .$$

This will be true for As supplied as As$_2$ or As$_4$.

At present the reason for the difference in sticking coefficients for the group V elements is not clear. It has been suggested that it may relate to the different lifetimes of the species.[17,18] This explanation must be incorrect, however, since the lifetime of the As$_2$ dimer is much shorter than that the of P$_4$ tetramer. As$_2$ is nevertheless incorporated preferentially.

At higher temperatures, when rate of loss of the dimer by dissociation and re-evaporation is comparable to the supply rate, composition control becomes much more difficult.[18,19] The rate of loss of the dimer does not relate in any simple way to its sticking coefficients but is determined more by the thermodynamic vapor pressure over the alloy. For GaAsP, therefore, the As fraction will decrease with increasing temperature but for InAsP the reverse will be true.[19] No simple rule can be given therefore in this situation. For Sb containing alloys similar behaviour is expected.[18]

2.   Alloys With Mixed Group III Elements

Both InGaAs and AlGaAs have been studied in some detail. At low temperatures with an excess group V flux the situation is quite straightforward. Both group III elements have a unity sticking coefficient and the growth rate is determined by the total cation flux reaching the surface whilst the composition depends on the relative concentrations of the group III elements.

For In containing alloys of this type the rate of loss of group V species from the surface at high temperatures has been shown both by desorption measurements using MBMS techniques and RHEED studies of the transition temperature from As stable (2x4) to Ga stable (4x2) growth to be similar to that observed in InAs or InP.[21] To compensate for the loss of As_2 it is necessary therefore to use rather high group V fluxes. For AlGaAs, however, it has been suggested that the opposite behaviour is observed and that the alloy grows more As rich than GaAs at high temperatures. Evidence for this statement is based entirely on RHEED studies, but the surface reconstructions observed for AlAs are quite different from those seen in GaAs which may confuse the issue.[22] A much more comprehensive study combining RHEED with MBMS studies is required to resolve this question.

In both InGaAs and InGaP evidence for the segregation and subsequent desorption of the higher vapor pressure element, In, has been observed.[21] This produces both a change in film thickness and composition compared with growth at low temperatures using the same fluxes.

Similar segregation behaviour has been reported in Al containing alloys.[23-25] The evaporation of Ga during the growth of AlGaAs at high temperatures has been considered from a thermodynamic standpoint by Heckingbottom.[26] For a fixed As overpressure the apparent activation energy of evaporation of Ga is about 4.6eV. This has no fundamental significance and will depend upon the particular conditions used. This number is significantly larger, however, than the observed heat of evaporation under equilibrium conditions for Ga over Ga[27] or Ga over GaAs[28-30] which is about 2.7eV. At low temperature, evaporation during growth is occurring under As rich conditions whereas at higher temperatures there is a gradual transition to Ga stable growth. This results in a change in the apparent activation energy. This implies that both the growth rate and composition of AlGaAs alloys will be altered by an increase in As overpressure and that the rate of loss of gallium for fixed temperature and As flux will depend upon the Al flux reaching the surface during growth.

RHEED oscillations corresponding to layer by layer sublimation of GaAs have been observed at high temperature.[31,32] The GaAs re-evaporation rate depends upon the As flux reaching the surface as expected thermodynamically[26] and it was also established that a few monolayers of AlAs could totally suppress the loss of underlying GaAs. This last observation suggests that the observed segregation of Ga during the growth of AlGaAs[23-25] does not always occur.

Direct measurements of the rate of loss of Ga during the growth of GaAs and AlGaAs using RHEED oscillations[33,34] showed no dependence upon the As flux or the Al fraction. This might be expected if Ga is lost from a mobile surface state prior to incorporation into the growing film. It is entirely consistent with the earlier work of Arthur[9] who observed a finite temperature dependent lifetime for Ga. It is also reasonable since the binding energy of Ga in its mobile state (about 2.5-2.7 eV) is much lower than the measured activation energies for the re-evaporation of GaAs 4.7eV[31] and 4.6eV[32] from the RHEED sublimation studies and the value calculated by Heckingbottom 4.6eV.[26] Direct measurements of Ga loss calculated from measurements of film thickness have produced two contradictory results, Alexandre et al[35] found that the Ga re-evaporation rates from GaAs and AlGaAs were similar whereas Fischer et al[36] found that the rate of loss of Ga from AlGaAs was much lower than the corresponding rate from GaAs. Recent mass spectrometric measurements of the Ga desorption rate have shown a reduction due to the presence of Al[37] even though the binding energy for Ga is apparently reduced by the arrival of Al.[38]

At present therefore there are two conflicting views of how to predict the loss of the more volatile group III element during the growth of alloys such as AlGaAs, InGaAs and InAlAs at high temperatures. The thermodynamic arguments suggest that the rate of loss will depend upon both the As flux and the alloy fraction and this will certainly occur for a situation where the loss occurs after growth of the film. If the group III element is lost from the mobile precursor state before reacting with the group V element no strong dependence upon either the anion or less volatile cation flux is expected.

To clarify this situation we have recently been studying this problem using MBMS techniques. In this study the desorption rate of Ga has been measured at high temperatures both under Langmuir evaporation conditions and during growth. Both As_2 and As_4 were used in order to try to explain some of the apparent inconsistencies in the existing literature. The Ga and Al fluxes used were set using the RHEED oscillation technique[3] to give a growth rate of 0.5 MLs^{-1} at low temperatures for GaAs and AlAs. The As flux was also determined using the RHEED oscillation method by predepositing several MLs of Ga.[39] This was set to approximately 5 x 10^{14} cm^{-2} s^{-1}. It should be noted that using this method assumes a unity incorporation rate for the As which may not be true at all temperatures. Relative fluxes of Al, Ga, As_2 and As_4 in the beam monitoring ion gauge were about 0.5:1:5:10 under this set of conditions. The substrate temperature was monitored using a thermocouple shielded to prevent direct radiation from the heaters and over the whole temperature range agreed with estimates from a radiation pyrometer (operating in the $2\mu m$ wavelength range with an emissivity setting of 0.62 for a Si doped substrate) to better than 5C.

Figure 3 shows the data obtained using As_2. From this we see that there is little difference between the rate of loss of Ga during Langmuir evaporation and growth. The measured activation energy is similar to that observed previously for Ga evaporation from Ga[27] and Ga over GaAs[28-30] under Knudsen conditions. In this we also find good agreement with our earlier RHEED data.[31,32]

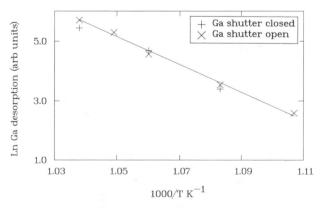

Fig. 3 Measurements of the rate of loss of Ga from GaAs under Langmuir and MBE growth conditions using As_2.

Figure 4 shows the data obtained using As_4. Here at low temperatures we do find a higher desorption rate for Ga during growth compared to the corresponding curve during Langmuir evaporation. This perhaps implies a higher concentration of Ga in a mobile more weakly bound precursor state when growing with As_4 than is the case for growth with As_2. At high temperatures, however, there is no significant difference and the desorption rates during growth and Langmuir evaporation are identical to those observed using As_2 within experimental error. It is interesting to note that the data points observed on reducing temperature (which fall below the solid lines) do not follow the behaviour observed with a progressive increase in sample temperature. This is not we believe due to any temperature error but because the nature of the GaAs surface has been changed by the thermal cycling process (see below).

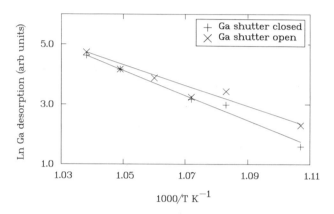

Fig. 4 Measurements of the rate of loss of Ga from GaAs under Langmuir and MBE growth conditions using As_4.

Figure 5 shows data from an experiment where the desorption rate of Ga is measured first during Langmuir evaporation with only As arriving at the surface. By opening the Ga shutter the corresponding loss rate during growth is obtained. The Al shutter is then opened to determine the rate of loss of Ga during the growth of $Al_{0.5}Ga_{0.5}As$. To determine the reproducibility of the data the shutters are again closed in the reverse order. At high temperatures (690C) we find that there is no difference in the Ga desorption rate for the growth of GaAs compared to $Al_{0.5}Ga_{0.5}As$. At lower substrate temperatures the results are inconsistent, the Ga desorption rate can be apparently increased or decreased by the presence of Al reaching the surface.

At present it is not clear why the results observed at low substrate temperatures are inconsistent or why the results we obtain differ from those recently published.[37,38]

It is perhaps worth noting that in the temperature range where our data is irreproducible and where the results in the literature diverge, it is well established that the morphology of AlGaAs is known to be rough, whereas GaAs surfaces grown at the same temperature are smooth. The desorption rate from the surface locally usually follows a cosine distribution. The signal detected by the mass spectrometer may depend, therefore, on the surface topography. Changes in morphology due to the

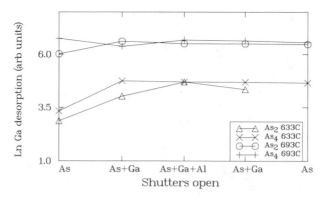

Fig. 5 The effect of Al on the desorption rate of Ga from (Al)GaAs using As$_2$.

arrival of Al may produce an apparent change in desorption rate which is merely an artefact of the measurement technique.

Work on this problem is continuing and the detailed results of this and further experiments will be published elsewhere.[40]

CONCLUSION

In summary, at present, there is conflicting data in the literature concerning the loss of the group III elements from the surface during growth by MBE. The indications from our present work suggest that under the conditions we have used the rate of loss of Ga does not depend upon the As species or the rate of supply of Al to the surface.

REFERENCES

1. J. R. Arthur, J. Appl. Phys. 39, 4032 (1968).
2. C. T. Foxon, M. R. Boudry and B. A. Joyce, Surf. Sci. 44, 69 (1974).
3. P. J. Dobson, B. A. Joyce, J. H. Neave and J. Zhang, J. Crystal Growth, 81, 1 (1987).
4. R. Heckingbottom, in Molecular Beam Epitaxy and Heterostructures, L. L. Chang and K. Ploog eds. (Martinus Nijhoff, Dordrecht, Holland, 719 (1967)).
5. A. Madhukar, Surf. Sci. 132, 344 (1983).
6. H. H. Farrell, J. P. Harbison and L. D. Peterson, J. Vacuum Sci. Technol. B5, 1482 (1987).
7. R. H. Jones, D. R. Olander, W. J. Siekhaus and J. A. Schwarz, J. Vac. Sci. Technol. 9, 1429 (1972).
8. J. A. Schwarz and R. J. Madix, Surf. Sci. 46, 317 (1974).
9. J. R. Arthur, J. Appl. Phys. 39, 4032 (1968).
10. J. R. Arthur, Surf. Sci. 43, 449 (1974).
11. C. T. Foxon and B. A. Joyce, Surf. Sci. 64, 293 (1977).
12. C. T. Foxon and B. A. Joyce, Surf. Sci. 50, 434 (1975).
13. C. T. Foxon, J. Vac. Sci. Technol. B1, 293 (1983).
14. J. H. Neave, P. K. Larsen, J. F. van der Veen, P. J. Dobson and B. A. Joyce, Surf. Sci. 133, 267 (1983).

15. J. H. Neave, P. Blood, and B. A. Joyce, Appl. Phys. Lett. 36, 311 (1980).

16. G. Duggan, P. Dawson, C. T. Foxon and G. W. 't Hooft., J. de Phys. C5, 129 (1982).

17. C. T. Foxon, B. A. Joyce and M. T. Norris, J. Cryst. Growth. 49, 132 (1980).

18. Chin-An Chang, R. Ludeke, L. L. Chang and L. Esaki, Appl. Phys. Lett. 31, 759 (1977).

19. K. Woodbridge, J. P. Gowers and B. A. Joyce, J. Cryst. Growth 60, 21 (1982).

20. J. S. Johannessen, J. B. Clegg, C. T. Foxon and B. A. Joyce, Physica Scripta 24, 440 (1981).

21. C. T. Foxon and B. A. Joyce, J. Cryst. Growth 44, 75 (1978).

22. R. Z. Bachrach, R. S. Bauer, P. Chiaradia and G. V. Hansson, J. Vac. Sci. Technol. 19, 335 (1981).

23. R. A. Stall, J. Zilko, V. Swaminathan and N. Schumaker, J. Vac. Sci. Technol. B3, 524 (1985).

24. J. Massies, J. F. Rochette, and P. Delescluse, J. Vac. Sci. Technol. B3, 613 (1985).

25. J. Massies, F. Turco and J. P. Contour, Semicond. Sci. Technol. 2, 179 (1987).

26. R. Heckingbottom, J. Vac. Sci. Technol. B3, 572 (1985).

27. R. E. Honig and D. A. Kramer, RCA Rev. 30, 285 (1969).

28. J. R. Arthur, J. Phys. Chem. Sol. 28, 2257 (1967).

29. C. T. Foxon, J. A. Harvey and B. A. Joyce, J. Phys. Chem. Sol. 34, 1693 (1973).

30. C. Pupp, J. J. Murray and R. F. Pottie, J. Chem. Therm. 6, 123 (1974).

31. T. Kojima, N. J. Kawai, T. Nakagawa, K. Ohta, T. Sakamoto and M. Kawashima, Appl. Phys. Lett. 47, 286 (1985).

32. J. M. Van Hove and P. I. Cohen, Appl. Phys. Lett. 47, 725 (1985).

33. C. T. Foxon, J. Vac. Sci. Technol. B4, 867 (1986).

34. C. T. Foxon, in Heterojunctions and Semiconductor Superlattices, G. Allen, G. Bastard, N. Boccara, M. Lannoo and M. Voos, eds. Springer-Verlag, 216 (1986).

35. F. Alexandre, N. Duhamel, P. Ossart, J. M. Masson and C. Meillerat, J. de Phys. C5, 483 (1982).

36. R. Fischer, J. Klem, T. J. Drummond, R. E. Thorne, W. Kopp, H. Morkoc and A. Y. Cho, J. Appl. Phys. 54, 2508 (1983).

37. A. J. SpringThorpe and P. Mandeville, J. Vac. Sci. Technol. B6, 754 (1988).

38. K. R. Evans, C. E. Stutz, D. K. Lorance and R. L. Jones, J. Vac. Sci. Technol. B7, 259 (1989).

39. J. H. Neave, B. A. Joyce, and P. J. Dobson, Appl. Phys. A34, 179 (1984).

40. E. M. Gibson, C. T. Foxon, J. Zhang and B. A. Joyce, to be published.

SILICON MOLECULAR BEAM EPITAXY

T. Sakamoto, K. Sakamoto, K. Miki, H. Okumura, S. Yoshida
and H. Tokumoto

Electrotechnical Laboratory, 1-1-4 Umezono
Tsukuba, Ibaraki, 305 Japan

ABSTRACT

Aspects of the molecular beam epitaxy of Si and Ge on Si(001) are described. For Si-on-Si, step formation and growth mechanisms on vicinal surfaces are considered, using RHEED and STM. Ge-on-Si heteroepitaxy is discussed, including growth mode and lattice relaxation. Strained-layer superlattices have been fabricated by phase-locked epitaxy; analysis suggests that Ge_4/Si_{12} has a direct band gap.

INTRODUCTION

Reflection high-energy electron diffraction (RHEED) has provided much information about crystal surface during molecular beam epitaxy (MBE). In the past few years damped intensity oscillations of the specular beam spot in RHEED patterns have been widely studied in GaAs MBE growth.[1-3] The RHEED intensity oscillation technique has been used to study growth dynamics including heterostructure formation in MBE, and has been extended to the precise control of short period superlattice structure.[4-7] The RHEED intensity oscillation has been observed widely such as with III-V and II-VI compound semiconductors, Si, Ge, metal and high T_c superconductor materials. In this paper we will focus on the RHEED intensity oscillations during Si and Ge growth.

In the first part of the paper, we describe the RHEED observation of Si homoepitaxial growth. We first illustrate the effect of high-temperature substrate annealing on the RHEED intensity oscillations of well-oriented Si(001) surface. Then the growth kinetics and the step formation on the (001) vicinal surfaces are investigated using both RHEED and scanning tunneling microscopy (STM).

In the second part of the paper we describe heteroepitaxy of Ge/Si. The initial growth kinetics of Ge/Si(001) heteroepitaxy is discussed at first. We fabricated Ge_m/Si_n strained-layer superlattice (SLS) structures by phase-locked epitaxy (PLE) using RHEED intensity oscillations. A Ge_4/Si_{12} SLS on Si(001) substrate showed intense photoemission and sharp absorption in the energy region of 0.8eV. These results suggest that the Ge_4/Si_{12} sample has a direct band gap.

Kinetics of Ordering and Growth at Surfaces
Edited by M. G. Lagally
Plenum Press, New York, 1990

1.   Experimental

Growth and measurements were performed in an ion-pumped MBE system (base pressure 5×10^{-8}Pa). A Si molecular beam was evaporated from a high-purity single-crystalline Si. The Si(001) substrates used were typically p-type, 2 ohm-cm, well-oriented (<0.08°) and tilted by 0.5°, 1° and 4° toward the [110] azimuth.[8] Samples were cut into pieces of $47\times10\times0.5$mm^3 and subjected to standard chemical cleaning and loaded onto a sample holder by Ta clips. The substrate was then heated by passing a direct current through its length. A two inches lens-shaped substrate having a radius of curvature of 1625 mm was also used which allows growth on all surface orientations within 0.8° of the (001) plane. A thin oxide film was decomposed with an assist of a low-flux Si beam (3×10^{13} atoms/cm$^2\cdot$s) at the substrate temperature of 800°C.[10] Deposition of a Si buffer layer at 700°C and a subsequent annealing at 1000°C were repeated in order to eliminate initial surface roughness and carbon contaminations on the as-cleaned surface.[11] Total buffer layer thickness was typically 200 nm. A 40 keV RHEED system was used for surface analysis. A glancing angle was typically 15 mrad. Detail of the RHEED monitoring system has been described elsewhere.[12]

2.   Results and Discussion

2.a   Si-on-Si(001), Well-Oriented Substrate

In case of well-oriented Si(001) a high-temperature annealing following a buffer layer growth was found to be very effective to get an atomically smooth surface. Figure 1 shows the intensity oscillations of the RHEED specular beam spot ((00) spot) observed in the [110] azimuth during homoepitaxial growth on Si(001) at a substrate temperature of 500°C. The substrate was subjected to various number of annealing cycles before growth. The annealing process consisted of a buffer-layer growth of 20 nm in thickness and a high temperature annealing at 1000°C for 20 min. It can be clearly seen that the amplitude of RHEED intensity oscillations increases with repetition of the annealing process. The initial intensity of the specular beam spot also increases and its FWHM decreases after the cycles of the annealing process. These facts suggest that the annealing process enlarges the size of terraces.

Moreover, during this annealing process, a change of the surface reconstruction was found. That is, a double-domain structure (2x1+1x2) which is usually observed on as-cleaned or as-grown Si(001) changed to a 2x1 single-domain structure after proper high temperature annealing.[13] The single-domain structure on well-oriented Si(001) has later been observed by other groups using the methods of reflection electron microscope (REM),[14] micro-probe RHEED[15] and low-energy electron diffraction (LEED).[16] Then, it has been revealed that the annealed surface consisted of large terraces and surrounding stepbands which contain totally even number of monatomic-layer steps. The reason why the single-domain structure can be obtained on the well-oriented Si(001) surface has not been clear yet but it has been pointed out that the DC current passing through the wafer would play an important role in determining which domain will grow.[16] That is on the surface with very small misorientation ($\theta < 0.1°$), the stable domain is determined not only by the relation between steps and dangling bonds but also by the polarity of the DC current.

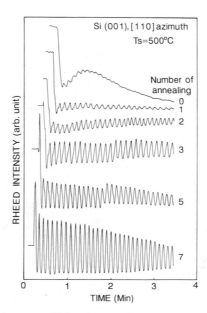

Fig. 1   RHEED intensity oscillations obtained from Si on well-oriented
Si(001): [110] azimuth, repeated annealing process. Each time,
1000°C annealing for 20 min followed a buffer layer growth 20 nm
thick.

The 2x1 single-domain structure has previously been reported for
vicinal surfaces. Kaplan[17] pointed out in his studies using LEED that the
steps of biatomic-layer height could be observed in a vicinal surface
(6°-10° off from (001)). More recently, Aizaki and Tatsumi[18] observed the
single-domain (001)-2x1 structure on a 0.5° off (001) surface only during
the growth. However, to the authors' knowledge, our paper[13] is the first
report concerning a stable single-domain (001)-2x1 structure on the
well-oriented Si (001) surface.

It has been observed that alternating surface reconstructions
between 2x1 and 1x2 appear during growth on the single-domain surface.[19]
In this well-oriented case, both of the surface reconstructions can
stably exist at the lower growth temperature range (400°C to ～ 500°C),
since the influence of the steps on the surface reconstructions in the
large size terraces is considered to be weak. The oscillation mode of the
RHEED specular beam spot strongly depend on the electron-beam incidence
azimuth, as shown in Fig. 2. Though a biatomic-layer mode oscillation is
observed in <110> azimuth, other azimuths than <110> show monatomic-layer
mode oscillation.[20] This distinctive feature can be explained as follows.
In the single-domain structure, there exists large difference in
reflectivity between the 2x1 and the 1x2 surfaces when the electron beam
is directed along the <110> azimuth. In addition, the reflectivity
difference would be enhanced by the anisotropic island formation during
growth. As a result, if the reflectivity difference caused by the surface
reconstruction change is larger than that caused by the "smooth and
rough" surface transition which occurs at every monatomic-layer growth,
the biatomic-layer mode oscillations become observable. This result has
been theoretically explained using multiple-scattering theory.[21]

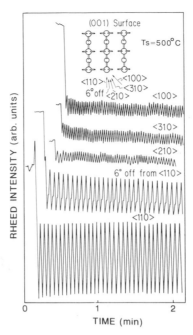

Fig. 2   RHEED intensity oscillations taken from various azimuths on
Si(001): T_s = 400°C, θ_i = 15 mrad.

2.b   Si-on-Si(001), Tilted by Large Angle (4°)

When the surface is misoriented from the (001) plane toward the
<110> direction, steps generated by tilt run along one of the two <110>
azimuths. Before describing the experimental results, it would be
convenient to classify the terrace-and-edge combinations of the stepped
Si (001) surface by the dangling bonds configurations. Figure 3 shows the
two kinds of dangling bond configurations at [110]-oriented
monatomic-layer steps. On both the top and the third atomic-planes,
dangling bonds of Si atoms run along the step edge. This terrace is
called a type-A terrace (terminology after Kroemer).[22] A type-B terrace
is the second atomic plane. Surface atoms of this terrace belong to the
other sublattice, whose dangling bonds point perpendicularly to the step
edge. The surface structure of a Si(001) substrate tilted by 4° toward
the [110] azimuth was double-domain just after the removal of the oxide
film. After high-temperature annealing at 1000°C followed by the growth
of a buffer layer, a single-domain structure was observed.

Figure 4 shows RHEED patterns taken from (a) [110], (b) [110] and
(c) [100] azimuths. In the [110] incident azimuth; (a), parallel to the
step edges, half-order spots diffracted from the type-B terraces
disappear in the zeroth-order Laue zone (L_0). On the other hand, the
half-order Laue-ring ($L_{1/2}$) diffracted from the type-A terraces is
observed outside of L_0. In the [110] incident azimuth; (b), looking down
the staircase, fractional-order streaks (0 m/2) diffracted from type-A
terraces are clearly observed between integral-order streaks. In both (b)
and (c), well-developed streaks can be observed.   These are due to
surface disordering caused by the steps.

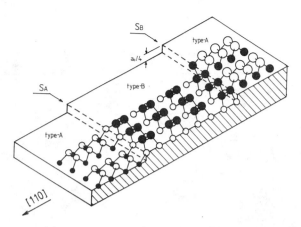

Fig. 3   Schematic illustration of a Si(001) vicinal surface tilted
          toward the [110] azimuth.   Two types of steps, S_A and S_B, and
          two types of terraces, type-A and type-B, are indicated.

Figure 5 shows the RHEED pattern obtained from the [110] azimuth.
Integral-order diffraction spots (m 0) were split into two or three spots
due to the interference between waves scattered from staircase terraces.
A mean separation of terraces was estimated to be 4 nm from the splitting
distance. This value is equal to that obtained through an assumption of a
uniformly stepped surface with biatomic-layer high steps. We conclude
that the vicinal Si(001) surface tilted by 4° toward the [110] azimuth
has the 2x1 single-domain structure with ordered biatomic-layer high
steps which separate the type-A terraces. This type of biatomic-layer
step is called D_B-step.

Figure 6 shows the RHEED intensity evolutions taken from the [110]
azimuth during growth on the Si(001) 4° off substrate at several
substrate temperatures.   At a substrate temperature above 580°C, no
significant change was observed in the RHEED pattern and the specular
intensity. This result indicates the occurrence of step growth in the
biatomic-layer unit. On the other hand at substrate temperatures below
450°C, RHEED intensity oscillations were observed. The oscillations,
however, damped rapidly and then the RHEED pattern changed to a streaky
one and the half-order streaks diffracted from type-B terraces became
observable. We conclude from these results that the growth occurs in
monolayer-by-monolayer fashion at least at the initial stage of the
growth.

2.c   Si-on-Si(001), Tilted by Small Angle (0.3°-1°)

The surface step structure of samples tilted by a small angle
(0.3°-1°) is quite different from that of the well-oriented substrate or
4°-off substrate. Using STM we have investigated anisotropic step-edges
on the Si(001) 0.3°-off surface towards [110].[23] In this study, we used
an ultra-high vacuum (UHV)-STM system[24] with a mechanically ground
Pt-tip.   An Sb doped n-type Si(001) wafer (10 ohm·cm) was cleaned using
standard chemical cleaning methods.[9] The sample was thermally flashed
several times at 1200°C for 20-30 sec in a pressure of 10^{-7} Pa. After
cooling the substrate down to a room temperature, STM measurements were

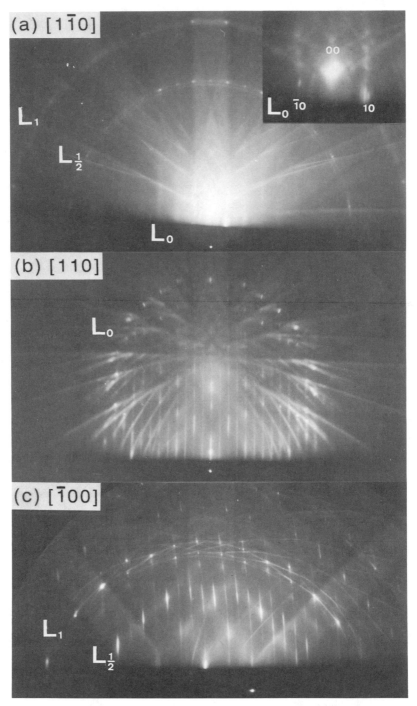

Fig. 4   RHEED patterns obtained from a Si(001) vicinal surface tilted by 4° toward [110] after high-temperature annealing: (a) [1$\bar{1}$0] azimuth, (b) [110] azimuth and (c) [$\bar{1}$00] azimuth.

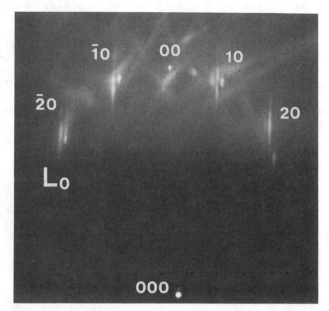

Fig. 5 RHEED pattern obtained from Si(001) 4° off substrate after high-temperature annealing: [1̄10] azimuth, θ_i = 33 mrad.

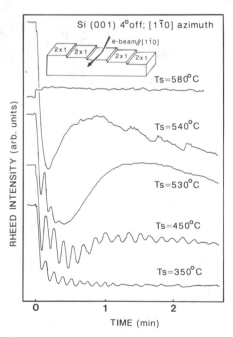

Fig. 6 Substrate temperature dependence to the RHEED intensity oscillations of the specular beam taken from [1̄10] azimuth during growth of Si on Si(001) 4° off toward [110]: θ_i = 7mrad.

carried out in a variable current mode. A typical 210 x 210 nm^2 STM image is shown in Fig. 7. We can see terraces with almost uniform size and edges nearly parallel to each other. From the ratio of the monatomic-step height to the mean terrace width (30nm), the misorientation from the Si (001) surface is estimated at 0.3°. In addition to the above observation, the STM image shows the existence of two types of characteristic step edges, straight and kinked step edges, which are indicated in the photograph by A and B respectively, alternating with each other. Microscopically, the straight and the kinked step edges correspond to Chadi's steps[25] as follows: the straight step edges A are mainly composed of an S_A-type step and the indents and juts of the kinked step edges B are composed of both S_A- and S_B-type steps.

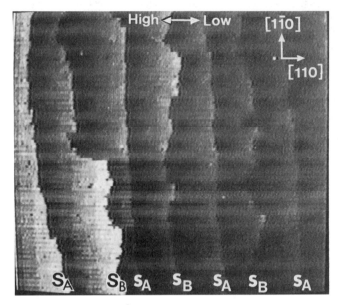

Fig. 7   A typical 210x210 nm^2 STM image of the Si(001) 0.3° off clean surface.

From the viewpoint of step formation kinetics, we shall discuss the difference in roughness between the two type of step edges observed in STM images. Firstly, according to Chadi's calculation, the step formation energy of S_A-steps is considerably smaller than that of S_B-steps.[25] So at the step edge A, the kinks which are composed of S_B-steps are energetically expensive to form while at the step edge B the kinks which are composed of S_A-steps are relatively inexpensive to form. Then, the step edges A are rather stable while the step edges B are very active in thermal equilibrium conditions. As a result, step edge A tends to become straight while step edge B hardly tends to become straight.

A typical RHEED pattern taken in the [$1\bar{1}0$] azimuth of a (001) vicinal surface tilted by 1° after high-temperature annealing is shown in Fig. 8a. Unlike the sample tilted by 4°, half-order spots (1/2 0) and (1/2 0)) on L_0 diffracted from the type-B terraces were clearly seen. The half-order Laue-ring ($L_{1/2}$) diffracted from the type-A terraces was observed at the same time (not shown in Fig. 8). The surface reconstruction was thus assigned to the double-domain structure. The

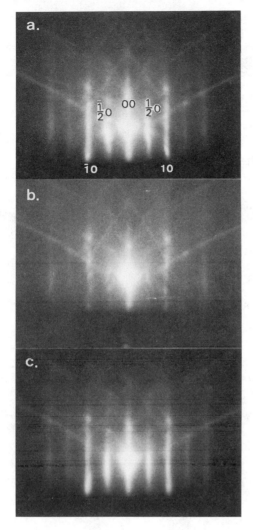

Fig. 8   RHEED patterns obtained from [110] azimuth of a Si (001)
substrate tilted by 1° toward [110] azimuth; (a) before, (b)
during, and (c) after growth: T_s = 500°C.

adjacent terraces were therefore considered to be primarily separated by
the monatomic-layer steps as observed in the STM image in Fig 7. The
substrate was subjected to various annealing conditions, but no
significant change of the RHEED pattern was observed. We therefore
concluded that the Si(001) surface tilted by 0.3°-1° has no tendency
toward step doubling through prolonged high-temperature annealing.
However, the mechanism of coexistence of the both types of steps under
thermal equilibrium condition is not clear yet.

Figure 8b shows a RHEED pattern during growth. The substrate
temperature was 500°C and the growth rate was 1 nm/min. The half-order
spots on the L_0 disappeared. The surface structure during growth was
identified as the 2x1 single-domain structure with the type-A terraces.

The surface structure, however, returned to the double-domain after the growth (Fig. 8c). The single-domain structure was observed during growth at medium-range temperatures (450°C-550°C). Our observation agrees with the earlier work of the 0.5°-tilted samples by Aizaki and Tatsumi.[18] It should be noted here that the single-domain structure during growth is not necessarily due to step doubling. Aizaki and Tatsumi had proposed two possibilities: (a) step doubling and the biatomic-layer step growth, (b) monatomic-layer growth which selectively destroyed the reconstruction of the type-B terraces, but they could not determine the validity of each model.

The single-domain structure during growth was further examined by RHEED intensity analysis. Figure 9 shows a typical RHEED intensity evolution of the (1/2 0) spot in the [1̄10]-incident azimuth. The intensity corresponds to the terraces labeled type-B in Fig. 3. We paid attention to the initial decrease of the intensity and its recovery after growth.

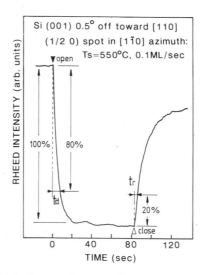

Fig. 9   A typical RHEED intensity evolution of the (1/2 0) spot in the [110] pattern.   Definitions of characteritic parameters discussed in the text are also shown.

Figure 10 shows the deposited layer thickness vs substrate temperatures when the intensity decreased by 80% to the first minimum. At that thickness the surface was almost covered with the type-A terraces due to the step doubling. As is shown in the figure the step doubling was mostly completed after less than one monolayer deposition and was independent of the growth rate. This indicates that since S_B are very active and advance much faster than S_A, eventually S_B-steps become very close to S_A-steps and D_B-steps might be partly formed. The selective growth behavior on the vicinal surface has recently been directly observed by another group using STM.[26] The results are in good agreement with our proposed model. This phenomenon is probably due to the large difference in the step formation energy between two steps as mentioned before.

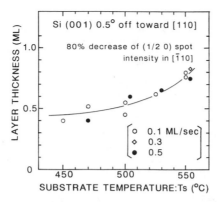

Fig. 10 Substrate temperature dependence of deposited layer thickness when the intensity of the (1/2 0) spot decreased by 80% to the first minimum.

On the other hand, the transition of surface reconstruction from single-domain to double-domain after growth seems to be a sophisticated process. We have no argument yet whether after growth type-A and type-B terraces come back eventually to their size before growth. We will discuss here only the initial stage of the dissociation process of the steps into the double-domain structure.

The recovery of the (1/2 0) spot intensity, which reflected the increase of type-B terraces, was very sensitive to substrate temperatures. We characterized the initial step dissociation process by t_r which is the period of intensity recovery by 20% to the value before growth. Figure 11 is the Arrhenius plot of $1/t_r$. The activation energy of

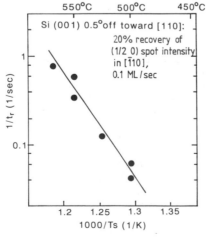

Fig. 11 Arrhenius plot of the inverse of the period t_r needed for the intensity to recover by 20% to the value before the growth.

intensity recovery E_r was 2.4eV. To understand the kinetics of this intensity recovery process, we made the following speculation according to the step formation energy. The step edge of S_A would be straight and the S_A-to-terrace transfer rate of atoms would be small because of its small step energy. The S_B, which has larger step energy, on the contrary, would have many kinks and the S_B-to-terrace transfer rate would be large when the growth was stopped. The STM image shown in Fig. 7 supports the above speculation. Therefore, the adatom density n_B in equilibrium with S_B is supposed to be larger than n_A, in equilibrium with S_A. We considered the following model to explain the initial stage of the increase of type-B terraces which are considered to be small enough during the growth. Si atoms are detached from the kinks of S_B. Adatoms diffuse from S_B to S_A on the type-B terrace driven by the adatom density difference n_B-n_A and are captured by the kinks of S_A. According to this model the temporal evolution of the type-B terrace-width (L_B) after the growth is expressed as

$$L_B = 2(D_s \Delta n/n_0)^{1/2} t^{1/2} , \qquad\qquad (1)$$

where D_s, Δn, and n_0 represent the diffusion coefficient in the dimerization direction, n_B-n_A, and the area density of lattice sites, respectively. It should be noticed here that this equation is valid for only the very initial stage of dissociation, and cannot be applied throughout the intensity recovery process. From Eq. (1), the inverse of 20% recovery time, $1/t_r$, is proportional to $D_s \Delta n$. Therefore the intensity recovery energy E_r is expressed as $E_r = E_d + W_s$ where the activation energies E_d and W_s represent barriers for surface diffusion and kink-to-terrace transition, respectively. Since there is no experimental data on W_s for Si, we roughly estimate it from data on fcc crystals, i.e. 1/3 of the sublimation energy.[27] From the value of S, i.e., 4.5eV for sublimation energy of Si, we deduce W_s as 1.5eV. Using E_r of 2.4eV and W_s of 1.5eV, we obtain E_d of 0.9eV. This estimated value is in good agreement with recent measurements.[28-30]

Ge-ON-Si(001) HETEROEPITAXY

1.   Experimental

Experiments were made in the same MBE system as described in 2.1. Ge was evaporated from a PBN Knudsen cell. The typical growth rate of Ge was about 0.05 ML/sec. Si substrates used were well oriented (001) (<0.08°). The substrates were prepared to form the Si(001)-2x1 single-domain structure. The incidence angles were 4 mrad for the RHEED intensity analyses and 15 mrad for measuring the in-plane lattice constant. The in-plane lattice constant was determined by measuring the spacing of RHEED spots. RHEED patterns during the growth were recorded by a VTR and analyzed by an image processor with 256 x 256 pixels.

2.   Results and Discussion

2a.   Growth Kinetics of Ge/Si(001)

Figure 12 show the intensity variations of the RHEED specular beam spot in the [100] azimuth during Ge/Si(001) heteroepitaxy.[31] The growth temperature was varied from room temperature to 600°C. One period of the RHEED intensity oscillation observed in this azimuth corresponds to a monatomic-layer growth as in the case of Si homoepitaxy. As seen in Fig. 12, at a growth temperature above 350°C, 5 or 6 periods of oscillation were observed clearly. A two-dimensional (2-D) streak RHEED pattern was observed during this period. At the point of peak a, the pattern started to change from 2-D to 3-D (three-dimensional). And then at the point b, where a small shoulder was observed, it became a perfect 3-D pattern. These results indicate that the growth mode of Ge/Si(001) can be classified as the Stranski-Krastanov type.

Figure 13 shows the mean in-plane lattice constant during the growth at 400°C which was measured from the RHEED spot spacing between (10) and (10) spots in the [110] azimuth. The vertical axis represents the relative in-plane lattice constant of the Ge heteroepilayer to bulk Si as follows;

$$\Delta = (a_{surf} - a_{si}) / a_{si} . \tag{2}$$

These data suggest the pseudomorphic growth of Ge is up to about 6 monatomic layers. This layer thickness corresponds well to the period while 2-D layer-by-layer growth was observed as shown in Fig. 12. The

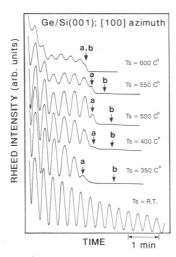

Fig. 12 Growth temperature dependence of the RHEED specular intensity variation during Ge/Si(001) heteropeitaxy. Above 350°C, oscillations were observed up to 5 or 6 periods, and after that, the pattern started to change from 2-D to 3-D at the point a and perfect 3-D pattern was observed at the point b.

peak a, where the surface morphology started to change to 3-D, also corresponds to the initiation of lattice relaxation. Beyond the point b, where the perfect 3-D RHEED pattern was observed, the in-plane lattice constant of Ge layer rapidly turned to the lattice constant of bulk Ge. It is conceivable that misfit dislocations are mostly induced around the point b. A lattice constant minimum was observed between points a and b. Similar behavior was reported in case of strained-layer growth of InAs/GaAs(001),[32] however, a clear explanation has not been given. These anomalies may essentially be related to the fashion of strain relaxation. Further investigation remain to be done. Bevk et al[33] reported that Ge layers were pseudomorphic if the thickness was less than 6 monatomic-layers and were relaxed if they were thicker than 10 monatomic-layers using a method of RBS measurement. Our result is in good agreement with their result.

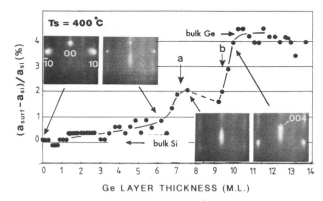

Fig. 13 Mean in-plane lattice constant variation measured from the RHEED
spot spacing between ($\bar{1}0$) and (10) in the [$1\bar{1}0$] azimuth during
the heteroepitaxial growth of Ge/Si(001) at T_s - 400°C.

2b. Ge$_m$/Si$_n$ Strained-Layer Superlattices

Taking into account the experimental results mentioned above,
Ge$_m$/Si$_n$ SLSs were fabricated using the phase-locked epitaxy (PLE)
method.[34] In the PLE method, the RHEED intensity oscillation was
monitored to control precisely the growth period of Si and Ge.

Figure 14a shows the RHEED intensity oscillation observed during
the growth of a Ge$_1$/Si$_4$ SLS. A Ge source shutter was opened for growth,
and closed at the first peak of the oscillation. Then a Si source shutter
was opened, and at the fourth peak of oscillation it was closed. This
controlling procedure was repeated sequentially. Although the amplitude
of the RHEED oscillation decreased as the Ge$_1$/Si$_4$ growth was repeated, we
could observe oscillation clearly throughout 49 superlattice periods.
Figure 14 (b) shows the RHEED intensity oscillation observed during the
growth of a Ge$_4$/Si$_{16}$ SLS. The sequential oscillation damping and recovery
were repeated, and the oscillation was clearly observed throughout 12
superlattice periods. From these results, we confirm that using the PLE
method the Ge$_m$/Si$_n$ SLSs can be fabricated up to the total thickness of
240-245 monatomic-layers (33nm).

The Ge$_m$/Si$_n$ SLS structure was examined using X-ray diffraction and
Raman scattering spectroscopy. We fabricated Si$_{20}$/(Ge$_1$/Si$_4$)$_{100}$/Si(001)
and Si$_{20}$/(Ge$_4$/Si$_{16}$)$_{60}$/Si(001) SLSs. Since these SLS layers are too thick
to apply the PLE method throughout the growth of such a thickness, a
time-control method was also employed after the RHEED oscillation had
been damped. That is, both growth periods of Si and Ge were calibrated
from the RHEED intensity oscillation during the PLE growth period.

The X-ray diffraction spectra of the Ge$_4$/Si$_{16}$ SLS are shown in Fig.
15.[34] At angles of (a) 1.6°, 3.15°, 4.78° and (b) 30.48°, 32.41°, 34.08°,
diffraction peaks are observed. These peaks are attributed to (001),
(002), (003) and (00 18), (00 19), (00 20) diffraction peaks originated
from the Ge$_4$/Si$_{16}$ superlattice structure, respectively.

Figure 16 shows the Raman spectra below 200 cm^{-1} for (Ge$_n$/Si$_{4n}$)
(n = 2, 4 and 6) superlattices.[35] The Raman spectra were measured in the
backscattering configuration with a photon counting system at room

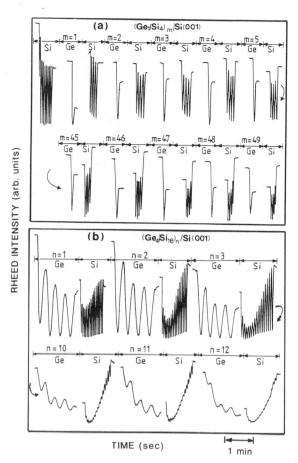

Fig. 14 RHEED intensity oscillations obtained during the growth of (a) Ge_1/Si_4 SLS and (b) Ge_4/Si_{16} SLS.

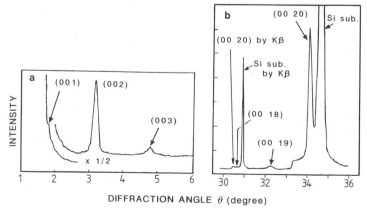

Fig. 15 X-ray diffraction spectra of $(Ge_4/Si_{16})_{60}$ SLS: (a) lower angle region (<6°) and (b) around the Si(004) diffraction peak (30°-36°).

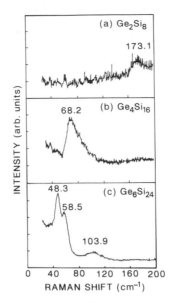

Fig. 16 Raman spectra below 200 nm^{-1} for (a) Ge_2/Si_8, (b) Ge_4/Si_{16} and (c) Ge_6/Si_{24} SLSs.

temperature using the 488 nm line of an Ar^+ laser. The spectra exhibit the peaks assigned to such zone-folded modes. Although the observed peaks are quite broad and some peaks do not split, the measured peak frequencies agree with the calculated ones quite well. This result indicates that the long range periodicity due to the (Ge_n/Si_{4n}) (n = 2, 4 and 6) superlattice structures is properly formed as designed, and that a folded acoustic phonon structure can be explained with the elastic model even for highly-mismatched superlattices, assuming lattice strain.

Photoluminescence and optical absorption measurements were carried out at 4.2K and 2.8K, respectively. In the photoluminescence measurements, the 488 nm line of an Ar^+ laser was used as an excitation light, and signals were detected by a Ge detector. In the optical absorption measurements, a W-lamp light source and a PbS cell detector were used.

In Fig. 17a, the photoluminescence spectra measured for the Ge_4/Si_{12} sample are shown. The emission between 750 and 1000 meV with multiple structures is due to the signal from the SLS layer. Intense sharp peaks are observed at 800 and 865 meV. The energy region where the emission is observed is between the band gap energy of Si and that of Ge lattice-matched to Si. The emission peak at the lower energy side has the stronger intensity. Ge_m/Si_n SLSs with other superlattice structures exhibit only a weak broad emission band, as shown by spectrum (b), in the energy region between 700 meV and 900 meV, although they were prepared by similar procedures in the same MBE system. Compared with spectrum (b), the emission intensity of spectrum (a) is more than one order of magnitude higher, and the peak widths are quite small. If the band gap of

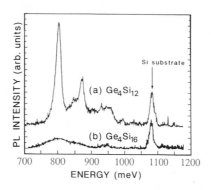

Fig. 17 Photoluminescence spectra of (a) $(Ge_4/Si_{12})_{69}$ and (b) $(Ge_4/Si_{16})_{60}$ SLSs measured at 4.2K. The peaks of about 1100 meV are due to the emission from Si substrates.

the Ge_4/Si_{12} sample is direct type as predicted theoretically, an increase of transition probability and enhancement of emission intensity should be expected, with which the result is consistent.

The optical absorption spectrum measured for the Ge_4/Si_{12} sample is shown in Fig. 18. Above around 790 meV an increase of the absorption coefficient α is seen. Ge_m/Si_n SLSs having other superlattice structures

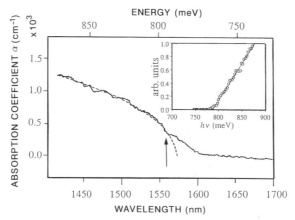

Fig. 18 Optical adsorption spectrum of $(Ge_4/Si_{12})_{69}$ SLS measured at 2.8K. The arrow indicates the energy position of the strongest photoluminescence emission line shown in Fig. 17. Relation of $(h\nu \cdot \alpha)^2$ vs $h\nu$ is shown in the inset.

279

did not exhibit such a increase in the measured spectral region. The energy corresponding to the strongest emission peak in Fig. 17a is located just above the rising position of α. Considering these results together, the rising point of the absorption coefficient is considered to correspond to the band gap energy. Above this rising point, the measured values of the absorption coefficients are in the order of 10^3 cm^{-1}. These values are too large for band-to-band transitions in optical indirect semiconductors. On the contrary, as large α as actually observed is expected for optical direct semiconductors, because transitions can occur without phonon assistance. In the case of direct transitions, the exponent of energy dependence of α is generally $1/2$. In the inset of Fig. 18, the relation between $(h\nu \cdot \alpha)^2$ and $h\nu$ is shown. The linear relation shown in the inset indicates that the observed α follows the energy dependence typical of direct transitions quite well, and that the optical direct transition occurs in the Ge$_4$/Si$_{12}$ SLS sample above 790meV.

Thus, the results described above suggest that the Ge$_4$/Si$_{12}$ SLS sample has a direct band gap. The observation of a direct transition means that the band structure is converted from indirect to direct on account of the zone-folding effect of the Ge$_m$/Si$_n$ SLS structure, considering that the conduction band minima of Si are located along the [001] direction in k-space.

CONCLUSIONS

We have used RHEED to investigate the surface step structure and growth kinetics on Si(001) well-oriented substrates and tilted ones from 0.3° to 4° toward the [110] azimuth. It was found that high-temperature annealing at 1000°C is very effective to obtain an atomically smooth surface. In case of the well-oriented (001) substrate, large terraces with 2x1 single-domain structure were obtained. This dramatic change of surface structure greatly improves the RHEED intensity oscillations during the growth. It was found that 1x2 and 2x1 surface reconstructions appeared in turn during the growth on the 2x1 single-domain surface. The alternative surface reconstruction produces biatomic-layer mode RHEED intensity oscillations along the <110> azimuth while other azimuths show monatomic-layer mode oscillations.

In case of 4°-off substrates toward [110] azimuth, a single-domain structure with biatomic-layer high steps was observed. A step growth with biatomic-layer units was observed at substrate temperatures above 580°C.

For surfaces tilted by 0.5°-1°, a double-domain structure was observed after annealing. However, a single-domain structure was observed only during the growth in the substrate temperature range from 450°C to 550°C. Using STM alternate terraces of double-domain structure showing straight and kinked step edges were observed on 0.3° off surface.

These results are attributed to a large difference in formation energy between the two types of monatomic steps.

The studies of step structure and the growth mechanism on a vicinal surface are very important from the viewpoint of heteroepitaxial growth such as GaAs, SiC and silicide metal on Si substrate. The investigation of the step formation and the growth kinetics on the vicinal surface tilted toward other azimuths than <110> remains to be done.

We reported the dynamic RHEED observation during Ge/Si(001) strained-layer heteroepitaxy over a wide range of growth temperatures. Above 350°C, pseudomorphic growth was observed up to 5 or 6 monatomic-layers, and after that, the growth mode changed from 2-D to 3-D, accompanying a lattice relaxation. This can be classified as the Stranski-Krastanov type growth. The growth mode and the lattice relaxation strongly relate with each other and both are affected by the growth temperature.

Using the PLE method, Ge_m/Si_n SLSs were fabricated. It was confirmed that the PLE method makes possible to fabricate Ge_m/Si_n SLSs up to a total thickness of about 240 monatomic layers. These sample were examined using X-ray diffraction, Raman scattering spectroscopy, photoluminescence and optical absorption measurements. Superlattice periods were confirmed for the samples with $m \geq 2$. It was found that the structure of folded acoustic phonon branches is described with the elastic model including lattice strain. The minimal limit of the formation of Ge layer is 2 monatomic-layers in terms of the superlattice phonon structure. We observed for the first time that a Ge_4/Si_{12} SLS sample demonstrates the luminescent and optical absorption properties typical of direct band gap transitions. The structural conditions for producing such properties should be investigated in further studies, in order to clarify the possible occurrence of direct band gap transitions caused by zone-folding.

Acknowledgements

The authors would like to thank Prof. T. Kawamura of Yamanashi University for helpful discussions and Drs. T. Tsurushima and K. Tanaka for continuous encouragement and S. Nagao, G. Hashiguchi, K. Kuniyoshi, and N. Takahashi for their cooperation in the experimental work.

REFERENCES

1. J. H. Neave, B. A. Joyce, P. J. Dobson and N. Norton, Appl. Phys. A31, 1 (1983).
2. J. M. Van Hove, C. S. Lent, P. R. Pukite and P. I. Cohen, J. Vac. Sci. & Technol. B2, 741 (1983).
3. B. F. Lewis, F. J. Grunthaner, A. Madhukar, T. C. Lee and R. Fernadez, J Vac. Sci. & Technol. B3, 1317 (1985).
4. T. Sakamoto, H. Funabashi, K. Ohta, T. Nakagawa, N. J. Kawai, T. Kojima and Y. Bando, Superlattices and Microstructures 1, 347 (1985).
5. T. Sakamoto, H, Funabashi, K. Ohta, T. Nakagawa, N. J. Kawai and T. Kojima, Jpn. J. Appl. Phys. 23, L657 (1984).
6. N. Sano, H. Kata, M. Nakayama, S. Chika, H. Terauchi, Jpn. J. Appl. Phys. 23, L640 (1984).
7. F. Briones, D. Golmaya, L. Gonzalez and A. Ruiz, J. Cryst. Growth 81, 19 (1987).
8. K. Sakamoto, T. Sakamoto, K. Miki and S. Nagao, J. Electrochem. Soc. 31, 2705 (1989).
9. A. Ishizaka and Y. Shiraki, J. Electrochem. Soc. 133, 666 (1986).
10. K. Kugimiya, Y. Hirafuji and M. Matsuo, Jpn. J. Appl. Phys. 24, 564 (1985).
11. T. Sakamoto, T. Kawamura, S. Nagao, G. Hashiguchi, K. Sakamoto and K. Kuniyoshi, J. Cryst. Growth 81, 59 (1987).
12. T. Sakamoto, K. Sakamoto, S. Nagao, G. Hashiguchi, K. Kuniyoshi and Y. Bando, Thin Film Growth Techniques for Low-Dimensional Structures, R. F. C. Farrow, S. S. P. Parkin, P. J. Dobson, J. H. Neave and A. S. Arrott eds. (Plenum Publishing Co.) 225

(1987).

13. T. Sakamoto and G. Hashiguchi, Jpn. J. Appl. Phys. 25, L78 (1986).
14. N. Inoue, Y. Tanishiro and K. Yagi, Jpn. J. Appl. Phys. 26, L298 (1987).
15. T. Doi and M. Ichikawa, J. Cryst. Growth. 95, 468 (1989).
16. Y. Enta, S. Suzuki, S. Kono, and T. Sakamoto, Phys. Rev. B39, 5524 (1989).
17. R. Kaplan, Surf. Sci. 93, 145 (1980).
18. N. Aizaki and T. Tatsumi, Surf. Sci. 174, 658 (1986).
19. T. Sakamoto, T. Kawamura and G. Hashiguchi, Appl. Phys. Lett. 48, 1612 (1986).
20. T. Sakamoto, N. J. Kawai, T. Nakagawa, K. Ohta, and T. Kojima, Appl. Phys. Lett. 47, 617 (1985).
21. T. Kawamura, T. Sakamoto and K. Ohta, Surf. Sci. 171, L409 (1986).
22. H. Kroemer, J. Cryst. Growth 81, 1983 (1987).
23. K. Miki, H. Tokumoto, T. Sakamoto and K. Kajimura, Jpn J. Appl. Phys. 28, L1483 (1989) and ibid L2107 Errata.
24. H. Tokumoto, K. Miki, H. Murakami, N. Morita, H. Bando, A. Sakai, S. Wakiyama, M. Ono, and K. Kajimura, to be published in J. Microscopical Society.
25. D. J. Chadi, Phys. Rev. Lett. 59, 1691 (1987).
26. A. J. Hoeven, J. M. Lenssinck, D. Dijkkamp, E. J. van Loenen and J. Dieleman, Phys. Rev. Lett. 63, 1830 (1989).
27. W. K. Burton, N. Cabrera and F. C. Frank, Philos. Trans. Roy. Soc. (London) A243, 299 (1951).
28. E. Kasper, Appl. Phys. A28, 129 (1982).
29. S. S. Iyer, T. F. Heinz and M. M. T. Loy, J. Vac. Sci. Technol. B5, 709 (1987).
30. M. Ichikawa and T. Doi, Appl. Phys. Lett. 50, 1141 (1987).
31. K. Miki, K. Sakamoto and T. Sakamoto, Proc. Mat. Res. Soc. Symp. 148, 323 (1989).
32. H. Munakata, L. L. Chang, S. C. Woronick and Y. H. Kao, J. Crystal Growth 81, 237 (1987).
33. J. Bevk, J. P. Mannaerts, L. C. Feldman, B. A. Davidson and A. Ourmazd, Appl. Phys. Lett. 49, 286 (1986).
34. Kazushi Miki, Kunihiro Sakamoto and Tsunenori Sakamoto, J. Cryst. Growth 95, 444 (1989).
35. H. Okumura, K. Miki, K. Sakamoto, T. Sakamoto, K. Endo and S. Yoshida, Appl. Surf. Sci. 41/42, 548 (1989)
36. H. Okumura, K. Miki, S. Misawa, K. Sakamoto, T. Sakamoto and S. Yoshida, Jpn. J. Appl. Phys. 28, L1893 (1989)

Si MOLECULAR BEAM EPITAXY: A COMPARISON OF SCANNING TUNNELING MICROSCOPY

OBSERVATIONS WITH COMPUTER SIMULATIONS

E. J. van Loenen, H. B. Elswijk, A. J. Hoeven, D. Dijkkamp,
J. M. Lenssinck, and J. Dieleman

Philips Research Laboratories
P. O. Box 80.000
5600 JA Eindhoven, The Netherlands

ABSTRACT

Initial stages of Si Molecular Beam Epitaxy (MBE) on Si(001) surfaces have been studied using Scanning Tunneling Microscopy (STM) and Monte Carlo computer simulations.

Regular step distributions have been obtained on samples which were cleaned in UHV at 1500 K. For samples, misoriented by 0.5° towards [110], steps parallel to the dimer rows on the upper terrace are straight (type A), whereas the steps perpendicular to the dimer rows are rough (type B), indicative of a strong interaction between dimers in the row, and a weak interaction between dimers in different rows. STM images of such surfaces after growth of about 0.5 ML of Si at 750 K showed that growth occurs preferentially at the type B steps, resulting in a single domain surface with biatomic steps. For higher coverages island formation is observed.

These observations have been compared with Monte Carlo computer simulations, and the characteristic activation energies involved have been deduced.

INTRODUCTION

Si Molecular Beam Epitaxy (MBE) is a technique for growth of advanced semiconductor device structures and has been the subject of a large number of experimental and theoretical investigations. However, in order to understand the influence of step structures and surface defects on the growth process, information is required about the actual surface structure on an atomic scale. With the advent of the Scanning Tunneling Microscope (STM), it has become possible to observe surfaces at different growth stages in real space with atomic resolution.[1-6] In the present paper, a STM study of the growth of Si on stepped Si(001) surfaces is presented and a direct comparison with results of Monte Carlo computer simulations will be made. As a model system we have chosen the growth at 750K on Si(001) samples with a surface misorientation of about 0.5° towards [110], because in this system almost all aspects of growth can be observed, including step roughening, step flow and island nucleation. It will be shown that all phenomena can be described using a Solid On Solid (SOS) model with only 3 parameters: an activation energy for surface

Kinetics of Ordering and Growth at Surfaces
Edited by M. G. Lagally
Plenum Press, New York, 1990

diffusion, and effective binding energies of atoms in the directions perpendicular and parallel to the dimer row, respectively.

EXPERIMENTAL

The experimental set-up consists of a UHV chamber equipped with a home-built STM,[7,8] a LEED system, Si sublimation sources and a transfer rod with sample heating facilities.

Si(001) samples (20x5x0.3 mm^3, 1 Ωcm, n-type) were cut from commercially available wafers with a misorientation directed predominantly towards [110]. The samples used for this study have a misorientation of 0.52° towards [110] and 0.09° towards [110][9,10] as determined with x-ray diffraction.[9,10]

Before insertion into the UHV chamber, the samples were rinsed in ethanol only. The surfaces were cleaned in situ by heating to 900 K for typically 1 hour to outgas the holder and subsequent heating to 1500 K for 1-2 minutes to remove both oxide and carbon from the surface. Samples annealed at lower temperatures do not exhibit a regular step distribution, because they still contain SiC impurities[11,12] which act as step pinning centres.[9,10]

The samples were resistively heated on the transfer rod and subsequently inserted into the STM. After determination of the clean surface structure, the samples were transferred to the growth position in front of a Si evaporation source which consists of a resistively heated Si strip. The deposition rate was calibrated using Rutherford Backscattering Spectrometry (RBS) analysis of thin layers deposited on C substrates.

The sample was held at 750 K during deposition, and quenched to room temperature immediately after stopping the growth, in order to freeze in the actual state of the growing surface.[4-6] The sample was then transferred back to the STM for analysis.

RESULTS

Scanning Tunneling Microscopy

After cleaning at 1500 K the Si(001) surface exhibits a regular array of monatomic steps, as shown in Fig. 1. The average terrace width of 16 nm corresponds to a misorientation of 0.48° in good agreement with the misorientation determined with x-ray diffraction (0.52°). From the angle between the average step direction and the actual [110] dimer row direction (10°) the misorientation towards [110] has been determined to be 0.09° again in excellent agreement with the x-ray analysis. This indicates that the step array is uniform across the sample surface.

The Si(001) surface exhibits two inequivalent kinds of steps: steps parallel to the dimer rows on the upper terrace (A-type), and steps perpendicular to the dimer rows (B-type). As can be observed in Fig. 1 the A-type steps are straight, whereas the B-type steps are rough. This means that the kink formation energy of an A-type step (E_{kA}) is much higher than that of a B-type step (E_{kB}). We denote the total binding energy between dimers in the direction along the dimer row by $2E_{//}$ ($E_{//}$ per unit length) and in the direction perpendicular to the dimer row by $E_\perp$. On a simple cubic lattice with nearest neighbour interactions only, moving a dimer from a kink site to a site further along the step

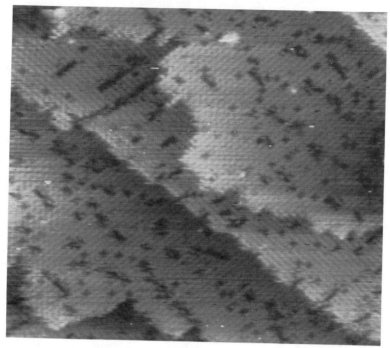

Fig. 1 STM image of a 50x50 nm^2 area of a Si(001) sample after cleaning at 1500 K.

corresponds to the formation of 2 additional kinks. For the A-type step this involves breaking bonds with total energy $2E_{/\!/}$, so

$$E_{kA} = E_{/\!/} \quad . \tag{1}$$

For a B-type step bonds with energy $E_\perp$ are broken, so

$$E_{kB} = 1/2 \; E_\perp \; . \tag{2}$$

Since we have seen that $E_{kA} > E_{kB}$ it follows that

$$2E_{/\!/} > E_\perp \quad . \tag{3}$$

During growth at sufficiently low temperatures, the stronger interaction in the direction along the dimer rows as compared to that perpendicular to the rows, should result in longer residence times for adatoms at B-steps. Therefore, with a supersaturation during growth, the adatoms have a larger probability of being incorporated into the terrace, leading to a faster growth of B steps, provided island nucleation is prevented. This is indeed the case for the growth conditions discussed here, i.e. T = 750 K, a deposition rate of 0.6 ML/min. and average terrace widths of 16 nm.

Figure 2 shows an STM image of the stepped Si(001) surface after the growth of about 0.7 ML of Si. Growth has occurred preferentially at the B-steps, which has resulted in the formation of a single domain surface with double steps. Steps separated by one or more dimer rows are also

Fig. 2 STM image of a 50x45 nm^2 area of a Si(001) sample after deposition of ~0.7 ML of Si at 750 K.

found, particularly near kinks in the underlying step. These structures are very similar to those observed on highly misoriented Si(001) surfaces.[13]

Several observations can be made from Fig. 2. Firstly, the B-type step has moved entirely across the A-terrace (defined as the upper terrace adjacent to the A-type step) after deposition of about 0.7 ML of Si. This means that almost all atoms which have been deposited on the B terrace must have diffused down the adjacent B-step and/or across an A-step.

Secondly, upon completion of the double steps, some islands have nucleated in the middle of the terrace.

Finally, at some points the rapid growth of the dimer rows has been blocked, resulting in the troughs in the growth front. The nature of the blocking centers is unclear at present, but they likely consist of impurities deposited on the surface during growth. The missing dimers and other vacancies, which are always present in the clean Si(001) surface,[14] apparently do not block the growth. This can be concluded from the fact that the number of troughs is much smaller than the number of missing dimer defects.

Monte Carlo Computer Simulations

Computer simulations have been performed, in order to obtain a more quantitative description of the growth processes described above.[15]

A fast Monte Carlo simulation algorithm is used.[16] In this algorithm rates are calculated for deposition and migration events, depending on the total surface configuration. Therefore, the process can be coupled to a true time scale which allows a direct comparison with the experimental conditions.

In the growth model identical "average atoms" are used, with anisotropic nearest neighbor interactions to simulate the anisotropy induced by the surface dimer reconstruction, as described by Clarke et al.[17-20] The dimer formation itself is not included. In the model, the surface is represented by a 160 x 60 simple cubic lattice with periodic boundary conditions, where the edges correspond to the Si[110] and Si[1$\bar{1}$0] directions, that is, the unit distance a corresponds to 0.384 nm. The surface contains an A-type and a B-type step, initially both straight and spaced 80 unit distances apart. This corresponds to a smaller misorientation than used in the experiments, however similar simulation results were obtained for the actual 0.5° misorientation.[15]

Surface migration is included by calculating a hopping probability $k(E,T)$ for each surface atom:

$$k(E,T) = k_o \exp(-E/k_b T) ,\tag{4}$$

where k_o is a frequency factor (assumed to be 10^{13} s^{-1}), k_b is Boltzmann's constant and T is the sample temperature (750 K). The energy barrier E consists of a surface contribution E_s plus a term representing the number and type of nearest neighbour bonds formed:

$$E = E_s + mE_\parallel + nE_\perp \ (m,n = 0,1,2) ,\tag{5}$$

where m and n are the number of bonds formed in the two orthogonal directions respectively. The direction of the anisotropy rotates over 90° on successive layers. Overhangs and buried vacancies are not allowed. In accordance with our STM observations, migration across steps can occur, i.e., an atom desorbing from a step has equal probability of moving to the upper or the lower terrace. The diffusion is assumed to be isotropic.[3]

Deposition consists of a random site selection for additional atoms, at a rate of 0.6 ML/min corresponding to the actual deposition rate in the experiments. For the temperature used here, we have not included an additional local search for an optimal position immediately after arrival.[17-20] Evaporation is negligible at 750K.

The only free parameters in the simulation are the three energy terms E_s, $E_\parallel$ and $E_\perp$. The whole growth sequence observed at 750 K, i.e., roughening of the B-type step, double step formation and subsequent island nucleation, has been simulated with $E_s = 1.0$ eV, $E_\parallel = 0.5$ eV and $E_\perp = 0.05$ eV.

The results are shown in Fig. 3. Starting out from straight steps, annealing at 750 K results in spontaneous roughening of the B-type step, due to the very small interaction energy parallel to that step edge (Fig. 3a). During growth (Fig. 3b-3m), the B-type step remains rough and grows much faster than the straight A-type step, due to the much higher binding energy and correspondingly higher residence time and incorporation probability for adatoms. It is important to notice that during this stage of the growth, no island formation occurs. The adatom density is so small that the mean free path is large enough for all atoms to reach a step before nucleating into islands with other adatoms.

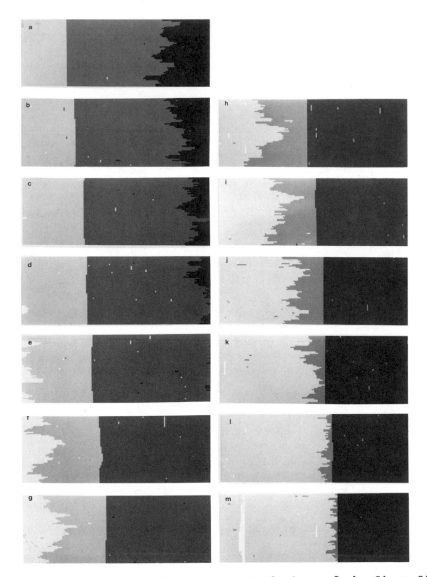

Fig. 3 Results of Monte Carlo computer simulations of the Si on Si(001)
        growth process.   (a) After annealing only, to 750K for 20 sec.
        (b)-(m) After incremental depositions of 0.1 ML.   T = 750 K,
        deposition rate = 0.6 ML/min., E_s = 1.0 eV, $E_\parallel$ = 0.5 eV and
        $E_\perp$ = 0.05 eV.

        During the formation of the double step the preferential adsorption
sites at the B-type step begin to disappear and the adatom density starts
to increase. Islands then nucleate and grow (Fig. 3m). Since the adatom
density is highest far away from the steps (sinks), the islands nucleate
preferentially in the middle of the terrace, in excellent agreement with
the experimental observations (Fig. 2).

Since there is a finite probability for atoms to grow at the A-type step, deposition of more than 0.5 ML is required for the B-step to reach the A-step. Starting from a 55% coverage of the surface with B terraces (Fig. 3c), as observed experimentally, it takes about 0.9 ML to form a single domain surface with the present set of parameters, which is slightly more than observed in the experiments.

The ratio of $E_{/\!/}$ and $E_{\perp}$ chosen here (10:1) gives a qualitative description of the observed shape of small islands and of the B-step roughness. However, the anisotropy in binding energies is only directly reflected in the equilibrium shape of islands. During growth the length to width ratio of the islands is kinetically determined: it depends on the growth velocities of the A- and B-type island edges, and therefore also depends on the diffusion energy as well as the temperature and supersaturation. In future studies[15] parameter sets with a binding energy anisotropy closer to (3:1) will be included in order to be able to describe the elongated growth shape as well as the equilibrium shape of islands.[3]

The surface diffusion energy $E_s = 1.0$ eV is in good agreement with other observations.[21,22] The simulations show island nucleation before the double steps are formed when higher values are chosen for E_s. Clarke et al. used a value of 1.3 eV to describe RHEED oscillations.[17-20] The difference is probably due to the absence in our simulations of the "free diffusion" steps immediately after arrival of an atom.[17-20]

It is tempting to regard the effective energies $E_{/\!/}$ and $E_{\perp}$ as true binding energies. The step formation energies in the model would then simply correspond to $1/2E_{\perp} = 0.025$ eV/a for an A-type step and $1/2E_{/\!/} = 0.25$ eV/a for a B-type step. These values are in fact in fairly good agreement with those calculated by Chadi,[23] who found 0.01 eV/a for an A-type step, and 0.15 eV/a and 0.31 eV/a for the two different type-B step edge configurations which are observed experimentally.[24]

However, $E_{/\!/}$ and $E_{\perp}$ are merely effective bond energies for the hypothetical identical atoms in the simulation. Modification of the model to include the formation of (strong) dimer bonds will strongly affect the values for $E_{/\!/}$ and $E_{\perp}$. In such a modified model a dimer bond energy will have to be included, as well as an additional term to ensure aligning of the dimers in rows,[25] which significantly increases the number of free parameters.

CONCLUSIONS

In conclusion, with STM we have observed the formation of a single domain Si(001) surface during MBE growth. The A-type step remains straight, whereas the B-type step roughens. Growth occurs preferentially at the B-type step, that is at the ends of the dimer rows, until a single domain surface is formed after which islands are nucleated. These phenomena indicate that the interaction between dimers in the direction parallel to the dimer rows is much stronger than in the direction perpendicular to the rows. The entire growth process at 750 K can be simulated with a relatively simple SOS model, using an activation energy for surface diffusion $E_s = 1.0$ eV, and effective bond energies parallel and perpendicular to the dimer rows of $E_{/\!/} = 0.5$ eV and $E_{\perp} = 0.05$ eV per atom respectively.

REFERENCES

1. U. Köhler, J.E. Demuth and R. J. Hamers, J. Vac. Sci. Technol. A7, 2860 (1989).
2. R. J. Hamers, U. K. Köhler and J. E. Demuth, Ultramicroscopy 31, 10 (1989).
3. M. G. Lagally, R. Kariotis, B. S. Swartzentruber and Y. W. Mo, Ultramicroscopy 31, 87 (1989).
4. A. J. Hoeven, J. M. Lenssinck, D. Dijkkamp, E. J. van Loenen and J. Dieleman, Phys. Rev. Lett. 63, 1830 (1989).
5. A. J. Hoeven, E. J. van Loenen, D. Dijkkamp, J. M. Lenssinck and J. Dieleman, Thin Solid Films, Proc. of the E-MRS Conference, Strasbourg 1989, in press.
6. A. J. Hoeven, D. Dijkkamp, E. J. van Loenen, J. M. Lenssinck and J. Dieleman, J. Vac. Sci. Technol. A8, 207 (1990).
7. E. J. van Loenen, D. Dijkkamp and A. J. Hoeven, J. Microscopy 152, 487 (1988).
8. A. J. Hoeven, D. Dijkkamp, E. J. van Loenen and P. J. G. M. van Hooft, Surface Sci. 211/212, 165 (1989).
9. D. Dijkkamp, E. J. van Loenen, A. J. Hoeven and J. Dieleman, J. Vac. Sci. Technol. A8, 218 (1990).
10. D. Dijkkamp, A. J. Hoeven, E. J. van Loenen, J. M. Lenssinck and J. Dieleman, Appl. Phys. Lett. (1989), to be published.
11. Y. Ishikawa, Y. Hosokawa, I. Hamaguchi and T. Ichinokawa, Surface Sci. 187, L606 (1987).
12. T. Ichinokawa and Y. Ishikawa, Ultramicroscopy 15, 193 (1984).
13. P. E. Wierenga, J. A. Kubby and J. E. Griffith, Phys. Rev. Lett. 59, 2169 (1987).
14. R. J. Hamers and U. K. Köhler, J. Vac. Sci. Technol. A7, 2854 (1989).
15. H. B. Elswijk et al., to be published.
16. P. A. Maksym, Semicond. Sci. Technol. 3, 594 (1988).
17. S. Clarke, M. R. Wilby, D. D. Vvedensky and T. Kawamura, Phys. Rev. B40, 1369 (1989).
18. S. Clarke, M. R. Wilby, D. D. Vvedensky, T. Kawamura and T. Sakamoto, Appl. Phys. Lett. 54, 2417 (1989).
19. S. Clarke and D. D. Vvedensky, Phys. Rev. B36, 9312 (1987).
20. S. Clarke and D. D. Vvedensky, J.Appl. Phys. 63, 2272 (1988).
21. E. Kasper, Appl. Phys. A28, 129 (1982).
22. K. Sakamoto, E-MRS Conference, Strassbourg 1989.
23. D. J. Chadi, Phys. Rev. Lett. 59, 1691 (1987).
24. R. J. Hamers, R. M. Tromp and J. E. Demuth, Phys. Rev. B34, 5343 (1986).
25. S. A. Barnett, A. Rockett and R. Kaspi, J. Electrochem. Soc. 136, 1132 (1989).

SCANNING TUNNELING MICROSCOPY STUDY OF Ge MOLECULAR BEAM EPITAXY ON Si(111) 7 X 7

O. Jusko, U. Köhler, and M. Henzler

Institut für Festkörperphysik
Universität Hannover
D3000 Hannover 1, Federal Republic of Germany

ABSTRACT

Scanning tunneling microscopy is used to study the nucleation and epitaxial growth of germanium on Si(111) substrates at elevated temperature. Different growth regimes of germanium are observed for coverages from the submonolayer range to about 10 ML. For small coverages germanium grows pseudomorphic. Triangular shaped islands and large reconstructed areas of (5 x 5) or (7 x 7) are observed. For coverages of about 5 ML the misfit of germanium to silicon leads to strain, which results in a quite disordered layer. For coverages of more than 10 ML the layer relaxes and assimilates the germanium bulk behavior. Here dislocations have been observed.

INTRODUCTION

We report on the observation of nucleation and growth phenomena of germanium MBE on Si(111) with tunneling microscopy. This surface has already been studied by LEED,[1,2] RHEED[3,4] and STM.[5,6,7] But atomic scale information of in situ grown germanium layers is still not available. The goal of this investigation is to follow the various stages of the epitaxial layer as a function of the deposited amount. Here it is of special interest, whether Ge heteroepitaxy shows a significantly different growth behavior than the better known homoepitaxy of silicon on silicon.[9]

EXPERIMENTAL

The experiments were performed in a UHV double chamber (base pressure : < 10^{-10} mbar) containing the STM, LEED and a germanium - evaporator. The sample can be moved through the chamber by a transfer rod. A loadlock is used to introduce new samples without breaking the vacuum. The STM itself and the mechanical sample tip approach are described elsewhere.[10]

The samples were cut from highly doped Si(111) wafers, and were in situ cleaned by thermally removing the native oxide at 950°C followed by a 1250°C flash. Heating is done resistively. Before this procedure the sample and its holder were thoroughly degassed at 700°C for several

Kinetics of Ordering and Growth at Surfaces
Edited by M. G. Lagally
Plenum Press, New York, 1990

hours. The germanium was deposited in situ from a heated carbon tube. The chosen deposition rate was 0.8 ML/min. Here we define 1 ML as the number of atoms required to form a complete bilayer. Absolute coverage calibration was performed by counting the covered fraction of the surface after depositing less than 2 ML of Ge. Most images were obtained at a bias voltage of +2V (tunneling into empty surface states) and a tunneling current of 1 nA. The size of the Si - (7 x 7) unit cell is used to calibrate lateral distances to scan voltages.

RESULTS

In the temperature range from 300°C to 600°C, where all experiments took place, we observed always epitaxial growth. Depending on the deposited amount four different growth regimes could be observed.

1.   Submonolayer Range

All deposited germanium grows in complete bilayers (step height 3.2 Å). We never observed smaller step heights. After depositing 0.3 ML to 2 ML we found triangular shaped islands on the surface. There are three preferred directions of growth reflecting the threefold symmetry of the silicon substrate [Fig. 1]. In this low coverage regime already some main results of the germanium epitaxy can be derived. So it could be observed, that domain boundaries in the substrate or larger point defects act as nucleation centers. If such a nucleation center is formed, further atoms from the incoming flux will preferentially stop at the barrier formed from a covered domain boundary leading to a agglomeration of germanium at the defect and a depletion in the surrounding area. This could be called 'defect decoration'. A large fraction of grown islands has a less perfect structure than the substrate. Quite often they are partly covered by second layer, even if the deposited amount is fairly small. This is understandable, because the faults in the first layer act also as preferred nucleation centers.

The substrate in between the islands seems to be unaffected. A (7 x 7) superstructure on an germanium island can not be distinguished from the substrate. No contrast between silicon and germanium or a possible alloy of both has been observed.

If the deposition temperature is about 500°C to 600°C we observed large surface areas covered by (5 x 5) or (7 x 7) germanium reconstruction. The structure of the Ge - (5 x 5)[6] is very similar to the well known DAS - (7 x 7) - silicon reconstruction.[9] In this growth regime germanium behaves very similar to silicon homoepitaxy except the fact, that the (5 x 5) structure, which also was observed at silicon growth, is more stable.

2.   Growth at Steps

Some samples used as substrates had a random miscut of about 0.5°. After annealing they showed a nearly perfect regular step array with straight and parallel steps. If now germanium is deposited onto such a surface the completion of nucleation and diffusion processes on the terraces and at the steps can be studied nicely. In the upper left area of Fig. 2 there is a large terrace with islands on top. The smaller terraces in the image do not show separated islands or the grown step edges have captured one while propagating. Islands are only stable on large terraces. If the terrace width is smaller than an effective diffusion length, the grows will prefer the step edges. The main step direction is determined by the macroscopic miscut, but because there are

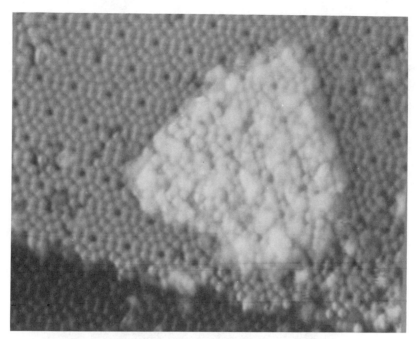

Fig. 1 Triangular shaped germanium island on the silicon substrate. On top is a (7x7) reconstruction and some less ordered Ge. Deposition: 1/3 ML at 500°C.

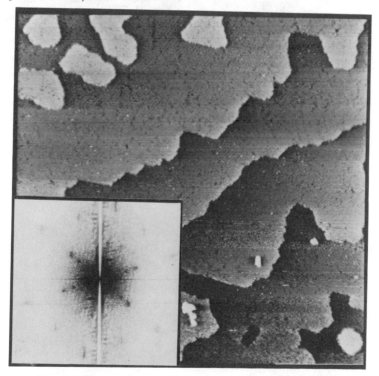

Fig. 2 A step array after deposition of 2 ML of Ge at 500°C. The image reflects the competition between step propagation and island nucleation. Size: 3500 x 3500Å^2. The inset shows the corresponding Fourier transform with 1/5 and 1/7 spots similar to a LEED pattern.

preferred directions of growth, the steps start to meander locally. If Fig. 2 is examined closely, two different types of domains can be distinguished. One has a periodic structure and the other one a finer or somewhere not resolved structure. The power spectrum of a 2D fourier transformation of the image can be seen in the inset in Fig. 2. It shows that these structures are (5 x 5) and (7 x 7). They appear as 1/5 and 1/7 'diffraction spots'.

3. Critical Thickness

At a thickness of about 5 ML the germanium layer has lost its silicon similarity. Very few areas can be found, where DAS - structures are to be seen. The significant strain, which is caused by the 4 % misfit of germanium to silicon, leads to an outermost layer exhibiting only short range order. Small domains of 100 Å to 300 Å diameter of various superstructures like (2 x 2) and c(2 x 4)[5] can be found. We even found a local (2 x 1)- reconstruction, which is normally only stable on cleaved germanium crystals. Figure 3 shows this reconstruction for two different bias voltages where the surface is not reconstructed, adatoms take

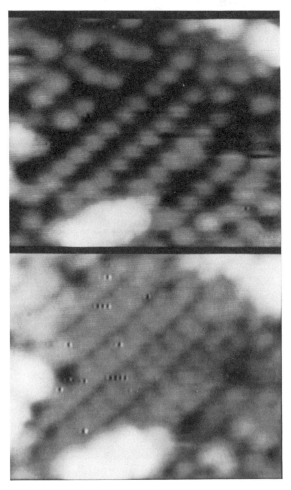

Fig. 3 A 2x1 reconstruction on a 5 ML germanium film. Top: U = - 1V, bottom: U = - 2V.

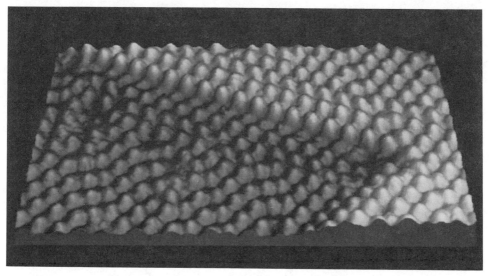

Fig. 4 A germanium bulk dislocation to be seen at the surface. Step
        height ≈ 1Å.

various possible positions to saturate the dangling bonds of the deeper
germanium layer.

    5 ML seem to be the intermediate region, where the germanium layer
has the highest strain. This regime is called 'critical thickness'. See
Sakamoto et al.[8] for a more detailed work on this.

4.   Relaxed Layer

    At a thickness of about 10 ML the strained germanium layer relaxes
by forming dislocations. In Fig. 4 we can see a germanium dislocation.
The step height between the two visible areas is about 0.8 Å. Therefore
it can not be a monatomic step. On these 'thick' layers germanium already
looks like a bulk crystal forming a (2 x 8) reconstruction. For a more
detailed description of the 2 x 8 structure, see Ref. 5.

CONCLUSION

    Substrate defects play a major role in the first nucleation.
Preferred growth directions could be observed. Again it should be
emphasized that at the examined bias voltages germanium and silicon could
not be distinguished by STM. The local density of states seems to be very
similar if the structures are, too. Coming spectroscopic data might give
more reliable information about the electronic states of the surface.
Strained areas in the germanium layer show a strong tendency to form
metastable reconstructions. Thicker layers lose their stress and relax by
forming dislocations.

REFERENCES

1. H. J. Gossmann, L. C. Feldman and W. M. Gibson, Surf. Sci. 155, 413 (1985).
2. J. Krause, diploma thesis, Hannover 1987.
3. P. M. Marée, K. Nakagama, F. M. Mulders and K. L. Kavangh, Surf. Sci. 191, 305 (1987).
4. M. A. Lavin, O. P. Pchelchyakov, L. V. Sokolov, S. I. Stenin, and A. I. Toropov, Surf. Sci. 207, 418 (1989).
5. R. S. Becker, J. A. Golovchenko, B. S. Swartzentruber, Phys. Rev. Lett. 54, 2678 (1985).
6. R. S. Becker, J. A. Golovchenko, B. S. Swartzentruber, Phys. Rev. B32, 8455 (1985).
7. R. S. Becker, B. S. Swartzentruber, J. Vac. Sci. Technol. A6, 472 (1988).
8. T. Sakamoto, K. Sakamoto, K. Miki, H. Okumura, S. Yoshida, and H. Tokumoto, this volume.
9. U. Köhler, J. E. Demuth and R. J. Hamers, J. Vac. Sci. Technol. A7, 2860 (1989).
10. J. E. Demuth, R. J. Hamers, R. M. Tromp and M. E. Welland, J. Vac. Sci. Technol. A4, 1320 (1986).

GROWTH KINETICS ON VICINAL (001) SURFACES: THE SOLID-ON-SOLID MODEL OF

MOLECULAR-BEAM EPITAXY

D. D. Vvedensky, S. Clarke, K. J. Hugill,
A. K. Myers-Beaghton, and M. R. Wilby

The Blackett Laboratory
and
Indisciplinary Research Centre for Semiconductor Materials
Imperial College
London SW7 2BZ, United Kingdom

ABSTRACT

Growth kinetics during molecular-beam epitaxy (MBE) on vicinal surfaces are simulated with a kinetic solid-on-solid model. Comparison is made between the simulated step-density and the specular intensity of reflection high-energy electron-diffraction (RHEED) measurements. In addition to identifying the kinetic mechanisms giving rise to observed phenomena, this similarity provides considerable insight into the sensitivity of specular RHEED to specific features of morphological sensitivity involving two-point surface correlations. Applications of the model encompass both III-V and group-IV semiconductors. Examples for GaAs(001) MBE include: (i) a discussion of growth modes on vicinal surfaces as a function of substrate temperature, with (ii) a direct determination of model parameters from a comparison of step-density and RHEED evolutions; and (iii) the fabrication of quantum wires. For growth on Si(001) surfaces, we discuss (iv) the role of monatomic and biatomic steps and the dimer reconstruction in determining the mode of growth, and (v) the coverage of 2 x 1 and 1 x 2 domains during growth and after recovery.

INTRODUCTION: THE SOLID-ON SOLID MODEL

Many aspects of crystal surfaces may be described in terms of lattice models. The most commonly used is the so-called solid-on-solid (SOS) model,[1-3] whereby the lattice is simple cubic and vacancies and overhangs are forbidden. Each surface configuration may then be described by an array H whose entries $h_{i,j}$ contain the numbers of atoms in columns perpendicular to the (001) plane. Kinetic processes such as adsorption, migration, and evaporation that characterize surface dynamics can then be described in terms of rules for the addition or subtraction of atoms in the columns. Once the rules are specified, the evolution of the surface can be described by a Master equation and solved either with Monte Carlo methods or, in certain cases, analytically.[3] This approach

Kinetics of Ordering and Growth at Surfaces
Edited by M. G. Lagally
Plenum Press, New York, 1990

has been successfully applied to crystal growth from solution and to the investigation of structural features such as nucleation of clusters of atoms, the dynamics of step edges, and the influence of screw dislocations upon growth rates.

A growth technique of particular current interest is molecular beam epitaxy (MBE),[4] where growth proceeds by ballistic deposition of atoms or molecules onto a heated substrate. Interest in MBE stems from the technological importance of semiconductor and metallic structures fabricated, as well as the opportunity to study growth under controlled far-from-equilibrium conditions. The ultra-high vacuum environment also enables the in-situ application of modern surface analysis tools. Foremost among these is reflection high-energy electron-diffraction (RHEED) due both to simplicity of the experimental setup and the oscillations in the specular intensity having been first observed for MBE of GaAs.[5,6] These oscillations provide the most direct evidence of the layer-by-layer growth mode associated with MBE, and may be used as a probe of the growth kinetics and the surface morphology.[7]

In the kinetic SOS model of MBE, growth is initiated by random deposition of atoms onto the substrate with a rate FA, where F is the beam flux, and A is the area of the substrate. Surface atoms migrate by nearest-neighbor hopping, the probability of which is given by

$$k(E,T) = k_o \exp(-E/k_B T), \tag{1}$$

where k_o is the adatom vibrational frequency, E is the diffusion barrier, k_B is Boltzmann's constant, and T is the substrate temperature. The diffusion barrier E is a site-dependent quantity including a substrate term E_S and a contribution E_n from nearest-neighbor bonding parallel to the plane of the surface:

$$E = E_S + E_n . \tag{2}$$

The precise form of E_n is determined by the system studied, generically $E_n = nE_N$, where E_N is the contribution per bond formed, and n the number of such bonds formed (n = 0,...,4). Unlike the parameterization of the SOS model used to study, say, the roughening transition, where $E_S = E_N$, we find that a choice where $E_S \gg E_N$ is more appropriate for describing epitaxial growth of both metals and semiconductors. Thus, the epitaxial growth process may viewed approximately as a competition between mobility (E_S) and ordering (E_n).

We describe below the application of the SOS model to MBE. Unlike growth from solution, MBE is a far-from-equilibrium process, due to the mass and energy currents from the incident particle beams, so there are few analytic theories with which simulations can be compared. However, considerable progress has been made based upon comparisons between simulations and RHEED measurements.[8-15] Since full dynamical RHEED calculations from a disordered surface are not yet practical, such comparisons have been based upon kinematic diffraction calculations,[8-11] or upon quantities such as the step density[11-15] or "smoothness".[16] Our studies have shown that the step density provides an excellent qualitative representation of many features of RHEED profiles both during growth[11-13] and in the recovery phase upon cessation of growth.[14,15] This has led to many useful insights into the kinetics of growth and recovery in MBE, as well as highlighting the intrinsic morphological sensitivity of RHEED.

STEP DENSITY, KINEMATIC SCATTERING, AND SURFACE CORRELATIONS

The probability of a particular configuration H after a total of m atoms have been deposited is denoted by P(H;m). From the complete probability distribution it is possible to derive several useful quantities in terms of reduced probabilities. We first introduce the following shorthand notation for summation over all configurations of heights:

$$\sum_{h_{1,1}=0}^{\infty} \sum_{h_{1,2}=0}^{\infty} \cdots \sum_{h_{2,1}=0}^{\infty} \cdots \sum_{h_{\sqrt{N},\sqrt{N}}=0}^{\infty} = \sum_{\{h_{i,j}\}=0}^{\infty} .$$

The probability that a site [i,j] has height $h_{i,j}$ at time t is given by the one-site reduced probability $p(h_{i,j};t)$. Similarly, the two-site reduced probability that two sites a [i,j] and [i',j'] have heights $h_{i,j}$ and $h_{i',j'}$ at time t is given by the two-site reduced probability $p(h_{i,j},h_{i',j'};t)$. These quantities are given in terms of P by:

$$p(h_{i,j};m) = \sum_{\substack{\{h_{k,\ell}\}=0 \\ [k,\ell]\neq[i,k]}}^{\infty} P(H;t),$$

$$p(h_{i,j}, h_{i+1,j};t) = \sum_{\substack{\{h_{k,\ell}\}=0 \\ [k,\ell]\neq[i,j],[i',j']}}^{\infty} P(H;t) . \tag{3}$$

Generalizations to higher multi-site reduced probabilities are straightforward. Thus, for example, the average height at time t, <h(t)>, is given by

$$<h(t)> = \frac{1}{N} \sum_{i=1}^{\sqrt{N}} \sum_{j=1}^{\sqrt{N}} \sum_{h_{i,j}}^{\infty} h_{i,j} p(h_{i,j};t) . \tag{4}$$

The surface step density, S(t), is calculated as the number of steps formed parallel to the surface by neighboring sites of different heights and, from Eq. (3), is given by

$$S(t) = \frac{1}{N} \sum_{i=1}^{\sqrt{N}} \sum_{j=1}^{\sqrt{N}} \sum_{h_{i,j}=0}^{\infty} \left[\sum_{\substack{h_{i+1,j}=0 \\ h_{i+1,j}\neq h_{i,j}}}^{\infty} p(h_{i,j},h_{i+1,j};t) \right.$$

$$\left. + \sum_{\substack{h_{i,j+1}=0 \\ h_{i,j+1}\neq h_{i,j}}}^{\infty} p(h_{i,j},h_{i,j+1};t) \right] . \tag{5}$$

To formulate the RHEED intensity in the kinematic approximation, we write the diffracted intensity as $I(\underline{S}) = N \sum_u C(\underline{u}) \exp(-i\underline{S}\cdot\underline{u})$, where N is the number of scattering centers, $\underline{S}$ is the momentum transfer to the surface, $\underline{u}$ is a vector which spans over all surface positions, and $C(\underline{u})$ is the two-point correlation function of the surface. For a simple cubic lattice with a lattice constant of a, we write the displacement vector $\underline{u}$ as $\underline{u}=\underline{x}+\underline{z}$, where $\underline{x}$ is a lattice vector parallel to the surface, $\underline{x}=am_x\hat{x}+am_y\hat{y}$, and $\underline{z}$ is the height in the vertical direction, $z = al\hat{z}$. Thus, $C(\underline{u}) = C(\underline{x},\underline{z})$. Since $C(\underline{u})$ is the probability of finding two scatterers separated by a vector $\underline{u}$, we may write it in terms of the two-site probability Eq. (3) as follows:[17]

$$C(x,1) = \begin{cases} \delta_{\ell,0}, \quad \sqrt{N} & \text{if } |x|=0; \\ (1/N) \sum_{i,j=0}^{\infty} \sum_{h_{i,j}=0}^{\infty} p(h_{i,j},h_{i+m_x,j+m_y} = h_{i,j}+|\ell|;t), & \text{if } |x|<\infty; \\ \sum_{h=0}^{\infty} p(h;t)p(h+|\ell|;t), & \text{if } |x|=\infty. \end{cases} \tag{6}$$

Note that the lack of correlation as $|x|\to\pm\infty$ means that the probabilities of finding scatterers at the various levels for very large separations are independent and equal to the product of one-site reduced probabilities. Using Eq. (6) for the correlation function, and recognizing that the kinematic diffracted intensity is the Fourier transform of the correlation function, the general form of the intensity is given by:[17]

$$I(\underline{S},t) = N(2\pi/a)^2 \sum_{\underline{G}_{||}} \delta(\underline{S}_{||}-\underline{G}_{||}) \left[1 - 2 \sum_{\ell=1}^{\infty} \sum_{h=0}^{\infty} p(h;t)p(h+\ell;t) \right.$$

$$[1-\cos(S_z\ell a)]\Big]$$

$$+ 2N \sum_{x} \sum_{\ell=1}^{\infty} e^{-i\underline{S}_{||}\cdot\underline{x}} [1 - \cos(S_z\ell a)] \left[\sum_{h=0}^{\infty} p(h;t)p(h+\ell;t) \right. \tag{7}$$

$$\left. - \frac{1}{N} \sqrt{N} \sum_{i,j=0}^{\infty} \sum_{h_{i,j}=0}^{\infty} p(h_{i+m_x,j+m_y} = h_{i,j}+\ell;t)(1 - \delta_{x,0}) \right] .$$

The first term is a lattice of delta functions at the Bragg spots $\underline{G}_{||}$, whose magnitudes depend only on the one-site reduced probabilities. Consequently, the Bragg spots alone give no information about the microscopic arrangement of atoms at the surface, only their distribution among the various layers. The second term is the diffuse intensity and is distributed throughout S-space. It contains information about the microscopic disorder of the atoms on the surface through the two-site reduced probability. This has important consequences for making comparisons with measured RHEED intensities, since only by integrating the diffuse intensity over some specified region about the specular beam is there information about surface configurations rather than simply coverage. In contrast, the step density (5) is given entirely in terms of reduced two-site probabilities and so is an intrinsically configurational quantity. This is most evident in the decay of the RHEED intensity at half-layer completion, which is also a feature of the step-density evolution (see below), but is present in the kinematic intensity only if a integration over the diffuse pattern is included.[11]

GROWTH AND RECOVERY ON GaAs(001)

The pioneering work of Neave et al.[7] demonstrated the existence of temperature dependent growth modes on vicinal surfaces during GaAs MBE. By recording the temporal evolution of the RHEED intensity at increasing

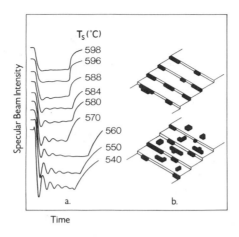

Fig. 1 (a) RHEED measurements showing eventual disappearance of intensity oscillations with increasing substrate temperature, T_S. (b) Schematic illustration of the transition from growth by two-dimensional nucleation to step propagation (after Ref. 7).

substrate temperatures (Fig. 1), they observed a transition from low-temperature growth that proceeds through the formation and coalescence of clusters on terraces (oscillating RHEED intensity) to high-temperature growth characterized by "step propagation" (constant RHEED intensity), whereby arriving atoms have sufficient mobility to be incorporated at the step edges.

In Fig. 2 we display the step-density data for growth on a 180x180 lattice, comprising ten 18-atom terraces - the same terrace width as in the experiments. The growth rate, F, = 0.368 ML/s, also coincides with that of the experiment. The form of E_n has been chosen to reflect the approximately isotropic mobility of Ga (see, however, Refs. 18-21)

$$E_n = n \, E_N, \quad n = 0, \ldots, 4 \, . \tag{8}$$

The substrate contribution, E_S has been arbitrarily set at 1.3eV, with E_N chosen such that E_N/E_S = 0.2 or 0.3. The simulated step densities are seen to reproduce the gross features of the experimental RHEED data, including the decay in the oscillations at both layer completion and half-layer completion for low-temperature growth, a decrease in the definition and number of oscillations with increasing substrate temperatures, and a decrease in the mean step-density with increasing substrate temperature. In addition, the recovery stage in the step-density upon cessation of the molecular beam flux also shows the same trends as the RHEED data with increasing substrate temperature, namely, the initial stage of recovery occurring over a progressively shorter time span as surface adatom mobility increases.

The comparison between Figs. 1 and 2 also provides one way of parameterizing our model. We consider first the determination of E_S. The transition to step flow occurs when the thermal energy imparted to arriving atoms by the substrate is sufficient to promote incorporation at

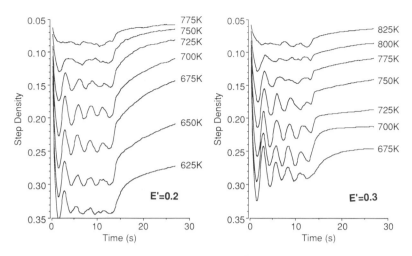

Fig. 2 Simulated step densities during growth and recovery on a vicinal
surface for the the following model parameters and growth
conditions: 180x180 surface with 10 steps, F=0.368 ML/s, E_s=1.3eV.
The growth rate and vicinal angle were chosen to correspond to
those used by Neave et al.[7] (Fig. 1), thereby allowing a direct
comparison to determine the transition temperature. (Note that
the ordinate has been inverted).

step edges at the expense of cluster formation on terraces. The
temperature T_{exp} at which step flow begins to dominate is therefore
determined to a large extent by E_S. Thus, since the terrace length is
known, a simulation can be performed to produce a transition at the
simulated temperature T_{sim}, which immediately yields E_S =
$E_{S_{sim}}(T_{exp}/T_{sim})$. The value of E_N could then be determined from a
quantitative analysis of recovery data.[22] An alternative basis for
making comparisons between simulations and experiment is by using
kinematic diffraction theory including the contribution to the diffuse
intensity. These calculations will be reported elsewhere.

At present, the best that can be done is a direct, but qualitative,
comparison between simulation and experiment for growth on vicinal
surfaces with different values of $E' \equiv E_N/E_S$ (Figs. 1 and 2). The
simulations with $E' = 0.2$ are seen to best reproduce the measured RHEED
profiles both during growth and upon recovery (Fig. 1). The data with
the higher value of $E' = 0.3$ show marked differences in both the growth
and the recovery processes, since the comparatively large
nearest-neighbor bond promotes cluster formation on terraces, which
inhibits both the transition to step flow growth as well as the
subsequent recovery of the surface. Alternatively, for lower values of
E' (not shown) cluster lifetimes are shorter, leading to weaker
step-density oscillations and recovery which is too facile.[21] By
adjusting E_S to align the simulated and measured temperature where the
transition temperature occurs for $E' = 0.3$, we estimate the appropriate
models parameters as

$$E_S \approx 1.45eV, \quad E_N \approx 0.30eV \tag{9}$$

though we must emphasize that the effective barriers E_S and E_N depend
upon the group-V flux, and there is the underlying assumption that the

kinetics of the group-V species do not affect the growth in a
rate-limiting way.

QUANTUM-WELL WIRES

Recent advances in MBE have led to reports of the successful
fabrication of unusual low-dimensional structures, the most exotic being
one-dimensional quantum-well wires (QWWs).[23-25] The basic principle
behind one method for growing QWWs stems from the work of Neave et al.
described in the preceding section. The high-temperature growth mode may
be exploited for growing (Al,Ga)As QWWs by arranging the substrate
temperature, Ga and Al fluxes, and misorientation angle to favor growth
by step propagation for both group-III species. To simulate the growth
of low-dimensional III-V alloy structures, we have initially considered
equally mobile atoms, setting both surface barrier energies at 1.3eV and
the three nearest-neighbor bond energies at 0.25eV, as for GaAs. Our
results will illustrate the difficulty of growing even the simplest
system, as well as highlighting problem areas in fabricating QWWs by MBE
in more general settings.

The simulations presented were performed on 120 x 40 lattices, with
4 steps of terrace width 10. In Fig. 3 we present QWWs simulated by
alternately depositing one-half monolayer of type-B followed by one-half
monolayer of type-A for a total of 10 monolayers, upon a type-A
substrate, thus forming an $A_{0.5}B_{0.5}$ QWW. The shading of each box in the
image indicates the percentage of type-B atoms present in the 120 atom
column perpendicular to the cross-section: the darker the shading the
higher the percentage. These data suggest that there is an optimum
temperature for the growth of these structures, the wire definition being
noticeably better at T = 750 K than for those grown at lower or higher
temperatures. The black rectangle in Fig. 3(c) marks out the area that
would be completely black if the highlighted QWW were perfect, the area
outside then being the same grey as the substrate. We have used this
region to define a quantity that is a measure of the QWWs' quality. An
alternative, more refined definition of "quality" could be obtained from
the connectivity of the QWW.[26]

We present in Fig. 4 the "quality" of QWWs averaged over five
independent simulations, as a function of substrate temperature for F=0.4
ML/s and F=0.6 ML/s. Of particular significance is the maximum in
quality as a function of both the flux and the temperature. For example,
the highest-quality wire with a flux of 0.4 ML/s is obtained for a
temperature of approximately 750 K. On the other hand, for a flux of
0.6ML/s, the growth quality is maximized at somewhat higher temperatures.
It appears, therefore, that there is an optimum substrate temperature for
a particular flux. Alternatively, this can be interpreted as indicating
an optimum mobility for a fixed substrate temperature and flux, which
highlights the intrinsic kinetic limitations imposed upon QWW growth with
two group-III species having very different mobilities.

The poor quality at low temperatures ($\sim$ 710K) is simply due to the
low mobility of surface atoms. Growth proceeds not purely by step
propagation, but has a significant degree of cluster growth. Thus, there
is very little correlation between the quality of neighboring
quantum-wire structures. At high temperatures ($\sim$ 780K) the poor quality
is due to enhanced thermal fluctuations in the step-edge structure, which
cause large differences in terrace lengths moving across a given step.
Thus, the decreasing quality can be seen in correlated lateral shifts of
darkened regions for adjacent wires (Fig. 3).

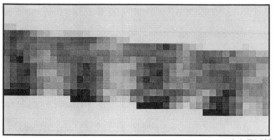

(a) 710K

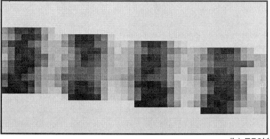

(b) 750K

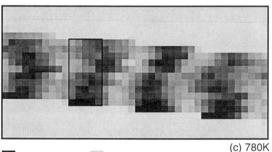

(c) 780K

■ 100% "B"   ░ 0% "B"

Fig. 3 Quantum well wires grown with a flux F = 0.4 ML/s for
temperatures: (a) T = 710K, (b) T = 750K, and (c) T = 780K. The
shading represents the density of "B" atoms found perpendicular to
the plane of the diagram, the higher the density the darker the
shading.

The identification of a narrow window within which the quantum-well
wire is optimized, even when both materials have the same kinetic
properties, has important implications for experimental realization of
these structures. Unless there is a substantial overlap between the
"growth windows" of the two materials employed, it may not be possible to
grow high-quality structures using this method, unless a technique such
as growth interruption is used to bypass this limitation.

GROWTH AND RECOVERY ON Si(001)

The influence of steps upon growth acquires an even more profound
significance for growth upon the Si(001) surface. When terminated at a
vacuum interface the (001) plane of Si reconstructs into ordered rows of

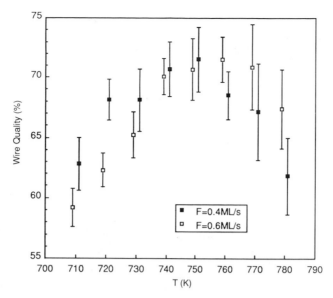

Fig. 4 Plots of wire quality as a function of temperature for F=0.4ML/s and F=0.6ML/s.

surface dimers with 2 x 1 configurations. On a surface comprising monatomic steps this leads to the coexistence of 2 x 1 and 1 x 2 domains, with antiphase boundaries formed at the step edges. Prolonged annealing of highly vicinal surfaces (2°-4°) at high temperatures, however, yields a single-domain 2 x 1 reconstruction with terraces separated by biatomic steps.[27,28] These structural changes have a significant influence upon the behavior of RHEED intensity evolutions during Si(001) homoepitaxy. On single-domain surfaces an azimuthal dependence is observed in the oscillation period, a period corresponding to bilayer deposition being observed in an azimuth parallel to a dimer axis and monolayer period in an azimuth bisecting the two dimer axes.[28] On double-domain surfaces the period is monolayer in all azimuths.[29]

Evaluation of monatomic step energies yield a pronounced anisotropy: steps in which the upper terrace dimers are perpendicular to the step edge, S_A, are considerably lower in energy than those in which the dimers are aligned parallel to the steps, S_B.[30] In accord with these calculations we introduce an anisotropy into surface bonding with the following properties. On a given layer, nearest-neighbor bonding is enhanced perpendicular to the dimer bonds. This accounts for the preferential step energies after reconstruction and subsequently must rotate through 90° with each successive monolayer. These effects are introduced into the simulation through the nearest-neighbor term:

$$E_n = mE_{/\!/} + nE_{\perp}, \quad m = 0, 1, 2 \text{ and } n = 0, 1, 2 \tag{10}$$

where $E_{/\!/}$ and $E_{\perp}$ are nearest-neighbor contributions parallel and perpendicular to the enhancement direction, and m and n are the associated numbers of bonds formed, respectively. We employ the following parameters: $E_S = 1.3eV$, $E_{/\!/} = 0.45eV$, and $E_{\perp} = 0.05eV$.[31] This ratio of 9:1 for $E_{/\!/}$: $E_{\perp}$ should be compared with the value of 10:1 determined by van

Loenen et al.[32] from comparisons between topographies generated by this model with those using scanning tunneling microscopy and the value of $\lambda(S_B):\lambda(S_A) = 15 : 1$ calculated by Chadi[30] for the energies per unit length λ for S_A and S_B steps.

In Fig. 5 we show step density evolutions, resolved in the $\phi = 45°$ and $\phi = 90°$ directions, for the growth of twenty monolayers at T = 700 K and flux F = 1ML/s. The top panel data was obtained from a surface with four monatomic steps, representative of coexistent 2 x 1 and 1 x 2 domains, while that in the lower panel is from a surface with two biatomic steps, representative of a single-domain surface. Addressing the single-domain evolutions first, we observe an azimuthal dependence in the step density calculation, a monolayer period in the $\phi = 45°$ direction (<100> azimuth), however a bilayer mode in the $\phi = 90°$ direction (<110> azimuth). We also note the enhanced stability of the bilayer oscillations, prolonged oscillations for 300 monolayers being readily observable. This behavior may be explained by examination of the simulated surface. Figure 6 shows monatomically and biatomically stepped monatomically and biatomically stepped surfaces after 0.25ML of growth. The elongation of the surface islands in the direction of nearest-neighbor bond enhancement is evident. On the single domain surface (Fig. 6b) with a single axis of elongation, a probe resolved along one of the nearest-neighbor bond directions will observe a bilayer period in the surface structure, as the axis of elongation changes with

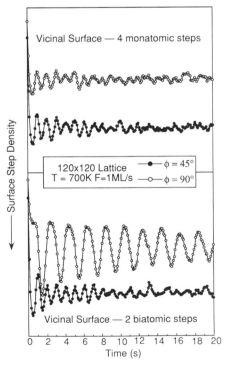

Fig. 5 Step-density evolutions projected along the indicated azimuths for growth on a two-domain surface (top) and a single-domain surface (bottom).

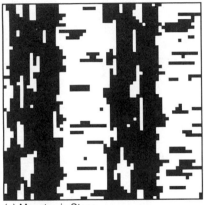

(a) Monatomic Steps

(b) Biatomic Steps

Fig. 6 60 x 60 sample substrates showing elongated clusters on stepped
substrates. In order to clarify the domain structure of the
surface atoms even-numbered terraces are shaded. Note the
alternating elongation in the upper figure and the constant
orientation for a biatomically stepped surface.

bilayer periodicity. However, when both domains are present (Fig. 6a), a
monolayer period will be measured irrespective of the azimuth.

We now proceed to examine the implications of the model for the
structure of the monatomically stepped Si(001) surface in its relaxed
state and during growth. In order to do so we must invoke a further
monitor of the surface: the coverage of atoms in even layers,
corresponding to the proportion of the surface covered by a 2 x 1 versus
1 x 2 domain. This also has a ready experimental analogue in RHEED
measurements, namely, the intensity of the second-order spots.[33] In Fig.
7 we compare "coverage" data from the simulation with experimental RHEED
measurements.[34] Prior to growth, the coverage converges towards an
equilibrium density of $\sim$0.3. However, when the Si beam is turned on the
surface immediately switches from odd-layer dominance to even-layer
dominance, remaining in such a state until the flux is terminated,
whereupon it reverts to its relaxed value. The experimental data follow

307

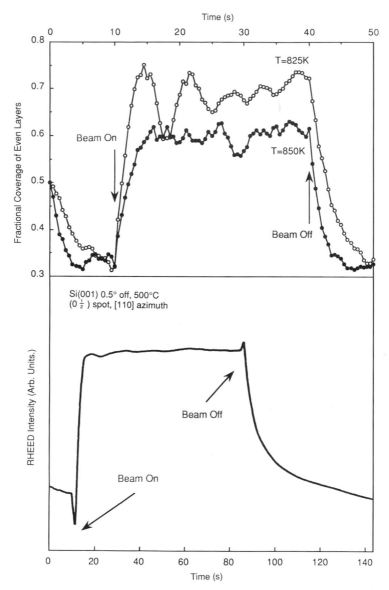

Fig. 7 Upper panel: Fractional coverage by even-numbered layers during
annealing, growth and subsequent recovery on a monatomically
stepped vicinal substrate. The surface used in this simulation
was a 120 x 120 site array, with 12 monatomic steps, exposed to a
flux F = 0.25 ML/s (during growth). Lower panel: Half-order RHEED
intensity for a growth regime corresponding to that of the upper
panel (after Ref. 34).

the same trend, yielding an intensity corresponding to a mixed 2x1/1x2 domain coverage before and after growth, but upon opening the Si shutter the intensity rapidly rises to a value indicative of a single-domain 2 x 1 surface.

This behavior has its origins in the relative stabilities of the step edges in our model. If the bond enhancement is parallel to the step, the step is very stable, and does not easily capture migrating adatoms. Conversely, if the bond enhancement is perpendicular then the step is unstable and hence disordered, but very "sticky", i.e., an atom adjacent to such a step has a high nearest-neighbor bond energy. With no external adatom flux the unstable steps decay, the lost atoms nucleating at the stable step edges (Fig. 8a). However, during the growth phase newly arriving atoms have a longer residence time at the "sticky" steps, which when combined with the higher entrapment area, ensures that most new growth nucleates at these steps (Fig. 8b). This alternating rough and smooth step structure is in direct accord with recent scanning tunneling microscope studies of misorientated Si(001) surfaces.[33,35,36]

Such detailed correspondence between simulation and experiment allows us to make several observations concerning the microscopic kinetics of Si(001) homoepitaxy. We conclude that the flux dependent transition from 1 x 2 to 2 x 1 dominance of the substrate is driven by the relative stabilities of the two monatomic step types, i.e. S_A steps are more stable than S_B steps but present a lower barrier to migrating surface atoms. This also gives rise to the formation of islands

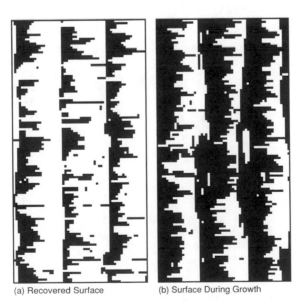

(a) Recovered Surface   (b) Surface During Growth

Fig. 8 "Domain maps" of (a) an equilibrated surface after 10s annealing at T = 850K, and (b) the same surface following 10s growth. The lattices are those used for the data in Fig. 1. To highlight the domain structure of the samples we have shaded even layer sites black and odd white. The steps descend from left to right. Each is a 60 x 120 subsection of a 120 x 120 lattice.

elongated in the direction of the bond anisotropy, and accounts for the observed dependencies in the specular RHEED intensity oscillations upon the azimuthal angle and upon substrate preparation.

SUMMARY

We have demonstrated the versatility of the kinetic SOS model studying MBE of a variety of systems. In particular, simple modifications to the basic model allow the treatment of both III-V and group-IV semiconductors. In each of these cases we find that the growth kinetics are controlled by the interplay between ordering E_n and surface mobility E_S. Detailed inclusion of surface structure appears to be unnecessary in a regime dominated by surface diffusion kinetics on the length and time scales currently accessible by in-situ probes. This will not be the case however, for GaAs, if the As incorporation kinetics enter into the growth kinetics in a rate-limiting fashion, such as may occur at very low As pressures. The simplification of the growth description also adds to our ability to treat large scale phenomena by making possible the study of mesoscopic sample sizes. Lattices of up to 500 x 500 sites are easily accommodated by a desktop computer such as a micro-VAX.

Of considerable value to subsequent studies of MBE is our ability to parameterize the model by reproducing characteristic features of experimental studies, a development that would be substantially enhanced by an evaluation of the As-flux dependence of the step-flow transition temperature. As well as providing an insight into fundamental aspects of growth, our approach is particularly amenable to the study of microfabrication, wherein we are concerned with the confinement of atoms within a prescribed region. For the case of quantum-well wires, even the simplest choice of materials leads to major difficulties in ensuring a high degree of uniformity in wire composition. Thus, simulations can offer the prospect of modelling processing procedures on a computer before attempting expensive and time-consuming trials in a growth chamber.

<u>Acknowledgements</u>

S. C. would like to thank the U. K. Science and Engineering Research Council for provision of financial support. A.K.M-B would like to acknowledge the provision of a NATO fellowship to pursue this work. We would all like to acknowledge the following individuals for valuable discussions and assistance concerning the results included in this paper: Prof. B. A. Joyce, Dr. T. Kawamura, Dr. T. Sakamoto and Dr. K. Sakamoto.

REFERENCES

1. T. L. Hill, J. Chem. Phys. <u>15</u>, 761 (1947).
2. D. E. Tempkin, Sov. Phys. Crystall. <u>14</u>, 344 (1969).
3. J. D. Weeks and G. H. Gilmer, Adv. Chem. Phys. <u>40</u>, 157 (1979).
4. B. A. Joyce, Rep. Prog. Phys. <u>48</u>, 1637 (1985).
5. J. H. Neave, B. A. Joyce, P. J. Dobson, and N. Norton, Appl. Phys. <u>31</u>, 1 (1983).
6. J. M. Van Hove, C. S. Lent, P. R. Pukite, and P. I. Cohen, J. Vac. Sci. Technol. <u>B1</u>, 741 (1983).
7. J. H. Neave, P. J. Dobson, B. A. Joyce, and J. Zhang, Appl. Phys. Lett. <u>47</u>, 100 (1985).
8. J. Singh, S. Dudley, and K. K. Bajaj, J. Vac. Sci. Technol. <u>B4</u>, 878 (1986).
9. M. Thomsen and A. Madhukar, J. Crystal Growth <u>80</u>, 275 (1987).

10. S. A. Barnett and A. Rockett, Surf. Sci. 198, 133 (1988).
11. S. Clarke and D. D. Vvedensky, J. Appl. Phys. 63, 2272 (1988).
12. S. Clarke and D. D. Vvedensky, Phys. Rev. Lett. 58, 2235 (1987).
13. S. Clarke and D. D. Vvedensky, Phys. Rev. B36, 9312 (1987).
14. S. Clarke and D. D. Vvedensky, Appl. Phys. Lett. 51, 340 (1987).
15. S. Clarke, D. D. Vvedensky, and M. W. Ricketts, J. Crystal Growth 95, 28 (1989).
16. S. V. Ghaisas and A. Madhukar, J. Appl. Phys. 65, 3872 (1989).
17. A. K. Myers-Beaghton and D. D. Vvedensky, Surf. Sci. (in press).
18. P. R. Pukite, G. S. Petrich, S. Batra and P. I. Cohen, J. Cryst. Growth 95, 269 (1988).
19. K. Ohta, T. Kojima, and T. Nakagawa, J. Cryst. Growth 95, 71 (1988).
20. T. Shitara and T. Nishinaga, Jpn. J. Appl. Phys. 28, 1212 (1989).
21. T. Shitara, D. D. Vvedensky, S. Clarke, and B. A. Joyce, to be published.
22. D. D. Vvedensky and S. Clarke, Surf. Sci. (in press).
23. J. M. Gaines, P. M. Petroff, H. Kroemer, R. J. Simes, R. S. Geels, and J. H. English, J. Vac. Sci. Technol. B 6, 1378 (1988).
24. M. Tsuchiya, J. M. Gaines, R. H. Yan, R. J. Simes, P. O. Holtz, L. A. Coldren, and P. M. Petroff, Phys. Rev. Lett. 62, 466 (1989).
25. M. Tanaka and H. Sakaki, Jpn. J. Appl. Phys. 27, L2025 (1988).
26. K. J. Hugill, S. Clarke, D. D. Vvedensky, and B. A. Joyce, J. Appl. Phys. 66, 3415 (1989).
27. T. Sakamoto, N. J. Kawai, T. Nakagawa, K. Ohta, and T. Kojima, Appl. Phys. Lett. 47, 617 (1985).
28. T. Sakamoto and G. Hashiguchi, Jpn. J. Appl. Phys. 25, L78 (1986).
29. J. Aarts, W. M. Gerits, and P. K. Larsen, Appl. Phys. Lett. 48, 931 (1986).
30. D. J. Chadi, Phys. Rev. Lett. 59, 1691 (1987).
31. S. Clarke, M. R. Wilby, D. D. Vvedensky, and T. Kawamura, Phys Rev. B40, 1369 (1989).
32. T. Sakamoto, T. Kawamura, and G. Hashiguchi, Appl. Phys. Lett. 48, 1612 (1986).
33. E. J. van Loenen, A. J. Hoeven, D. Dijkkamp, J. M. Lenssinck, H. Elswijk, and J. Dieleman, this volume.
34. K. Sakamoto, T. Sakamoto, K. Miki, S. Nagao, G. Hashiguchi, K. Kuniyoshi, and N. Takahashi J. Electrochem. Soc. (in press).
35. E. J. Van Loenen, A. J. Hoeven, D. Dijkkamp, and H. Lenssinck. Paper presented at the 3rd International Symposium on Silicon Molecular Beam Epitaxy, Strasbourg, France on 30th May 1989; and E. J. van Loenen, private communication.
36. M. R. Wilby, S. Clarke, D. D. Vvedensky, T. Kawamura, K. Miki, and H. Tokumoto, to be published.

STRUCTURE BIFURCATIONS DURING SURFACE GROWTH

R. Kariotis

Department of Materials Science and Engineering
University of Wisconsin
Madison, Wisconsin 53706 USA

G. Rowlands

Department of Physics, University of Warwick
Conventry, CV4 7AL, UK

ABSTRACT

Surface structure transitions are described and shown to take place as a function of deposition rate. This behavior is seen to occur as a result of two competing time scales, one determined by the deposition rate, and the other by the rate of lateral diffusion. Discussion is given of the physical mechanism behind this effect, and its relation to non-equilibrium roughening.

INTRODUCTION

Non-equilibrium roughening transitions are of importance in the field of Molecular Beam Epitaxy (MBE).[1-4] Transitions of this sort have been discussed recently[5-7] and are distinguished from the more familiar equilibrium roughening transition[8] as is discussed below. To describe this phenomenon, consider a two-dimensional substrate with an external source depositing adatoms at a given rate R and define the interface width as the number of incomplete layers at a given instant. This will be denoted as Z. In the case (usually associated with zero temperature) that the adatoms arrive on the surface and stick to where they land it is easy to show that in the long time limit $Z \sim t^{\alpha}$ where the T = 0 value is $\alpha = 1/2$. Generally deposition can be monitored in the laboratory by means of diffracted-intensity oscillations[1-3] and a simple interpretation of the observed results requires that the oscillations die out as the surface grows increasingly rough. The diffracted intensity in the out-of-phase condition is given in the kinematic approximation by

$$I(s_z) = \int d^2r_1 d^2r_2 < \exp[is_z(\phi(r_1) - \phi(r_2))] > .$$

(1)

$\phi(r)$ is the column height at lattice site r and s_z is evaluated at the out-of-phase condition ($s_z = \pi/d$, d is the vertical lattice spacing). In many laboratory experiments, the observed behavior is such that as $t \to \infty$, $I \to 0$. As discussed later, there are cases where this condition is not met, but the surface is still regarded as rough; in some situations this distinction is one of experimental limitations[9] however, from a

Kinetics of Ordering and Growth at Surfaces
Edited by M. G. Lagally
Plenum Press, New York, 1990

theoretical point of view the loss of diffracted intensity due to surface disorder is a result of competing length scales, laterally and vertically. In the literature[8] the quantity often of theoretical interest characterizing the surface is the column-column height correlation function

$$G(r) \equiv G(|r_1 - r_2|) = \langle [\phi(r_1) - \phi(r_2)]^2 \rangle , \qquad (2)$$

which is the theorist's means of defining Z. For Gaussian noise (used to describe the random deposition of adatoms) the scattered intensity is a simple functional of $G(r)$. However, we will also want to make use of the two-point correlation function

$$\Gamma(r,t) \equiv \Gamma(|r_1 - r_2|,t) = \langle \phi(r,t)\phi(0,0) \rangle . \qquad (3)$$

Later we will say that the surface is bounded if Γ is bounded, and unbounded if Γ is unbounded.

In the following paragraphs we will describe the two limiting cases, the Poisson and Diffusive models which are, respectively, always and never rough. Then we compare with the Binary model which interpolates between them as a function either of temperature or deposition rate. The Binary model does not include re-evaporation and shadowing, which generally are of importance and may influence the behavior of the system.

An important conclusion of this paper is that the two time scales of the process, fixed by the deposition and diffusion rates respectively, determine the macroscopic structure of the interface width. When the deposition rate is greater than the diffusion rate, the surface is rough (Poisson-like). When the diffusion rate is greater, the surface is smooth (diffusive). When the two time scales are comparable, the structure will undergo a transition from bounded to unbounded. The nature of this transition apparently is described by a sequence a bifurcations, adding more and more discontinuities, or kinks, to the surface, as the deposition rate is increased (or the temperature lowered). A secondary conclusion is that it is not sufficient to state whether a surface is rough or smooth without giving a lateral length scale over which this property is supposed to appear. This "roughening length" may be zero, finite or infinite. In the equilibrium case the interface is infinite above the roughening temperature T_R, that is, the interface width diverges logarithmically with column-column separation. This is not sufficient to result in a total loss in diffracted intensity. In the kinematic approximation, the diffracted intensity has two parts, the incoherent part, (of order N for a surface of N columns) and a coherent part (of order N^2 for a flat surface). The incoherent part is always present. It is the strength of the coherent part which is a measure of the surface roughness. There is a lateral length scale in the problem which determines the diffusion distance over which atoms must be able to migrate in order that the inteface be bounded. This is the property which we will now discuss.

TWO LIMITING CASES

The simplest case is that of Poisson statistics, where the atoms are deposited at random, but at a fixed average rate R, and in each case, the atom sticks where it lands and does not move. Each column grows independently of all others, and we can simply look at the rate of growth of a representative column. Calling the height of the column above the surface ϕ, the time rate of change of this height is

$$\frac{\partial \phi}{\partial t} \equiv \phi_t(x,t) = f(x,t) \quad , \tag{4}$$

where $f = 0,1$ is a random function that is completely uncorrelated in time and space. The result is that the variable ϕ is a Poisson process, and is distributed such that at time t, the probability of $\phi = n$ is

$$P(n,t) = \frac{(Rt)^n e^{-tR}}{n!} \quad , \tag{5}$$

where $<f> = R$. The diffracted intensity at time t in the out-of-phase condition is

$$I(S_z = \pi/a) = \sum_{n=0}^{\infty} P(2n,t) - P((2n+1),t) \tag{6}$$

$$= [\cosh(Rt) - \sinh(Rt)] e^{-tR}.$$

There are no oscillations and the diffracted intensity decays exponentially with time to zero. In the Poisson case, the two-point function of Eq. 3 is $\Gamma = R|tt'|$, somewhat pathological since it is not translationally invariant, but it is clear that for most separations it increases without bound. This is interpreted as an indication that the interface width grows without bound.

The opposite limit is the Diffusive case[10] where the deposition is given as a Gaussian process with zero mean (which implies that the column heights are given with respect to the moving interface) and atoms which land at a given column have a certain probability of dropping down to neighboring columns if the neighbors are less tall. In the continuum limit this can be approximated by adding a diffusive term to the equation of motion given above for the column heights

$$\phi_t(x,t) - J\nabla^2\phi(x,t) + f(x,t) \quad . \tag{7}$$

This equation describes the evolution of a surface above the x,y plane, and is easily solved in terms of the external driving force

$$\phi(x,t) = \int d^2x' dt' G_o(x-x',t-t') f(x',t') \quad , \tag{8}$$

where $G_o(r,t)$ is the Green's function in two dimensions. The disorder is described by the Gaussian random variable f distributed by

$$P[f(x,t)] = \exp\left\{-\frac{1}{2\sigma}\int d^2x dt \, f^2(x,t)\right\} \quad . \tag{9}$$

Although this has somewhat different properties than the distribution used in the Poisson case, it suffices for our purposes here with the reminder that there is a constant term to the equation which represents the average deposition rate. The diffracted intensity is

$$I(s) = \int d^2r < \exp(is_z[\phi(r,t) - \phi(0,t)]) > \quad . \tag{10}$$

Performing the implied integration we have

$$I(s) = \int d^2r \, \exp\left[-\frac{s_z^2}{J} \, Q(r,t)\right] , \tag{11}$$

where

$$Q(r,t) = \int dx^2 dt [G_0(r,t) - G_0(0,t)]^2 \sim \log |r| . \tag{12}$$

as $t \to \infty$. This suggests that for a finite system, the diffracted intensity decays to a non-zero value in the long-time limit. Also of interest is the two-point correlation

$$\Gamma(r,t) \equiv \int_\epsilon^\Lambda \frac{dk}{k} \, e^{-Jk^2 t} \, J_0(rk) \tag{13}$$

$$\leq e^{-tJ\epsilon^2} \log\left(\frac{\Lambda}{\epsilon}\right) ,$$

where $\epsilon = \pi/L$, $\Lambda = \pi/a$, L is the system size, and a is the lattice spacing. Γ decays monotonically to zero, in constrast to the Poisson case, and we interpret this as a signature of a bounded interface width. Unlike $G(r)$, the two-point function is not, to our knowledge, directly related to any observable quantity; however, in the next section we will find it a useful means of interpolating between bounded and unbounded interface growth.

In conclusion, the rate of lateral diffusion in the Poisson case is too small, and that in the Diffusive limit is too large for either of these models to describe both bounded and unbounded interface width growth. The Poisson model is always rough, the Diffusive model is never rough.

BINARY MODEL

We now consider a model which appears to interpolate between the two cases described in the previous section. We start with a generalized version of Eq. (7):

$$\phi_t(x,t) = V[\phi(x,t)] + f(x,t) \tag{14}$$

and instead of the diffusive term used above, choose for the interaction a modified version called the Binary model

$$V[\phi(x)] \equiv - J_o \sum_\alpha \tanh(H_\alpha) , \tag{15}$$

where $J_o > 0$, and

$$H_\alpha \equiv D_1[\phi(x) - \phi(x + \alpha)]. \tag{16}$$

The differences are summed over the nearest neighbors indexed by α. $P[f]$ is given by Eq. (9). In this model the columns exchange atoms at a fixed, threshold rate as long as they are not of equal height (provided D_1 is sufficiently large). This model is particularly appropriate to chemisorbed systems where an important mechanism of exchange between columns is that of diffusion down the side of the column. Since this proceeds at a finite and constant rate as long as the columns differ in height, the model should be qualitatively accurate. (It can be seen from this why the Diffusive model is incorrect since the exchange rate between columns is proportional to the column height difference, and thus grows large if the height difference is large. This is enough to guarantee that the interface width is always bounded in the sense defined previously.) Also this model conserves particle number during the exchange between columns.

Because we assume that the distribution of $f(x,t)$ is known and that the relation between f and ϕ is given by Eq. (14), it is possible to study the nature of the resulting system by looking at its average behavior. The change of variables can be done in the continuum limit without too much trouble

$$Df(x,t) \; P[f] = D\phi(x,t)J(f;\phi)P[\phi] = D\phi(x,t)e^{-H} \tag{17}$$

where the Jacobian of the transformation is written

$$J(f;\phi) = \mathrm{Det}\left|\frac{\delta f(x,t)}{\delta \phi(y,\tau)}\right| \tag{18}$$

and the effective "action" is given by

$$H = \frac{1}{2\sigma} \int d^2x dt [\phi_t - V]^2 - \int d^2x dt \, \log |J(f;\phi)| \tag{19}$$

$$= \frac{1}{2\sigma} H_1 - H_2 .$$

To investigate the average behavior of the system, we apply the method of steepest descent, or the "saddle point" method. The saddle point equation is found by minimizing the effective action with respect to the field variable ϕ and its derivatives. The full equation can be expressed in terms of functional derivatives with respect to $\phi(y_1) = \phi_1$ if H_1 and H_2 are properly symmetrized in terms of near neighbor variables. Thus the equation of motion for the saddle point is most easily obtained before taking the continuum limit

$$\phi_{tt} = \frac{\delta H_1}{\delta \phi_1} - \sigma \frac{\delta H_2}{\delta \phi_2} . \tag{20}$$

These functions are easily evaluated as

$$\frac{\delta H_1}{J_o \delta \phi_1} = V_1 \left[\cosh^{-2}(\vartheta_{10}) + \cosh^{-2}(\vartheta_{12}) \right] \tag{21}$$

$$- V_2 \cosh^{-2}(\vartheta_{21}) - V_o \cosh^{-2}(\vartheta_{01}),$$

while the second term is

$$\frac{\delta H_2}{\delta \vartheta_1} = \tanh(\psi_{10}) \cosh^{-2}(\vartheta_{10}) + \tanh(\vartheta_{12}) \cosh^{-2}(\vartheta_{12}) , \qquad (22)$$

where $\vartheta_{10} = D_1/2 \ (\phi_1 - \phi_0)$. We have also abbreviated the discrete form of the column-column interaction as

$$V_j = J_0 [\tanh(\vartheta_{j,j+1}) + \tanh(\vartheta_{j,j-1})]. \qquad (23)$$

The full equation can be integrated numerically, or studied in part analytically.

SADDLE POINT STRUCTURE

First we consider the linear analysis of the SP equation. It turns out to be relatively simple to obtain the linear dispersion relation to all orders in k, giving

$$\omega^2 = k \sin(k) [\sigma - J_0 4 \sin^2(k)] , \qquad (24)$$

which is shown for $\sigma/J_0 = 3$ in Fig. 1. For $\sigma/J_0 = 0$ all modes are damped and all surface structures decay. For $\sigma/J_0 > 4$ no modes are damped and all excitations persist indefinitely. The two-point function defined in Eq. 3 in the quadratic case is

$$\Gamma(r,t) = \int_\epsilon^\Lambda \frac{dk}{k} e^{-i\omega(k)t} J_0(rk) . \qquad (25)$$

In the simple Fick's law type diffusion model, $\omega = \pm i J_0 k^2$, where only the negative solution is chosen. (The saddle point equation is second order in time, requiring two initial conditions, unlike Eq. 14, allowing us to eliminate the unphysical solutions.) Now we note that with a real function $\omega(k)$, the decay of Γ to zero is no longer obtained, and as σ/J_0 increases past 4, none of the modes are damped and the interface width grows without bound.

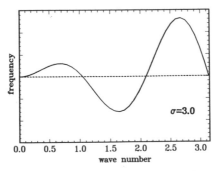

Fig. 1   The dispersion relation for the saddle point equation at a deposition rate of $\sigma = 3$.

The full non-linear equation can be numerically integrated forward in time. This was carried through, and we have found what appears to be a sequence of increasingly rough, or bifurcated, surface structures as a function of increasing σ. In Fig. 2a, we show the integration of the SP equation for small σ, starting with a sinusoidal configuration and resulting in a squared-off top, as the system comes to its most favored state. Results like this have been seen previously.[11,12] As the deposition rate is increased, the structure bifurcates, as shown in Fig. 2b, where $\sigma = 2$ and, apparently, bifurcates again as shown in Fig. 2c, where $\sigma = 4$. The belief is that there exists a systematic sequence of transitions of increasing number of kinks with increasing deposition rate, however, it is not yet clear what the overall form of this sequence is.

In a somewhat different approach, we have found that the saddle point equation to first order in the non-linearity can be put into the form of the modified Korteweg-deVries equation,[13] thus suggesting that the time evolution of the deposit takes the form of soliton-like structures. It was also found that the number of solitons emitted from an initial disturbance increased with increasing σ, which would imply that increasing number of solitons in a structure corresponded to increased roughness. Further we have shown that this equation can be used to describe the periodic ring-like structures seen in the deposition of Gold on silicon,[14] however, work remains to be done before the connection between these non-dissipative excitations and the transition to roughness can be understood.

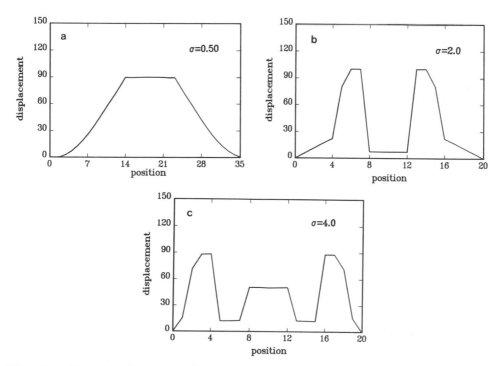

Fig. 2   The displacement for three different values of the deposition rate, after integrating forward in time 100 integration steps, dt = 0.01.

CONCLUSION

In summary, we have shown that the familiar Poisson and Diffusive models are always and never rough, respectively, and that the Binary model interpolates between these two. Our analysis used the two-point correlation function in order to draw these conclusions. We showed that the linear analysis of the Binary model is smooth for small σ, rough for large σ, and the transition from one to the other takes place as σ is increased from 0 to $4J_0$. We have also integrated the full non-linear equation in an attempt to determine the influence of the non-linear terms and found that the transition from smooth to rough appears to follow a sequence of bifurcations, with kinks in the surface structure appearing as σ is increased.

Acknowledgements

This work was supported by a grant from the NSF Solid State Chemistry Program (No. DMR 86-15089). We also would like to thank Mike Schacht and Tom Wuttke for assistance with the numerical analysis.

REFERENCES

1.   For a review, see A. Madhukar, Surface Science 132, 344 (1983).
2.   S. Clarke and D. D. Vvedensky, Phys. Rev. Lett. 58, 2235 (1987).
3a.   S. Das Sarma, S. M. Paik, K. E. Khor, and A. Kobayashi, J. Vac. Sci. Tech. B5, 1179 (1987).
3b.   S. V. Ghaisas and A. Madhukar, J. Vac. Sci. Tech. B3, 540 (1985).
3c.   J. Singh, S. Dudley and K. K. Bajaj, J. Vac. Sci. Tech. B4, 878 (1986).
3d.   J. Singh and A. Madhukar, J. Vac. Sci. Tech. B1, 305 (1983).
4a.   M. Schneider, I.K. Schuller, and A. Rahman, Phys. Rev. B36, 1340 (1987).
4b.   P. Meakin, P. Ramanlal, L.M. Sander, and R. C. Ball, Phys Rev. A34, 5091 (1986).
5a.   M. Kardar, G. Parisi, and Y.-C. Zhang, Phys. Rev. Lett. 56, 889 (1986).
5b.   R. Bruinsma and G. Aeppli, Phys. Rev. Lett. 52, 1547 (1984).
6.   P. Nozieres and F. Gallet, J. de Phys. 48, 353 (1987).
7a.   R. P. U. Karunasiri, R. Bruinsma, and J. Rudnick, Phys. Rev. Lett. 62, 788 (1989).
7b.   R. Kariotis, J. Phys. A22, 2781 (1989).
8.   J. D. Weeks and G. H. Gilmer, Adv. Chem. Phys. Vol. 40; eds. I. Prigogine and S. A. Rice, Wiley-Interscience, New York (1979).
9.   M. G. Lagally and R. Kariotis, Appl. Phys. Lett. 55, 960 (1989).
10.   S. F. Edwards and D. R. Wilkinson, Proc. R. Soc. A381, 17 (1982).
11.   W. Selke and J. Oitmaa, Surf. Sci 198, L346 (1988).
12.   F. Lançon and J. Villain this volume.
13.   R. Kariotis and G. Rowlands (in preparation).
14.   T. Ichinokawa, I. Hamaguchi, M. Hibino, and J. Kirschner, Surface Sci. 209, L144 (1989).

RHEED INTENSITIES AND OSCILLATIONS DURING THE GROWTH OF IRON

ON IRON WHISKERS

A. S. Arrott, B. Heinrich, and S. T. Purcell*

Surface Physics Laboratory, Department of Physics
Simon Fraser University
Burnaby, B. C., Canada V5A 1S6
*Present address: Philips Research Laboratories
5600JA Eindhoven, The Netherlands

ABSTRACT

The homoepitaxial growth of Fe on (001) surfaces of Fe whiskers is used to study Reflection High Energy Electron Diffraction. Variations in the intensity of the specular spot are followed during growth for temperatures from 100 K to 600 K. Mean field models of growth are used for the classification of types of RHEED oscillation patterns. Simple ideas of diffraction are employed to partially account for the effects of attenuation and surface disorder on the RHEED intensities.

INTRODUCTION

Reflection High Energy Electron Diffraction is certainly a convenient tool for the study of epitaxial growth in ultra high vacuum. Because the electron beam impinges on the sample at a small angle there is minimal interference with the growth process. RHEED allows a continuous picture of the evolution of surface structure, the distribution of atoms on the surface and their density. One can surmise that high quality surfaces have been grown and that sharp interfaces between multilayers have been achieved. While progress has been made in the calculation of intensities, a full interpretation of the RHEED observations is not readily available to the experimentalist.[1] Under these circumstances it is reasonable to study a well-behaved system in order to increase our abilities to draw conclusions from the readily obtained data. For this purpose we have used the homoepitaxial growth of iron on iron whiskers, partly because of our experience with them and partly because our basic interest is in ferromagnetism.

We would like to leave the impression that iron whiskers provide very useful substrates for surface science. Fe on Fe grows mostly layer by layer, even at low temperatures. There are changes in the atom-atom correlation function with temperature of growth that show up in the RHEED patterns and can be extracted by analysis of spot profiles. The work reported here focuses on the measurement of specular intensity from the (001) surface of Fe whiskers. We study the effect of surface preparation. We detect the diffusion of atoms on the surface during and after growth. We discuss how to separate the transient effects of the

Kinetics of Ordering and Growth at Surfaces
Edited by M. G. Lagally
Plenum Press, New York, 1990

first cycle from the subsequent cycles of growth and how to distinguish between changes in correlation function on a given layer, i.e., clustering, and interference effects between adjacent layers.

We attempt to model the growth process and to say something about the RHEED intensity. Modeling is necessary, for there is little likelihood of unique interpretations even if we had a full theory of RHEED. As it is, even if we have a full description of the surface, we do not yet know how to calculate the RHEED intensities. We are optimistic that increasing interest in the subject of RHEED will attract the theoretical talent needed to provide us with increased understanding. At the higher energies of electron microscopy[2] and at the lower energies of LEED[3] the theories are much better developed. Here we speculate on the problem of diffraction from a partially covered surface, understanding of which is needed for the interpretation of the time evolution of RHEED intensity oscillations during growth.

Fe Whiskers

The Fe whiskers show a variety of morphologies. The ideal shapes are the <100> and <111> whiskers. The <100> whiskers have (100) faces and form as long bars with approximately square cross sections (typically 1 to 7 cm in length and 10 to 500 μm in width). The <111> whiskers have (110) faces on the six sides of long bars with hexagonal cross sections. We have not yet studied growth on the <111> whiskers. The whiskers are prepared by the decomposition of $FeCl_2$ powder in an iron boat placed in a stream of wet hydrogen at $\sim$725 C. After three decades of growing Fe whiskers, we still do not know the details of the growth mechanism. We know that the starting $FeCl_2$ powder is critical. Carefully dried, finely ground light-grey powder does not work as well as coarser material with a pinkish cast. Important variables in growth are the convection currents arising from deliberately created temperature gradients from the middle to the ends of the iron boat. The variety of morphologies permit the selection of many of the low index planes and a wide range of step densities. These have yet to be investigated.

Surface Preparation

The whiskers are kept in their boats in a dry atmosphere until mounting on the substrate holder. The whiskers have been found to oxidize rapidly in air on the Atlantic coast of North America, but not in Vancouver where the moisture content of the atmosphere is controlled by the Pacific Ocean. Minor oxidation can be removed in the UHV system by sputter cleaning and annealing at 700 C. As this is close to the original growth conditions, it is not surprising that the surfaces are remarkably good after this treatment. Even a grey dull surface became bright and shiny during this in situ preparation.

An important part of the surface preparation is the suppression of accumulation of S or C at the surface. The initial treatment is a prolonged sputtering at 700 C to deplete the S in the near surface region. Auger analysis at 500 C shows no evidence of S or C, after sputtering, but on slow cooling to ambient temperature the diffusion of C to the surface can produce as much as 0.05 coverage. This is suppressed by clamping the sample holder between liquid nitrogen cooled jaws, reducing the cooling time from two hours to ten minutes. The rate of passage through the range near 200 C seems to be important for the suppression of surface segregation of C. This is consistent with studies of C in Fe whiskers by ac susceptibility measurements.[4] The bulk concentration of C is 10-100 ppm. For a surface grown at 330 C and rapidly cooled to 20 C, there is no detectable C.

The epitaxial growth of Fe on Fe whiskers below 200 C leads to changes in the flatness of the surfaces. This is evident from the RHEED patterns and from the changes in the RHEED oscillation during growth. The initial surfaces can be recovered by annealing above 500 C or by carrying out growths above 300 C.

RHEED Intensities

The RHEED patterns for the annealed Fe whiskers show sharp spots corresponding to the intersections of reciprocal lattice rods with the Ewald sphere of reflection.[5] The spots appear on a phosphorescent screen. The positions of the spots are recorded photographically or by video camera or by focusing the light onto a photomultiplier. Analyses of the intensity profiles of these spots indicate that the whiskers are flat on the scale of several thousand Å.[5] The intensity of the specular spot, the simplest of diffraction conditions, is of particular interest as it varies greatly with angle of incidence θ_i and the condition of the surface.

We report our observations of the variation of the intensity of the specular spot with θ_i, (see Fig. 9, discussed below) but make no pretense of understanding. The intensity depends on the preparation of the surface, distribution of terraces, density of steps and edges, as well as the fundamental dynamical problem of electrons interacting with the periodic structure, which includes the role of Kikuchi lines, inelastic scattering, and the diffraction geometry. At very low angles the whisker intercepts only part of the beam.

At 300 K the growth appears to include the formation of island clusters with a statistically preferred distance between centers. In reciprocal lattice space these correlated clusters create hollow cylinders that enclose the rods, and that is how they appear in LEED patterns.[6] In RHEED the hollow cylinders intersect the steeply descending Ewald sphere, to produce the pattern shown in Fig. 1. The splitting of the streaks is a measure the correlation of the one-atom-high islands. In Fig. 1 the average distance between island centers is ∿18 atom spacings. The sample is oriented in the [10] azimuth after the growth of 9 layers at 20 C.

RHEED OSCILLATIONS

Our main interest is in the oscillations of RHEED intensity during growth of Fe on Fe whiskers. To be able to say something about these data we adopt the position that there is some truth in the application of simple ideas of interference.[7] The interference argument says that if one has a flat surface that is covered by a partial monolayer, then the electrons that are reflected after reaching the partial monolayer will have a phase difference $\delta\phi$ with respect to those that are reflected from the uncovered part of the original surface. This accounts for the observation that during growth the intensity of the specular spot periodically drops to the background level for some angles of observation, particularly for the condition that the phase between the electrons reflected from adjacent surface levels differs by 3π. (For Fe(001) with 10 keV electrons, this is at $\theta_i = 65$ mrad. The wave length of the incoming beam is $\lambda = 0.124$ Å.) The simple interference argument seems to work well at this angle, see Fig. 2, except that the initial drop is too large. That there is more to RHEED oscillations than simple interference becomes evident from observations of RHEED at smaller angles. For example, the minimum intensity of the oscillations does not drop to the background for the angle at which $\delta\phi$ should be π.

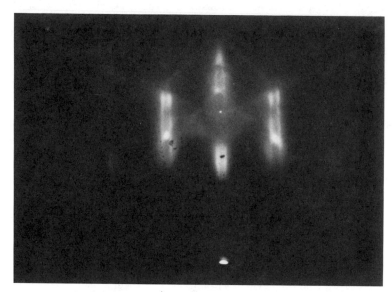

Fig. 1   The diffraction pattern in the [10] azimuth after growth of 9
layers of Fe on Fe whiskers at 20 C.   The split streaks measure
the correlation of the one-atom-high islands; in this case an
average of ∿18 atom spacings between centers.

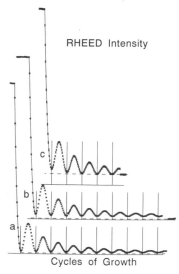

Fig. 2   The time dependence of the intensity of the specular spot during
growth at 20 C.   Each cycle of growth takes ∿50s.   The angle of
incidence is θ_i = 65 mrad, for which the phase difference
between two adjacent layers is 3π.   Curve (a) was taken after
sputtering and annealing at 650C.   Curve (b) was taken after
growing at 300C.   Curve (c) was taken after growing at 330C.

Usually we observe the RHEED oscillations for the (001) surface after rotating the surface about its normal by 0.1 rad from having the electron beam along the [11] azimuth. This is to avoid the complications arising from the changes in the intensity of the Kikuchi lines which cross along the principle azimuths. (There is one Kikuchi line, that is parallel to the surface, which one cannot avoid by rotation, see Fig. 9.) We distinguish several ranges of temperature because they show different RHEED oscillations and different dependences of the RHEED oscillations on the angle of observation. At 100 K one might expect the atoms to stick where they hit. From the point of view of interference this would lead to a rapid fall of intensity of the specular spot and no oscillations. We do observe oscillations at 100 K, see Fig. 3, and this leads us to adopt the view[8] that the incoming atoms are mobile at low temperatures because they deposit their heat of condensation locally. At temperatures above 650 K, where oscillations are not observed, it is assumed that atoms migrate sufficiently rapidly that there is no change in the correlation function of the atom positions with time. At 600 K the oscillations are large and sustained. At lower temperatures the oscillations are damped as growth continues.

The interference model of RHEED intensity oscillations predicts no oscillations when the electrons from adjacent layers are in phase with one another, but this is not what is observed at θ_i = 0.043 mrad, where this is expected to be the case because $\delta\phi \approx 2\pi$. As can be seen in Fig. 4, the intensity increases by a factor of almost five during growth. The

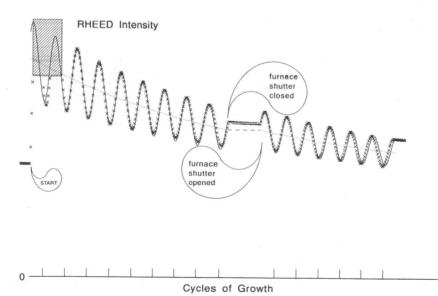

Fig. 3 Time dependence of the intensity of the specular spot during growth at 100 K. θ_i = 12 mrad. The solid line and dashed lines are empirical fits using Eq. (1). Each cycle of growth takes $\sim$52s. A horizontal white line has been drawn through the data points where the furnace shutter is closed to emphasize that there is a small decrease in the signal with time. There was an accidental loss of data near the beginning of the measurement.

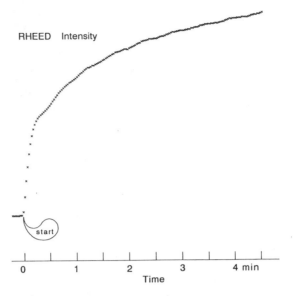

RHEED   Intensity

start

| | | | | | | | | | | |
0   1   2   3   4 min
              Time

Fig. 4   The time dependence of intensity of the specular spot at 20C.
          θ_i = 43 mrad, for which the phase difference between two
          adjacent layers is $\sim 2\pi$. From the simple interference argument
          one does not expect the intensity to change. It is likely that
          the surface reflectivity increases when the surface is covered
          with small one-atom-high islands.

nature of the surface is surely changing. The formation of the islands
leads to greater scattering of electrons into the specular spot[2,7]. We
know that the growth is basically layer by layer from observations at
other angles. We also believe that surface roughness affects the RHEED
intensity. Thus we are not surprised to observe small periodic
variations superimposed upon the general increase in Fig. 4.

Sometimes one can obtain an empirical fit to the intensity observed
during the growth. This is shown in Fig. 5 for θ = .018 rad for growth
at 300 K. The empirical fit

$$I = A \exp\left[-\frac{t}{\tau_2}\right] \cos\left[2\pi \frac{t-t_o}{\tau_1}\right] + B \exp\left[-\frac{t}{\tau_3}\right] + C \qquad (1)$$

has an oscillation of period τ_1 with an initial amplitude A that decays
exponentially with time constant τ_2 superimposed upon an average
intensity that decays from an initial value of B+C to a final value of C
with a time constant τ_3. The phase of the oscillation is set by t_o. In
Fig. 5 the damping ratios τ_2/τ_1 = 0.15 and τ_3/τ_1 = 0.16 are almost the
same. The data was taken as a function of time. It was confirmed, by
measuring with a quartz thickness gauge, that the flux of Fe atoms was
constant during the deposition. From the fit in Fig. 5 we would conclude
that, if the peaks in the intensity correspond to "filled layers", then

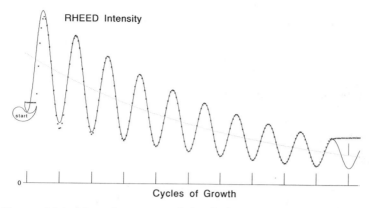

Fig. 5   The solid line is an empirical fit using Eq. (1) to the time
         dependence of the intensity of the specular spot at 20 C. $\theta_i=18$
         mrad.   Each cycle of growth takes ∿44s.   The first cycle clearly
         differs from succeeding cycles.   The dotted line is the last two
         terms of Eq. (1).

the starting condition is that of almost "filled layers." The initial
change in intensity then represents a change in the correlation function
of the atoms in the plane during the initial stage of deposition.   We
believe that there is a greater reflectivity as the surface develops a
high density of one-atom-high islands, producing lots of edges.   A step
down at the edge of an island allows the electron beam to emerge, as
shown for reflection electron microscopy.[9]

      In an attempt to understand some aspects of these types of results
we have investigated mean field models of growth and analyzed them in
terms of surface roughness and the intensity of RHEED oscillations for
simple models of diffraction.

MODELING GROWTH

      At low temperatures one might have expected to obtain correspondence
with the "stick-where-it-hits" model of growth.   For nondiffusive growth
on a low index plane, the fractional coverage $\theta_n(t)$ of the nth layer
obeys the differential equation[10]

$$\frac{d\theta_n(t)}{dt} = \frac{\theta_{n-1}(t)-\theta_n(t)}{\tau} , \qquad (2)$$

where $\theta_{n-1}(t)-\theta_n(t)$ is the portion of the area of the n-1 layer that is
exposed and   is the deposition time for one monolayer.   If one starts
with a flat surface, the intensity of the specular spot for $\delta\phi = (2n+1)\pi$
decreases as $I = I_0 \exp(-4t/\tau)$.   The surface roughness, defined[11] as

$$\Delta^2 = \sum_{n=0}^{\infty} (n-t/\tau)^2 \ [\theta_{n-1}(t)-\theta_n(t)] \ , \tag{3}$$

oscillates between 0 to 1/4 for perfect layer by layer growth. The time in Eq. (3) is measured from the start of the first layer. For nondiffusive growth the roughness increases as $\Delta^2 = t/\tau$.

The measurements of RHEED oscillations at 100 K show clearly that the non-diffusive model is not appropriate for Fe on Fe growth.

Simple models of diffusive growth on a low index surface can be obtained by adding one term to Eq. (2) for net transfers from n+1 to n and for subtracting one for net transfers from n to n-1. For one such model[10,12]

$$\frac{d\theta_n(t)}{dt} = \frac{\theta_{n-1}(t)-\theta_n(t)}{\tau} + \alpha_n[\theta_n(t)-\theta_{n+1}(t)] - \alpha_{n-1}[\theta_{n-1}(t)-\theta_n(t)]. \tag{4}$$

Of the $[\theta_{n-1}(t)-\theta_n(t)]/\tau$ incoming atoms that land per unit time on top of the n-1 layer, a fraction $(1-\tau\alpha_{n-1})$ join the nth layer and a fraction $\tau\alpha_{n-1}$ are transferred to the n-1 layer. If the n-1 layer is filled, $\alpha_{n-1}=0$. The transfer coefficients α_n depend upon the effective perimeters $d_n(\theta_n)$ of the layers. The greater the perimeter of the n-1 layer the more likely the atoms that land on top of it will find a site at its perimeter rather that joining a cluster on top of the n-1 layer. We take α_{n-1} to be proportional to the fraction of the available perimeter $d_{n-1}(\theta_{n-1})+d_n(\theta_n)$ that belongs to the n-1 layer, that is

$$\alpha_{n-1} = \left(\frac{A}{\tau} + B\right) \frac{d_{n-1}(\theta_{n-1})}{d_{n-1}(\theta_{n-1})+d_n(\theta_n)} \ , \tag{5}$$

where the proportionality constant has two parts: A/τ, proportional to the rate of incoming atoms, and B which represents the redistribution rate of atoms already joined to the layers. If $B = 0$, this model distributes the atoms as they arrive on each surface. If $B > 0$, the atoms continue to redistribute in the absence of flux ($\tau = \infty$). If $A = B = 0$, one recovers the non-diffusive growth. If $1-A-\tau B = 0$, there is no growth of the second layer before the first is filled. The richness of this model arises from the range of choices for the dependence of d_n upon θ_n. Let us assume that

$$d_n(\theta_n) = \theta_n^{P1} (1-\theta_n)^{P2} \ , \tag{6a}$$

or that

$$d_n(\theta_n) = \begin{cases} \theta_n^{P1} & \text{for } \theta_n < \theta_c \\[2ex] (1-\theta_n)^{P2} & \text{for } \theta_n > \theta_c \ , \end{cases} \tag{6b}$$

where

$$\theta_c^{P_1} = (1-\theta_c)^{P_2} .$$

The results of the calculation depend mainly on how the perimeter grows with θ_n near zero and how it shrinks with θ_n near one. Therefore both of these forms produce equivalent behaviors. One can view the exponents p_1 and p_2 as fractal descriptions of the perimeters. For fixed sized islands that just increase in numbers as the growth proceeds, $p_1 = 1$. For a fixed number of nuclei that increase in size, $p_1 = 0.5$. We show examples of two growths for which the behavior is quite different even though both yield similar increases of the roughness parameter with deposition time.

In Fig. 6a we have chosen $p_1 = 0.6$, $p_2 = 0.5$, $A = 0.99$ and $B = 0$. In Fig. 6b we have chosen $p_1 = 1$ $p_2 = 1.5$, $A = 0.99$ and $B = 0$. In the first case, Fig 6a, the upper layer starts easily and at a later time the lower layer has an easy time filling. In the second case, Fig. 6b, the initiation of the upper layer is discouraged, but, once it does starts, the final filling of the lower layer is slow. If we use the interference model for the RHEED intensities, in the first case the oscillations damp rapidly and in the second case they reach an almost steady state. If one makes some association of the RHEED intensity with the surface roughness, one also would get stronger oscillations for the second case than for the first. The difference is that in the first case there are two layers growing at once and in the second case the growth is mainly one layer at a time. Both of these would be classified as good growths. In either case, if the growth were stopped on a maximum of the intensity, the final structure is that of two layers, one almost full and one almost empty, with the second case being better than the first, consistent with the roughness parameter, Eq. 3, being lower for the second case at the maximum of the intensity. In the second case the deviations from flatness are mainly holes in the almost filled surfaces. The number of holes tends to saturate, and the same state repeats at each maximum in intensity. In the first case the deviations are islands on top of the filled surfaces. The area of the islands that have formed at each filling of the lower layer increases with cycle number.

High-Temperature Oscillations

At high enough temperature and low enough growth rates each atom could diffuse along the surface until it reached a step edge. In this case the step edges would move across the surface at rates determined by the areas of the surfaces from which the edges collect atoms.[13]

Pukite[14] has worked out the case where the atoms diffuse on the surface while the edges move forward to collect atoms. This leads to spurts of edge motion. When the edges move faster, there is less time for the atom density on the terrace in front of the edge to build up. When the density of atoms in front of the edges is lower, they move more slowly. But, the slower they move, the more time for the atom density to build up on the terrace in front of the edge, and the greater the density of atoms to be collected. This has the effect that the density of atoms on the terraces away from the edges oscillates with time, and this should be observed in the RHEED intensity. We have obtained similar results with a set of differential equations that are a generalization of Eqs. (4-6) to include the effects of steps and to take into account specifically the diffusion to the steps, the sticking at the steps and the formation of clusters on the terraces. One result that we think has some general validity is that competition between clustering on the terraces and growth from the step edges leads to RHEED oscillations for

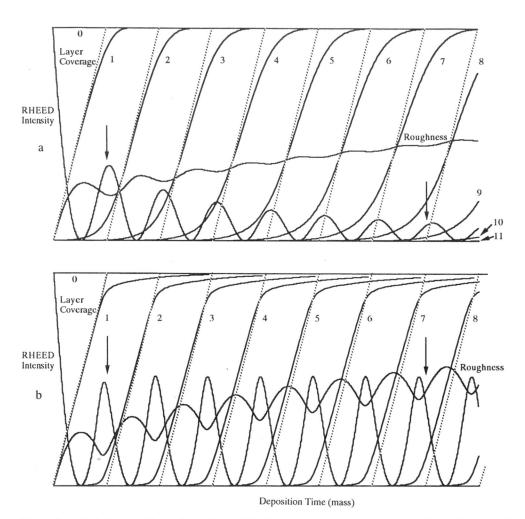

Fig. 6   Model growth using Eqs. (4-6) to generate the time dependence of the layer coverage $\theta_n(t)$, starting from $\theta_0(t) = 1$, and $\theta_n(0) = 0$ for $n > 0$. The coverages are labeled by their layer numbers. The roughness parameter is calculated using Eq. (3). The RHEED intensity is calculated using the column approximation for $\delta\phi=(2n+1)\pi$. In (a) $p_1 = 0.6$, $p_2 = 0.5$, $A = 0.99$ and $B = 0$. In (b) $p_1 = 1$, $p_2 = 1.5$, $A = 0.99$ and $B = 0$.

which the minimum intensity increases with time for a damped oscillation.[15] This effect often has been observed for vicinal surfaces.

SCATTERING FROM A PARTIALLY FILLED SURFACE

It seems clear that atoms sticking up from a flat surface should scatter differently than those which are sequestered in a filled surface. Consider the scattering from a few atoms sitting above a filled surface.

330

Sitting on a flat plane, these atoms would contribute to the specular spot even if there where no scattering from the filled surface. The amplitude of the wave diffracted from them would be calculated using the atomic scattering power of isolated atoms in a 10 keV electron beam. If the upper atoms were not there, the diffracted wave from the flat filled surface would depend only on the normal component of the incoming electron wave vector, that is, it would be the same for LEED or for RHEED if they had the same normal component. But the diffracted wave for the upper atoms would be different for LEED and RHEED because of their much different energies.

We assume that the scattering from the upper atoms is more effective mainly because of the low angle of incidence. We incorporated this into a picture of RHEED, that is not much more complicated than the simplest kinematic theory, by considering the effects of attenuation of the beam. The coherent elastic scattering contributes to the specular spot in competition with other processes, both elastic and inelastic. An atom in a filled surface layer scatters with a cross section b^2, but the amplitude of the wave that gives rise to the scattering is attenuated by $\exp(-\beta a_1/2)$ where $a_1/2$ is the distance to the center of that atom and b is a measure of the degree to which all scattering processes decrease the amplitude of the incoming wave. For an atom in the second layer the incoming wave amplitude is attenuated by $\exp(-3\beta a_1/2)$ because the wave has to go all the way through the first layer and partially through the second layer in order to scatter. For unfilled layers it will be legislated that $\beta(n) = \beta\, C(n)$, where $C(n)$ is the coverage of the nth layer. The outgoing waves scattered from each atom would also be attenuated by the same factors. The specular intensity from such a simple extension of kinematic theory is given by:

$$I = \left| \sum_{n=-\infty}^{\infty} \Psi(n)\, \exp(in\delta\phi) \right|^2 , \qquad (7)$$

where $\Psi(n)$ is the scattered wave amplitude from the nth layer and $\delta\phi$ is the phase difference from layer to layer of the contributions to the out-going wave. The amplitude from the nth layer is

$$\Psi(n) = b\, C(n)\, \exp(-\beta L(n)), \qquad (8)$$

where b is the scattering length of an atom. $L(n)$ is the effective attenuation length from the surface to an atom in the nth layer and back to the surface, given by

$$L(n) = \left\{ C(n) + 2 \sum_{m=n+1}^{\infty} C(m) \right\} a_1 , \qquad (9)$$

where the layer $n = 0$ is buried deep below the surface and the layers are counted positive from there toward the surface. The intensity is readily calculated from Eqs. (7-9), if the layers are complete to the nth layer and only the n+1 layer is partially filled to coverage C. As we are concerned only with the question of the dependence of the intensity upon C, we show the intensity normalized to the intensity for $C = 0$ as

$$\frac{I}{I(C=0)} = \frac{1+2Ce^{C\beta a_1}\left\{e^{\beta a_1}\cos\delta\phi - e^{-\beta a_1}\right\}+2\left\{Ce^{C\beta a_1}\right\}^2\left\{\cos 2\beta a_1 - \cos\delta\phi\right\}}{e^{4C\beta a_1}} . \qquad (10)$$

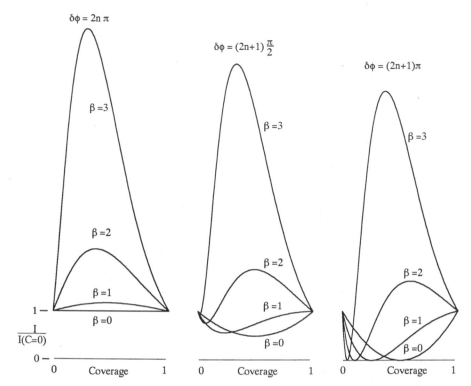

Fig. 7 Intensity as a function of coverage for layer-by-layer growth
for three choices of the phase difference $\delta\phi$ between adjacent
layers. The effect of attenuation is calculated using Eq. (11)
with several values of the parameter β. For $\beta = 0$ one has the
same dependence on coverage as for the column approximation.

This expression for the specular intensities makes no reference to the
dependence of $\delta\phi$ or β on ϕ_i, the angle of incidence of the incoming beam.
We expect β to increase a low angles as $1/\sin\phi_i$. As phenomenology, the
dependence of $I/I(C = 0)$ upon C for several values of β and $\delta\phi$ is shown
in Fig. 7. For $\beta = 0$ one recovers the results of the simplest kinematic
theory which considers only the phase difference between the top two
exposed layers. For large values of β, this model makes the top layer a
much more efficient scatterer because all the atoms of the lower layer
are partially shielded by the atoms on the top layer as well as by all
their neighbors in the lower layer. The point of the calculation is to
show that for a model where the top layer of atoms reflect more
effectively, the nature of the RHEED oscillations could be quite
different than expected from the simple interference argument. Depending
on the choices of β, the intensity can increase or decrease rapidly at
the initiation of growth and the phases of the maximum and minima are
shifted.

OBSERVATIONS OF RHEED OSCILLATIONS ON Fe WHISKERS

We have observed the intensity oscillations of the specular spot for several choices of the angle of incidence of the 10 keV electron beam and at several growth temperatures. The above ideas were given to provide a framework for discussion of some of the experimental observations.

Growth Studies at $\delta\phi \simeq 5\pi/2$

A particularly striking observation was made on a whisker that was sputtered at 700 C, annealed at 650 C, and rapidly cooled to 20 C. At an angle corresponding to a $5\pi/2$ phase difference between adjacent layers, there was one position on the sample where the specular intensity decreased to well below the small background intensity. A phase difference of $5\pi/2$ is neither the Bragg condition nor the anti-Bragg condition for interference between two layers. We postulate that the incoming beam was reflected from a portion of the surface that had a small number of very flat regions , which had just the right ratio of areas to lead to complete cancellation. Possibly as few as three levels might be involved. In Fig. 8 we show a sequence of growths where the intensity of the specular spot as a function of time was measured starting from this special condition.

When the shutter in front of the iron source was opened, the RHEED intensity grew rapidly, see Fig. 8a. This should be expected on the most general grounds just because we started at a special condition of vanishing specular intensity. Just after it reached a maximum (as determined from the recorder trace, which is continuous, whereas the data in all the figures was collected in equal time steps), the shutter was closed. There was a subsequent decrease in intensity over a time scale of minutes. The intensity maximum can not have corresponded to perfectly filled layers, otherwise it would be difficult to account for the decrease with time after the shutter was closed. We could explain this if the topmost layers were sparsely occupied. The migration of atoms on the topmost layers could lead to larger clusters with fewer edges to contribute to the intensity.

The specimen was then annealed at 650 C and cooled rapidly to 20 C. The intensity of the specular spot for the same settings of angles and sample position was slightly reduced from before annealing, that is, from the end of Fig. 8a to the beginning of Fig. 8b. One might have expected the intensity to decrease after opening the shutter, but instead it increased much further, see Fig. 8b, going through a maximum in about the same time as before. The shutter was closed as the intensity went through the subsequent minimum. There followed a decrease in specular intensity with time even though this time the shutter was closed on the minimum rather than the maximum. Again we could say that the rearrangement of atoms on the surface produces a decrease in reflectivity. At the start of a growth we would say that the dispersion of the atoms on the top surface produces an increase in reflectivity.

The sample was reannealed a second time. The intensity after annealing was less than after the first reanneal. This correlates with the fact that the first reanneal took place after stopping at a maximum while the second reanneal took place after stopping on a minimum.

Having stopped on a minimum, one might expect the intensity to rise at the beginning of growth 8c, but it drops and then very quickly rises as if the growth started just before a minimum. This is similar to the behavior calculated in Fig. 7.

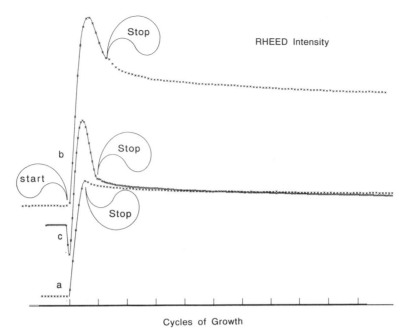

RHEED Intensity

Cycles of Growth

Fig. 8   The time dependence of intensity of the specular spot at 20 C.
θ_i = 54 mrad, for which the phase difference between two
adjacent layers is $\sim 5/2$.   The three curves were taken
sequentially with an anneal at 650 C between each.   Curve (a)
was taken after sputtering and annealing at 650 C starting from
an anomalous position where the specular spot intensity is below
the background signal.   In each case the signal decreases with
time after closing the furnace shutter.   The data points in (c)
are taken every 2s rather than every 4s as in (a) and (b).

One may surmise from these experiments (1) that the surface does
remember something about the number of atoms deposited on the surface
previously to the 650 C anneal, (2) that there are some dramatic changes
in RHEED intensity due to the surface condition, and (3) the atoms
continue to move for several minutes on the surface of Fe whiskers at 20
C.

Observations of Growth at $\delta\phi \approx 3\pi$

A sequence of growths at 20 C with three different preparations of
the surface are shown in Fig. 2 with the specular spot observed at $\delta\phi \approx$
3π.   In preparation for the growth shown in Fig. 2a the whisker was
sputter-cleaned for 50 minutes at 650 C and annealed for 5 minutes before
rapidly cooling to 20 C.   Fig. 2a shows the first 10 of a series of 25
oscillations.   By the 25th oscillation the peak intensity was 2% of the
first maximum.   The growth was continued as the temperature was increased
to 300 C.   At 300 C the oscillations had increased in amplitude,
exceeding the first maximum by 25%.   The growth was stopped on a peak of
the 300 C oscillations.   The sample was then rapidly cooled to 20 C.   The
initial part of the second growth is shown in Fig. 2b.   After checking
the background, as shown in Fig. 2b at the end of the trace, the angle of

observation was changed to $\delta\phi \simeq \pi/2$. At this low angle the oscillations are only slightly damped after 25 cycles. Increasing the temperature through 330 C increased the amplitude of these oscillations by more than a factor of two, as shown in Fig. 11a. The growth was stopped at the minimum of an oscillation near 340 C. The whisker was then cooled rapidly to 20 C, followed by the third growth shown in Fig. 2c. Clearly the time dependence of the RHEED intensity is quite the same in each of the three growths despite the differences in sample preparation.

The rapid drop in the first cycle is a general feature of all the observations of the specular spot with $\delta\phi \simeq 3\pi$. The simple interference argument predicts a minimum in the intensity of the specular spot from a flat surface for this angle. We have measured the dependence of the specular spot intensity upon the angle of incidence as shown in Fig. 9. The intensity is rising as the angle goes through $\delta\phi \simeq 3\pi$. We suspect that this is because of the proximity to the Kikuchi line. If electrons can leave at this angle, they can also penetrate at this angle. If channeling accounts for the high intensity before growth, perhaps the grown surface inhibits channeling.

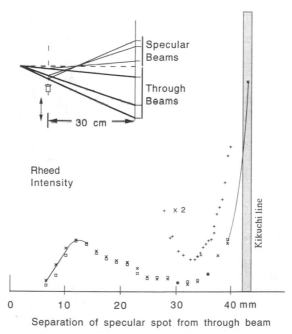

Fig. 9 The dependence upon angle of the specular intensity from the (001) surface of an iron whisker for three preparations by annealing at 650 C. The inset shows the geometry of RHEED in the PHI-400 MBE facility, where the sample must be translated and the beam redirected for each choice of incident angle. The incoming electron beam makes an angle of 0.1 rad with respect to the [11] azimuth. The data indicated by +'s were taken at many positions across the surface of the whisker. The ordinate has been scaled by a factor of 2 for the +'s.

Observations at Low Angles

As mentioned above, the oscillations at 20 C seem to persist much longer when observed at lower angles than at higher angles. If flat-topped islands form, the electron beam would not see the valleys between them at low angles. Even though the appearance of oscillations would signal layer by layer growth on the islands, one would not obtain a true picture of the roughness parameter.

Low-Temperature Growth

The diffraction patterns in the [10] azimuth and the [11] azimuth after growth of 15 layers of Fe on Fe whiskers at 100 K are shown in Fig. 10. The Kikuchi patterns remain after growth. The split line structure seen in Fig. 1 at 20 C is not so obvious in Fig. 10, but it is discernable.

The RHEED intensity oscillations in Fig. 3 were measured at θ=.0175 rad. This is the largest angle at which we could observe the low temperature growth because of obstruction of the RHEED beams by the cooling mechanism. The sample had been annealed at 650 C after growing at 330 C and stopping near a maximum (the end of Fig. 11b.) We did not properly anticipate the large initial increase in intensity. As a result we saturated the voltmeter in the data acquisition system, losing some data in Fig. 3. This was corrected after the second oscillation. The average intensity is greater and the amplitude of the oscillations smaller at 100 K than at 20 C. The greater intensity may be attributed to a smaller size of one-atom-high islands formed during growth. The reduction in amplitude of oscillation may be ascribed to a rougher surface for the low temperature growth. The dampings per cycle, found by fitting with Eq. (1), are $r_2/r_1 = 0.07$ and $r_3/r_1 = 0.11$ at 100 K. Both are smaller than observed at 20 C. This might be used as a argument against a rougher surface, but this difference is more likely the result of the lower angle of observation.

It may be that the heat of condensation provides most of the energy for rearrangement of atoms once they hit. But, there is some indication in Fig. 3 that the atoms may continue to move for a minute or more after the shutter is closed. We have not ruled out that this may be due to some instrumental drift.

Observation of the Recovery of Oscillations at 330 C

After each of two growths at 20 C, the controller of the temperature of the substrate holder was set to 330 C to observe the recovery of the oscillations. The temperature rose rapidly to 320 C and then slowly increased through 330 C. The results for the recovery of the specular spot at $\theta_i = 14$ mrad and $\theta_i = 65$ mrad are shown in Fig. 11. In both runs the oscillations go through a maximum as the temperature goes through 330 C. For $\theta_i = 65$ mrad, the maximum intensity at 330 C matches the intensity at the beginning of the growth at 20 C. (Fig. 11b is a continuation of the experiment shown in Fig. 2b.) At one point in Fig. 11b the furnace shutter was closed at the maximum of an oscillation and at another before the maximum was reached. It appears that the rearrangement of the atoms while the shutter is closed leads to a change in intensity in the same direction as it would if the deposition had been continued. This may be interpreted as the migration of atoms, if we assume that the changes in intensity follow the arrival of atoms at their final resting positions.

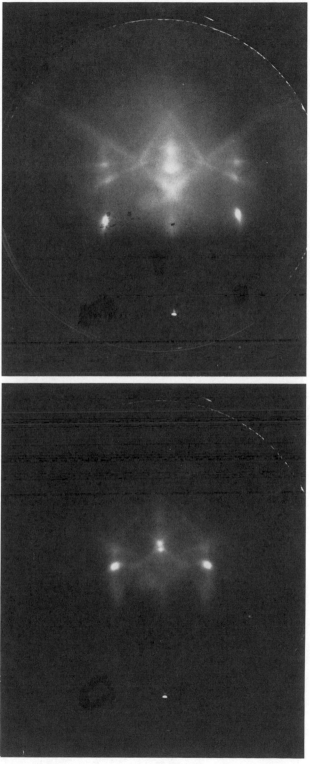

Fig. 10 The diffraction patterns in the top: [10] and bottom: [11] azimuths after growth of 15 layers of Fe on Fe whiskers at 100 K.

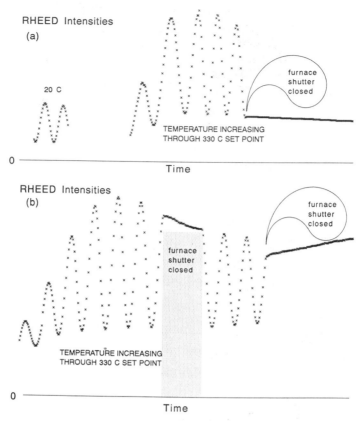

Fig. 11 Recovery of RHEED oscillations on heating. The data points are taken in 2s intervals. The specular spot is observed at (a) θ_i=14 mrad and (b) θ_i = 65 mrad in separate experiments. In each case the substrate heater control was set to 330 C, but this produced a 10-20 C overshoot at the thermocouple. In (b) the furnace shutter was closed at the maximum of an oscillation at one time and then just before the maximum was reached at a later time.

Effects of Temperature on Growth

An obvious effect of increasing the substrate temperature is an increase of the atom mobility. The split streak structure shown in Fig. 1 depends on the growth temperature. At 80 C the split lines are more sharply defined and closer together than at 20 C, indicating greater mean distance between the center of the one-atom-high islands and a narrower distribution of the separation distances. Above 200 C, the split line pattern is absent. The comparison of 20 C RHEED patterns with the 100 K patterns leads us to conclude that even at 20 C the heat of condensation plays an important role in determining the surface correlation function.

Oscillations of the specular spot show quite different damping depending upon the angle of observation and the temperature of observation. RHEED intensities for growths at 20 C, 200 C, and 350 C are

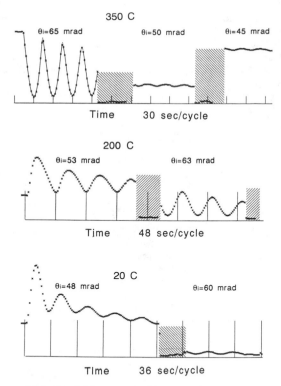

Fig. 12 Growth at 350 C, 200 C and 20 C. The specular spot is observed for several choices of the angle of incidence as indicated. At 350 C the oscillations approach that expected for perfect layer by layer growth on a flat surface with large amplitude oscillations when the specular spot is observed at θ_i = 65 mrad for which $\delta\phi$ = 3π. When observed at θ_i = 45 mrad or θ_i = 50mrad the intensity is almost constant. For growth at 200 C the amplitudes of the oscillations are almost the same at θ_i = 53 mrad and θ_i = 63 mrad. There is a similarity in the oscillations at 200 C and 20 C but the attenuation with time is greater for the lower temperature. The hatched areas correspond to times during which the angle of observation was being changed while the growth continued.

compared in Fig. 12. In each case a sequence of angles of incidence is selected during continuous depositions. Only the beginning of the sequences are shown in Fig. 12. The hatched areas are times in which the angle of observation was being changed while the growth continued.

The data for $\delta\phi \cong 3\pi$ at 350 C appears to be close to perfect layer by layer growth on a flat surface. In the simplest interference interpretation of RHEED oscillations, one would obtain a series of parabolas for perfect growth. The maximum intensities at several angles for growth at 350 C are consistent with the intensities measured on the as prepared surfaces at 20 C as shown in Fig. 9. There is a strong contrast between the data at 200 C and 350 C. While the angles being compared are not quite the same, it seems that the dependence on angle is reversed between 200 C and 350 C. There is more similarity between the 200 C and 20 C runs.

These studies of homoepitaxial growth lead to some ideas of the effect of temperature on the degree of layer by layer growth obtained. Clearly the best growths are obtained by going to temperatures in the 300 - 400 C range, but this is not very helpful in guiding the choice of growth temperatures for heteroepitaxial growth. For these temperatures will produce mixing of one transition metal element on another. For heteroepitaxial growths, it is a challenge for surface science to determine the trade off between the flatness of the layers achieved by elevated temperatures and the suppression of mixing achieved at low temperatures. For this problem RHEED oscillations should provide part of the information.

CONCLUSIONS

Fe whiskers provide excellent substrates for the study of growth. This growth is a strong function of temperature. The intensity of the specular spot is of particular interest as it varies greatly with θ_i and the condition of the surface. So far the studies have been of growth on the (100) surface. Presumably there will be surprises when we study other types of whisker surfaces.

Growths of Fe overlayers near ambient temperature produce surfaces with small one-atom-high islands of rather uniform separations, which depend upon growth temperature. Studies of the evolution of this structure and its dependence on growth temperature and how it is affect by annealing should be carried out.

From the rather good growths at low temperature, e.g. 100 K, it seems reasonable to assume that the energy from the heat of condensation contributes to the migration of atoms on the surface, even at room temperature.

The recovery of RHEED oscillations during growth at 330 C and subsequent measurements of growth at lower temperatures show that one obtains essentially the same surfaces by this method as by annealing at higher temperatures.

Mean field models of growth appear to be useful for the classification of types of RHEED oscillation patterns. Simple ideas of diffraction can be extended to partially account for the effects of attenuation and surface disorder on the RHEED intensities.

Acknowledgements

The assistance of K. Myrtle and Z. Celinski in the measurements and X.-Z. Li in the preparation of Fe whiskers is much appreciated. The modeling of growth has been carried out in stimulating collaboration with P. I. Cohen and P. R. Pukite. This work is supported by grants from the Natural Sciences and Engineering Research Council of Canada.

REFERENCES

1. A review of active research in RHEED is to be found in Reflection High-Energy Electron Diffraction and Reflection Electron Imaging of Surfaces, P. K. Larsen and P. J. Dobson eds. (Plenum Press, New York, NATO ASI Series vol. 188, 1989).
2. Z. L. Wang, Dynamical Calculations for RHEED and REM Including the Plasmon Inelastic Scattering, Surface Science 215, 201-216, and

217-231 (1989); and references therein; A Multislice Theory of Electron Inelastic Scattering in a Solid, Acta Cryst. $\underline{A45}$, 636-644 (1989).

3. M. A. Van Hove and S. Y. Tong, <u>Surface Crystallography by LEED</u>, (Springer, Berlin, 1977).

4. B. Heinrich, A. S. Arrott and S. D. Hanham, The Magnetic After-effect in 180 Degree Magnetically Bowed Bloch Walls, J. Appl. Phys. $\underline{50}$, 2134-2136 (1979); B. Heinrich and A. S. Arrott, Interactions of Mobile Impurities with a Single Domain Wall in Iron Whiskers, "3-M-1976", A.I.P. Conf. Proc. $\underline{34}$, 119-121 (1976).

5. S. T. Purcell, A. S. Arrott, and B. Heinrich, Reflection High-energy Electron Diffraction Oscillations during Growth of Metallic Overlayers on Ideal and Non-ideal Metallic Substrates, J. Vac. Sci. Technol. $\underline{B6}$, 794-797 (1988).

6. R. Hahn and M. Henzler, LEED Investigations and Work-function Measurements of the First Stages of Epitaxy of Tungsten on Tungsten (110), J. Appl. Phys. $\underline{54}$, 2079-2084 (1980); M. Henzler, Spot Profile Analysis (LEED) of Defects at Silicon Surfaces, Surface Sci. $\underline{132}$, 82-89 (1983); also this volume.

7. J. M. Cowley and H. Shuman, Electron Diffraction from a Statistically Rough Surface, Surface Sci. $\underline{38}$, 53 (1973); P. R. Pukite, P. I. Cohen and S. Batra, The Contribution of Atomic Steps to Reflection High-Energy Electron Diffraction from Semiconductor Surfaces, in <u>Reflection High-Energy Electron Diffraction and Reflection Electron Imaging of Surfaces</u>, P. K. Larsen and P. J. Dobson eds. (Plenum Press, New York, NATO ASI Series vol. 188, 1988), 427-447.

8. W. F. Egelhoff and I. Jacob, Reflection High Energy Electron Diffraction (RHEED) Oscillations at 77K, Phys. Rev Lett $\underline{62}$, 921-924 (1988).

9. Z. L. Wang, J. Liu, P. Lu and J. M. Cowley, Electron Resonance Reflections from Perfect Crystal Surfaces and Surfaces with Steps, Ultramicroscopy $\underline{27}$ 101-112 (1989); J. M. Cowley, Reflection Electron Microscopy, in <u>Surface and Interface Characterization by Electron Optical Methods</u>, A. Howie and U. Valdre eds. (Plenum Publishing Corp., 1988) pp 127-157.

10. P. I. Cohen, G. S. Petrich, P. R. Pukite, G. J. Whaley and A. S. Arrott, Birth-Death Models of Epitaxy, Surface Science $\underline{216}$, 222-248 (1989).

11. M. Horn, U. Gotter and M. Henzler, LEED Investigations of Si MBE onto Si(100), in <u>Reflection High-Energy Electron Diffraction and Reflection Electron Imaging of Surfaces</u>, P. K. Larsen and P. J. Dobson eds. (Plenum Press, New York, NATO ASI Series vol. 188, 1989), pp 463-473, Eq. (14).

12. A. S. Arrott and B. Heinrich, Crystallographic and Magnetic Properties of New Phases of Transition Elements Grown in Ultra Thin Layers by Molecular Beam Epitaxy, in <u>Metallic Multilayers and Epitaxy</u>, (M. Hong, S. Wolf and D. C. Gerber eds., The Metallurgical Society, Inc. Warrendale, Pa., 1987) pp 147-166.

13. Y. Tokura, H. Saito and T. Fukui, Terrace Width Ordering Mechanism During Epitaxial Growth on a Slightly Tilted Substrate, J. Cryst. Growth $\underline{94}$, 46-52 (1989).

14. P. R. Pukite, G. S. Petrich, G. J. Walley and P. I. Cohen, RHEED Studies of Diffusion and Cluster Formation During Molecular Beam Epitaxy in <u>Diffusion at Interfaces - Microscopic Concepts</u>, (M. Grunze, H. J. Kreuzer, J.J. Meimer eds., Springer Series in Surface Science 12, Berlin, 1988).

15. G. S. Petrich, P. R. Pukite, A. M. Wowchak, G. J. Whaley, P. I. Cohen and A. S. Arrott, On the Origin of RHEED Intensity Oscillations, J. Cryst. Growth $\underline{95}$, 23-27 (1989).

ION SCATTERING STUDIES OF SURFACE MELTING

J. F. van der Veen, B. Pluis and A. W. Denier van der Gon

FOM-Institute for Atomic and Molecular Physics
Kruislaan 407
1098 SJ Amsterdam, The Netherlands

ABSTRACT

Recent ion scattering investigations of the thermal disordering of Pb and Al surfaces are discussed. The results of these experiments are compared with predictions of a Landau-Ginzburg type theory of surface melting.

INTRODUCTION

The notion that the melting of a solid might commence at the surface was accepted long ago.[1-3] Since these first predictions of surface melting, various experimental searches for the effect have been made, many of them on surfaces of ice.[4,5] Often, the evidence brought forth in these investigations is not too convincing because the measurement technique used either lacked surface sensitivity or did not guarantee surface cleanliness. It is only very recently that surface melting was directly observed on a microscopic scale on a well-characterized, atomically clean surface.[6] The latter investigation was performed on the (110) surface of a Pb single crystal, using medium-energy ion scattering (MEIS) in conjunction with shadowing and blocking. Substantial advances have also been made with the development of, mostly phenomenological, theories of surface melting.[7-12]

In this paper we briefly discuss the most recent ion scattering investigations on surfaces of Pb[13-15] and Al[16] crystals and compare these results with the predictions of a phenomenological theory of surface melting.

The term "surface melting" has raised some confusion in the recent literature, so we first explain what we mean by it. Surface melting is, by our definition, the formation, below the triple point T_m, of a quasiliquid skin on the surface of a solid, which thickens as the temperature comes closer to T_m and finally converts the bulk solid into a melt at T_m. In this context, "surface melting" is just an abbreviation of "surface-induced melting of a solid". It proceeds strictly under thermal equilibrium, with the surface temperature being exactly equal to the temperature of the bulk. As such, it has nothing to do with the non-equilibrium melting of a crystal surface that may occur upon irradiation by e.g. a short laser pulse. It is not a nucleation and

Kinetics of Ordering and Growth at Surfaces
Edited by M. G. Lagally
Plenum Press, New York, 1990

growth phenomenon and there are no noticeable kinetic effects, at least not on the time scale of an experiment. In these respects, surface melting is fundamentally different from many of the other surface disordering phenomena discussed in this workshop. Finally, note our using the term "quasiliquid" layer instead of just "liquid" layer. This is to stress that there must be residual lattice order in a layer of finite thickness, which is imposed by the crystal with which it is in contact. Formally speaking, the layer has properties between those of a solid and a pure liquid. It is only in the limit of $T \rightarrow T_m$ that the quasiliquid layer attains macroscopic thickness and becomes a true liquid. The phenomenon of surface melting with diverging layer thickness can be considered to be a case of complete wetting, i.e. a complete wetting of the solid by its own melt.[17]

EXPERIMENTAL

The divergence of the disordered layer thickness can directly be observed by the use of MEIS in combination with shadowing and blocking. The principles of MEIS have been discussed elsewhere[18]. Here we briefly recapitulate some aspects relevant to surface melting.[4]

First suppose the surface of the crystal is well-ordered. If a parallel beam of medium-energy (e.g. 100 keV) protons is aligned with a low-index direction in the crystal, a shadowing effect arises along the incoming path within cones as schematically shown in Fig. 1a. When, in addition, the detection direction of backscattered protons is aligned with atomic rows, then the (near)-surface atoms along the rows block the backscattered protons on their way out and thus a blocking effect arises along the exit path (Fig. 1a). The combined effect of shadowing and blocking is to substantially reduce the backscattered yield of protons from subsurface layers relative to the yield from the topmost layers. The energy spectrum of backscattered protons then shows, at the elastic

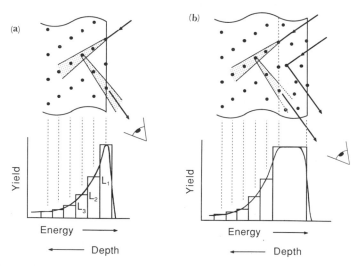

Fig. 1 Schematic representation of shadowing and blocking in the surface region of (a) a well-ordered crystal and (b) a crystal with a disordered surface layer. The corresponding energy spectra are shown as well.

backscattering energy, a distinct peak ("surface peak") against a low background, which arises from protons backscattered from the non-shadowed and non-blocked atoms in the topmost layers. For the hypothetical case that thermal vibrations in the crystal are completely absent, the shadowing and blocking would be perfect and the surface peak would contain only protons backscattered from the top layer only. The corresponding detection probability L_1 for that layer (Fig. 1a) is equal to one and those for the deeper layers $\{L_i\}$ (i > 1) would be zero. In reality however, the atoms are thermally vibrating, which makes the shadowing and blocking less effective. The $\{L_i\}$ for the deeper layers now become nonzero as indicated in Fig. 1a. Consequently, the surface peak becomes asymmetrically broadened toward lower energies (i.e., greater depths). With the use of a standard calibration procedure,[18] the integrated surface peak area can be converted into the number of monolayers H_{ord} that are visible to ion beam and detector:

$$H_{ord} = \sum_i L_i , \tag{1}$$

with $L_1 = 1$ and L_i decreasing with layer number i. H_{ord} increases smoothly with temperature because the thermal vibration amplitude increases, making shadowing and blocking less effective.

Now consider a crystal terminated by a thin disordered layer as schematically indicated in Fig. 1b. The atoms in the layer are displaced from the atom rows and hence do not contribute to shadowing or blocking. This makes these atoms fully visible to beam and detector and the surface peak broadens correspondingly (Fig. 1b). The surface peak area becomes much larger and the difference in area $H_{dis}-H_{ord}$ with respect to the yield that an ordered surface would have at the same temperature equals the number of disordered monolayers (or equivalently the number of disordered atoms per unit surface area). At temperatures where surface melting occurs, H_{ord} is of course experimentally inaccessible but it can readily be extrapolated from out of the lower temperature range where the surface is still ordered. In summary, with MEIS we have a technique at hand, with which we can quantitatively determine the temperature-dependent thickness of the disordered layer with submonolayer sensitivity.

MEIS studies of surface melting have been performed on differently oriented crystal faces of Pb[13-15] and Al.[16] Close-packed surfaces of, e.g., (111) and vicinal orientations were found to remain ordered up to T_m, while faces with relatively open packing [e.g. Pb(110), (112), (113) etc.] exhibit a pronounced melting effect. The strong crystal-face dependence in the melting behaviour is illustrated in Fig. 2 which shows a selection of backscattering energy spectra from the (110) and (111) surfaces of Pb for increasing temperature. For the (110) surface, the surface peak area increases dramatically as the temperature is raised above ~580 K, and close to T_m = 600.70 K it broadens in the way shown in Fig. 1b. By contrast, the surface peak area for the (111) surface increases with temperature at just the slow rate expected for a surface which vibrates with increasing amplitude but otherwise remains well-ordered.

The areal density N = $H_{dis}-H_{ord}$ of disordered atoms at the melted Pb(110) surface is shown in Fig. 3 as a function of the temperature difference T_m-T. Clearly, N is seen to diverge at T → T_m. Hence, we are dealing with a genuine surface-induced melting effect. The disordered layer thickness increases first logarithmically with ~$\ln[T_m/(T_m-T)]$ (regime I, Fig. 3a), then with a power law ~$(T_m-T)^{-1/3}$ (regime II, Fig. 3b).[13] The logarithmic and power-law growth result from short-range and

345

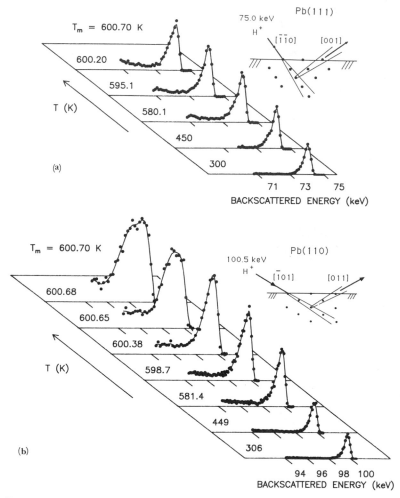

Fig. 2 Energy spectra of backscattered protons for increasing temperature, from (a) Pb(111) and (b) Pb(110). The insets show the shadowing/blocking geometry in which these spectra were taken.

long-range atomic interactions, respectively, see the next section.

The (111) and (110) oriented crystal faces of Al (T_m = 933.52 K) behave qualitatively the same way as the respective surfaces of Pb.[16] There is no sign of surface melting on Al(111), while there is a logarithmically diverging melt layer on Al(110). The disordered layer thickness increases logarithmically at least up to 0.3 K below T_m. As for Pb, a changeover to a power law increase is expected to occur very close to T_m, but this growth regime proved experimentally inaccessible because of temperature drifts in the sample.

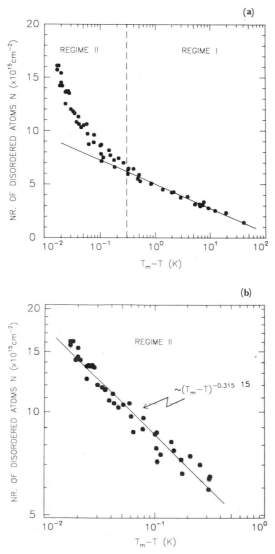

Fig. 3 (a) The number of disorderly positioned atoms per unit area of the Pb(110) surface, derived from the measured surface peak areas, as a function of T_m-T. The solid line represents the best fit of Eq. (16) to the data in regime I. (b) The data of regime II, reproduced on a double-log plot. The solid line represents the best fit of Eq. (20) to the data. From Ref. 13.

LANDAU-GINZBURG THEORY OF SURFACE MELTING

The experiments discussed in the preceding section show that, in a surface melting transition, the crystalline order at the surface transforms continuously into complete disorder as T approaches T_m. In the bulk of the crystal, however, the degree of crystalline order does not decrease continuously to zero, but instead jumps abruptly to zero at T_m, as expected for a first-order transition. It thus appears that a semi-infinite system which undergoes a first-order transition in the bulk, may show critical behavior at the surface. This property of the system can be understood within the framework of Landau-Ginzburg theory.[7,8]

Let us introduce, as a measure of the crystallinity, an order parameter $M(z)$ which is assumed to be only a function of the coordinate z perpendicular to the surface (the surface plane is at $z = 0$). A suitable order parameter to which the MEIS technique is sensitive, is the Fourier component of the density which corresponds to the smallest nonvanishing reciprocal lattice vector. The semi-infinite system is described by a Landau free-energy functional of the form[7]

$$F[M] = f_1(M_s) + \int_0^\infty [f(M) + \frac{1}{2} J \left(\frac{dM}{dz}\right)^2] dz , \qquad (2)$$

where $f(M)$ is the free energy per unit volume for homogeneous bulk material and $f_1(M_s)$ is the surface contribution to the free energy, per unit area. $M_s = M(z = 0)$ is the order parameter at the surface. The second term in the integral represents the energy cost associated with gradients in the order parameter M. For the function $f(M)$ we choose the "double-parabola" form

$$f(M) = \frac{1}{2} \alpha M^2 + \Lambda \qquad \text{for } M \leq M*, \qquad (3a)$$

and

$$f(M) = \frac{1}{2} \alpha (M_{cr} - M)^2 \qquad \text{for } M \geq M*, \qquad (3b)$$

with $\alpha > 0$ and $M*$ the order parameter where the parabolas intersect. The parameter Λ represents the free energy difference between solid and liquid: $\Lambda = \mathcal{L}(1 - T/T_m)$ with $\mathcal{L}$ the latent heat of melting per unit volume. The function $f(M)$ is sketched in Fig. 4a. With this choice of $f(M)$, the bulk has the proper melting characteristics; when the temperature crosses T_m, the equilibrium order parameter jumps discretely from $M = M_{cr}$ to $M = 0$, and at T_m itself the crystalline and liquid phases are in coexistence. For the surface term $f_1(M_s)$ in Eq. (2) we assume

$$f_1(M_s) = C + \frac{1}{2} \alpha_1 M_s^2 , \qquad (4)$$

with $\alpha_1 > 0$ and C constants (Fig. 4b). By this choice of $F_1(M_s)$, the surface at $z = 0$ has the tendency to be in a disordered state.

The equilibrium order parameter profile $M(z)$ is obtained by minimizing $F[M]$ using the variational principle and solving the resulting Euler equation[7,19]

$$\frac{\partial f(M)}{\partial M} = J \cdot \frac{d^2 M}{dz^2} , \qquad (5)$$

together with the boundary conditions

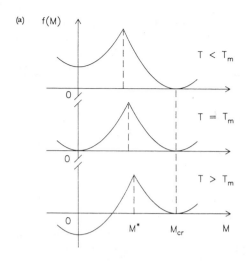

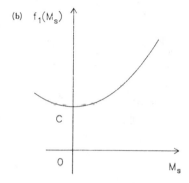

Fig. 4 (a) The bulk term f(M) in the Landau free energy for a first-order
transition at T_m, for temperatures $T < T_m$, $T = T_m$ and $T > T_m$. (b)
The surface contribution $f_1(M_s)$ to the Landau free energy.

$$\left.\frac{dM}{dz}\right|_{z=0} = \frac{1}{J} \cdot \frac{\partial f_1(M_s)}{\partial M_s} \tag{6a}$$

and

$$\left.\frac{dM}{dz}\right|_{z \to \infty} = 0 \quad . \tag{6b}$$

Henceforth, we assume $T < T_m$, that is $\Lambda > 0$. Solving Eqs. (5) and (6)
leads to the order parameter profile

$$M(z) = \frac{1}{2} M_{cr} e^{\kappa(z-z^*)} - \frac{\Lambda}{\alpha M_{cr}} \cdot e^{-\kappa(z-z^*)} \qquad \text{for } z \leq z^*, \tag{7a}$$

and

$$M(z) = M_{cr} \left[1 - \frac{1}{2} e^{-\kappa(z-z*)} \right] - \frac{\Lambda}{\alpha M_{cr}} \cdot e^{-\kappa(z-z*)} \qquad \text{for } z \geq z*, \qquad (7b)$$

provided

$$\alpha_1 > (J\alpha)^{1/2} \text{ and } \Lambda \leq \Lambda_{max} = \frac{\alpha M_{cr}^2 [\alpha_1 - (J\alpha)^{1/2}]}{2[\alpha_1 + (J\alpha)^{1/2}]}. \qquad (8)$$

Here, $\kappa = (\alpha/J)^{1/2}$ and $z*$ is given by

$$z* = \frac{\kappa^{-1}}{2} \ln \left(\frac{\Lambda_{max}}{\Lambda} \right). \qquad (9)$$

The profile $M(z)$ is sketched in Fig. 5 for increasing T. Clearly, this solution for the order parameter profile represents the case of surface melting. As $T \to T_m$ ($\Lambda \to 0$), the order parameter at the surface $M_s = M(z=0)$ decreases continuously to zero, while the disordered layer thickness $z*$ diverges logarithmically.

The equilibrium order parameter profile given by Eq. (7) minimizes the Landau free energy, for which one finds

$$F(z*) = f_1(M_s) + \int_0^\infty J \left(\frac{dM}{dz} \right)^2 dz$$

$$= C + \frac{1}{4} M_{cr}^2 (J\alpha)^{1/2} + z*\Lambda + \frac{1}{4} M_{cr}^2 (J\alpha)^{1/2} \frac{\alpha_1 - (J\alpha)^{1/2}}{\alpha_1 + (J\alpha)^{1/2}} e^{-2\kappa z*} \qquad (10)$$

after substitution of Eq. (7) in the integral over z and use of Eqs. (8) and (9). The constants C, M_{cr}, J, α and α_1 are all input parameters of the theory. We can relate these parameters to the (semi-) empirically known interfacial free energies per unit area, γ_{sv}, γ_{lv} and γ_{sl} for the respective solid-vapor, liquid-vapor and solid-liquid equilibria. This is done[19] as follows. First consider the limit of infinite melt thickness $z* \to \infty$ for $T \to T_m$ ($\Lambda \to 0$). The corresponding free energy per unit area for this macroscopically wet surface must be equal to

$$F(z* \to \infty) = \gamma_{lv} + \gamma_{sl}. \qquad (11)$$

Letting $z* \to \infty$ and $\Lambda \to 0$ in Eq. (10) and considering the fact that the parameter C represents the surface term $f_1(M_s = 0)$ in the Landau free energy, we obtain

$$C = \gamma_{lv} \qquad (12)$$

and

$$\frac{1}{4} M_{cr}^2 (J\alpha)^{1/2} = \gamma_{sl}. \qquad (13)$$

Now consider the opposite limit of zero melt thickness $z* = 0$ ($\Lambda = \Lambda_{max}$), which corresponds to having a completely dry solid surface. Then one must have

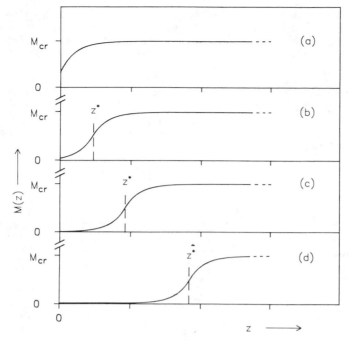

Fig. 5 Order parameter profiles M(z) for a surface melting transition, calculated with the use of Eq. (17) for increasing temperature $T \to T_m$. The position of the quasiliquid interface is at $z = z*$.

$$F(z* = 0) = \gamma_{sv} \, . \tag{14}$$

Taking $z* = 0$ in Eq. (10), one derives the equality

$$C + \frac{1}{2} M_{cr}^2 (J\alpha)^{1/2} \frac{\alpha_1}{\alpha_1 + (J\alpha)^{1/2}} = \gamma_{sv} \, . \tag{15}$$

Using Eqs. (12), (13) and (15) one can rewrite Eqs. (9) and (10) in the following useful form:

$$z* = \frac{\xi_d}{2} \ln \left[\frac{2T_m(\gamma_{sv} - \gamma_{lv} - \gamma_{sl})}{\mathcal{L}(T_m - T)\xi_d} \right] \tag{16}$$

and

$$F(z*) = \gamma_{lv} + \gamma_{sl} + \mathcal{L}(1 - T/T_m)z* + (\gamma_{sv} - \gamma_{lv} - \gamma_{sl}) \, e^{-2z*/\xi_d}, \tag{17}$$

where we have substituted $\mathcal{L}(1 - T/T_m)$ for Λ. $\xi_d = \kappa^{-1}$ is the correlation length in the disordered phase, i.e., the distance over which the order parameter M(z) changes by a factor e. It is readily verified that the equilibrium thickness $z*$ as given by Eq. (16) minimizes F. Replace $z*$ in Eq. (17) by an arbitrary thickness ℓ. Then $dF(\ell)/d\ell = 0$ for $\ell = z*$.

351

It is important to note that the interfacial free energies appearing in Eqs. (16) and (17) cannot be equilibrium quantities at temperatures for which the surface is wetted by a quasiliquid layer of <u>finite nonzero</u> thickness. It is only in the limit of <u>infinite</u> liquid layer thickness that γ_{lv} and γ_{sl} were introduced earlier and this limit is reached only at $T = T_m$. Likewise, γ_{sv} is an equilibrium quantity only in the limit of zero quasiliquid layer thickness. However, out-of-equilibrium values for γ_{sv} can still be estimated by extrapolating the equilibrium values known at lower temperatures to the temperature region of interest. There is no need for a downward extrapolation of γ_{lv} and γ_{sl} values, since surface melting occurs relatively close to T_m.

Surface melting will occur, if the following condition is satisfied:

$$\gamma_{sv} - \gamma_{lv} - \gamma_{sl} \equiv \Delta\gamma > 0 . \tag{18}$$

This condition follows from the requirement that the argument of the logarithm in Eq. (16) be positive. The physical meaning of $\Delta\gamma > 0$ is that energy is gained by replacing the dry solid surface by one wetted by a liquid layer. This provides a driving force for surface melting. If, on the other hand, $\Delta\gamma < 0$, then the solid surface will remain dry up to T_m.

In the foregoing we have discussed the case of short-range atomic interactions dominating the melting behavior. Typical for the short range are the exponentially decaying surface order parameter $M_s \approx 1/2\, M_{cr}$ $\exp(-z*/\xi_d)$ and the excess free energy term $\Delta\gamma \exp(-2z*/\xi_d)$ in F. However, in addition to short-range interactions there are always long-range Van der Waals interactions, here assumed to be non-retarded[8,14]. These interactions give rise to a pair potential energy which, for large interparticle separation r, has an attractive tail of the form $-\epsilon(r/\lambda)^{-6}$. The corresponding contribution to the excess free energy is given by $W = \pi\rho_\ell(\rho_s-\rho_\ell)\epsilon\lambda^6/12$. ρ_s and ρ_ℓ are the particle number densities in solid and melt, respectively. Adding this contribution to the free energy F, one obtains

$$F(\ell) = \gamma_{lv} + \gamma_{sl} + \mathcal{L}(1-T/T_m)\, \ell + \Delta\gamma^{SR}\, e^{-2\ell/\xi_d} + W/\ell^2 , \tag{19}$$

where $\Delta\gamma^{SR}$ represents the short-range contribution to the total $\Delta\gamma$. For very large quasiliquid layer thickness, the long-range interactions will eventually dominate. In that case, the short-range contribution to $F(\ell)$ can be neglected and minimization of $F(\ell)$ then yields for the equilibrium thickness of the quasiliquid layer

$$z* = \left[\frac{(T_m-T)\mathcal{L}}{2T_mW}\right]^{-1/3} , \tag{20}$$

provided $W > 0$. The layer thickness is seen to increase with T_m-T according to a power law with exponent $-1/3$. A necessary condition for surface melting is that $W > 0$, i.e. $\rho_s > \rho_\ell$. For Pb and Al, this condition is fulfilled.

DISCUSSION

The MEIS results for Pb and Al nicely confirm the predictions of the Landau-Ginzburg theory.

First consider our experimental finding that the (110) face melts, but the (111) face not. It is well known that the free energy of a solid crystal surface is anisotropic, i.e., dependent on the surface orientation. In general, crystal surfaces with open atom packing have a higher surface energy γ_{sv} than close-packed surfaces. For an f.c.c. crystal:

$$\gamma_{sv}^{\{110\}} > \gamma_{sv}^{\{111\}} \ . \tag{21}$$

A relatively small anisotropy in γ_{sv} can lead to a dramatic change in melting behavior. From measurements of the equilibrium shape of Pb crystals, Heyraud and Métois[20] have derived a value for the anisotropy: $\gamma_{sv}^{\{110\}}/\gamma_{sv}^{\{110\}0} \approx 1.06$. Using this value and known values[13] for γ_{lv}, γ_{sl} and $\gamma_{sv}^{\{111\}}$, it can be shown that $\Delta\gamma^{\{110\}} > 0$ and $\Delta\gamma^{\{111\}} < 0$ for Pb. This explains why Pb(110) melts but Pb(111) not. For Al, the anisotropy of γ_{sv} is not known and the γ values are not accurate enough for a reliable prediction of the sign of $\Delta\gamma$.

The temperature dependence of the quasiliquid thickness is also in accordance with the predictions of Landau-Ginzburg theory. Up to 0.3 K below T (regime I in Fig. 3a), the layer thickness on Pb(110) increases logarithmically, which is in agreement with the prediction for the case that short-range atomic interactions are dominant. Fitting Eq. (16) to the data yields $\Delta\gamma^{\{110\}} = 21$ mJ/m^2 and $\xi_d = 0.63$ nm.[14] Here, ℓ was fixed to the value of $\ell = 2.5 \times 10^8$ J/m^3 known for bulk melting of Pb and use was made of $N = z^* \rho_\ell$ for the conversion of quasiliquid layer thickness z^* to N, the number of disorderly positioned atoms per unit area as determined by MEIS. The best-fit value of $\Delta\gamma$ is close to the value obtained from semi-empirical estimates.[13] For the correlation length ξ_d in the disordered phase we find, as expected, a value which is in the range of 1.5-2 times the atomic diameter D ($\xi_d/D = 1.6$). For Al(110) we find best-fit values of $\Delta\gamma^{\{110\}} = 29$ mJ/m^2 and $\xi_d/D = 1.5$, rather close to those for Pb(110).

In our MEIS experiments on Pb(110) we were able to enter the asymptotic melting regime where long-range atomic interactions become dominant. From $T_m-T = 0.3$ K onwards, a departure from the logarithmic growth law is observed (regime II in Fig. 3). Apparently, the melt thickness at that temperature has become sufficiently thick that short-range interactions are screened out. The temperature dependence of the melt thickness in regime II can be fitted to a power law (Fig. 3b). Treating the exponent r of the power law $\sim(T_m-T)^{-r}$ as a free parameter of the fit, we obtain r = 0.315 ± 0.015, in excellent agreement with the value of 1/3 expected for non-retarded Van der Waals interactions. Fitting Eq. (20) with r fixed to 1/3, we obtain for the Hamaker constant the not unreasonable value of W = (0.40 ± 0.05) × 10^{-21} J.

Acknowledgements

This work is part of the research program of the Stichting voor Fundamenteel Onderzoek der Materie (Foundation for Fundamental Research on Matter) and was made possible by financial support from the Nederlandse Organisatie voor Wetenschappelijk Onderzoek (Netherlands Organization for the Advancement of Research).

REFERENCES

1. G. Tammann, Z. Phys. Chem. __68__, 205 (1910); Z. Phys. __11__, 609 (1910).
2. M. Volmer and O. Schmidt, Z. Phys. Chem. __B35__, 467 (1937).
3. I. N. Stranski, Z. Phys. __119__, 22 (1942); Naturwissenschaften __28__, 425 (1942).

4. J. F. van der Veen, B. Pluis and A. W. Denier van der Gon, in: <u>Chemistry and Physics of Solid Surfaces VII</u>, Springer Ser. in Surface Sci., Vol. 10, R. Vanselow and R. F. Howe eds., (Springer, Heidelberg, 1988), p. 455, and refs. therein.
5. G. Dash, Contemporary Physics <u>30</u>, 89 (1989).
6. J. W. M. Frenken and J. F. van der Veen, Phys. Rev. Letters <u>54</u>, 134 (1985); J. W. M. Frenken, P. M. Marée and J.F. van der Veen, Phys. Rev. <u>B34</u>, 7506 (1986).
7. R. Lipowsky and W. Speth, Phys. Rev. <u>B28</u>, 3983 (1983).
8. R. Lipowsky, Ferroelectrics <u>73</u>, 69 (1987).
9. A. A. Chernov and L. V. Mikheev, Phys. Rev. Letters <u>60</u>, 2488 (1988).
10. A. Levi and E. Tosatti, Surface Sci. <u>189/190</u>, 641 (1987).
11. E. Tosatti, in: <u>The Structure of Surfaces II</u>, Springer Ser. in Surface Sci., Vol. 11, J. F. van der Veen and M. A. Van Hove eds., (Springer, Heidelberg, 1988) p. 535.
12. A. Trayanov and E. Tosatti, Phys. Rev. <u>B38</u>, 6961 (1988).
13. B. Pluis, A. W. Denier van der Gon, J. W. M. Frenken and J. F. van der Veen, Phys. Rev. Letters <u>59</u>, 2678 (1987).
14. B. Pluis, T. N. Taylor, D. Frenkel and J. F. van der Veen, Phys. Rev. <u>B40</u>, 1353 (1989) 1353.
15. A. W. Denier van der Gon, B. Pluis, R. J. Smith and J. F. van der Veen, Surface Sci. <u>209</u>, 431 (1989).
16. A. W. Denier van der Gon, R. J. Smith, J. M. Gay, D. J. O'Connor and J. F. van der Veen, submitted to Surface Sci.
17. S. Dietrich, in: <u>Phase Transitions and Critical Phenomena</u>, Vol. 12 (Academic, London), in press.
18. J. F. van der Veen, Surface Sci. Rept. <u>5</u>, 199 (1985).
19. B. Pluis, D. Frenkel and J. F. van der Veen, to be published.
20. J. C. Heyraud and J. J. Métois, Surface Sci. <u>128</u>, 334 (1983).

THE ROUGHENING TRANSITION OF VICINAL SURFACES

J. Lapujoulade, B. Salanon, F. Fabre
Service de Physique des Atomes et des Surfaces

B. Loisel
Section de Recherches en Metallurgie Physique
C.E.N. Saclay
91191 Gif-sur-Yvette, France

ABSTRACT

On a vicinal surface the steps are expected to become rougher and rougher as the temperature is increased. Using a TSK (Terrace, Step, Kink) model it is shown that a transition of the Kosterlitz-Thouless type occurs. Then above $T = T_R$ the correlation function of step positions diverges logarithmically with distance. This behavior is experimentally studied by TEAS (Thermal Energy Atom Scattering). A comparison between the experimental peak profiles and exact Monte Carlo calculations on TSK models allows to evaluate important physical quantities such as the kink energy formation and the step-step interaction energy.

Various vicinal surfaces of copper have been studied:

Cu(113), Cu(115), Cu(1,1,11), Cu(331), Cu(310) in order to check the influence of step-step distance, step structure, terrace structure. The following trends are observed:

- At the same temperature the roughness increases with the distance between steps as a result of the weakening of the step-step interaction.
- Changing the terrace structure from (100) to (111) does not significantly affect the roughness. - Steps parallel to the loosely packed direction [100] are very much rougher than those parallel to the close-packed direction [110].
- Freezing of the roughness due to low surface mobility always occurs below room temperature.
- The large terrace surface Cu(1,1,11) deviates from the prediction of the Kosterlitz-Thouless model. A tentative explanation is given.

INTRODUCTION

The concept of the roughening transition was introduced in 1949 by W. K. Burton and N. Cabrera[1] in order to explain the crystal growth regimes which were experimentally observed (slow layer by layer regime or fast tridimensional). They introduced a special TSK (terrace, step, kink) model which was later called the SOS model which still remains the basic

Kinetics of Ordering and Growth at Surfaces
Edited by M. G. Lagally
Plenum Press, New York, 1990

one. Starting from an analogy with the Ising model they predicted that the interface width will diverge above a transition temperature T_R. However it was recognized later that the SOS model does not fall into the Ising class but into a new class: the Kosterlitz-Thouless class.[2] This has been proved establishing formal equivalences between the SOS model, the XY-Heisenberg model and the Coulomb lattice gas.[3,4] Moreover Monte-Carlo simulations carried out on the SOS model have shown that for close-packed surfaces T_R is even higher than the melting point. So the observation of the roughening transition on such surfaces is difficult and indeed it has only been achieved in 1977 on C_2Cl_6 and NH_4Cl,[5] in 1980 at the interface liq.He/sol.He[6] and for Au, In and Pb surfaces.[7] These experiments have stimulated many theoretical works but it is not easy to make the connection since these experimental data only rely upon the observation of the macroscopic equilibrium crystal shape: facetted below T_R and rounded above T_R. One does need to know the surface height-to-height correlation function. Its measurement was readily achieved by TEAS (Thermal Energy Atom Scattering) on vicinal surfaces: $Ni(115)$[8] in 1985 and $Cu(115)$[9] in 1986. Then the comparison of the experimental data with numerical exact calculations allows the estimation of energetic parameters like the kink formation energy and the step-step interaction energy which were not previously known. This last point is presently an object of discussions in the physics community.

THEORETICAL APPROACH

It is not our purpose to present here the theoretical background of the roughening transition, which can be found in previous papers devoted to the roughening transition in TSK models,[10] Sine-Gordon models[11] or to the equilibrium shape of crystals.[12] The special case of vicinal surfaces is only considered.

The basic ideas were developed in 1985 by Villain et al.[13] A vicinal surface is schematically represented in Fig. 1. At zero temperature the steps of the vicinal structure are straight, parallel and equidistant. When the temperature is raised the steps meander at large scale under the influence of thermal fluctuations. At a microscopic scale this is due to the formation of kinks along the steps. The elementary excitation is a detour as shown in the figure; its energy is

$$E = 2W_0 + w_1(\ell)L , \tag{1}$$

where W_0 is the kink formation energy, $w_1(\ell)$ is the energy per unit length needed when one step is displaced by one atomic distance between its two neighbors, ℓ is the distance between steps, and L is the length of the detour.
    The introduction of the step interaction $w_1(\ell)$ is crucial since this term is responsible for the stability of vicinal surfaces. An isolated step is well known to be rough at all temperatures. Physically the interaction may be due either to the elastic repulsion between two steps of the same sign or to the electrostatic repulsion between dipoles. There is also another source of interaction which has a statistical origin: it comes from the impossibility for two steps to cross each other. All these interactions are proportional to d^{-2} so that

$$w_1(\ell) \approx (\ell+a)^{-2} + (\ell-a)^{-2} - 2\ell^{-2} \approx \ell^{-4} , \tag{2}$$

where a is an elementary step displacement. Except for very close steps one can assume $w_\ell \ll W_0$ so that for usual temperatures $w_\ell < kT < W_0$.

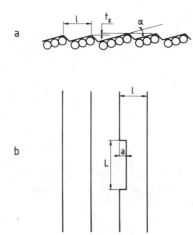

Fig. 1 a) Schematic view of a vicinal surface; b) Elementary excitation.

Under the effect of their mutual repulsion the detours have a tendency to group into domains as illustrated in Fig. 2. These domains have the same structure as the original surface but they are shifted by an integer multiple of a vector t; its normal component is

$$t_z = a \sin \alpha \,, \tag{3}$$

where α is the terrace angle as defined in Fig. 1.

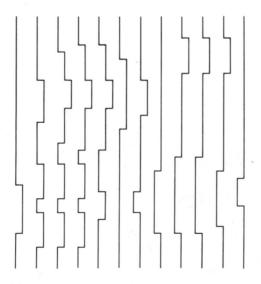

Fig. 2 Domain organization of the elementary excitations.

Having settled the physical basis of the problem one has now to write down the Hamiltonian of the surface and then to calculate the partition function from which all thermodynamical quantities can be derived. This cannot be achieved analytically in most cases and numerical methods such as Monte Carlo computations are generally required. However useful informations can be derived from a Sine-Gordon approach which can be handled analytically by renormalization methods.

1. The Sine-Gordon approach[13]

The stepped surface is modeled by a deformable array of lines (Fig. 3). Its deformations in a free regime are governed by anisotropic surface stiffnesses: η_x in a direction perpendicular to the steps and η_y parallel to the steps. The discreteness of the step displacements imposed by the atomic structure is introduced through a localization potential U. Then one can write a Sine-Gordon Hamiltonian:

$$H = (1/2)\Sigma_q(\eta_x\ell^2q_x^2+\eta_ya^2q_y^2)|u_q|^2 + U \Sigma_{m,y} [1-\cos(2\pi u_m(y))], \qquad (4)$$

where $u_m(y)=(1/\sqrt{N}) \underset{\substack{q_x<q_{cx} \\ q_y<q_{cy}}}{\Sigma} u_q\exp(iq_xm\ell+iq_yy)$.

$u_m(y)$ is the displacement of the mth step at the ordinate y, and q_{cx},q_{cy} are appropriate cut-offs. Such a formulation has the great advantage to be solved analytically by a renormalization method. The following results are obtained:

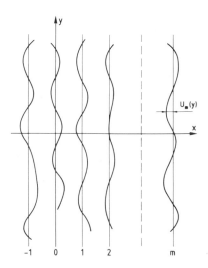

Fig. 3 The array of deformable lines used in the Sine-Gordon model.

i.   there exists a critical temperature T_R

    -for $T<T_R$ the localization potential is renormalized to a finite value U*. Thus the correlation function of step displacements $<(u_m(y)-u_0(0))^2>$ remains finite at any scale and the surface is smooth.

    -for $T\geq T_R$ U* vanishes in the renormalization process. Thus the correlation function diverges at large scale and the surface is rough. The behaviour is logarithmic:

$$<(u_m(y)-u_0(0))^2> = cte + A(T)\ln\rho , \qquad (5)$$

where

$$\rho^2 = y^2 \ (\eta^*_x/\eta^*_y)^{1/2}+m^2a^2 \ (\eta^*_y/\eta^*_x)^{1/2} , \qquad (6)$$

$\eta^*_x(T),\eta^*_y(T)$ are the renormalized stiffnesses, and

$$A(T) = (T/\pi\sqrt{\eta^*_x\eta^*_y} \quad .$$

    The correlation function is anisotropic. The anisotropy is due first to the intrinsic anisotropy of the structure geometry and secondly to the anisotropy of the surface stiffness. But it must be emphasized that the latter acts in the opposite direction to the former as $\eta^*_x<<\eta^*_y$.

ii. At $T = T_R$ one finds a universal relation between the renormalized stiffnesses η^*_x,η^*_y and T_R:

$$\sqrt{\eta^*_x\eta^*_y} = (\pi/2)T_R \quad . \qquad (7)$$

    The main advantage of this method is that the parameters η^*_x, η^*_y are easily obtained from a diffraction experiment as shown below but its drawback is that their physical significance is not obvious especially in terms of the previous model.

2.   The TSK approach[14,15]

    This approach is more in agreement with the model presented at the beginning of this section. The general form of the Hamiltonian is

$$H = \sum_{m,n} W_0(u_{m,n+1}-u_{m,n})^2 + \sum_{m,n} f(u_{m+1,n}-u_{m,n}) \quad . \qquad (8)$$

    In the first term W_0 is again the kink formation energy. The quadratic dependence upon step displacements is introduced in order to account for the lower probability of double kinks with respect to simple ones. The second term describes the step-step interaction. Its exact form depends upon the vicinal face geometry. Taking for instance a (11n) face of a FCC crystal (with $\ell = na/2$) one can write:

$$f(\delta u) = \begin{vmatrix} 0 & \text{for } \delta u\geq 0 \\ w_p(\ell) & \delta u=-1,-2,\ldots,-p,\ldots,(n-1)/2 \\ \infty & \delta u<-(n-1)/2 \quad . \end{vmatrix} \qquad (9)$$

    The last condition excludes step crossings. We have of course $w_1(\ell)<w_2(\ell)<w_3(\ell)$. This is a generalization of the model presented at the beginning of this chapter.

In order to derive the statistics of this Hamiltonian a numerical computation using a Monte-Carlo technique is needed. Such calculations have been performed which confirm the previous results ie. the existence of T_R and the logarithmic behaviour of the correlation function at large distances. But in addition the relation between the energetic parameter $W_0, w_p(\ell)$ and η^*_x, η^*_y allowing their derivation from the experimental data as shown below. It must be emphasized here that the approximate relations which were given by Villain et al. are only valid when $w_p(\ell) \ll W_0$ otherwise they can be in error by a factor two. Thus when the steps are close together like in (133) or (115) faces any reasonable estimation of W_0, $w_p(\ell)$ from the experimental data must use an exact numerical computation and not such approximations.

It is worth to note the similarity of this TSK approach with the usual SOS models: here the columns lie flat on the terraces but their interaction is now anisotropic.

In order to illustrate this section we present in Fig. 4 the results of a Monte-Carlo calculation which displays the configuration of the surface below, at and above T_R. This figure shows clearly how progressive is the transition. Indeed at T_R the surface is not so "rough" at a microscopic scale. It is quite impossible to define T_R by eye inspection. A careful determination of the correlation function is needed in this respect.

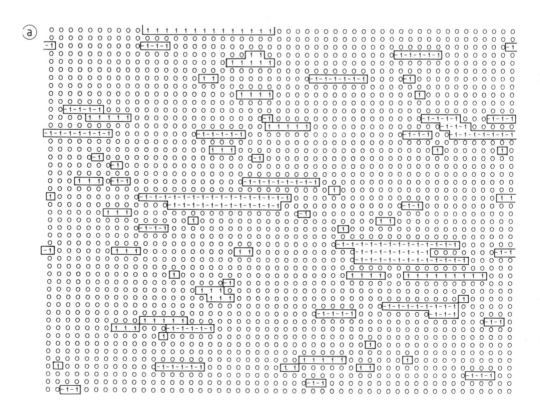

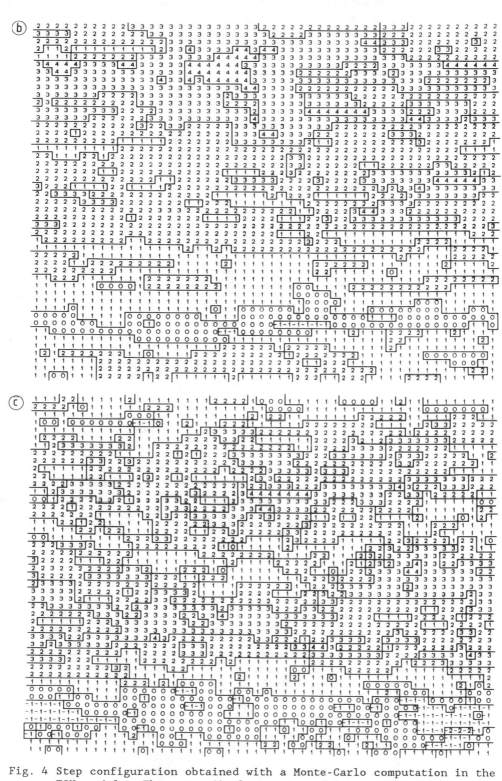

Fig. 4 Step configuration obtained with a Monte-Carlo computation in the TSK model. The step displacements $u_m(y)$ are indicated by the numbers. The solid lines indicates the kink locations. Anisotropy $w_1/W_0 = 0.1$ for Cu(115). a) $T = 0.8T_R$; b) $T = T_R$; c) $T = 1.4 T_R$.

361

EXPERIMENTAL MEASUREMENTS

The first observations of the roughening transition of close-packed surfaces were achieved by measuring the equilibrium shape of macroscopic crystals. However only very small crystals can be equilibrated and their vicinal areas are too small to be accurately observed. So the experiments are carried out on macroscopic metastable vicinal surfaces on which only the local equilibrium of the step configurations is achieved. The roughness is then characterized by the correlation function of the step positions which is the quantity to be measured. The easiest way to do that is to perform a diffraction experiment in which the peak profiles can be monitored. Helium atoms, x-rays, electrons beams can be used in this aim; indeed Helium was first used and more recently x-rays but in our knowledge HRLEED has not been used yet.

The exact calculation of the diffracted intensity can be difficult because multiple scattering may be important. It is always the case for Helium or electron scattering but even for x-rays this problem has to be considered when very glancing incidence is required in order to get enough surface sensitivity. A great simplication arises if the surface is assumed to be made of shifted domains as previously shown in Fig. 2 and if the domain size is large enough. Then the multiple scattering between domains can be neglected and the intensity can be expressed as a Fourier transform of the distribution of the domain phase shifts[16]

$$I(\delta K) = I_0(\delta K) \sum_{m,n} \exp(i\delta K \cdot R_{m,n}) \langle \exp(i\delta k \cdot t(u_{m,n} - u_{0,0}) \rangle , \qquad (10)$$

where $I_0(\delta K)$ is a form factor which only depends upon the structure of the vicinal surface unit cell; all multiple scattering effects are contained in this factor. All step statistics are contained in the second term.

A further simplification occurs if the distribution of the step displacements can be assumed to be Gaussian. This approximation has been checked to be valid in any practical situation.[17] Then one obtains:

$$I(\delta K) = \sum_{m,n} I_0(\delta K) \exp(i\delta K \cdot R_{m,n}) \exp(-\langle (u_{m,n} - u_{0,0})^2 \rangle f(\delta k \cdot t)) , \qquad (11)$$

where $\langle (u_{m,n} - u_{0,0})^2 \rangle$ is the correlation function of step displacements and $f(\delta k \cdot t)$ is a periodic function of the phase shift.

Two important situations may appear:

i. in-phase condition: $\delta k \cdot t = 2p\pi$ (p integer)

Waves scattered from the various domains interfere constructively. The disorder is not seen by the beam and the scattering pattern is made of δ-functions just like the smooth surface.

ii. anti-phase condition: $\delta k \cdot t = (2p+1)\pi$

Waves interfere destructively. Each diffraction peak is broadened by the step disorder and the effect is maximum.

The sequence of in-phase and anti-phase conditions characterizes the discreteness of the domain heights and it is thus possible to distinguish between step or terrace roughening.

For $T<T_R$ the correlation function remains finite at large distances, thus the above formula shows that the scattered intensity contains a coherent part made of an array of δ-functions superimposed to an incoherent diffuse scattering.

For $T \geq T_R$ the correlation function diverges logarithmically at large distances thus the coherent part of the scattering vanishes. Introducing the logarithmic form Eq. (5) into Eq. (11) gives

$$I(\delta k) = I_0(\delta k) \times \sum_{p,q} \Phi(\tau)[(\eta^*_y/\eta^*_x)^{1/2}(a_y\delta k_y+2\pi p)^2$$
$$+(\eta^*_x/\eta^*_y)^{1/2}(a_x\delta k_x+2\pi q)^2]^{-1+(\tau/2)} \tag{12}$$

with $\tau = A(T)f(\delta k)$ (13)

and $\Phi(\tau) = \Gamma(2-\tau)/(\Gamma(\tau/2)\Gamma((3-\tau)/2))$. (14)

In the derivation of Eq. (12) it has been supposed that $\tau<2$. Thus the scattering is made of anisotropic power law peaks centered at Bragg positions which reduce to δ-functions for in-phase conditions. The observation of this power law dependence is thus the signature of the roughening transition.

At $T = T_R$ in the anti-phase condition the exponent of the power law decay takes the universal value[17]

$$\tau(T_R) \approx 1 . \tag{15}$$

The measurement of the exponent universal value is the most accurate way to determine T_R.

The broadening of the diffraction peaks above T_R brings about a strong decrease of the intensity in the Bragg position. However it could be very misleading to deduce a roughening temperature from such an accident on the intensity curves since it can arise from other causes like enhanced inelastic scattering due to surface enhanced thermal motions (anharmonicity).[25]

A fit of the above formula on the experimental peak profiles gives the values of the renormalized surface stiffnesses η^*_x, η^*_y. It is, of course, very important to include carefully in this fit the other source of peak broadening such as instrumental width, mosaic structure of the crystal or surface plane miscut. As discussed above the correlation function can be computed numerically from TSK models. A logarithmic fit of its long distance behavior gives also η^*_x, η^*_y. Then it is possible to adjust the parameters of the TSK model in order to fit the calculated values to the experimental ones. If this program can be achieved one obtains a complete picture of the surface roughening. Moreover the knowledge of the kink formation energy and step interaction energy is by itself very interesting for the fundamental understanding of crystal structure and growth.

RESULTS

To date Ni(113),[18] Ni(115),[8] Cu(113),[15] Cu(115),[9] Cu(1,1,11),[19] Cu(331),[20] Cu(310)[20] have been studied by TEAS. X-rays were also used for Cu(113)[21] and very recently STM images of Ag(115)[22] have been obtained. The most complete analysis has been done for copper so we shall focus our presentation on copper results.

363

Cu(113), Cu(115) and Cu(1,1,11) present (100) terraces and [110] close-packed steps. Cu(331) has (111) terraces and [110] steps. Cu(310) has (100) terraces with loosely packed [100] steps.

The power law decay of the peaks was observed in all TEAS experiments except for Cu(1,1,11) the case of which will be discussed later. The peak profiles were not analyzed in the X-ray experiment. The values of the roughening temperatures listed on Table 1 are deduced from the universal value of the measured exponent. The peaks are always broader in q_x than in q_y which is the opposite of the behavior expected from geometric anisotropy so this is due to the energetic anisotropy and clearly demonstrate that $w_1(\ell)<W_0$. This observation rules out a model proposed by den Nijs et al. which does not take the energetic anisotropy into account.[8]

Table I

Surface	Roughening Temperature T_R (K)	Energy Of Creation Of a Step W_o (K)	Energy Of Interaction Between Steps $w_1(\ell)$ (K)
Cu(113)	720	800	560
Cu(115)	380	850	120
Cu(1,1,11)	<300	?	≈5
Cu(331)	650	≈1000	≈500
Cu(310)	<300	≈400	≈300

The following trends are found:

i.   At the same temperature the roughness of the different surfaces depends upon the step separation: the larger this distance, the rougher the surface. This is in agreement with the elastical origin of the step repulsion which is expected to decreases strongly with step separation.

ii.   Changing the terrace structure from (100) to (111) does not affect the roughness very much.

iii. The roughening is strongly dependent upon the step structure. Steps parallel to the loosely packed direction [100] are much rougher than those parallel to the close-packed one [110]. This is the expected behavior since in the latter case one bond between two nearest neighbour atoms has to be broken in order to make a kink while none is affected in the former case.

iv.   Below room temperature we observe a freezing of the roughness. This is due to a lack of surface mobility which prevents the achievement of the thermodynamical equilibrium. Only Cu(113) can be observed at low temperature in a smooth state because its roughening temperature is high enough. For Cu(1,1,11) and Cu(310) T_R is lower than the freezing temperature thus it cannot be determined. The kinetics of the approach to equilibrium has been directly observed in the case of Cu(310) in the temperature range 200-250 K.[20]

Figure 5 displays the evolution of the pre-factor A(T) of the logarithmic law Eq. (5) and of the anisotropy as a function of temperature for Cu(113), Cu(115), Cu(331) and Cu(310). Comparison with

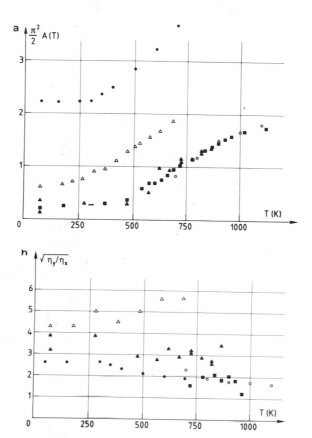

Fig. 5 Pre-factor of the exponential law A(T) multiplied by $\pi^2/2$ (a) and anisotropy η_x/η_y (b) vs. temperature Experimental data: ■ Cu(113), + Cu(115), x Cu(331), • Cu(310). o Monte-Carlo computation for Cu(113) (W_0 = 800K, w_ℓ = 560K). In the anti-phase condition $(\pi^2/2)A(T)$ is also the exponent of the power law. It was directly measured for Cu(113), Cu(115) and Cu(331). For the very rough Cu(310) surface an intermediate condition was used and then a correction using Eq. (13) has been applied in order to derive A(T) from the exponent τ.

model TSK calculations lead to the estimations of W_0, w_ℓ indicated in Table 1. The calculated values of the pre-factor are also plotted in Fig. 5 for Cu(113).

Kink formation energy: for a close-packed [110] step we find: W_0 = 900±100 K. This value is to be compared to theoretical estimations. Only one nearest neighbour is lost when a kink is formed thus in a broken bond model W_0 = $E_c/12$ (E_c: cohesive energy); for copper E_c = 3.4 eV so W_0 = 3400 K. This model is thus clearly too crude. A more refined estimation based on band structure[23] leads to W_0 = 2300 K which is still too large. More recently a calculation based on a pair potential carefully fitted on bulk and surface properties (lattice parameter, elastic constant, melting point, surface relaxation...)[27] leads to W_0 = 1500 K which is still twice the experimental value. A possible source of discrepancy may arise from the interaction between kinks which is neglected in the W_0 calculation and in the data analysis as well. But,

365

on the other hand, it is not the energy W_0 but the free energy W_F of creation of an isolated kink which has to be introduced in the model. Indeed there is a vibrationnal entropy contribution to this free energy. Then the value which is fitted to the experimental data is W_F:

$$W_F = W_0 - TS .$$

The vibrational entropy for the creation of a vacancy has been calculated by P. Wynnblatt[24] to be of the order of k_B. As W_F is measured in the range 400-800 K this entropy term thus explains at least part of the discrepancy.

For the loosely packed direction [100] W_0 is smaller as expected but there is no realistic calculation to be compared with. The broken-bond model which gives $W_0 = 0$ is clearly useless!

Step interaction energy: $w_1(\ell)$ decreases with increasing step distance as expected. The dependence over ℓ^{-4} is rather well verified for Cu(115) and Cu(1,1,11). We emphasized that this is the first experimental attempt to measure the step-step interaction.

However the same law predicts for Cu(113) $w_1(\ell) = 1500$ K, which is much larger than the experimental value, $W_1(\ell) = 560$ K. This discrepancy can be easily understood: on the (113) face the interaction is no longer purely elastic or electrostatic. When a step moves towards its neighbour by one atomic length it forms with it a (111) facet which has a lower surface energy.

1. The Cu(1,1,11) Anomalous Behaviour[19]

For this face the peak shape is strongly anisotropic as a consequence of the low value of $w_1(\ell)$. In the direction perpendicular to the steps the peak is very broad and surprisingly does not follow a power law but rather a Lorentzian profile. In the other direction the peak is only little broadened with respect to the instrumental residual shape and it can be fitted by a power law or a Lorentzian as well. The Lorentzian shape suggests a linear instead of logarithmic behavior of the correlation function but calculations with a TSK model have shown that the correlation function is always logarithmic as far as w_1 is not strictly zero. One way to remove the discrepancy is to suppose that the surface has only a finite extension L_S in the direction parallel to the steps. When $w_1 \ll W_0$ the average length L of the elementary excitations along the steps becomes large; if L is larger than L_S then calculations shows that one recovers the linear behavior which for $L_S = \infty$ was only obtained for $w_1(\ell) = 0$. The finite size of the surface is possibly imposed by the residual steps present on the surface as a consequence of macrogeometrical defects. The crossings of these residual steps with the normal steps of the vicinal structure act as traps for the kinks.

CONCLUSION

As shown above the roughening transition on vicinal surfaces of metals is now quite well characterized theoretically as well as experimentally. One of the main conclusion of these researches is that, except for a few of them, metal vicinal surfaces are always rough. The kink density is always high on such surfaces and this fact have to be taken into account in any study of chemical reactivity.

Kink formation energies are deduced from the experimental data which should stimulate theoretical efforts in order to be completely understood.

The interaction between steps is also estimated: it is not inconsistent with a dependence in d^{-2}. This is in a close connection with the problem of the equilibrium shape of crystals in the vicinity of a facet. A step interaction proportional to d^{-2} leads to a cubic term in the development of the surface energy with respect to surface inclination. However a quadratic term seems to help fitting the experimental data[7] in the case of In. Recent work on the equilibrium shape of the interface He liquid-He solid suggests the same trend.[26] However the physical significance of this quadratic term remains unclear. So more experimental effort in this field should be welcomed.

REFERENCES

1. W. K. Burton and N. Cabrera, Discuss. Faraday Soc. 5, 33. (1949).
2. J. M. Kosterlitz and D. J. Thouless, J. Phys. C6, 1181 (1973), J. M. Kosterlitz, J. Phys. C7, 1046 (1974).
3. J. V. Jose, L. P. Kadanoff, S. Kirkpatrick and D. R. Nelson, Phys. Rev. B16, 1217 (1977).
4. S. T. Chui and J. D. Weeks, Phys. Rev. B14, 4978 (1976).
5. K. A. Jackson and C. E. Miller, J. Crystal Growth 40, 169 (1977).
6. J. E. Avron, L. S. Balfour, G. G. Kuper, J. Landau, S. G. Lipson and L. S. Shulman, Phys. Rev. Lett. 45, 814 (1980); S. Balibar and B. Castaing, J. Physique Lettres (Paris) 41, L329 (1980); K. O. Keshishev, A. Ya. Parshin and A. B. Babkin, Sov. Phys. JETP 53, 362 (1981).
7. J. J. Metois and J. C. Heyraud, J. Cryst. Growth 50, 571 (1980); Acta Metall. 28, 1789 (1980); Surf. Sci. 128, 334 (1983), J. C. Heyraud and J. J. Metois, J. Cryst. Growth 57, 487 (1982).
8. M. den Nijs, E. K. Riedel, E. H. Conrad and T. Engel, Phys. Rev. Lett. 55, 1689 (1985), and erratum 57, 1279 (1986), E. H. Conrad, R. M. Aten, D. S. Kaufman, L. R. Allen, T. Engel, M. den Nijs and E. K. Riedel, J. Chem. Phys. 84, (1986), 1015 and erratum 85, (1986), 4856.
9. J. Lapujoulade Surf. Sci. 178, 406 (1986), F. Fabre, D. Gorse, J. Lapujoulade and B. Salanon, Europhys. Lett. 3, 737 (1987), F. Fabre, D. Gorse, B. Salanon and J. Lapujoulade J. Physique (Paris) 48, 1017 (1987).
10. J. D. Weeks in: Ordering of Strongly Fluctuating Condensed Matter T. Riste, ed. (Plenum, New York, 1980) H. van Beijeren and I. Nolden in: Structures and Dynamics of Surfaces, W. Schommers and P. von Blanckenhagen eds. (Springer, Heidelberg, 1987).
11. P. Nozieres and F. Gallet, J. Physique (Paris) 48, 353 (1987).
12. H. J. Schultz, J. Physique (Paris) 46, 257 (1985).
13. J. Villain, D. R. Grempel and J. Lapujoulade, J. Phys. F: Metal Phys. 15, 809 (1985).
14. W. Selke and A. M. Spilka, Z. Phys. B62, 381 (1986), W. Selke J. Phys. C20, L455 (1987).
15. B. Salanon, F. Fabre, J. Lapujoulade and W. Selke, Phys. Rev. B38, 7385 (1988).
16. G. Armand and B. Salanon, Surf. Sci. 217, 317 (1989).
17. B. Salanon and J. Lapujoulade, Vacuum in press.
18. E. H. Conrad, L. R. Allen, D. L. Blanchard and T. Engel, Surf. Sci. 187, 265 (1987).
19. F. Fabre, B. Salanon and J. Lapujoulade, Sol. St. Com. 64, 1125 (1987).
20. B. Loisel Thesis, Université Paris VII

21. K. S. Liang, E. B. Sirota, K. L. d'Amico, G. H. Hughes and S. K. Sinha, Phys. Rev. Lett. 59, 2447 (1987),

22. J. W. M. Frenken, R. J. Hamers and J. E. Demuth, 4th Int. Conf. on STM, Iberaki, Japan, 9-14 July 1989.

23. G. Allan and J. Wach in: Proc. 4th Int. Conf. on Solid Surfaces Cannes, France (1980) D. A. Degras and M. Costa, ed. Le Vide, les Couches Minces Suppl. n°201, (1980), 11.

24. P. Wynnblatt, Phys. Sta. Sol. 36, 797 (1969).

25. P. Zeppenfeld, K. Kern, R. David and G. Comsa, Phys. Rev. Lett. 62, 63 (1989).

26. A. V. Babkin, D. B. Kopeliovich and A. Ya. Parshin, Sov. Phys. JETP 62, 1322 (1985), O. A. Andreeva, K. O. Keshishev and C. Yu. Ossipian to be published.

27. B. Loisel, D. Gorse, V. Pontikis and J. Lapujoulade, Surf. Sci. 221, 365 (1989).

DYNAMICS OF A CRYSTAL SURFACE BELOW ITS ROUGHENING

TRANSITION

Frédéric Lançon and Jacques Villain

Département de Recherche Fondamentale
C. E. N. G.
85X
F-38041 Grenoble Cédex, France

ABSTRACT

The equations of the smoothening dynamics of a crystal surface below the roughening transition are written. It is assumed that only ledges separating smooth planes are mobile. The equations are applied to the classical experiment in which the decay of parallel grooves is studied. The formation of facets, which has been observed, is argued to be possible only if the average surface makes a small, but non-vanishing angle α with a high symmetry direction. If $\alpha = 0$ the profile has sharp maxima and minima with singularities.

1.   INTRODUCTION

A usual technological problem is: "how to obtain a smooth surface ?" Mechanical polishing is not always practicable, for instance in the case of a very thin film obtained by molecular beam epitaxy (M.B.E.). Then the appropriate method is annealing. This method works only if the temperature T is below the so-called roughening transition temperature T_R of the surface (Burton and Cabrera 1949, Weeks 1980). For $T > T_R$ the surface is rough at equilibrium and annealing does not lead to a smooth surface. On the other hand, T should be high enough to have a sufficiently fast dynamics. The appropriate order of magnitude is about $2T_M/3$, where T_M is the melting temperature. This temperature is sufficiently lower than the roughening temperature of the (001) and (111) faces of most of cubic materials, so that even local roughness is negligible at equilibrium. In most of this article we therefore address a surface of a cubic material, the average orientation of which is close to the (001) or (111) orientation, and the temperature of which is about $2T_M/3$. The z axis will be chosen perpendicular to this orientation (Figs. 1 and 2).

The condition $T < T_R$ implies that the surface tension $\sigma(z'_x, z'_x,)$ is not an analytic function of $z'_x = \partial z/\partial x$ and $z'_y = \partial z/\partial y$ for $z'_x = z'_y = 0$. The orientation z = Const is called singular. As a consequence of this non-analyticity, the surface dynamics are described by non-linear equations which, to our knowledge, were not solved until 1989. It will be assumed throughout this note that all other orientations of the surface are non-singular, i.e. σ is analytic and $T > T_R$.

Kinetics of Ordering and Growth at Surfaces
Edited by M. G. Lagally
Plenum Press, New York, 1990

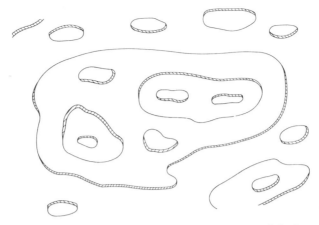

Fig. 1 A rough surface with terraces and ledges.

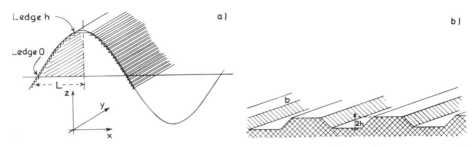

Fig. 2 a) Initial state of the profile, with a schematic representation of the ledges separating lattice planes. b) Profile observed after some time for faces of high symmetry (the trapezoidal shape is exaggerated).

A surface prepared by M.B.E. has a microscopic roughness, at least for an ideally large and thick sample. It may be viewed (Fig. 1) as consisting of terraces of atomic height, although their radius becomes large after a short annealing time. This case will be briefly discussed in the next section. The remainder of this paper is devoted to the decay of a macroscopic profile (Fig. 2). This type of experiment has been widely used (Mullins 1957-59, Yamashita et al 1981) as a tool for the investigation of surface diffusion.

2. DECAY OF MICROSCOPIC FLUCTUATIONS

This Section provides us with the opportunity to introduce, following Burton, Cabrera and Frank (1951), the description of a surface with a singular orientation in terms of terraces separated by ledges (Fig. 3). For the sake of simplicity the analysis will be limited to the case of a surface which has a self-similar structure on length-scales

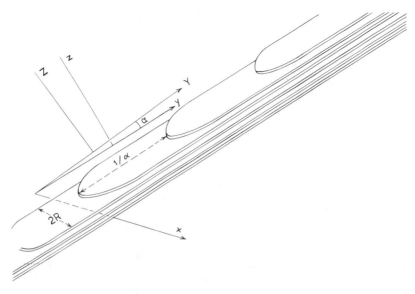

Fig. 3 The top of a "miscut" profile, showing ledges.

larger than some length $R_1(t)$. This is the case of a surface quenched from above T_R or T_M to a temperature below T_R. It is also the case of a surface which has been roughened by the growth process (see e.g. Wolf 1989). $R_1(t)$ is essentially the size of the smallest terraces, so the t is the lifetime of a terrace of size $R = R_1$. This can be calculated from the assumption (which can be justified) that dynamics are due to diffusion on the surface of isolated atoms or surface vacancies, while the motion of the terrace as a whole, or its decomposition in two parts, are assumed to be negligible. The atoms go from the terrace edge (where the chemical potential is μ_1) to a neighboring terrace where the chemical potential is $\mu_2 > \mu_1$. Surface vacancies follow the opposite way and the result is the same, namely the terraces with higher chemical potential (and valleys with lower chemical potential) decay first. The chemical potential of a terrace edge of radius R turns out to be

$$\mu = \mu_o + g/R \ , \tag{1}$$

where μ_o is the chemical potential of the bulk (or of the flat surface) and g is the ledge free energy per unit length. The unit length is assumed throughout this paper to be the atomic distance. Relation (1) can be obtained from the thermodynamic relation

$$\mu = \delta F/\delta N \tag{2}$$

where δF and δN are the respective changes in free energy and particle number due to a change of the surface shape. This change can be chosen to be a change of the radius R into $R+\delta R$. Then, for circular terraces, $\delta F = 2\pi g \delta R$ and $\delta N = 2\pi R \delta R$ and Eq. (2) follows. The assumption of circular terraces is easily seen to be unessential.

It is seen from (1) that smaller terraces have the higher chemical potential and therefore decay first. The mass M of a typical terrace decays (Villain 1986) according to the formula

```
dM/dt = Total flux of atoms from the terrace
     = Current density times ledge length 2πR
```

$$= K \; \frac{\text{Chemical potential difference } \delta\mu}{\text{Distance } \Delta \text{ to the next terrace or valley}} \; 2\pi R$$

$$\approx \frac{g/R}{R} \; KR \approx Kg/R \;,$$

where K is a kinetic coefficient. The fact that $\Delta \approx R$ is implied by self-similarity. For flat terraces, the mass M is proportional to R. More generally, one can assume (Wolf 1990) that fluctuations of radius R have a height proportional to R^α, so that $M \sim R^{2+\alpha}$. Insertion into the above formula yields $R^{1+\alpha} \dot{R} \sim -1/R$ and

$$R^{3+\alpha}(t) \approx R^{3+\alpha}(0) - Kgt \;.$$

Thus a terrace of size R decays after a time $\approx R^{3+\alpha}/Kg$. It follows that

$$R_1(t) \approx (Kgt)^{1/(3+\alpha)} \;. \tag{3}$$

For $\alpha = 0$, this is the same result as in the classical theory of Ostwald (1908), elaborated by Lifshitz and Slyozov (1961), which describes the last stage of nucleation in a supersaturated solution. The value $\alpha = 0$ is also relevant for thermal roughness (Villain 1986, Uwaha 1988) while the decay of adsorbed clusters corresponds to $\alpha = 1$ (Chakraverty 1967). The decay of clusters is more complicated because the geometry is not self-similar.

3. MACROSCOPIC PROFILE: VARIOUS TYPES OF DYNAMICS

In contrast with the previous Section, the decay of a macroscopic profile is governed by deterministic equations.

1. T>T_R. (Mullins 1963)

a) In the case of *surface diffusion* (addressed in the previous Section) there is a continuity equation

$$\dot{z} = -\text{div} \; \vec{j} = - \partial j_x/\partial x - \partial j_y/\partial y \;, \tag{4}$$

where the two-dimensional current $\vec{j}$ is given by the phenomenological transport equation

$$\vec{j} = -K \vec{\nabla} \mu \quad \text{or} \quad j_\alpha = -K \partial_\alpha \mu \quad (\alpha = x,y) \;, \tag{5}$$

where K is the same transport coefficient as in the previous Section and μ is the chemical potential. The set of equations needs to be closed by an expression of μ as a function of the profile $z(x,y)$. This will be done quite generally in the next Section. However, when all orientations are non-singular, it follows from Eq. (2) that μ is an analytic function of the derivatives of z. If these derivatives are not too large, the expansion of μ may be limited to the linear terms. Now, there can be no term proportional to the first derivatives because such term would imply that the equilibrium of a crystal depends on the orientation of its

372

surface. It is not so if the surface is plane. Therefore

$$\mu = A \left[\partial^2_{xx} z + \partial^2_{yy} z \right] = A \left[z''_{xx} + z''_{yy} \right] ,$$ (6)

where A is a constant. Insertion of Eq. (6) into Eqs. (5) and (4) yields a linear equation. An initially sinusoidal profile keeps the sinusoidal shape

$$z(x,t) = h(t) \sin (\pi x/2L) ,$$ (7)

and h decays exponentially:

$$h(t) = h(0) \exp(-t/\tau) ,$$ (8)

with

$$\tau = 16L^4/\pi^4 \; AK .$$ (9)

b) Another type of dynamics is *evaporation dynamics*. It is particularly simple if the chemical potential of the vapor is assumed to be always uniform and equal to the value μ_0 of the bulk solid (in order to avoid growth or decay of the crystal). The evaporation rate at a point of the surface where the chemical potential is p is proportional to the difference $(\mu-\mu_0)$.

$$\partial z/\partial t = \dot{z} = -B \; (\mu-\mu_0) \sqrt{1+z'^2_x + z'^2_y} ,$$ (10)

where B is a temperature-dependent coefficient. This formula is good for a fluid since the evaporation rate per unit surface area is the same for all orientations. For a crystal the linearized form $\dot{z} = -B(\mu-\mu_0)$ is equally good if the derivatives are small, as always assumed in this paper.

Insertion of Eq. (7) yields a linear equation again, so that there are sinusoidal solutions satisfying Eqs. (7) and (8). Equation (9) is replaced by

$$\tau = L^2/(4\pi^2 \; AB) .$$ (11)

The assumption of a uniform chemical potential in the vapor is acceptable only if the mean free path of the molecules through the vapor is larger than the distance of interest (for instance the wavelength in the case of a periodic profile as considered by Mullins 1959).

c) In the case of transport through the bulk, or through the vapor if the mean free path is shorter than the distance of interest, one has to write the equations of three-dimensional diffusion, which are complicated because of the surface. In the limit of very long distances, this process is in principle dominant because the current density between two points at distance R is still of order $K\delta\mu/R$ but the total flux is obtained by multiplying by R^2 rather than R. $\delta\mu$ is the chemical potential difference. Note that the time necessary to equilibrate a surface on large length-scales can be extremely long.

In the non-singular case, it can be shown that an initially sinusoidal surface decaying according to the process (c) remains sinusoidal and decays exponentially with a relaxation time proportional to the cube of the wavelength (Mullins 1959). In the remainder of this article the simple cases (a) and (b) will only be considered.

2. Singular Case

When one (and only one) orientation of the surface (namely $z =$ Const) is singular it is appropriate to consider the surface as consisting of terraces $z =$ Const separated by ledges. The treatment of subsection 1 fails for the following reasons. The growth or decay on a terrace far from its edge is an activated process described by nucleation theory and not by Eqs. (5) or (10). Indeed, there is no expression of μ as a function of the profile, which can close the set of equations. This is because, on a terrace, μ is a function of the size of the terrace according to Eq. (1). Therefore it is a nonlocal function of the profile $z(x,y)$ rather than a function of the local derivatives z'_α, which, on a terrace, are both zero.

The difficulty is partially overcome by the reasonable assumption that nucleation is negligible. Thus the whole dynamics are due to ledge motion. In that case it is sufficient to know the chemical potential μ on ledges, which will be calculated in the next section. Furthermore Eq. (10) must be multiplied by the density of ledges since evaporation takes place only at ledges. Thus the equation which corresponds to evaporation dynamics and $T < T_R$ is

$$\dot{z} = - \Lambda\ (\mu - \mu_o)\ \sqrt{z'^2_x + z'^2_y}\ . \tag{12}$$

In this formula μ is the chemical potential on the ledge at the point of interest or near this point. If there is no ledge in the neighborhood, this means that the point is on a terrace, so that the square root vanishes and $\dot{z} = 0$.

In the case of surface diffusion, Eq. (4) is of course valid even below T_R. Equation (5) is also correct at least qualitatively. The current between two ledges with distance ℓ and chemical potentials μ_1 and μ_2 is $K(\mu_1 - \mu_2)/\ell$ and can generally be replaced by Eq. (5). However, in certain circumstances, addressed below, this replacement is only qualitative.

4. CHEMICAL POTENTIAL

The chemical potential can be evaluated through Eq. (2) if the free energy F is known. The latter can be written as the sum of a volume integral (which is not interesting here) and a surface term.

$$F = \mu_o \iint z(x,y)\ dx\ dy + \iint \psi(z'_x,\ z'_y)\ dx\ dy\ , \tag{13}$$

where $z(x,y)$ is the height of the surface, assumed to be a uniform function of x and y. This formula neglects possible effects from the curvature of the surface, which would enter through higher derivatives. For a sufficiently smooth surface, these are presumably negligible. The function ψ is related with the surface tension σ (free energy per unit surface area) by

$$\psi(z'_x,\ z'_y) = \sigma(z'_x,\ z'_y)\ \sqrt{1 + z'^2_x + z'^2_y}\ . \tag{14}$$

Above T_R, μ can be obtained as follows (Herring 1952, Mullins 1963, Villain 1989). Equation (13) yields the change δF in free energy produced by a localized change $\delta z(x, y)$ of z. After integrating by parts, Eq. (2) yields

$$\mu = - \frac{\partial}{\partial x} \left(\frac{\partial \psi}{\partial z'_x} \right) - \frac{\partial}{\partial y} \left(\frac{\partial \psi}{\partial z'_y} \right) . \tag{15}$$

This formula can be found in the textbook by Landau and Lifshitz (1967) as a parameter which is constant on a surface at equilibrium [Eq. (143.3)]. This property is not sufficient to prove that the parameter μ (called -2λ by Landau and Lifshitz) is the chemical potential. But it is!

Above T_R, $\psi \sim z'^2_x + z'^2_y$. Insertion into Eq. (15) yields Eq. (6).

If the orientation $z = $ Const is singular, Eq. (15) yields a delta function and can therefore not be used. Then it is appropriate to write the surface free energy density as $\psi = \psi_1 + \psi_2$, where ψ_1 is the sum of the free energies of independent ledges and ψ_2 results from interactions. Assuming the line tension g of ledges to be isotropic (for a general formula see Villain 1989) one finds

$$\psi_1 = g \sqrt{z'^2_x + z'^2_y} , \tag{16}$$

which has no second derivative for $z'_x = z'_y = 0$. The interaction free energy density ψ_2 corresponds, in the case of a short range interaction energy, to an entropy loss due to the mutual hindrance of ledges. Neglecting anisotropy again one obtains for small z'_x. and z'_y (Gruber, Mullins 1967, Pokrovskii and Talapov 1979)

$$\psi_2(z'_x, z'_y) = 1/3 \, G_3 \, (z'^2_x + z'^2_y)^{3/2} , \tag{17}$$

where G_3 is a temperature-dependent constant. In the case of interactions through elastic strains it may be argued that the form Eq. (17) is still acceptable though, strictly speaking, there is no local free energy density ψ.

The chemical potential is $\mu = \mu_1 + \mu_2$, where μ_2 is given by Eq. (15), with ψ being replaced by ψ_2. This makes sense, because Eq. (4) has enough derivatives. According to Eq. (1), the part μ_1 arising from ψ_1 is $\mu_1(x,y) = g/\rho(x,y)$, where ρ is the radius of curvature of the ledge which goes though the point (x,y). Finally the chemical potential on ledges is given by

$$\mu = \frac{g}{\rho} - \frac{\partial}{\partial x} \left(\frac{\partial \psi_2}{\partial z'_x} \right) - \frac{\partial}{\partial y} \left(\frac{\partial \psi_2}{\partial z'_y} \right) . \tag{18}$$

5. DECAY OF A ONE-DIMENSIONAL PROFILE

As a first application of the equations obtained in the previous sections one may consider the decay of an artificial profile having at $t = 0$ the form Eq. (7). In this section the average orientation of the profile will be assumed to coincide exactly with a singular orientation, in practice (001) or (111). Then ledges are straight lines (Fig. 2) and the first term at the r.h.s. of Eq. (18) vanishes. The y axis can be chosen along the ledge direction, so that the third term also vanishes and one obtains from Eqs. (18) and (17)

$$\mu = -2G_3 z' z'',$$ (19)

with $z' = \partial z/\partial x$ and $z'' = \partial^2 z/\partial x^2$. In the case of dynamics governed by two-dimensional diffusion, Eqs. (19), (4) and (5) yield

$$\dot{z} = -2KG_3 \frac{\partial^2}{\partial x^2} (z' z'') \quad .$$ (20)

This equation differs from the one used by Bonzel et al (1982) who applied Eq. (15) instead of Eqs. (18) or (20), and approximated the singular delta part by a regular, but sharply peaked function. It will be seen in the next section that this procedure is in practice justified if the surface is slightly miscut. The purpose of Bonzel et al was to explain the formation of a trapezoidal profile observed experimentally (Fig. 2b) after an appropriate waiting time (Yamashita et al 1981) before complete decay of the profile.

Since the experimentally observed profile is flat at its top, it is of interest to check whether this feature can be reproduced by Eq. (20). Firstly, no facet can be present since it would satisfy $z' = z'' = 0$ and therefore $\dot{z} = 0$ according to Eq. (20). A plausible Ansatz is that the function $\delta z(\delta x, t) = z(L+\delta x, t) - z(L, t)$ behaves near its top (i.e., for $\delta x \approx 0$) as

$$\delta z = |\delta x|^b \quad .$$ (21)

Inserting Eq. (21) into Eq. (19) and assuming μ to be strictly positive and finite for $x = L$, one finds $b = 3/2$. Thus, the profile is sharper than a parabola at its top (and at its bottom as well). A nonvanishing value of $\dot{z}$ can only be obtained if higher-order terms in $\dot{z} = \lambda |\delta x|^b + \nu |\delta x|^{b'}$ are taken into account. One easily finds $b' = 7/2$.

In the case of *evaporation dynamics* Eqs. (12) and (19) yield

$$\dot{z} = \text{Const.} \ z'^2 \ z''$$ (22)

and the Ansatz Eq. (21), together with the condition that $\dot{z}$ be finite and strictly negative for $x = L$, yields now $b = 4/3$. Again the profile is found to be sharper than a parabola, at least after some time. It is remarkable that, if the initial profile is parabolic at its top (or bottom), it remains parabolic at short times as remarked by Jim Langer (private communication). Indeed if the profile is assumed for small times to have the form $z \approx h_o - 1/2 \ A \ \delta x^2$, insertion into (22) yields $-\dot{A}/A^3 = \text{Const.}$ Thus, $A(t) \sim 1/\sqrt{(t-t_c)}$, where the integration constant t_c depends on initial conditions. The form Eq. (21), with $b = 4/3$, becomes presumably valid for $t > t_c$.

Thus, in all cases and in contradiction with experiment, the profile is predicted to become sharper than a parabola. The explanation is probably that the average surface is not exactly cut parallel to a singular orientation, as seen in the next Section.

6.   CASE OF A MISCUT PROFILE

In this Section it is assumed that the direction Z perpendicular to the average orientation of the surface does not coincide exactly with the singular orientation z, which in practice is (001) or (111). Let Y be the direction of the grooves and X the direction perpendicular to Y and Z. y may be chosen in the YZ plane. The relation between the coordinates at the surface are

$$x = a_x X + a_y Y + a_z Z(X) \tag{23a}$$

$$y = b_y Y + b_z Z(X) \tag{23b}$$

$$z = c_x X + c_y Y + c_z Z(X) \; . \tag{23c}$$

Z(X) is the profile of the surface. In the previous section the diagonal coefficients a_x, etc. were equal to 1 and off-diagonal coefficients a_y, etc. were zero. As stated earlier we need the chemical potential on ledges, defined by z = n where n is an integer.   Since the problem is invariant under translations along Y, all ledges are identical and the ledge n = 0 can be chosen.   The system (23) yields for z = 0 after elimination of Y

$$x = KX + HZ(X) \tag{24a}$$

$$y = LX + MZ(X) \; , \tag{24b}$$

where

$$K = a_x - a_y \, c_x/c_y \tag{25a}$$

$$H = a_z - a_y \, c_z/c_y \tag{25b}$$

$$L = -b_y \, c_x/c_y \tag{25c}$$

$$M = b_z - b_y \, c_z/c_y \; . \tag{25d}$$

The evaluation of ρ in Eq. (18) requires the calculation of $y' = dy/dx$ and $y" = d^2y/dx^2$ along the curve defined parametrically by Eq. (24).   The result is

$$y' = \frac{L+MZ'(X)}{K+HZ'(X)} \tag{26a}$$

$$y" = \frac{KM-HL}{[K+HZ'(X)]^2} \, Z"(X) \; . \tag{26b}$$

Then the first term of Eq. (18) is

$$\mu_1 = g/\rho = g \, y" \, (1+y'^2)^{-3/2} \; . \tag{27}$$

For the sake of simplicity it will now be assumed that the axes x and X coincide. Then $a_x = a_z = c_x = H = L = 0$. Defining $c_y = \alpha$, one finds using Eqs. (18), (17), (27) and (26) that for small α (Villain and Lançon 1989)

$$\mu = \mu_o + \frac{\partial}{\partial x}\left[\frac{gZ'}{\sqrt{Z'^2+\alpha^2}} + \frac{\partial}{\partial Z'}\psi_2(Z',-\alpha)\right] = \partial\tilde{\psi}(Z')/\partial x \ . \tag{28}$$

This equation has the form (15) except that ψ is replaced by a function $\tilde{\psi}$ of $Z' = \partial Z/\partial x$. Equation (28) is essentially equivalent to that used and solved numerically by Bonzel et al (1984) in the case of two-dimensional diffusion. As stressed in the previous section, this equation is not correct for $\alpha = 0$ nor, presumably, for extremely small α. The flaw of the method for those very small values of the miscut angle is probably the following: it has been assumed in this section that $Z(X)$ does not depend on Y. This is macroscopically true, but this Ansatz neglects microscopic variations which may become essential for small α. A related problem is that the curves z = Const on the surface are not necessarily defined by an equation $y = f$ (x) in terms of a uniform function f.

Another source of quantitative error is the use of the continuous equation (5). At the top of the profile the distance between ledges becomes very large if α is large and the continuum approximation is only qualitatively valid.

This difficulty does not appear in the case of evaporation dynamics. The system of equations (12) and (28) has been solved numerically (Lançon and Villain 1989) and indicate facet formation (Fig. 4). The dependence of the amplitude vs time is shown by Fig. 5.

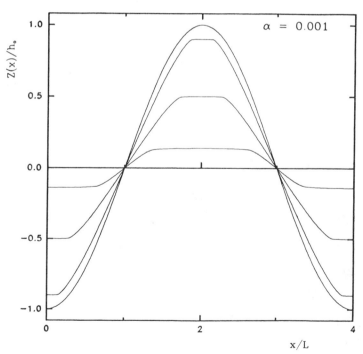

Fig. 4 Profile $Z(x,t)$ when the height $h(t)$ equals h_o (initial), $0.9h_o$, $0.5h_o$ and $0.14h_o$. α is the miscut angle and 4L is the wavelength.

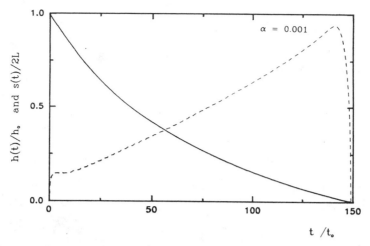

Fig. 5 Time evolution of the height h(t) and the width s(t) of the plateau of Z(x,t). h_o is the initial height, 4L is the wavelength and t_o is a time unit.

The smoothening of grooves has also been treated by the Monte-Carlo method by Jiang and Ebner (1989) and, in the case of evaporation dynamics, by Selke and Oitmaa (1988).

7. DECAY OF A BIDIRECTIONAL PROFILE

All experiments on the flattening of an artificial profile have been made on a unidirectional profile having initially the shape (7). It follows from the last two sections that the results are very sensitive to a small miscut angle. To avoid this drawback it might be interesting to use a bidirectional profile. Above T_R it follows from Eqs. (6), (4), (5), (10) that there are solutions of the type

$$z(x,y,t) = h(t) \sin (\pi x/2L) \sin (\pi y/2L) . \qquad (29)$$

When the average surface orientation is singular it was argued by Rettori and Villain (1988) that the surface takes after some time a profile shown by Fig. 6, reminiscent of the Arizona landscapes. We have confirmed this prediction by a numerical calculation (Fig. 7) in the case of two-dimensional diffusion, starting with the initial shape Eq. (29). In order to avoid the manipulation of the 3 variables x,y,t the ledges have been assumed circular, so that z is only a function of t and the ledge radius R. This Ansatz is qualitatively acceptable if the distance between hills and valleys is not too small. The calculation indeed shows that this distance is reasonably large, and this ensures the self-consistence of the method. The continuum equations (4) and (5) may only be used for z>0 or z<0. The current density through the plane z = 0 has been evaluated by the approximate equation $j = K\mu_1/(L-R_1)$ where R_1 and μ_1 are respectively the radius of the lowest ledge with positive height and its chemical potential.

Fig. 6 A bidirectional profile after some time.

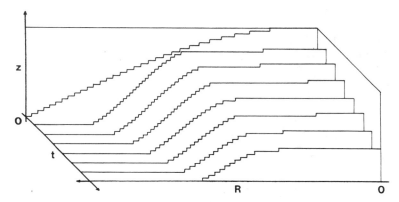

Fig. 7 Time evolution of a bidirectional profile. $R(z,t)$ is the average
ledge radius at given height z. The initial profile $z(R,t=0)$ is a
sine curve.

8. CONCLUSION

The questions addressed in this article belong to classical physics.
However, the nonlinear equations to be solved in the case of a singular
orientation do not seem to have been written until this year. While the
equations to solve are completely contained in the third and fourth
sections in the case of a macroscopic profile, their solution itself, as
proposed in Sections 5, 6, 7, is not completely satisfactory. Firstly, in
the numerical solutions given in sections 6 and 7, a simplification has
been used which is not completely safe; namely, the determination of the
function $z(x,y,t)$ has been replaced by the calculation of a function of 2
variables. In section 6, $Z(X,t)$ is actually a function of 2 variables
only if the miscut angle is not extremely small. In section 7 the
replacement of $z(x,y,t)$ by a function $z(R,t)$ is obviously a rough
approximation.

On the other hand the numerical solution does not say how the
lifetime depends on the wavelength. Tentative answers have been given
(Rettori and Villain 1988, Villain and Lançon 1989, Lançon and Villain
1990) using heuristic arguments which require confirmation.

It would be highly desirable to have an experimental check of the predictions made in this article. It is, however very difficult to cut a plane surface with the required angle, to give it the required initial profile and to analyze the shape of the decaying surface with a very good accuracy. Also, an ideal crystal has been assumed throughout this article, while dislocations are known to play an essential part in crystal growth (Friedel 1964) and might also be important for the smoothening dynamics.

Finally the decay of microscopic fluctuations is a difficult problem which requires more detailed investigations. In the second section only the calculation of the characteristic length $R_c(t)$ has been outlined. It would also be of interest to know, for instance, the distribution of terrace sizes for $R<R_c$.

Acknowledgement

It is a pleasure to thank Professors Jim Langer and Stephane Fauve for their helpful comments.

REFERENCES

H. P. Bonzel, E. Preuss, B. Steffen, Appl. Phys. A35, 1 and Surface Sci. 145, 20 (1984).

W. K. Burton, N. Cabrera, Discuss. Faraday Soc. 5, 33, (1949).

W. K. Burton, N. Cabrera, F. C. Frank, Phil. Trans. Roy. Soc. 243, 299 (1951).

B. K. Chakraverty, J. Phys. Chem. Solids 28, 2401 (1967).

Z. Jiang, C. Ebner, Phys. Rev. B40, 316 (1989).

J. Friedel, Dislocations (Pergamon Press, Oxford) (1964).

E. E. Gruber, W. W. Mullins, J. Phys. Chem. Solids 28, 875 (1967)

C. Herring, Structure and Properties of Solid Surfaces, R. Gomer and C. S. Smith ed. (University of Chicago Press, Chicago 1952).

L. Landau, E. Lifshitz, Physique statistique (Editions Mir, Moscou), (1967).

F. Lançon, J. Villain, Phys. Rev. Lett. 64, 293 (1990).

I. M. Liftshitz, V. V. Slyozov, J. Phys. Chem. Solids 19, 35 (1961).

W. W. Mullins, J. Appl. Phys. 28, 333 (1957).

W. W. Mullins, J. Appl. Phys. 30, 77 (1959).

W. W. Mullins, in: Metal Surfaces: Structure, Energetics and Kinetics, Am. Soc. Metals, Metals Park, Ohio p. 17 (1963).

W. Ostwald, Foundation of Analytical Chemistry, p. 22, Mc Millan, London, (1908).

V. L. Pokrovskii, A. L. Talapov, Phys. Rev. Lett. 42, 65 (1979).

E. Preuss, N. Freyer, H. P. Bonzel, Appl. Phys. A41, 137 (1986).

A. Rettori, J. Villain, J. Physique France 49, 257 (1988).

W. Selke, J. Oitmaa, Surface Sci. 198, L346 (1988).

M. Uwaha, J. Phys. Soc. Japan 57, 1681 (1988).

J. Villain, Europhysics Lett. 2, 531 (1986).

J. Villain, Comptes rendus de l'Academie des Sciences 309 II, 517 (1989).

J. Villain, F. Lançon, Comptes rendus de l'Academie des Sciences, 309 II, 647, (1989).

J. D. Weeks, in: Ordering in Strongly Fluctuating Condensed Matter Systems, T. Riste ed. (Plenum, New York) p. 293, (1980).

D. Wolf, this volume, (1990).

K. Yamashita, H. P. Bonzel, H. Ibach, Appl. Phys. 25, 231 (1981).

GROWTH OF ROUGH AND FACETTED SURFACES

Dietrich E. Wolf

Institut für Festkörperforschung, KFA Jülich
Postfach 1913
D-5170 Jülich, Federal Republic of Germany

ABSTRACT

   Scaling arguments have contributed a lot to the understanding of the growth of rough surfaces. Usually they are based on the assumption that the size of the surface provides the only characteristic length in the system. However in experiments as well as in computer simulations one generally has to take a second length into account. It is associated with the concept of an intrinsic width or short range roughness of the surface, to be distinguished from its long range fluctuations. Two physical manifestations of such a second length will be discussed: a) The intrinsic width is a major source of corrections to scaling and should be kept small if one wants to measure roughening exponents. In computer simulations this can be achieved by an algorithm called noise reduction. b) At a nonequilibrium roughening transition the second length may diverge, so that the intrinsic width determines the anomalous roughness right at the transition. This is demonstrated for a class of growth models, where analytical as well as numerical results are available.

INTRODUCTION

   Surfaces of growing aggregates are in general rougher than those in thermal equilibrium. A large number of idealized models[1] are being studied in order to extract the universal features of this roughness. From a theoretical point of view one is therefore mostly interested in the asymptotic long time, large wavelength behavior where a hydrodynamic approach can be used to classify the models. An overview will be given in the next chapter.

   On the other hand, what would be regarded as "small" scale features in these theories, can not always be ignored in laboratory or computer experiments. Being nonuniversal detailed predictions require a more sophisticated modelling of the microscopic dynamics. But still the investigation of simple models is useful. It has recently given new insight into the general nature of deviations from the universal behavior. The central notion is the intrinsic width[2,3] which will now be explained. Some of its physical implications will be discussed in the subsequent chapters.

Kinetics of Ordering and Growth at Surfaces
Edited by M. G. Lagally
Plenum Press, New York, 1990

Let us consider a film growing on a flat substrate. It can become arbitrarily thick but its linear size L parallel to the substrate should be restricted. No matter whether the growth is by vapor deposition, sedimentation, division of living cells, whether it is diffusion or reaction limited, in general the structure of the surface region becomes stationary for sufficiently thick films. Then the density profile (number of particles per unit area at a certain height h above the substrate) will just be shifted with a constant velocity as in Fig. 1. Typically one finds that the bulk density n of the film as well as the width w of the active zone, where the growth proceeds, scale with L,

$$n \sim L^{d_f - d} \quad , \quad w \sim L^\varsigma \quad , \tag{1}$$

where d_f is the fractal dimension of the aggregate, $d \geq d_f$ the space dimension and ς the roughening exponent.

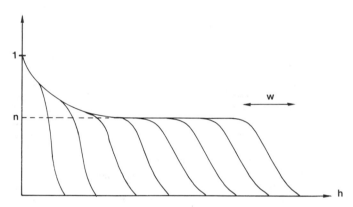

Fig. 1 Schematic time evolution of the density profile of a growing film.

Deviations from Eq. (1) indicate that in addition to L a second length ξ_r parallel to the substrate has to be taken into account. Its physical meaning depends on the model or the experimental setup. In the simplest cases it is just the lattice constant, but in more realistic situations it could be some correlation or diffusion length[4] and should in general vary e.g. with the film temperature. Then one can distinguish between a short range roughness or intrinsic width

$$w_i \sim \xi_r^{\varsigma'} \tag{2}$$

and the long wavelength fluctuations ($\sim L^\varsigma$). For example if $\varsigma' = 0$ the surface remains smooth within a distance ξ_r, and the roughness due to $\varsigma \neq 0$ can only be detected on much larger distances.

There are only few exceptions from the rule that the active zone develops a stationary structure. A simple example is random deposition.[5,6] In every time step the height h_i of the surface above a randomly chosen substrate site i is increased by 1. Trivially the only relevant length is the lattice constant ξ_r in this case, as there are no correlations between neighboring columns. The width of the active zone is not determined by L but is identical with the intrinsic width, the height fluctuation within the distance $\xi_r \ll L$. It increases like $t^{1/2}$ without ever reaching a stationary limit.

Usually the neighborhood does influence the growth at a site, however. For example a particle added to a column may seek to lower its potential energy by moving to the lowest column within a finite distance (Fig. 2a), or by diffusing along the surface until it finds a favorable adsorption site (holes, corners, or steps). Analytical calculations[5] as well as computer simulations[7] of random deposition with surface relaxation show that the intrinsic width now remains constant after a short time. In addition a long range roughness as in Fig. 1 builds up according to the power law

$$w \sim t^{\zeta/z} \tag{3}$$

until it reaches the stationary value Eq. (1) for $t \gg L^z$. The same is true if the particles are dropped onto the growing cluster along straight trajectories and stick to the first cluster site they touch (ballistic deposition,[8] Fig. 2b). Also if the growth is reaction limited like in the Eden model[9,10] (Fig. 2c), where particles are added to the cluster randomly at any perimeter site, the scaling (1) and (3) is found.

Equations (1) and (3) are limiting cases of the general scaling form[6]

$$w \sim L^\zeta f(t/L^z), \tag{4}$$

where $f(c) \sim c^{\zeta/z}$ for $c \ll 1$ and $f(c) = \text{const.}$ for $c \gg 1$. This scaling can only be observed if L is much larger than any other characteristic length, and if t is much larger than any correlation time in the system. In all cases considered in the following the intrinsic width may be regarded as constant because it builds up within a time interval independent of L which is much shorter than the time L^z until w reaches its stationary value (z > 0).

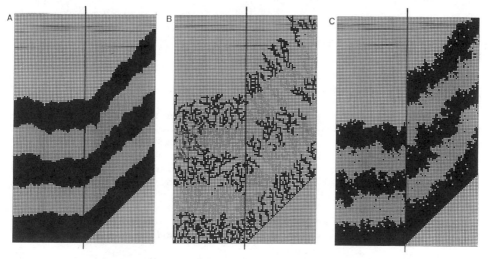

Fig. 2 a) Random deposition with relaxation of the atom to the lowest nearest neighbor column. Two simulations in strips of width $\ell = 80$ are shown, with periodic or helical boundary conditions, respectively, for the horizontal nad tilted substrate. Color changes indicate equal time intervals; b) Ballistic deposition, particles are dropped vertically; c) Eden model.

In this paper only growth processes with $d_f = d$ and $\zeta < 1$ will be considered. This excludes, e.g., diffusion limited deposits[11] as well as ballistic deposits at grazing incidence.[12] If ζ is strictly smaller than 1 the long wave-length surface configurations can be described by a height function $h(\underline{x},t)$, because looking at the surface on a scale ℓ with $\xi_r \ll \ell \ll L$ the width ℓ^ζ can be made arbitrarily small.

Edwards and Wilkinson[5] wrote down the following Langevin equation for the time evolution of $h(\underline{x},t)$:

$$\partial_t h = \lambda + \gamma \nabla^2 h + \eta, \tag{5}$$

where λ is the average height increase per unit time, γ is an effective surface tension and $\eta(\underline{x},t)$ is a white noise.[13] Kardar, Parisi and Zhang (KPZ) made the important observation[14] that λ in general depends on the local surface orientation, ∇h. This can be measured by simulating the growth on tilted substrates.[15,16]

Figure 2a shows a simulation of random deposition with surface relaxation on a horizontal[7] and on a tilted substrate. It is clear that the mass increase per unit time, dM/dt, is independent of the tilt: in each time step a particle is dropped in one of the ℓ columns and settles nearby. Also the bulk density $n = 1$ is independent of the tilt so that h increases with a constant average velocity

$$\lambda = dM/dt \quad 1/n\ell, \tag{6}$$

as can be seen clearly in Fig. 2a. This is the case solved by Edwards and Wilkinson.[5]

The situation is different for ballistic deposition[17] (Fig. 2b). Again, dM/dt does not depend on the tilt, but the bulk density n is smaller for oblique incidence as there are larger pores in the film than for normal incidence.[18] As the morphology does not depend on the sign of the tilt, an expansion of Eq. (6) for small tilt angles leads to a nonlinear term in Eq. (5):

$$\lambda \sim \lambda_1 + \lambda_2 (\nabla h)^2 . \tag{7}$$

Finally, in the Eden model (Fig. 2c) the bulk density $n = 1$ is independent of the tilt, however now the mass increases faster in the strip with the tilted substrate.[16] In order to use the same time scale one puts the particles one by one on sites picked at random from the total set of perimeter sites in both strips, as if they were added to a single cluster. The mass increases faster on the side where more perimeter sites are available. In fact the number of perimeter sites per surface area has the meaning of a normal growth velocity v_n in the Eden model. It changes very little with the orientation of the surface,[15] as can be seen in Fig. 2c. Since $\lambda = v_n (1 + (\nabla h)^2)^{1/2}$ one arrives at the same expansion Eq. (7) as for ballistic deposition. Therefore the long wavelength behavior of both models should follow from the KPZ-equation[14]

$$\partial_t h = \lambda_1 + \lambda_2 (\nabla h)^2 + \gamma \nabla^2 h + \eta . \tag{8}$$

This explains why the Eden model originally proposed to simulate the growth of cell cultures[9], can tell us something about the surface roughness of vapor deposits.

More generally one can consider a nonlinear term $|\nabla h|^\mu$ instead of $(\nabla h)^2$ in Eqs. (7), (8).[19] Below a simple example will be given where $\mu \neq 2$. It can be argued that the exponents ζ and z derived from Eq. (8) should be related[20-22,13] to each other by[19,12]

$$z = \min (2, \mu + \zeta(1-\mu)).$$ (9)

THE ROUGHENING EXPONENT IN THE EDEN MODEL

If one tries to determine ζ numerically in the Eden model from the mean square fluctuation of the perimeter sites about their average height, one has to cope with strong corrections to scaling.[10,23] Due to holes, overhangs and high steps the perimeter sites above any substrate site are scattered over an interval $\sim w_i$ about a mean local distance from the substrate (Fig. 2c). This distance varies along the surface according to a distribution characterizing the long wavelength fluctuations. It is therefore natural to assume[24] that the distribution of all perimeter sites is a convolution of an intrinsic one with width w_i and the scaling one with width (4). This implies that e.g. in the stationary limit Eq. (1) should be replaced by

$$w^2 = a^2 L^{2\zeta} + w_i^2 .$$ (10)

This has been confirmed[25] for d = 2 (Fig. 3) where[10,23,14] $\zeta = 1/2$. The convolution Ansatz does not only work in the stationary limit but represents the corrections to Eq. (3) as well.[25]

Diminishing w_i improves the scaling. This can be achieved by an algorithm called noise reduction[26,3]: A perimeter site becomes occupied only after the m'th attempt which leads to a preferential occupation of old perimeter sites so that overhangs and holes form less easily (Fig. 3).

Figure 3 also shows that ξ_r does not change significantly for large m. Otherwise L would be rescaled by $\xi_r(m)$ so that the coefficient a in Eq. (10) would not approach a constant value. This seems to be the case

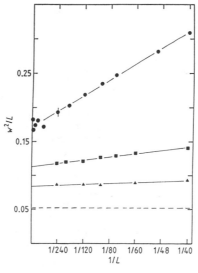

Fig. 3 Slopes are the intrinsic widths in the Eden model (d = 2) with noise reduction: m = 1($\bullet$), m = 2($\blacksquare$), m = 4($\blacktriangle$), m $\to \infty$ (---).

in related models[27,28] where noise reduction leads to a clustering of perimeter sites of similar age[22] and hence to a correlation length $\xi_r(m)$ increasing with m. In our case,[25] noise reduction only rescales the time by $\xi_t(m) \propto m$ which is the typical time (or cluster mass) until height fluctuations by one lattice constant occur.[22]

By means of noise reduction it was possible to show[24] that ζ is neither independent of the dimension d ("superuniversal")[29,30] nor is 0 for d ≥ 3 as in thermal equilibrium or for random deposition with surface relaxation[5]. The precise values are still unknown. Our simulation gave $\zeta(d = 3) \approx 0.33$ and $\zeta(d = 4) \approx 0.24$. Approximate analytical studies supported these values.[31,32] Based on their numerical investigation of a related model Kim and Kosterlitz[28] have recently conjectured that $\zeta = 2/(2 + d)$. Other simulations[33,34] seem to converge to a value between ours and theirs and indicate[34] that $\zeta > 0$ also for d = 5. The dynamical exponents measured independently are consistent with Eq. (8) and $\mu = 2$ in any case.

A NONEQUILIBRIUM ROUGHENING TRANSITION

ζ_r may diverge e.g. at a morphological transition between two phases with different roughening exponents. This leads to anomalous roughening at the transition. This phenomenon has been investigated[35] in a class of growth models with a transition between a smooth ($\zeta = 0$) and a rough ($\zeta > 0$) phase.[36,37]

A simple representative of this class is the RCH-model[38] (Roughening by Corrosion and Healing). Every growth step (t → t + 1) consists of D + u substeps. It begins with D corrosion steps

$$h_i(t + \delta t) = h_i(t) \pm 1 \quad \text{randomly}, \tag{11}$$

where $\delta t = 1/(D + u)$. During this corrosion phase all height variables perform independent random walks of D steps which corresponds to the application of uncorrelated noise with amplitude $D^{1/2}$. In the subsequent healing phase all surface steps move horizontally by u lattice constants provided they do not annihilate with a step of opposite sign before. In other words, the perimeter sites are filled in a deterministic fashion by applying the following cellular automat[19] for u times:

$$h_i(t + \delta t) = \max_j(h_i(t), h_j(t)), \tag{12}$$

where the maximum is taken over the nearest neighbors j of i.

Two properties are responsible for the nonequilibrium phase transition in the RCH-model. Any model with these properties will show the same type of transition. First, there is a maximal height H reachable within time t. In the RCH-model this height is H = Dt. Second, the mass increase dM/dt is tunable and may be sufficiently close to nLdH/dt. In the RCH-model this is the case as dM/dt → LD for u → ∞ and large L.

Now the mechanism for the kinetic roughening transition is almost tivial: If u is smaller than a critical value $u_c(D)$ the distance between H and the average surface height increases linearly in time. Then the surface does not feel the global height constraint H and becomes rough. On the other hand, if $u > u_c$, there is a finite probability that $h_i = H$. Hence the surface remains smooth.

388

The transition is triggered by directed percolation.[37] In order to reach the level H at every time the probability k must be big enough that somewhere on the uppermost parts of the surface the height variable increases by D lattice constants during the corrosion phase. This happens for a particular height variable with probability $p = 2^{-D}$. Suppose that at the uppermost level the surface consists essentially of isolated facets of the minimal size f at the transition. For instance, $f = 2u + 1$ for $d = 2$. Each of these facets remains at the level H with probability

$$1 - (1 - p)^f = k. \qquad (13)$$

However they are not really isolated: In the course of time they split and merge, and it is this collective behavior which determines the value of k at the transition. One expects it to be constant along the phase transition line.[35] In fact, the phase diagrams of the RCH-model in 2, 3 and 4 dimensions can be fitted nicely[38] with constant values of k: 0.896 ± 0.006 (d = 2), 0.81 ± 0.02 (d = 3) and 0.72 ± 0.06 (d = 4).

In the RCH-model it is not possible to investigate the critical behavior at the transition, because both model parameters are discrete. Therefore we studied a lattice version of the PNG-(Poly-Nucleation-Growth)-model[39] where the corrosion part is replaced by random nucleation of islands on facets: h_i is increased by 1 with probability p or else remains unchanged. Except that now $H = t$, the RCH-and the PNG-model are very similar so that Eq. (13) should be valid with the same k-values at the transition.

In the corresponding directed percolation problem for $u = 1$ all sites are filled with probability p. They are only connected by nearest or next nearest neighbor bonds pointing away from the substrate. The percolation threshold above which sites arbitrarily far from the substrate still are connected to it with a finite probability can easily be calculated by solving Eq. (13) for p. In dimensions 2, 3 and 4 the respective values for p_c are: 0.53 ± 0.01, 0.28 ± 0.02, 0.17 ± 0.03. The values in two and three dimensions have been confirmed by more precise measurements for the PNG-model, (0.539 ± 0.001[35] and 0.2723 ± 0.0003,[40] respectively).

Figure 4 shows some PNG-clusters. Sites nucleated at the current maximal height H(t) are left white. They constitute the clusters of the corresponding percolation problem. Below p_c they form "trees" of a typical height ξ_t and width ξ_r. In directed percolation these are the parallel and perpendicular correlation lengths.[41] In the growth of the rough surface ξ_t is a characteristic time over which a surface segment of length ξ_r remains facetlike correlated. Close to p_c these lengths diverge like[41]

$$\xi_t \sim |p - p_c|^{-\nu_t}, \quad \xi_r \sim |p - p_c|^{-\nu_r}. \qquad (14)$$

At p_c the trees form a fractal network with holes of all sizes. This is an example where the bulk density n of the cluster of white sites scales with an exponent $d_f < d$ (Eq. (1)). Above p_c there exist only holes up to a typical vertical size ξ_t and width ξ_r. In the growth of the smooth surface ξ_r means the typical size of holes in the facet at maximal height, and ξ_t their average lifetime.

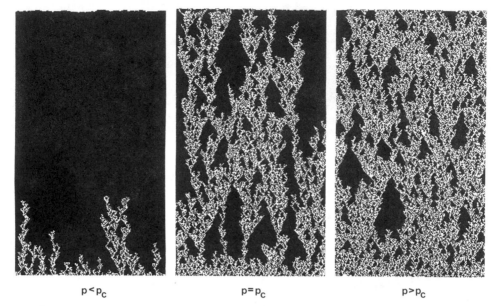

$$p < p_c \qquad\qquad p = p_c \qquad\qquad p > p_c$$

Fig. 4 PNG-clusters having grown for the same time at different nucleation probabilities p and u = 1. Sites nucleated at the maximum height H are marked white. The surface becomes rough where no nucleation at H occurs.

DEPENDENCE OF THE GROWTH VELOCITY ON THE TILT ANGLE

The KPZ-equation (8) can be analyzed exactly for d = 2 where it does not show any transition[14] of the kind we observed. Whereas in the rough phase one may expect the PNG-model to have the same universal properties as the Eden model, the smooth phase cannot be described by Eq. (8). Also at the transition a different hydrodynamic equation[42] is called for, as our results for the surface width and the growth velocity showed, which will be discussed in the next chapter. In order to make use of scaling concepts we studied the quantity[35]

$$V = d(H - \bar{h})/dt \ , \tag{15}$$

where $\bar{h}$ is the average surface height. In the PNG-model, $V = 1 - \lambda$. For a horizontal substrate V is nonzero in the rough phase, but vanishes above p_c. However it will be explained below that for a substrate tilted by an angle ϕ, Eq. (15) is finite everywhere, and is of the general form[42]

$$V = \xi_t^{-1} \ f_{\pm}(\phi \xi_r) \ , \tag{16}$$

where + (-) refers to $p > p_c$ ($p < p_c$).

The factor ξ_t^{-1} can best be understood for a horizontal surface in the rough phase: ξ_t is the time scale on which the surface lags behind the maximum height by one lattice constant.

In the smooth phase one has to consider $\phi \neq 0$. On a tilted surface there are no longer equally many up and down steps. The excess steps have an average distance $1/\phi$ and due to their motion give an extra contribution to the mass increase. Hence V is no longer 0 but should depend linearly on the step density ϕ. The proportionality factor is the effective step velocity, and one might expect it to be equal to u. However this is only correct for an isolated step separating perfect terraces. As the holes of size ξ_r live for a typical time ξ_t, the effective step velocity must be ξ_r/ξ_t, i.e. V ~ $\phi\xi_r/\xi_t$. By symmetry one has a cusp at $\phi = 0$, so that one arrives at Eq. (16) with

$$f_+(a) \sim |a| .\qquad(17)$$

This has been confirmed by our computer simulations.[38]

In the rough phase the excess steps are "screened" as far as the mass increase is concerned. There is no cusp in V. However the screening should only be valid if the number of excess steps within the range ξ_r of facetlike correlations is small. This explains why the argument of f_- is $\phi\xi_r$. For small a one expects

$$f_-(a) \sim a^2 + \text{const}.\qquad(18)$$

At the transition V remains finite for $\phi \neq 0$. Therefore it follows from Eq. (16) that[42]

$$f(a) \sim a^{\nu_t/\nu_r} .\qquad(19)$$

Equations (17)-(19) imply that the exponent 2 in Eq. (7) should be replaced by

$$\mu = \begin{cases} 1 & \text{for} \quad p < p_c \\ \nu_t/\nu_r & \text{for} \quad p = p_c \\ 1 & \text{for} \quad p > p_c \end{cases} .\qquad(20)$$

ANOMALOUS ROUGHENING

For the height fluctuations of the surface in the PNG-model we proposed the following scaling Ansatz:

$$w = w_i g_\pm \left(\frac{L}{\xi_r}, \frac{t}{\xi_t} \right) = \xi_r^{\zeta'} g_\pm \left(\frac{L}{\xi_r}, \frac{\tau}{\xi_r^{\nu_t/\nu_r}} \right) .\qquad(21)$$

For $L \gg \xi_r$ and $t \gg \xi_t$ one should see normal roughening with the appropriate exponents ζ_+, z_+ on either side of the transition. In the rough phase we obtained exponents ζ_- and z_- consistent with those for the Eden model in 2, 3 and 4 dimensions, confirming that $\mu = 2$. For facetted growth, $\zeta_+ = 0$ and $z_+ = 1$ in agreement with Eq. (9) and $\mu = 1$.

At the transition Eq. (21) implies anomalous roughening[35]

$$w = L^{\zeta'} g(t/L^{\nu_t/\nu_r}) .\qquad(22)$$

This is the same form as Eq. (4). The dynamical exponent is

$$z' = \nu_t/\nu_r = \mu ,\qquad(23)$$

so that Eq. (9) implies

$$\zeta' = 0, \tag{24}$$

which can also be derived independently.[35]

Both exponents were confirmed by our simulation of the two dimensional PNG-model.[35] The width was found to diverge logarithmically at the transition, (Fig. 5),

$$w \sim (\log L)^{1/2} . \tag{25}$$

Recent results show that the exponent on the right hand side of Eq. (25) is smaller (about 1/2.8) in three dimensions.[40]

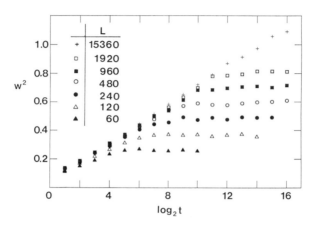

Fig. 5 Anomalous roughening in the d = 2 PNG-model at p = p_c.

CONCLUSION

Extensive numerical studies have given insight into how the asymptotic scaling Eq. (4) of a rough surface changes if a second length scale must be taken into account. In the Eden model the convolution picture Eq. (10) has been useful and in turn enabled us to extract new information about the asymptotic scaling. In a class of models with a nonequilibrium roughening transition the scaling Ansatz Eq. (21) gave us a valid description of anomalous roughening. It would be interesting to check whether the crossover function in this model again can be represented by a convolution Ansatz

$$w^2 = \xi_r^{2\zeta'} \left[a^2 \left(\frac{L}{\xi_r}\right)^{2\zeta} + b^2\right]. \tag{26}$$

The kinetic roughening transition in the above class of models is special because it is triggered by directed percolation. The scaling Ansatz Eq. (21) and the anomalous roughening should be valid for other nonequilibrium roughening transitions, too. For example the KPZ-equation (8) leads to such a transition[14] for d > 3. In contrast to Eq. (20) μ is 2 in the rough phase ($\zeta > 0$), the smooth phase ($\zeta = 0$) and at the transition where a recent ϵ-expansion[43] has given $\zeta' = 0(\log)$ and $z' = 2$ in agreement with Eq. (9). The smooth phase is not facetted in this case. Rather it is smooth like a fluid surface in thermal equilibrium above d = 3. One might go further and look for a growth model with two transitions: From a facetted phase ($\mu = 1$) to a smooth fluid like phase ($\mu = 2$) and finally to a rough one ($\mu = 2$)!

It would be most interesting to see whether there are further examples of nonequilibrium roughening transitions in the experimentally relevant dimensions d = 2 and 3. A final remark about the facetted phase in the two dimensional PNG-model: This is an example where due to the growth the surface is smoother than in thermal equilibrium, where it is well known that in the absence of long range interactions the surface is always rough at finite temperatures. Note however that in the growth model there is a long range kinetic constraint ($h_i < H$) which causes the surface to remain smooth.

Acknowledgement

Many of the results presented in this survey have been obtained in an enjoyable collaboration with J. Kertész. This work was supported by the Deutsche Forschungsgemeinschaft by SFB 341.

REFERENCES

1. An overview is given e.g. by T. Vicsek, Fractal Growth Phenomena (World Scientific, Singapore, 1989).
2. J. G. Zabolitzky and D. Stauffer, Phys. Rev. A34, 1523 (1986).
3. D. E. Wolf and J. Kertész, J.Phys. A20, L257 (1987).
4. R. Kariotis and M. G. Lagally, Surface Sci. 216, 557 (1989).
5. S. F. Edwards and D. R. Wilkinson, Proc. R. Soc. A381, 17 (1982).
6. F. Family and T. Vicsek, J. Phys. A18, L75 (1985).
7. F. Family, J. Phys. A19, L441 (1986).
8. M. J. Vold, J. Colloid Sci. 14, 168 (1959).
9. M. Eden, in: Symposium on Information Theory in Biology, H. P. Yockey ed. (Pergamon Press, New York 1958) p. 359.
10. R. Jullien and R. Botet, J. Phys. A18, 2279 (1985).
11. P. Meakin and F. Family, Phys. Rev. A34, 2558 (1986).
12. J. Krug and P. Meakin, Phys. Rev. A40, 2064 (1989); P. Meakin and J. Krug, Europhys. Lett. to be published (1989).
13. For correlated noise, see e.g. E. Medina, T. Hwa, M. Kardar and Y.-C. Zhang, Phys. Rev. A39, 3053 (1989).
14. M. Kardar, G. Parisi and Y.-C. Zhang, Phys. Rev. Lett. 56, 889 (1986).
15. R. Hirsch and D. E. Wolf, J. Phys. A19, L251 (1986).
16. D. E. Wolf, J. Phys. A20, 1251 (1987).
17. J. Krug, J. Phys. A22, L769 (1989).
18. D. Henderson, M. H. Brodsky and P. Chaudhari, Appl. Phys. Lett. 25, 641 (1974).
19. J. Krug and H. Spohn, Phys. Rev. A38, 4271 (1988).
20. P. Meakin, P. Ramanlal, L. M. Sander and R. C. Ball, Phys. Rev. A34, 5091 (1986).
21. J. Krug, Phys. Rev. A36, 5465 (1987).
22. D. E. Wolf and J. Kertész, Phys. Rev. Lett. 63, 1191 (1989).
23. M. Plischke and Z. Rácz, Phys. Rev. A32, 3825 (1985).
24. D. E. Wolf and J. Kertész, Europhys. Lett. 4, 651 (1987).
25. J. Kertész and D. E. Wolf, J. Phys. A21, 747 (1988).
26. J. Szép, J. Cserti and J. Kertész, J. Phys. A18, L413 (1985).
27. P. Devillard and H. E. Stanley, Phys. Rev. A38, 6451 (1988).
28. J. M. Kim and J. M. Kosterlitz, Phys. Rev. Lett. 62, 2289 (1989).
29. M. Kardar and Y.-C. Zhang, Phys. Rev. Lett. 58, 2087 (1987).
30. A. J. McKane and M. A. Moore, Phys. Rev. Lett. 60, 527 (1988).
31. T. Halpin-Healy, Phys. Rev. Lett. 62, 442 (1989); T. Nattermann preprint.

32. T. Halpin-Healy, Phys. Rev. Lett. <u>63</u>, 917 (1989).
33. D. Liu and M. Plischke, Phys. Rev. B <u>38</u>, 4781 (1988); P. Devillard and H. E. Stanley, Physica A to be published (1989); L. Tang and B. Forrest, private communication.
34. W. Renz, private communication.
35. J. Kertész and D. E. Wolf, Phys. Rev. Lett. <u>62</u>, 2571 (1989).
36. D. Richardson, Proc. Cambridge Philos. Soc. <u>74</u>, 515 (1973).
37. R. Durrett and T. M. Liggett, Ann. Probab. <u>18</u>, 186 (1981); R. Savit and R. Ziff, Phys. Rev. Lett. <u>55</u>, 2515 (1985).
38. D. E. Wolf, unpublished.
39. F. C. Frank, J. Cryst. Growth <u>22</u>, 233 (1974).
40. C. Lehner, N. Rajewsky, D. E. Wolf and J. Kertész, Physica A to be published.
41. W. Kinzel, in: <u>Percolation Structures and Processes</u>, G. Deutscher, R. Zallen and J. Adler, eds. Annals of the Israel Physical Society, Vol. 5 (Hilger, Bristol, 1983).
42. J. Krug, J. Kertész and D. E. Wolf, to be published.
43. T. Nattermann, private communication.

GROWTH AND EROSION OF THIN SOLID FILMS

R. Bruinsma, R. P. U. Karunasiri, and Joseph Rudnick

University of California, Los Angeles
Los Angeles, CA 90024 USA

ABSTRACT

Both growth and erosion of thin films lead to microstructures that are complex, exhibiting fractal-like features. Theories to describe these processes are reviewed. In particular, the Eden model and the shadow model are discussed.

INTRODUCTION

The process of sputtering is widely used for the deposition of thin amorphous films and for cleaning surfaces. For these reasons there is widespread interest among materials scientists in the properties of surfaces prepared by sputtering. In recent years condensed matter physicists have also found themselves drawn to the study of films and other structures produced via off-equilibrium processes. Some of these structures have been found to exhibit fractal or fractal-like features. Issues related to pattern formation and self-organization have also emerged. Indeed, such is the current interest in various sputtering processes that their study now constitutes a sizable subfield in condensed matter physics.

EXPERIMENTAL ASPECTS

1.   Sputter Deposition

Coatings are of great importance, both in daily life applications and in sophisticated technological processes.[1] Frequently, we wish to protect a surface from corrosion or abrasion which may be achieved by coverage with a suitable coat. Well known examples are Teflon coats and hardened cutting tools. The optical properties of thin films are exploited in for instance sunglasses. More sophisticated applications involve electron-beam microlithography for the production of integrated circuits.

Coatings can be made in a variety of ways, involving electrodeposition, chemical vapor deposition, thermal evaporation and, the subject of this talk, sputter deposition.[2] Sputter deposition

Kinetics of Ordering and Growth at Surfaces
Edited by M. G. Lagally
Plenum Press, New York, 1990

involves the deposition of fairly energetic atoms or ions on a growing surface. The incoming atoms move along ballistic trajectories, predominantly along the normal to the substrate surface. Once they reach the substrate, the atoms move around - by surface diffusion - and coalesce into small microcrystallites. Additional adsorbed atoms may lead to growth of the nuclei.

If the temperature is sufficiently low, then the microcrystallites will be small and the film will be amorphous for any reasonable deposition rate. Such non-equilibrium films develop fascinating surface morphologies. We can classify the various observed morphologies according to the so called structure zone model (SZM) (Fig. 1) originally proposed by Movchan and Demchishin[3] and subsequently expanded by Thornton.[4]

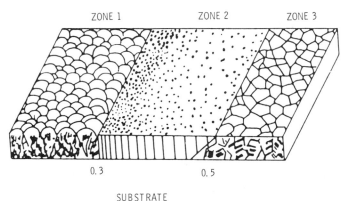

Fig. 1   Structure zone model.

The model distinguishes between four zones, each corresponding to a given ratio, T/T_M, of the temperature of the substrate, T, to the bulk melting temperature of the adsorbed material, T_M. In zone 1, we observe a columnar, fine-grained microstructure. The columns themselves are polycrystalline and are highly defected. In zone 2, surface diffusion acts as an annealing mechanism leading to less defected and larger grains. In zone 3, bulk diffusion and recrystallization dominate. The microstructure resembles that of materials grown via a melting process. The T zone is a transition region between 1 and 2.

The boundaries between the columns are regions of low adatom density and will thus play a significant role in determining the mechanical and transport properties of the film, as well as its durability and stability. It would thus appear that growth at elevated temperatures is most preferable. However, high temperatures also lead to interdiffusion between substrate and overlayer as well as thermal roughening.[5] In addition, it may be the goal of the film grower to produce mixtures that cannot be achieved in equilibrium, such as, finely spaced inclusions of impurities. Sputter deposition may also be utilized to grow heterostructures of alternating layers of different materials.[1] In both cases, temperatures must be low enough to keep diffusion from altering or destroying the desired film properties.

Columnar structures were first observed by Wade and Silcox[6] for sputtered Pd films using small angle electron scattering. Since then there have been many reports of columnar microstructure by microfractography,[7-15] by transmission electron microscopy,[7,16-26] and by small angle electron and X-ray scattering.[7,26-29] An example of a microfractograph[8] is shown in Fig. 2.

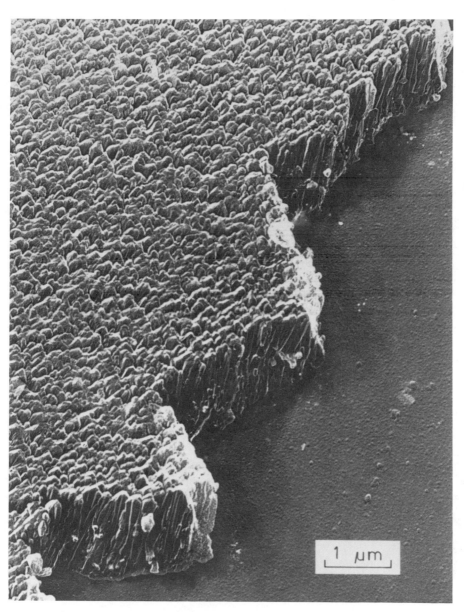

Fig. 2   Microfractograph of Al film.

The columns are observed to have an orientation which is always more nearly perpendicular to the substrate than the vapor beam direction. Nieuwenhuizen and Haanstra first reported a careful determination of the angles involved.[9] If a α the angle between the normal and the vapor beam, and β the angle of the columns, then to a good approximation one finds that $2 \tan\beta = \tan\alpha$, the famous tangent rule.

The columnar microstructure has a strong effect on the properties of the coating[49]. Magnetic,[10-15,17,32,33] optical,[16,34,35] electrical,[17,20-25,34-38] and mechanical[12,39] effects have been reported. The columnar structure also affects the surface topography,[16,40,41] the specific surface area,[42-44] the oxygen uptake[24,45-47] and the surface adhesion.[48]

The columnar structure is only observed for large angles of incidence. For near normal or random incidence a "cauliflower" structure is observed[50-52] shown in Fig. 3. Messier and Yehoda[50] found that the characteristic length scale D of the pattern ξ scales with the layer height h according to $D \propto h^x$ with $x \approx .72$ at low temperatures. In addition, they found that the surface topography is self-similar: if we expand a section of a micrograph then the basic features of the surface are not changed.[51] An example of this scaling is shown in Fig. 4.

A second characteristic microstructure encountered during coating is the "void." Its existence was first predicted by Moss and Graczyk[53] in amorphous Ge and Si, based on low-angle scattering and electron-spin-resonance.[54] Galeener[55] interpreted dielectric-constant measurements on low density films of Ge and Si as revealing the presence of oriented disk-like. Oriented disk-like voids connected together in a network of microcracks. Donovan and Heinemann[56] found that the voids tend to vanish if the substrate temperature is raised, which is reminiscent of the SZM model. The subsurface void-network appears to be correlated with the columnar grain structure of the coating. This was confirmed by Leamy and Dirks.[57] They found that the columnar regions observed in rare-earth/transition-metal thin films are surrounded by a network of less dense material. An example is shown in Fig. 5. It appears tempting to attribute void formation either to differential thermal contraction of the coat with respect to the film or to condensation of vacancies. The numerical work discussed in the next section indicates, however, that void formation is intrinsic to the non-equilibrium growth of overlayers.

Finally, a third type of object encountered is the nodule,[58] an isolated structure protruding from an otherwise smooth surface. They are believed to originate at small asperities of the substrate.

It has proven difficult to elucidate experimentally the underlying physical mechanism of columnar growth and void network formation. As we will discuss shortly, certain proposed explanations rely on atomic mechanisms. This would mean that we must be able to observe the columnar structure down to the 10 Å level or so. However, scanning electron microscopies offer a resolution of only 3-4 nm. Transmission electron microscopies go down to 0.1 - 0.2 nm resolution. Less direct, but more microscopic, are low-angle scattering probes. This has the advantage that the experiment can be performed in situ. However, it is difficult to reconstruct the "real-space" surface topography out of a diffraction image, especially in the absence of phase information. None of these methods is likely to be able to probe the early history of columnar growth.

398

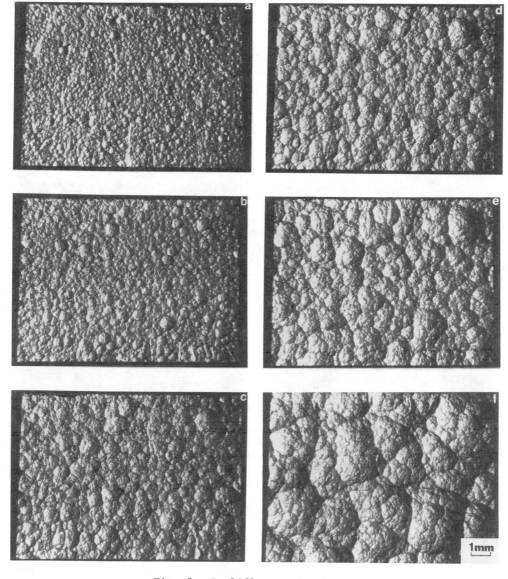

Fig. 3   Cauliflower structure.

2.   Sputter Erosion and Etching

The "inverse" of sputter deposition is sputter erosion.[60] This is a process of very considerable technical importance. Surfaces are frequently cleaned by exposing them to a bombardment of ions ("sputter cleaning"). Furthermore, in a number of applications we are interested in the erosion of surfaces under bombardment, e.g., the interior of a reactor vessel. Finally, etching by ion beams is used for the fabrication of micro-electronic devices[61] (in particular to make circuit patterns).

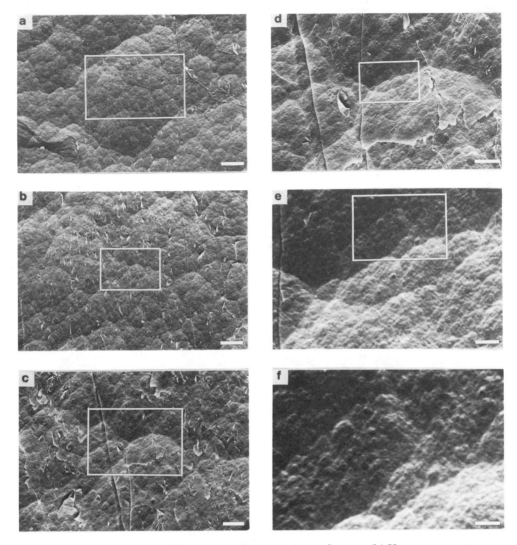

Fig. 4   Repeated magnification of a section of a cauliflower structure.

In discussing erosion, we must distinguish the macroscopic and microscopic features of the morphology. The evolution of macroscopic features during erosion and etching has been studied in much detail. Depending on the initial shape of the substrates, a number of features such as cones or ridges are seen. Eventually, though, the macroscopic surface features shrink and smooth surfaces are seen. Protruding surface features erode more rapidly than their neighborhood. Smith et al.[61-66] showed that if the surface evolves according to the "Huyghens Principle" (i.e., the Eden model discussed below), then one can make detailed predictions for the erosion of macroscopic features such as edges. On the microscopic level, the situation is less transparent. The ion bombardment will constantly create damage on the atomic level. Statistical fluctuations are likely to create considerable microstructure. At the same time, erosion and surface diffusion will reduce the size of the

Fig. 5   Subsurface groove network in GdCo amorphous films.   From Ref. 57.

microstructure. It is likely that a steady-state will develop where fresh damage is balanced by erosion.

THEORETICAL ASPECTS

1.   <u>Numerical</u>

The first numerical simulations of the growth of thin amorphous films were done by Henderson, Brodsky, and Chaudhari.[67] Their method has been the prototype of numerical simulations of growth under ballistic conditions. They assumed that incoming particles move along straight lines until they hit the surface with a sticking coefficient of unity. The incident particle is allowed to relax to the extent that it moves to the nearest pocket where it touches at least two other particles. This latter process mimics surface diffusion. A growth scenario of this type is called "ballistic aggregation". A typical example, for deposition at an angle is shown in Fig. 6. Gratifyingly, both columns and voids are visible. This indicates the existence of a universal mechanism for the microstructure formation, i.e., independent of the details of the kinetics and potentials. Extensive work by Leamy and Dirks[57] and Leamy, Gilmer, and Dirks[59] showed that the columns also obey the tangent rule.

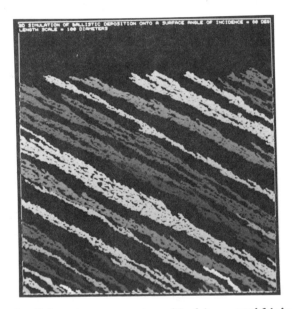

Fig. 6   Columnar aggregates (Meakin, unpublished).

A fundamental problem with the numerical work was pointed out by Kim et al.[68] If we compare Fig. 6 with, say, Fig. 2, then it is clear that the scale is all wrong. Whereas the numerical work finds voids and columns on the length scales of atoms, the observed length scale is many orders of magnitude larger ($\sim 0.1 \mu$). Apparently, there is far more relaxation than allowed by the ballistic aggregation model. The consequence is that the densities are much too low in the numerical simulations.

If we follow a particular aggregate of adsorbed particles during deposition, then it has the aspect of a growing tree. A very interesting question is whether these tree structures found in the computer experiments are fractals. Theoretical work by Ball and Witten[69] suggests they should not be. However, it required very large scale simulations to ascertain this result.[70,71] Large scale, two dimensional simulations by Meakin[71] showed that the growing trees compete with each other. As the layer thickness grows, large trees suffocate small trees (Fig. 7). Although the resulting structure is dense (i.e., non-fractal), it can still have an internal scale-invariant void network of the type first proposed by Messier and Yehoda (Fig. 8). In addition, Meakin also found deviations from the tangent rule.

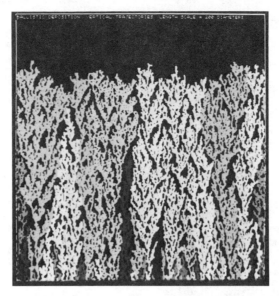

Fig. 7   Tree structure of ballistic aggregation.   From Ref. 71.

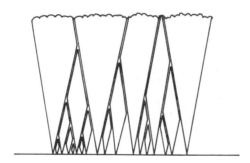

Fig. 8   Self-similar subsurface groove network.   From Ref. 50.

2.   Analytical Theories

We now turn to the various analytical growth laws which have been proposed to explain the observed microstructures. The numerical work on ballistic aggregation indicates that the basic physical ingredients are, at least in part, particularly present in the ballistic aggregation model, but that we must allow for far more relaxation of the surface than is possible with present day computer capabilities. We need a continuum theory because the observed surface structures are smooth on length scales less than a nanometer. We first must identify both the annealing mechanism and the destabilizing mechanism to construct such a continuum a model.

The dominant annealing mechanism at lower temperatures is surface diffusion. This was established by sintering experiments[72] and confirmed by scanning electron microscopy. Near the melting temperature, atoms can also evaporate and recondense in spots where they gain binding energy and thus smoothen the surface. The annealing mechanisms will smooth out roughness on atomic length scales.

2a. Local Self-Shadowing

The second ingredient in a continuum theory must be the destabilizing mechanism. The first proposed mechanism was the local "self-shadowing" effect of Leamy and Dirks[57] shown in Fig. 9. An atom sitting on the surface catches fewer atoms if the incident atoms make an angle with the surface. The tangent rule can be justified using this idea as shown in Fig. 9. Ramanlal and Sander[73] constructed a mean-field theory based on this argument and showed that it spontaneously develops a microstructure resembling columns. The local "self-shadowing" mechanism is essentially the same as the ballistic aggregation model mentioned before. In the absence of surface diffusion it will lead to the free structure of Fig. 7, which has too low a density. A variant of local self-shadowing which includes surface diffusion was discussed by Manor et al.[74] They found columnar structures whose characteristic length was temperature dependent through the surface diffusion constant.

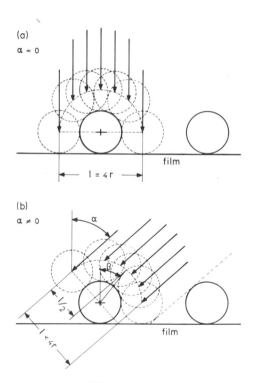

Fig. 9. Local Self-Shadowing.[57] A schematic illustration of possible impingement angles for two disks at (a) normal ($\alpha = 0$) and (b) oblique ($\alpha \neq 0$) incidence. At $\alpha = 0$ the capture length ℓ is 4r and the mean pair orientation is perpendicular. (b) shows that the capture length is shortened by self-shadowing at $\alpha \neq 0$ and that the mean pair orientation is consequently shifted away from α towards the substrate normal. The angle β defines the growth direction. It is related to α by $\tan \beta = 2 \tan \alpha$.

2b. Eden Model

An alternative mechanism is incorporated in the Eden model.[75] There, the surface is assumed to grow in the direction normal to the surface at a given rate (Fig. 10). The height h at a given site above the substrate thus increases more rapidly if we are on a slope ($\Delta h^{(1)}$), as compared to a mountain peak or valley bottom ($\Delta h^{(2)}$). The Eden model develops a "cusp structure" which could be related to the cauliflower geometry. All the proposed destabilizing mechanisms so far are local, i.e., the decision whether or not a given site grows faster or slower depends only on the local geometry.

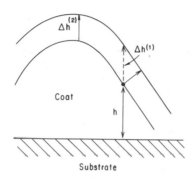

Fig. 10 Eden model growth. The growth direction is the surface normal.

We now must combine stabilizing and destabilizing mechanisms. In addition, we also must allow for noise. The incoming particle current will have an average value with statistical fluctuations on top ("shot noise"). Some of the effects of surface diffusion on an Eden-model type nonlinearity were discussed by Lichter and Shen.[76] A general analysis of "evaporation-recondensation" annealing and local growth in the limit of large length scales was done by Kardar, Parisi, and Zhang (KPZ).[77] In particular, they assumed a growth law

$$\frac{\partial h}{\partial t} = \sigma\nabla^2 h + J + \lambda(\vec{\nabla} h)^2 + \eta(\vec{r},t) .$$

(1)

Here, $h(r,t)$ is the height of the coat at a position r on the surface. This assumes the "solid-on-solid" approximation where the profile of the coat is assumed to have no overhangs. The first term on the right hand side has an annealing effect, i.e., it tries to remove height irregularities. As discussed by Edwards and Wilkinson,[78] it can be due to either evaporation/recondensation processes or surface reconstruction. We will see below that if surface diffusion is the dominant annealing mechanism, then the ∇^2 operator must be replaced by a ∇^4 operator. The second and third terms represent the deposition current. J is the average current while the third term is the lowest-order contribution of an Eden model type destabilizing effect. The last term in Eq. (1) represents the statistical fluctuations of the current (shot noise).

405

Equation (1) has been studied in much detail. The importance of the work of KPZ is that they recognized that $(\nabla h)^2$ is the lowest-order nonlinear term. It is expected to be, at long distances, the dominant destabilizing term. In the language of renormalization-group analysis, we say that it is the most relevant perturbation on a linear growth model. KPZ showed that, in d = 2, a sample of length L should have a noise-roughened surface of a width $W(L) \propto L^\chi$ with $\chi = 1/2$. The exponent X should not change if we add higher order nonlinear teams. KPZ argued that $W(L) \propto L^{1/2}$ should be a feature of all local ballistic growth theories. The computer simulations of Meakin generate surfaces with a roughness that is consistent with their prediction. Unfortunately, the KPZ model does not show columnar growth nor does it produce a self-similar network of voids, the two most prominent features of the observed microstructures.

2c.   The Shadow Model

If we look at the materials science literature instead of the physics literature, then it becomes clear that a different destabilizing mechanism is assumed to be operative (e.g., Ref. 1). The microstructure is believed to be due to shadowing but it is not just the atomic self-shadowing mechanism of Leamy and Dirks. Instead, the shadowing of a given site is due to the entire <u>mountain topography</u> of the structure near a particular site.[79] An example is shown in Fig. 11a. If the growth is by isotropic sputtering, then the exposure rate of a site is roughly proportional to the exposure angle θ of the site. The shadowing of valleys by peaks enhances the instability leading to roughness. The growth rate at a point on the surface is thus strongly influenced by the topology of nearby and not-so-nearby regions. This non-local growth mechanism will be relevant in particular if the incoming beam has a broad distribution of angles or, for a collimated beam, if it has a low incoming angle.   If, on the other hand, growth is by Molecular Beam Epitaxy (MBE) with the incoming atoms collimated in a beam in the normal direction (Fig. 11b) then we expect that only local instabilities play a role because the nearby mountains cannot shadow a particular site. Presumably, for MBE we should thus use a local growth model of the KPZ type discussed in the previous section.

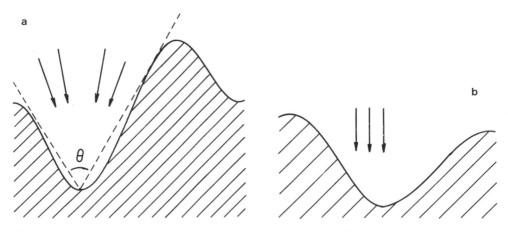

Fig. 11 a) Non-local shadowing (sputtering). b) Molecular-beam epitaxy.

The shadow model is thus based on the idea that "global" shadowing is the destabilizing mechanism of growth under sputtering conditions. The new mechanism is a powerful destabilizing effect. In fact, nonlocal shadowing can be shown to be, in renormalization group language, a relevant perturbation on the KPZ model. Note also that it is operative at macroscopic length scales. It is not a local nonlinearity whose characteristic lengthscale must "work itself up" from atomic lengths all the way to microns.

The shadowing effect must compete with surface annealing mechanisms. We will focus on the dominant one, surface diffusion. Following Herring,[72] we note that the surface diffusion current $\vec{J}_s$ along the surface of an amorphous solid is

$$\vec{J}_s = D\ \vec{\nabla}_s\ \kappa\ ,$$ (2)

with ∇_s the gradient along the surface, κ the local curvature, and

$$D = \frac{D_s \gamma \Omega^2 \nu}{k_B T}\ .$$ (3)

Here D_s is the surface diffusion constant, Ω the atomic volume, ν the number of surface atoms/unit area, and γ the surface energy/unit area. The growth velocity ϑ_n in the normal direction is simply the sum of $\nabla_s \cdot J_s$ and the deposition current:[80]

$$\vartheta_n = -D\ \nabla_s^2\ \kappa + \frac{R\tilde{\theta}}{\pi} + \eta(\{\tilde{\theta}\})\ .$$ (4)

Here, R is the deposition current and $R\tilde{\theta}/\pi$ is the average over all allowed incoming directions of the projection of the current on the normal to the surface. The shot-noise contribution η has an RMS width $\langle\eta^2\rangle^{1/2}$ proportional to θ. Note that the exposure angle θ is not equal to $\tilde{\theta}$. Erosion is described by Eq. (4) if we change the sign of R.

As presented, Eq. (4) indeed incorporates the combined physics of shadowing and surface diffusion. It presents though a formidable mathematical problem due to the nonlocal nature of θ. As a first step we analyze the resulting profile in d = 1, assuming no overhangs and setting $\tilde{\theta} = \theta$.

2c1. Solid-on-Solid Approximation (SOS)[79]

In the SOS approximation, Eq. (4) reduces to

$$\frac{\partial h}{\partial t} = -D\ \frac{\partial^4 h}{\partial x^4} + \frac{R\theta(\{h\})}{\pi} + \eta\ .$$ (5)

The nonlinearity in Eq. (5) lies in the functional dependence of θ on the height profile h(x,t). Equation (5) ignores any local nonlinearity, not because those effects are insignificant but because we are here interested in the asymptotic properties of a surface roughened by shadowing alone. If η is a Gaussian white-noise source and if we simply replace θ by its average $\langle\theta\rangle$, then the resulting surface has a mean height $R\langle\theta\rangle t$ and a width ω proportional to $(t^3/D)^{1/8}\ \langle\eta^2\rangle^{1/2}$. To go beyond this "mean-field" description we can do a linear stability analysis. Assume that the surface starts out as nearly flat. If, at t = 0, 0, the surface has a small periodic modulation $h(x,0) = h_k°\cos kx$, then the exposure angle $\theta = \langle\theta\rangle + \alpha k h_k°\cos kx$ with $\alpha = 2/\pi$. The perturbation grows exponentially, $h_k(t) = h_k° e^{\omega_k t}$, with a rate constant

$$\omega_k = \alpha Rk - Dk^4 . \qquad (6)$$

Note that for $k \to 0$, we can write ω_k as Uk with U proportional to the steady-state growth rate R. Short-distance details of the surface erode quickly ($\omega_k < 0$), while the amplitude of a long wavelength mode (with $w_k > 0$) grows. The mode with the highest growth rate has a wavevector $k* = (\alpha R/4D)^{1/3}$ and a growth rate $\omega* = 3/4\ \alpha R(\alpha R/4D)^{1/3}$.

If we follow the surface at late times numerically, then a columnar structure is found (Fig. 12). (Noise was not included). Note the similarity between Fig. 12 and Fig. 2.

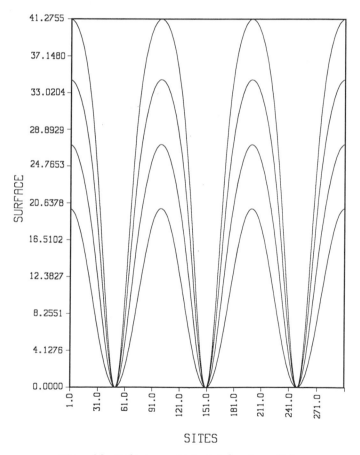

Fig. 12 Columns produced by Eq. (4).

For random initial conditions, a "self-similar" looking mountain landscape evolves[79] (Fig. 13). If we follow the coverage c(h), the number of atoms per unit area of an initially smooth surface, as a function of time, then initially $c(h) \simeq 1$ up to a height h*. At a time t*, of order $(\omega*)^{-1}$, the surface starts to roughen. A typical plot of c(h) for $t \gg t*$ is shown in Fig. 14a for $\omega < 10^{-4}$. The overlayer exhibits complete coverage up to h* beyond which the mountain landscape starts. For $h \gg h*$, h*, c (h) $\sim$ 1/h if D is small. The self-similar landscape is only seen for $h \gg h*$.

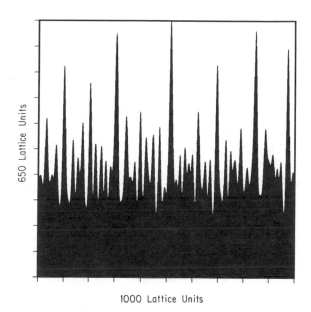

650 Lattice Units

1000 Lattice Units

Fig. 13 Mountain landscape produced by shadowing.

In Fig. 14b, we show the growth scenario for $D > 10^{-4}$, in this case $D = 10^{-1}$. For earlier times we again have a compact film (c = 1) whose thickness increases with time. The film reaches a maximum height h* such that at later times c(h) = 1 for $h < h*$ and c(h) < 1 for $h > h*$. The surface is very flat until it reaches h* while for $h > h*$ it again grows a mountain structure. The height profile c(h) now falls off more rapidly than a power law and it has a width which increases with time reminiscent of the "mean-field" picture.

This result has very interesting implications for thin-film growth. It is possible, in our model, to grow flat films if D/R > 0.1. However growth should be terminated, or rather R must be reduced, once h reaches the critical height h* because from then on the film only roughens. Can we compute h*? If the amplitude of the fastest growing mode is proportional to $\exp^{(\omega*t)}$, then roughening becomes important if $\omega*t$ is of the order one. The corresponding height h* $\sim$ R/$\omega*$ $\sim$ $(D/R)^{1/3}$. The actual critical height h* was found to depend on D roughly as $D^{0.3}$.

As a function of the surface diffusion constant D, we found a power-law dependence c(h) $\sim$ $1/h^{\chi}$ if D is small (D/R < 10^{-4}) with the exponent χ dependent on the value of D. For large D, (D/R > 10^{-4}) the

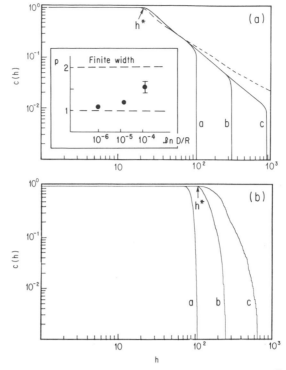

Fig. 14 Surface coverage c(h) versus height h, a) D=10^{-5}; b) D=10^{-1}. In each panel, curves a, b, c refer respectively to 111, 333, and 1000 time steps. The dashed line is the t = ∞ result for D = 0.

exponent χ depend on the value of D while the surface has the a more compact profile. This is reminiscent of the phase transition of two-dimensional X-Y model but more accurate numerical work will be necessary to establish whether or not there is indeed a phase transition at a critical value of D.

2c2. Beyond SOS

The SOS approximation is expected to be valid as long as the mountain landscape does not develop extremely steep slopes or overhangs. For large surface diffusion constants, this should be a good approximation because the landscape should be reasonably smooth. How about small surface diffusion constants with the shadow effect dominant? Recently,[80,81] Eq. (1) has been investigated numerically for small surface diffusion constants without resorting to the SOS approximation and without replacing $\tilde{\theta}$ by θ. A poster on the results was presented during the conference by S. Bales. The computed mountain landscape was found to consist of dome-like columns separated by narrow grooves. The columns originated from asperities on the starting surface. As in the SOS approximations, the columns appeared to compete with each other for space. However, contrary to the SOS approximation, the top film was relatively smooth outside the deep grooves. This could just as well have been a description of the Zone I morphology with its deep grooves and domed columns. The Shadow Model thus appears to provide us with an explanation of the morphology of amorphous films but many questions

410

remain: can we explain the scaling laws of Messier and Yehoda and how about the Zone II morphology? Yet, it is remarkable that the simple physics of the Shadow Model - surface diffusion versus shadowing - appears capable of producing realistic film morphologies without having to resort to the complicated microphysics of the impact of individual atoms on a surface.

REFERENCES

1. R. F. Bunshah and D. M. Mattox, Physics Today 50, May 1980.
2. For a review see: Thin Film Processes, Part 2, J. L. Vossen and W. Kern, eds. (Academic Press, New York, 1978).
3. B. A. Movchan and A. V. Demchishin, Phys. Met. Metallogr. 28, 83 (1969).
4. J. A. Thornton, Ann. Rev. Mater. Sci. 7, 239 (1977).
5. J. D. Weeks, in Ordering in Strongly Fluctuating Condensed Matter Systems (Plenum, New York, 1980), p. 293.
6. R. H. Wade and J. Silcox, Appl. Phys. Lett. 8, 7 (1966).
7. H. J. Leamy and A. G. Dirks, J. Phys. D 10, L95 (1977).
8. C. Kooy and J. M. Nieuwenhuizen, in Basic Problems in Thin Film Physics, R. Niedermayer and H. Mayer, eds. (Vandenhoeck and Ruprecht, Göttingen, 1966), p. 181.
9. J. M. Nieuwenhuizen and H. B. Haanstra, Philips Tech. Rev. 27, 87 (1966).
10. W. Metzdorf and H. E. Wiehl, Phys. Status Solidi 17, 285 (1966).
11. T. Hashimoto, K. Hara, and E. Tatsumoto, J. Phys. Soc. Jpn. 24, 1400 (1968).
12. K. Okamoto, T. Hashimoto, K. Hara, and E. Tatsumoto, J. Phys. Soc. Jpn. 31, 1374 (1971).
13. K. Okamoto, T. Hashimoto, K. Hara, H. Fujiwara, and T. Hashimoto, J. Phys. Soc. Jpn. 34, 1102 (1973).
14. K. Hara, H. Fujiwara, T. Hashimoto, K. Okamoto, and T. Hashimoto, J. Phys. Soc. Jpn. 39, 1252 (1975).
15. T. Hashimoto, K. Hara, K. Okamoto, and H. Fujiwara, J. Phys. Soc. Jpn. 41, 1433 (1976).
16. H. König and G. Helwig, Optik 6, 111 (1950) ; L. Reimer, Optik 14, 83 (1957).
17. D. O. Smith, M. S. Cohen, and G. P. Weiss, J. Appl. Phys. 31, 1755 (1960).
18. T. M. Donovan and K. Heinemann, Phys. Rev. Lett. 27, 1794 (1971).
19. N. G. Nakhodkin and A. I. Shaldervan, Thin Solid Films 10, 109 (1972).
20. J. J. Hauser and A. Staudinger, Phys. Rev. B8, 607 (1973).
21. W. Fuhs, H. J. Hesse, and K. H. Langer, in Amorphous and Liquid Semiconductors, J. Stuke and W. Brenig, eds. (Taylor and Francis, London, 1974), p. 79.
22. A. Barna, P. B. Barna, Z. Bodó, J. F. Pócza, I. Pózsgai, and G. Radnóczi, in Amorphous and Liquid Semiconductors, J. Stuke and W. Brenig, eds. (Taylor and Francis, London, 1974), p. 109.
23. D. K. Pandya, A. C. Rastogi, and K. L. Chopra, J. Appl. Phys. 46, 2966 (1975).
24. D. K. Pandya, S. K. Barthwal, and K. L. Chopra, Phys. Status Solidi A 32, 489 (1975).
25. A. Barna, P. B. Barna, G. Radnóczi, H. Sugawara, and P. Thomas, in Structure and Excitations of Amorphous Solids, G. Lucovsky and F. L. Galeener, eds. (American Institute of Physics, New York, 1976), p. 199.

26. N. G. Nakhodkin, A. I. Shaldervan, A. F. Bardamid, and S. P. Chenakin, Thin Solid Films 34, 21 (1976).
27. R. H. Wade and J. Silcox, Appl. Phys. Lett. 8, 7 (1966).
28. S. C. Moss and J. F. Graczyk, Phys. Rev. Lett. 23, 1167 (1969).
29. G. S. Cargill, III, Phys. Rev. Lett. 28, 1372 (1972).
30. M. L. Rudee, Philos. Mag. 28, 1149 (1973).
31. W. I. Kinney and G. S. Cargill, III, Phys. Status Solidi A 40, 37 (1977).
32. K. Okamoto, K. Hara, H. Fujiwara, and T. Hashimoto, J. Phys. Soc. Jpn. 40, 293 (1976).
33. T. Iwata, R. J. Prosen, and B. E. Gran, J. Appl. Phys. 37, 1285 (1966).
34. T. M. Donovan, W. E. Spicer, J. M. Bennett, and E. J. Ashley, Phys. Rev. B2, 397 (1970).
35. F. L. Galeener, Phys. Rev. Lett. 27, 1716 (1971).
36. P. Thomas, A. Barna, P. B. Barna, and G. Radnócni, Phys. Stat. Sol. A 30, 637 (1975).
37. B. A. Orlowski, W. E. Spicer, and A. D. Baer, Thin Solid Films 34, 31 (1976).
38. L. Kubler, G. Gewinner, J. J. Koulmann, and A. Jaéglé, Phys. Status Solidi B78, 149 (1976).
39. J. D. Finegan and R. W. Hoffman, J. Appl. Phys. 30, 597 (1959).
40. L. Holland, J. Opt. Soc. Am. 43, 376 (1953).
41. J. M. Pollack, W. E. Haas, and J. E. Adams, J. Appl. Phys. 48, 831 (1977).
42. J. A. Allen, C. C. Evans, and J. W. Mitchell, in Structure and Properties of Thin Films, C. A. Neugebauer, J. B. Newkirk, and D. A. Vermilyea, eds. (Wiley, New York, 1959), p. 46.
43. J. W. Swaine and R. C. Plumb, J. Appl. Phys. 33, 2378 (1962).
44. J. W. Geus, in Fast Ion Transport in Solids, W. van Gool, ed. (North-Holland Publ. Co., Amsterdam, 1973), pp. 331-371.
45. L. Holland, Br. J. Appl. Phys. 9, 336 (1958).
46. D. E. Speliotis, G. Bate, J. K. Alstad, and J. R. Morrison, J. Appl. Phys. 36, 972 (1965).
47. W. Ma and R. M. Anderson, Appl. Phys. Lett. 25, 101 (1974).
48. J. J. Garrido, D. Gerstenberg, and R. W. Berry, Thin Solid Films 41, 87 (1977).
49. For a review of the effect of the columns on the coating, see A. Dirks and H. J. Leamy, Thin Solid Films 47, 219 (1977), from which a number of references were borrowed.
50. R. Messier and J. E. Yehoda, J. Appl. Phys. 58, 3739 (1985).
51. J. E. Yehoda and R. Messier, Appl. Surf. Sci. 22/23, 590 (1985).
52. R. Messier and R. C. Ross, J. Appl. Phys. 53, 6220 (1982).
53. S. C. Moss and J. F. Graczyk, Phys. Rev. Lett. 23, 1167 (1969).
54. M. H. Brodsky and R. S. Title, Phys. Rev. Lett. 23, 581 (1969).
55. F. L. Galleener, Phys. Rev. Lett. 27, 1716 (1971).
56. Terence M. Donovan and Klaus Heinemann, Phys. Rev. Lett. 27, 1716 (1971).
57. H. J. Leamy and A. G. Dirks, J. Appl. Phys. 49, 3430 (1978).
58. Karl H. Guenther, Applied Optics 23, 3806 (1984).
59. H. J. Leamy, G. H. Gilmer, and A. G. Dirks, in Current Topics in Materials Science, E. Kaldis, ed. (North-Holland, Amsterdam, 1980), vol. 6.
60. For a review see: G. Carter, in Erosion and Growth of Solids Stimulated by Atom and Ion Beams, G. Kiriakidis, G. Carter, and J. L. Whitton, eds. (Martinus Nihjoff Publishers, 1986), pp. 70-97.
61. See for instance: H. P. Bader and M. A. Lardon, J. Vac. Sci. Technol. A3, 2167 (1985).
62. A. Smith and J. M. Walls, Philos. Mag. A 42, 235 (1980).
63. R. Smith, S. S. Makh, and J. M. Walls, Philos. Mag. A 47, 453 (1983).

64. R. Smith, G. Carter, and M. J. Nobes, Proc. R. Soc. London Ser. A407, 405 (1986).
65. R. Smith, M. A. Tagg, G. Carter, and M. J. Nobes, J. Mater. Sci. Lett. 5, 115 (1986).
66. R. Smith, S. J. Wilde, G. Carter, I. V. Katardjiev, and M. J. Nobes, J. Vac. Sci. Techn. B 5, 579 (1987).
67. D. Henderson, M. H. Brodsky, and P. Chaudhari, Appl. Phys. Lett. 25, 641 (1974).
68. S. Kim, D. J. Henderson, and P. Chaudari, Thin Solid Films 47, 155 (1977).
69. R. C. Ball and T. A. Witten, Phys. Rev. A 29, 2966 (1984).
70. P. Meakin, J. Colloid Interface Sci. 105, 240 (1985); D. Bensimon, B. Shraiman, and S. Liang, Phys. Lett. A102, 238 (1984).
71. P. Meakin, CRC Critical Reviews in Solid State and Materials Science 13, 143 (1987).
72. C. Herring, in Structure and Properties of Solid Surfaces, R. Gomer and C. S. Smith, eds. (University of Chicago, Chicago, 1953), pp. 5-72.
73. P. Ramanlal and L. M. Sander, Phys. Rev. Lett. 54, 1828 (1985).
74. A. Manor, D. J. Srolovitz, P. S. Hagen, and B. G. Bukiet, Phys. Rev. Lett. 60, 424 (1988).
75. M. Eden, in Symposium on Information Theory in Biology, H. P. Yockey, ed. (Pergamom Press, New York 1958).
76. S. Lichter and Jyhmin Chen, Phys. Rev. Lett. 56, 1396 (1986).
77. M. Kardar, G. Parisi, and Yi-Cheng Zhang, Phys. Rev. Lett. 56, 889 (1986).
78. S. F. Edwards and D. R. Wilkinson, Proc. Royal Soc. A381, 17 (1982)
79. R. P. U. Karunasiri, R. Bruinsma, and J. Rudnick, Phys. Rev. Lett. 62, 788 (1989).
80. G. S. Bales and A. Zangwill, Phys. Rev. Lett. 63, 692 (1989).
81. G. S. Bales, R. Bruinsma, E. A. Eklund, R. P. U. Karunasiri, J. Rudnick, and A. Zangwill, submitted to Science.

413

MORPHOLOGY OF CRYSTALS BASED ON CRYSTALLOGRAPHY,

STATISTICAL MECHANICS, AND THERMODYNAMICS:

APPLICATIONS TO Si AND GaAs

P. Bennema

RIM Laboratory of Solid State Chemistry
University of Nijmegen
Toernooiveld
6525 ED Nijmegen, The Netherlands

ABSTRACT

First a brief summary is given of the implications of the solid on solid lattice gas or Ising interface model. The key concept resulting form the theory and indirectly from simulations is the concept of roughening transition. This explains that crystals can grow with flat faces, provided the dimensionless temperature is below the roughening temperature.

The alternative one-layer interface model is also discussed. This model shows according to Onsager an order-disorder phase transition. The corresponding Ising temperature is close to the roughening transition temperature.

It is shown that the theory of Onsager can be generalized and applied to complex planar connected nets. The resulting Ising temperatures can be considered as a measure for the relative morphological importance of a face (hkl). The stronger the net of a face the higher the Ising temperature, the lower the growth rate, the higher the morphological importance. It is shown how an integration of the concept of connected nets showing a roughening transition and the crystallographic morphological Hartman Perdok theory can be carried out. The implications of the integrated network-roughening transition theory for reconstructed surfaces of Si and GaAs are discussed briefly.

INTRODUCTION

1a.   Ising Models, Crystal Surfaces and Crystal Growth

In order to apply statistical mechanical Ising models to the interface crystal-mother phase this whole interface is partitioned in cells of equal size and shape (see Fig. 1a) and the cells can have only two properties which are now solid (s) or fluid (f). Within such an interface model the solid on solid (SOS) condition can be introduced.[1-7] This implies that solid cells only can occur on top of solid cells and overhangs are ruled out. Also isolated fluid cells in the solid body and the reverse are ruled out. Looking on top of an SOS surface it can be seen that each surface configuration is characterized by the height of towers α_{xy} (see Fig. 1b). Monte Carlo simulations have been carried out on such surfaces, using as independent variables the quantities

$$\frac{\phi}{kT} = \frac{\phi^{sf} - 1/2\,(\phi^{ss} + \phi^{ff})}{kT} \qquad (1)$$

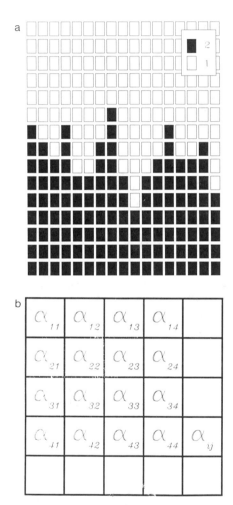

Fig. 1 (a) A cut through a SOS "landscape". Solid cells ■ (2); fluid cells □ (1). Solid cells only occur on top of other solid cells. This makes the solid phase a connected phase. The same holds automatically for the fluid phase. (b) the SOS surface seen from above. Due to the SOS restriction the whole interface can be characterized by a set of "towers", which are now given by the matrix elements α_{xy}. Theoretically spoken the number α_{xy} varies in a discrete way from $-\infty$ to $+\infty$ and characterizes the height of a tower referenced to a reference plane. The set of numbers α_{xy} characterize a surface configuration.

and

$$\beta' = \frac{\Delta\mu}{kT} = \frac{\mu^f - \mu^s}{kT} . \tag{2}$$

Here ϕ is a generalized bond energy, k is the Boltzman constant, and T the absolute temperature. ϕ represents the energy of formation of a solid-fluid (sf) bond in reference to half a solid-solid (ss) bond and half a fluid-fluid (ff) bond. The bond energies ϕ^{sf}, ϕ^{ff} and ϕ^{st} are supposed to be negative. In case the fluid phase is considered as vacuum

$$\frac{\phi}{kT} = -\frac{\phi^{ss}}{2kT} , \tag{2a}$$

which is a positive number. (In the following we will assume that ϕ is positive).

The quantity ϕ/kT defines the equilibrium properties of the surface such as roughness, etc. The quantity β' defines the non-equilibrium driving force for crystallization since μ^f is the chemical potential of a fluid cell and μ^s the chemical potential of a solid cell. The dimensionless driving force β' can be interpreted as being proportional to the supersaturation or undercooling.

In Fig. 2 the results of computer simulation studies are presented for different α values, where α is defined as

$$\alpha = \frac{4\phi}{kT} . \tag{3}$$

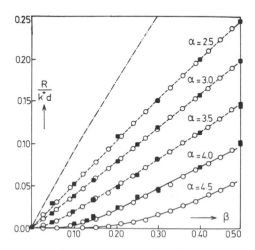

Fig. 2 Dimensionless growth rate versus β' for $0 < \beta' < 0.5$: (■) results of Gilmer and Bennema;[4] o special-purpose computer results (De Haan et al.;[6] solid lines: two dimensional birth and spread nucleation formula; dashed lines: empirical relation; uppermost line: maximal rate of growth (of a kinked or rough surface).

It can be seen that the higher α which means, keeping ϕ constant, the lower the temperature, the lower the growth rate. This is understandable because the lower the temperature the less rough the surface, the lower the sticking fraction. At about $\alpha \approx 3.2$ a change from a linear to non-linear R versus β' curve occurs. It is now well known that this change marks the so called roughening transition in an SOS or XY model. Independent of computer simulations the character of this phase transition is now clarified thanks to the work of theoretical physicists like Leamy and Gilmer,[7] van Beijeren,[8] Swendsen,[9] Müller-Krumbhaar,[10] van der Eerden and Knops[11] and Weeks and Gilmer.[12] For the practice of crystal growth the roughening transition can be defined as: at a SOS (Ising surface) a dimensionless temperature,

$$\theta^R = \left(\frac{2kT}{\phi}\right)^R , \tag{4}$$

exists such that, if for the actual dimensionless temperature, θ,

$$\theta < \theta^R, \ \gamma_{step} > 0 \tag{5}$$

and if

$$\theta > \theta^R, \ \gamma_{step} = 0 , \tag{6}$$

where we use the convention

$$\theta = \frac{2kT}{\phi} . \tag{7}$$

γ_{step} is defined as the edge free energy of a step. This is an order parameter playing a key role in theories on crystal surfaces and crystal growth. It will be clear that for $\theta < \theta^R$, or what comes to the same $\alpha > \alpha^R$, where α^R is defined as

$$\alpha^R = 4\left(\frac{\phi}{kT}\right)^R , \tag{8}$$

the surface remains, apart from statistical fluctuations, in essence flat over "infinitely" long distances. This is because large islands with a height of one atomic layer or holes cannot be formed due to the fact that $\gamma_{step} > 0$. If, however, $\theta > \theta^R$ surfaces roughen up and will lose their crystallographic orientation (hkl).

In case of growth ($\beta' > 0$) if $\theta > \theta^R$ or $\alpha < \alpha^R \approx 3.2$ the crystal face can grow barrierlessly, while if $\theta < \theta^R$ it has to grow by a layer mechanism, i.e., a two-dimensional nucleation mechanism or a spiral growth mechanism. The non-linear curves for $\alpha \geq 4$ can be fitted with a non-linear two-dimensional nucleation curve (Fig. 2 and Refs. 2,4,5,6 and references in Ref. 2). If the supersaturation becomes sufficiently high the size of a two-dimensional nucleus becomes equal to a few atoms (molecules). Then also roughening occurs. This is the so-called kinetical roughening caused by the driving force for crystallization. Also in this case crystal faces which at low supersaturations grow as flat faces below their roughening temperature lose their crystallographic orientation at higher supersaturation. This also depends on the "strength" of the connected net. The lower this strength the lower the edge free energy the lower the supersaturation for which kinetical roughening occurs.[13-16]

It is interesting to note that simulation of Ising models developed from 1966 till 1980 and in fact most essential points such as the roughening transition were clarified at the end of the seventies (see survey[18] and Refs. 8,9,11).

Yet in order to clarify specific theoretical and experimental issues still simulation studies on Ising models are carried out. As examples we mention studies of Liu Guang Zhao et al. on stress fields around dislocations,[19] modified Ising models,[20-21] and work carried out by Clarke et al.[22,23] and van Loenen et al. (see this volume) to simulate the growth of the reconstructed (001) face of silicon (see further section 3b of this chapter). We note that in the model of Clarke et al. cells and bonds are anisotropic, but the stacking to simulate the (001) reconstructed surface fulfills the space group symmetry $F^4/d\ 3\ ^2/m$ of the diamond structure. Similar simulations on anisotropic orthorhombic structures were carried out before.[24,25]

1b.   Two-Dimensional Interface Models

The change from a linear so called continuous crystal growth mechanism obeying a linear R versus β' law to a non-linear law is a key issue in crystal growth theories and was already studied in the pioneering paper by Burton, Cabrera and Frank[26] and by Jackson,[27] Mutaftschiev,[28] and Voronkov and Chernov.[29] In those papers the interface as presented in Fig. 1a was reduced to one or two mixed solid-fluid layers. Such a layer can be considered as a two-dimensional mixed fluid-solid crystal. And to such a two-dimensional mixed solid fluid model the theory of Onsager for an order-disorder phase transition applies.[30] The work of Onsager was another pioneering paper, because it showed that a real order-disorder phase transition occurred. Moreover, Onsager was able to calculate the dimensionless order-disorder phase temperature θ^C exactly. Although the SOS model is definitely a better model for the interface crystal mother phase, because it allows for more layers and hence more fluctuations, it can be noted that upon comparing the dimensionless roughening temperature θ^R and θ^C for simple interfaces as given in Table 1, it can be justified to assume that these values are close to each other so that

$$\theta^R \approx \theta^C . \tag{9}$$

Table 1

	θ^C	θ^R
SC(100)	2.26918...	2.56
BCC(110)	2.26918...	-
BCC(110)SOS	-	2.8855390...
FCC(100)	2.26918...	2.2
FCC(111)	3.64148	3.0-4.0
Diamond(111)	1.518652	1.48

The BCC(110)SOS value was taken from the special model introduced by van Beijeren,[8] which could be solved exactly. We note that the character of the two phase transitions, namely a roughening transition in an SOS or SOS like model and an order-disorder phase transition in a two-dimensional Ising model, is quite different. Yet the numerical values of the transition temperatures are rather close. This must be attributed to the fact that the fluctuations within the top layer give the highest contribution to the free energy of a step.

1c.   Integration of Statistical Mechanics and Crystallography

In the following the results of an exact theory will be discussed which allows for the calculation of θ^C for complex surfaces of real complex crystals. This can be considered as a kind of generalization of the Onsager theory. Resulting Ising temperatures can be used to calculate roughening transitions occurring on particular surfaces and to predict the shapes of crystal growth forms. We refer for a full treatment of this theory to the paper of Rijpkema, Knops, Bennema and Van der Eerden[31].

So in the following we replace as it were a SOS like surface model by a two dimensional Ising or lattice gas model. This replacement has an ad hoc character. This ad hoc replacement allows us to make a connection between the world of real crystal structures and statistical mechanical Ising models. This ad hoc replacement would be unacceptable if details of the character of the roughening transition (or related type of transition) of a particular interface crystal face (hkl)-mother phase were the subject of a theoretical and/or experimental study. In this case molecular dynamical studies are needed and the domain of validity of the theory of roughening transition, the theory of order disorder phase transition in a two-dimensional Ising model and the concept of surface melting (which is beyond SOS or two-dimensional Ising models) has to be studied (work in progress, see references at the end of subsection 2).

In this section it will be discussed how the dimensionless roughening temperature $\theta^R_{(hkl)}$ which will be replaced by the Ising temperature $\theta^C_{(hkl)}$ (assuming that $\theta^R_{(hkl)} \simeq \theta^C_{(hkl)}$) can be used as a measure for the relative morphological importance of faces (hkl) on the growth forms of crystals. In addition $\theta^C_{(hkl)}$ can be used to calculate the temperature at which a certain interface crystal-mother face becomes rough. For a demonstration how the theory works and an actual test of the predictions we refer to more than ten recent papers. In most cases the agreement between predictions and observations is quite satisfactory (see next subsection 2 where references are given in Section 2.b).

We note that the relative morphological importance MI of a particular crystal face (hkl) or better form {hkl} on crystal growth forms is a relative statistical measure of the relative frequency of occurrence of this form and the relative size of the faces of this form. The higher the frequency of occurrence and the higher the relative size on a set of crystals, the higher MI (A form {hkl} is a set of symmetrically equivalent faces (hkl)).

The interface crystal mother-phase may be considered as an interface with a thickness of one atomic layer. This layer may be considered as a two-dimensional Ising lattice (gas). How to make connections between the world of perhaps very complex crystal structures and their surfaces and the world of two-dimensional Ising models will be discussed in Section 2, when the so-called Hartman Perdok theory will be treated in a nutshell.

We note that till now crystals were mostly considered as structures consisting of cells having the two states: spin up, spin down or solid and fluid. Between the cells interaction energies due to "contacts" between cells occurred. In the following, cells will be reduced to centres of gravity or points and these points will be called atoms, ions or simply points. Contacts between cells will be reduced to covalent like chemical bonds represented by strings. The energy between the ith and the jth points is indicated by ϕ_{ij}. The atoms or points can have again two properties solid (s) or fluid (f). We will first treat a mean field model and next discuss the results of an exactly soluble model for the order-disorder phase transition. The mean field model has the advantage that the physics of this model can be easily understood and that first, second and in fact all-"neighbor" interactions can be taken into account.

1d. Mean Field Approximation and Its Implications

Let us suppose to have a two-dimensional net with a composition given by

$$A_{\nu A}B_{\nu B}C_{\nu C} \ . \tag{10}$$

All kinds of interactions between atoms A, B and C etc.: interactions in the first coordination sphere, second, third, etc. coordination spheres between the atoms A, B, C are supposed to occur. Assuming that a fraction x of the atoms A, B, C etc. are in a solid state and a fraction (1-x) in a fluid state, then as can be shown, using as a reference state a pure solid crystal consisting of xM solid "molecules" or stoichiometric units, $A_{\nu A} B_{\nu B} C_{\nu C}$ and a pure fluid crystal consisting of (1-x) M fluid "molecules", the total free energy of the mixed crystal is given by:

$$F_{tot} = \Delta F_{mix} + F_{ref} = \frac{M}{\beta} \ [(\Sigma \nu_k)(\alpha_k)x(1 - x) + x\ln x +$$

$$(1 - x)\ln(1 - x) + \Sigma_k \ \nu_k(x\mu_k{}^{\circ S} + (1 - x)\mu_k{}^{\circ f})] \ . \tag{11}$$

Equation (11) can be derived using the mean field, zeroth order or Bragg-Williams approximation. In Eq. 11 $\beta = 1/kT$. M is the number of stoichiometric units or molecules $A_{\nu A} B_{\nu B} C_{\nu C} \ldots$, ν_k represents the stoichiometic coefficient of the kth atom, where k represents A, B, C atoms etc., x is the fraction of the solid and (1-x) the fraction of fluid atoms in the crystal. $\mu_k{}^{\circ S}$ is the standard chemical potential of the kth type of atom if it is in the solid state and $\mu_k{}^{\circ F}$ the corresponding standard chemical potential of a fluid atom. The factor α_k of the kth type of atoms is defined in the following way

$$\alpha_k = \beta \ \underset{l_k}{\Sigma} \ (\phi 1_k{}^{sf} - \frac{1}{2} (\phi 1_k{}^{ss} + \phi 1_k{}^{ff})) \ ; \tag{12}$$

l_k refers to the lth interaction of an atom of the kth type of a first or next etc. neighboring atom, k = A, B or C etc.

Equation (11) can be derived in the following way without using statistical mechanics explicitly. Take as a reference a purely solid crystal consisting of xM molecules and a purely fluid crystal consisting of (1-x) M molecules. Then the reference free energy F_{ref} is given by

$$F_{ref} = M(\underset{k}{\Sigma} \nu_k) \; (x\mu_k°^S + (1 - x) \; \mu_k°^f) \; . \tag{13}$$

Introducing the ad hoc assumption, which is essential for Ising models, that the chemical potential (corresponding to the average internal energy) is independent of the ss, ff or sf bonds formed, we find after mixing the solid and the fluid crystal Eq. (11). This can be seen as follows: the essence of the very well known mean field, zeroth order or Bragg-Williams approximation can be summarized as follows: (i) an atom "sees" or "feels" in its direct environment the same composition of atoms as the average composition of the crystal (this implies that clustering is ruled out in our case) and (ii) fluid and solid crystal mix ideally, giving rise to the ideal entropy of mixing (in addition we mention that only pair interactions are presupposed and no three point, four point interactions etc.).

So every atom of the kth kind "feels" any sf pair interaction with a weight of (1-x), the fraction of fluid atoms, surrounding a given solid (or fluid) atom. The total number of ϕ^{sf} bonds formed due to mixing for the atoms of the kth type is given by ν^k Mx(1-x) and the total energy of these bonds is given by

$$M(\Sigma \nu_k) \; \underset{k}{\Sigma} \; \phi_{1_k}^{sf} \; x(1 - x) \; . \tag{14}$$

This explains the first term of the right hand part of Eq. (11). It is easy to see that for each sf bond formed, one half ss and one half ff bond must be broken, so the formation energy for each sf bond is given by

$$\phi_{1k}^{sf} = \phi_{1_k}^{sf} - \frac{1}{2} \; (\phi_{1_k}^{ss} + \phi_{1_k}^{ff}) \; . \tag{15}$$

This then explains Eq. (12) (see also Eq. (11)). Since an ideal entropy of mixing is presupposed we get Eq. (11).

After dividing Eq. (11) by the total number of atoms in the mixed solid-fluid crystal $M(\Sigma\nu_k)$ and multiplying by β we get

$$\overline{f} = \alpha x(1 - x) + x\ln x + (1 - x)\ln(1 - x) + x\overline{\Delta\mu°} + \mu°^f \; , \tag{16}$$

where $\overline{f}$ is the average free energy per atom in the mixed solid fluid crystal divided by kT. $\Delta\mu°$ is the average difference between the standard chemical potentials and μ of the average standard chemical potential of the fluid cells. α is an average over α_k's Eq. (12).

Employing Eq. (16) $\overline{f}$ can be plotted as a function of x for different α values (Fig. 3) according to the very well known plots generally used in crystal growth, first presented by Jackson.[27] In this case $\Delta\mu°$ has to be chosen to be zero (if this were not so the crystal would grow or dissolve in absence of a driving force for crystallization). The entropy function $x\ln x + (1-x)\ln(1-x)$ has a (flat) minimum for x = 1.2. For x → 0 and x → 1 the slopes become -∞ and +∞ respectively. The energy function $\alpha x(1-x)$ gives for $\alpha > 0$ a mountain parabola with a maximum x = 1/2. Adding the mountain parabola energy function and the entropy function, two minima occur corresponding to a

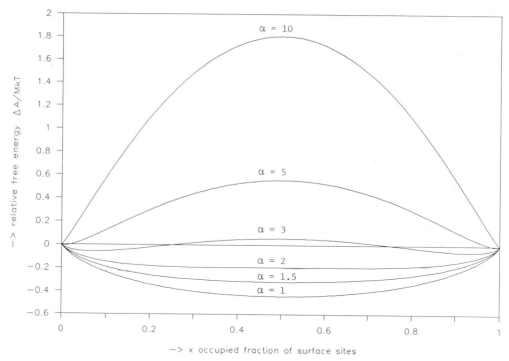

Fig. 3 Free energy f versus x for the single level Bragg-Williams approximation for different α values.

solid phase with a little fluid and a fluid phase with a little solid. If α decreases (i.e. the temperature increases if the bond energies are supposed to be independent of the temperature), the mountain parabola shrinks and the minima shift to the middle where x = 1/2. If α decreases further a critical point develops where only one phase, half solid half fluid occurs, with a composition x = 1/2. Due to the mean field approximation $\alpha^c = 2$.

The mean field approximation gives values of α^c which may be 100% wrong. It does give a good picture of the result of the competition between the ordering energy αx(1-x) (with α > 0) and the disordering entropy of mixing function, xlnx + (1-x)ln(1-x) leading to two phases below a certain critical temperature (or $\alpha > \alpha^c$) and one phase above this temperature (or $\alpha < \alpha^c$). In the following the results of an exact calculation will be discussed for the order-disorder phase transition, which also follows from the mean field approximation. It will then be discussed how the Ising temperature can be calculated precisely.

The results of the two-dimensional mean field model clearly demonstrate the phenomena of surface roughening (or order disorder phase transition). Below a certain critical temperature or ($\alpha > \alpha^c$), a two-dimensional layer consists of solid and fluid domains (phases), separated by steps having an edge free energy larger than zero. If $\alpha < \alpha^c$ for the two-dimensional layer, edge free energies of steps vanish because solid and fluid domains become one mixed solid fluid phase (note the analogy with surface roughening Eqs. (5),(6)). If a supersaturation is imposed on such an interface and if $\alpha > \alpha^c$ a layer growth mechanism occurs and if $\alpha < \alpha^c$ growth as a rough face occurs.

It is an advantage of the mean field approximation that all interactions: first, next-nearest neighbors, etc. to the interactions with atoms "infinitely far away" can be taken into consideration. This may be of importance if Coulomb interactions - as for most inorganic crystals - play a role in the bond structure of crystals. Results corresponding to Madelung like calculations and mean field approximations are compatible.

In the considerations given above we had in mind two-dimensional nets (crystals). But this was never used explicitly. So the considerations given above apply to three-dimensional crystals as well. We will not use this in this chapter.

We note that the mean field approximation applied to Ising like models finds a very wide application in all kind of branches of solid state physics, chemistry, solution chemistry (think about the regular solution theory), surface physics, etc.

1e. Exact Solution for the Calculation of the Ising Temperature θ^c for Complex Two-Dimensional Connected Nets

In the paper of Rijpkema, Knops, Bennema and van der Eerden a theory or method is developed to calculate dimensionless Ising temperatures for two-dimensional rectangular connected nets. We refer for the theory to the original paper.[31] The principle of the method goes back to the method of Kadanoff and Ceva[32] and can be explained as follows: In Fig. 4 a macroscopic picture of a connected net is presented. Within this connected net a monomolecular step is enforced by a screw dislocation which creates a step at A and a screw dislocation of opposite sign which annihilates it at B. Making a step means changing along the step the sign of the couplings J_{1k} or ϕ_{1k}. ϕ_{1k} has the shape of Eqs. (1,15). This comes to the same as breaking bonds along the step. Now the edge free energy step of the enforced step can be defined in the following way:

$$\frac{Z_{step}}{Z} = \exp(-\beta\gamma_{step}) = \exp[-\beta(F_k - F)] \ . \tag{17}$$

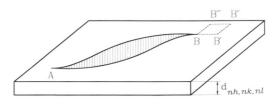

Fig. 4 Connected net with a screw dislocation (operator) to create a step at A and a screw dislocation of opposite sign to annihilate a step at B. B', B" and B''' are adjacent equivalent points of B. The thickness of the connected net is the interplanar distance corrected for the extinction conditions of the space group $d_{nh,nk,nl}$.

Here γ_{step} is the edge free energy, $\beta = 1/kT$. F_k is the free energy of a step of the system with the enforced step and F of the corresponding step-free system.

$$Z_{step} = \sum_{s_i} \exp(-\beta E_K) .$$ (18)

Z_{step} is the partition function of the system with step and Z of the system without a step. Equation (17) is based on the idea that the step free energy should be defined in reference to a step-free system. It is impossible to calculate the free energy of a step. It is, however, possible to define correlations between the step free energy starting at the same point A and ending in a point B and the three points B', B" and B''' adjacent to point B (see Fig. 4). At the critical point the fluid and the solid phases become one phase and the edge free energy vanishes (here again the same analogy with the vanishing of γ at the roughening transition occurs (Eqs. (5),(6)). Since at large distances correlations vanish at the Ising temperature a matrix equation can be found for each connected net and provided the ratio of bond energies is given, the Ising temperatures for each net can be calculated (within the accuracy of the computer program) exactly (see Ref. 31). A price has to be paid for this: contrary to the mean field approach where all second, third etc. neighbor interactions can be taken into account now only first-nearest neighbors can be taken into account.

It is amazing how well first nearest neighbor models seem to work in order to explain crystal growth and morphology data. A possible justification for this will be published elsewhere (work in progress).

1f.   Calculation of Advance Velocities of Steps with Different Crystallographic Directions

Information about structures of more or less complex nets can be used to carry out calculations on the advance velocities of steps with different crystallographic directions. Coupled sets of differential equations were developed using a pair approximation approach.[33,34] Relative bond energies for the system Yttrium Iron Garnet (YIG) were obtained using the proportionality hypothesis mentioned above, which implies that

$$\phi_i : \phi_j = \phi_i^{ss} : \phi_j^{ss} ,$$ (19)

where ϕ_i^{ss} and ϕ_j^{ss} are two (negative) bond energies referenced to vacuum and ϕ_i and ϕ_j the corresponding (positive) bond energies as given by Eqs. (1) and (15).

In order to find the absolute value of the bond energies the experimental fact was used that on a sphere of a single crystal of YIG growing from a lead oxide rich flux the weakest connected nets {332} still occur.[33,34] This implies that the actual temperature is close to the roughening temperature. Hence

$$\theta_{\{332\}}^R \approx \theta_{\{332\}}^C \approx \theta_{\{332\}}$$ (20)

where θ^R and θ^C are the dimensionless roughening and Ising temperatures and θ the actual dimensionless temperature of the interface (see Eq. (7)) respectively. Expressing θ's in reference to the strongest bond ϕ_{str}, if T is known it follows from Eq. (20) that the strongest bond energy ϕ_{str}

is known and due to the proportionality hypothesis (Eq. (19)) also all other ϕ's, ϕ_i, ϕ_j etc. can be calculated. Using this set of bond energies a very good agreement between observed relative growth rates of steps with different crystallographic directions with a height of about 4.5 to 9 Å and calculated values was obtained. It had to be assumed that Y^{3+} ions went into the lattice a thousand times faster than Fe^{3+} ions, which could be justified. Both the dominant faces of the forms {110} and {211} were investigated. Again it may be concluded that Ising models taking only first nearest neighbors into account give a satisfactory description of all kinds of crystal growth phenomena.

In the following we will apply the theory to silicon and gallium arsenide structures. Then quite a new problem has to be faced, namely reconstruction. But first we will discuss the crystallographic morphological Hartman-Perdok theory briefly.

THEORY OF HARTMAN AND PERDOK; CRYSTAL GRAPHS AND CONNECTED NETS

2a. Procedure to Derive Morphology

So far only Ising interface models were discussed. Except in Section 1.f no connection was made with the world of crystallography as described for example by the crystallographic morphological theory of Hartman and Perdok.[35-39] In order to predict important faces on equilibrium forms and crystal growth forms, using the Hartman Perdok theory, structural centre-of-gravity models are the best tools, together with drawings of structural models, projections, nets etc. made by computers.

In order to derive the theoretical morphology (habit) of crystals the following procedure consisting of three steps has to be carried out, according to the Hartman-Perdok theory:

2a.1 Determination of Growth Units and Bonds from the Crystal Structure

After studying carefully the supposedly known crystal structure, bond energies between growth units are calculated.

Growth units are ions, molecules, complexes, etc. occurring in the mother phase from which the crystal grows. Bonds within the growth units are not taken into consideration since only the overall bonds between growth units are relevant. These bonds may consist of many atom-atom pair potentials of atoms which are a part of the growth units. In principle, growth units are defined after considering the formation of the complexes occurring in the mother phase and comparing these complexes with the crystal structure. Very often growth units are not known and then ad hoc hypotheses concerning the growth units must be made. Also alternative growth units can be chosen which may lead to different morphologies. Comparing the alternative theoretical morphologies with the real one, conclusions concerning the structure of growth units in the mother phase may be drawn a posteriori.

After growth units are defined, interaction energies referenced to vacuum (ϕ_{1k}^{ss}) between growth units have to be calculated. Most sophisticated calculations are carried out starting with precise electron density maps (see Ref. 40). In case of crystal structures of organic crystals like naphthalene, paraffin, fat-crystals, etc. overall bonds between molecules are calculated by adding up all pair potentials such as C-C, H-H and C-H potentials, occurring between molecules, calculated for each of these pairs from reliable Lennard Jones like potentials, based on precise atomic distances which follow from structure determinations by x-rays (work in progress to be published, see also Section 2.b).

2a.2 Determination of Crystal Graph

Once the set of overall bonds between molecules or growth units is known a crystal graph has to be defined. A graph is a set of elements (points) with relations between them. A crystal graph is an infinite set of points fulfilling the symmetry of one of the 230 space groups with relations (bonds) between the growth units. In order to define the crystal graph of the crystal structure under consideration the molecules or growth units are reduced to centres of gravity or points and these form the set of points of the crystal graph. Next the strongest (first nearest neighbor) bonds are chosed as relations between the points.

2a.3 Determination of Connected Nets, F Faces

From the crystal graph defined in this way the so called connected nets have to be determined. For complex crystal structures this may be a tedious job. Connected nets have an overall thickness d_{hkl} which is the interplanar distance of the net planes, corresponding to the crystallographic plane (hkl), corrected for the extinction conditions of the space group of the crystal structure and crystal graph. A connected net is defined as a net where all points within the thickness d_{hkl} are connected by all kinds of arbitrary uninterrupted paths of bonds. Such connected nets show an order-disorder phase transition at a definite Ising temperature to be calculated with the method presented in Ref. 31. We note that the requirement of connectedness is essential for the occurrence of an order-disorder phase transition and roughening transition, because within a slice of thickness d_{hkl}, also non connected nets may occur and these according to Eqs. (5), (6) roughened up at T=0°K since in at least one direction $\gamma_{step} = 0$. This especially occurs if indices (hkl) are large and d_{hkl} small.

Within connected nets paths of uninterrupted bonds occur having a period [uvw] of the lattice. Such paths of bonds are called in the Hartman Perdok theory PBC's (Periodic Bond Chains). Connected nets consist of at least two different connected sets of parallel PBC's. We mention as an example the diamond structure or crystal graph with first nearest neighbor tetrahedron bonds. In this case the face diagonals <110> form zig zag chains or PBC's (see Section 3).

2b. The Role of the Interplanar Thickness d_{hkl} for Connected Sets

In the following we will explore the role of the interplanar thickness d_{hkl} in some more detail. Assume that we have a crystal graph of a particular crystal structure and that we want to determine the connected nets corresponding to a certain crystallographic direction (hkl). Let us assume for the time being that we can find an unambiguous cut corresponding to the elementary two-dimensional cell of the two-dimensional net having the surface area of the mesh and having the lowest cut energy of possible alternative cuts. It may well be that the surface changes due to reconstruction or interaction with the mother phase. But for the time being we will assume that somehow a "cheapest cut" corresponding to the lowest surface energy can be identified. Then due to the crystallographic periodicity, after a repeat distance d_{hkl} exactly a same cut occurs (which may for example be rotated due to a screw axis). If no connected net can be made in fact no unambiguous cheapest cut occurs and a stepped (S) or kinked (K) face occurs which is rough at T = 0°K. (We refer to discussions on this subject in Refs. 2 and 35-39). We note that for complex crystal structures alternative connected nets parallel to one face (hkl) are possible. After calculating the Ising temperature or E^{slice} (the energy per growth unit within a slice), it can be calculated which slice is the strongest and in this case it is assumed that the strongest slice will dominate the crystal growth form. Such a procedure was followed for garnet,[31,33,34,38,39] apatite[41,42] bismuth germanium oxide,[43] gypsum,[44] potassium titanyl phosphate (KTP),[45] 1,2,3, high-T_c crystals,[46] naphthalene (studying the phenomenon of kinetical roughening[13-16]), potassium hydrogenephtalate,[47] paraffin crystals,[48] fat crystals[49] and β-lactose.[50]

2c.   Determination of Ising Temperatures; Rectangularization of Connected Nets

Once the connected nets are determined Ising temperatures have to be calculated using a computer program giving a solution of the matrix equation which follows from the theory.

The formalism derived in Ref. 31 is a formalism developed for rectanglar nets. Very often connected nets are not rectangular. It is however, possible to transform non-rectangular connected nets into rectangular connected nets, having the same partition function as the original net by allowing bonds of infinite strength and zero energy. In order to "rectangularize" nets, points of the original net may be split into two with a bond of infinite strength connected these points. As an example we give the transformation of an hexagonal net consisting of three types of bonds into a rectangular net (see Fig. 5).

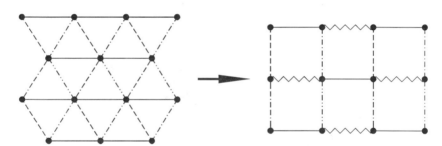

Fig. 5 Transformation of a non-rectangular net having also bonds with zero energy into a rectangular net by splitting nodes. The new points are connected by bonds of infinite strength. Wiggled bonds are bonds with an infinite strength.

Sometimes it is not possible to calculate Ising temperatures from connected nets, because the connected nets are not real planar in the sense that crossing bonds occur or that nets consist of subnets which are within the slice thickness d_{nhnknl} stacked upon each other and connected to each other. In this case by omitting (weak bonds) or adding extra (strong) bonds a weaker and a stronger rectangular net can be made giving an upper and a lower bound of the Ising temperature.

2d.   Construction of Crystal Growth Forms

Under well defined conditions tiny crystals, with dimensions of microns or less, will get in finite times of say hours, by statistical fluctuations the so called equilibrium shape. This corresponds to a minimum to the free energy of crystal and environment. In this case shapes of crystals will be unambiguous. Also from the theory of equilibrium such forms can be constructed unambiguously, if in the surface (free) energies of the different faces (hkl) were known, using the so called Wulff plot.

Contrary to equilibrium forms unambiguous crystal growth forms do not exist and all scientists working in the field of crystal growth know that each crystal is unique and also has a unique shape. It is, however, possible to replace the actual growth forms of ten or more crystals of a certain crystalline compound grown under the same conditions by an average growth form fulfilling the point group symmetry. It is also possible to formulate ad hoc rules for the construction of idealized (theoretical) crystal growth forms. So for example the hypothesis can be introduced: the higher the Ising temperature θ_{hkl}^{c} in a series θ_1^{c}, θ_2^{c}, θ_3^{c}, ... of connected nets, obtained from a crystal graph, the stronger the connected net, the lower its growth rate R_{hkl} and the higher its morphological importance (MI). MI is a statistical measure for the relative frequency of occurrence of a crystal face, parallel to a connected net, and its relative size. The larger the frequency and/or the size the larger MI.

So a crystal growth form may be constructed from a Wulff like plot, taking the vectors R_{hkl}, representing the growth rate of a face (hkl), proportional to $(\theta_{hkl}^{c})^{-1}$, or if the actual dimensionless θ can be calculated, taking $R_{hkl} \sim (\theta^c - \theta)_{hkl}^{-1}$. For complex nets θ and θ^c are defined as respectively:

$$\theta = \frac{2kT}{\phi_{str}}, \quad \theta^c = \left(\frac{2kT^c}{\phi_{str}}\right) \tag{21}$$

where ϕ_{str} is the strongest bond of the crystal graph. In order to obtain Ising temperatures of connected nets the ratios of bond energies must be known. According to the so called proportionality hypothesis, these are assumed to be proportional to the overall ϕ^{ss} bond energies (see Eqs. (1), (15) and (19)).

A more traditional way to construct growth forms is to take R_{hkl} proportional to E_{hkl}^{att}, where according to the concepts used in the Hartman Perdok theory E_{hkl}^{att} is the energy per growth unit which is cut by the boundary of the connected net. A better way to define E_{hkl}^{att} is to define it as the energy to remove a growth unit in its crystallographic position from a flat surface (hkl). E_{hkl}^{slice} is defined as the energy of a growth unit within a slice. Using these concepts of the Hartman Perdok theory it can be seen that

$$E_{hkl}^{att} + E_{hkl}^{slice} = E^{cr} \tag{22}$$

where E^{cr} is the crystallization energy, which is a bulk property; which is independent of the crystallographic face.

The justification that there is a parallel relationship between R_{hkl} and E_{hkl}^{att} was demonstrated in a paper by Hartman and Bennema.[51] In order to analyze possible growth forms in practice R_{hkl} is taken to be proportional to E_{hkl}^{att}. This recipe has been applied with great success by Dr. Z. Berkovitch Yellin from the group of Professor Lahav and Professor Leisserovitch from the Weissman Institute in Rehovot, Israel[40] and by Hartman and coworkers.[35-37] In the near future the recipe $R_{hkl} \sim (\theta_c - \theta)^{-1}$ will also be tried. (Work in progress).

429

2e. Morphology of Quasi Crystals and Modulated Crystals

Recently the classical law describing crystal faces by the law of rational indices leading nowadays to a description with Miller indices (hkl), was generalized for faces with four indices (hklm) occurring on modulated crystals.[52-57] It was tremendous success that Dam, Janner and Donnay could describe the remaining 82 crystal forms (out of the total number of the extremely large number of 92 crystal forms occurring on crystals of the mineral crystal calaverite ($Au\ Te_2$)), using again four integers {hklm}.[58,59] This was possible since calaverite turned out to be modulated.[52]

Recently it was shown that the theory of Hartman and Perdok could be in principle generalized to describe crystal faces occurring on quasi-crystals,[60] with six indices {hklmno}.

As noted above in Section 2.a in all complementary morphological theories the interplanar thickness d_{hkl} plays a key role. The recent findings that on modulated and quasi-crystals, crystal faces to be indexed with four or more indices (hklm....) occur, can be considered as a generalization of the oldest morphological theory of Bravais, Friedel, Donnay and Harker stating that the higher the interplanar thickness the higher the morphological importance (MI) of a face (hkl). From a modern point of view this law can be interpreted in the following way; the higher d_{hkl} the higher the energy content, the higher the roughening temperature, the lower the growth rate, the higher MI.

MORPHOLOGY OF SILICON AND GALLIUM ARSENIDE

3.a Morphology of Si and GaAs According to Hartman and Perdok

Let us apply the Hartman Perdok theory to Si and GaAs. As usual in the Hartman Perdok theory projections are made in the most important PBC directions corresponding to the lowest translation distances. Taking the only relevant first-nearest-neighbor interactions - the tetrahedron bonds-into account it can be seen that the well known zig zag chains with directions alternatively in the [$1\bar{1}0$] and [110] a/4 lower or higher in reference to each other are the only PBC's. (a is the dimension of the elementary cell). Within the proper slice thicknesses only the (111) faces correspond to connected nets. Both for silicon and GaAs all other faces do not correspond to connected nets and should grow as rough faces. As an example we mention that due to the extinction conditions of the space group the interplanar distance for the face (001) corresponds to a/4. d_{004} is a/4. Without using extinction conditions, this also can be seen at once by partitioning the projections of Figs. 6a,b into equal "cheapest cuts". (The spacegroups for Si and GaAs are $F^4 1/d$ and $\bar{3}\ 4/m$ and $F\ \bar{4}\ 3m$ respectively from which the extinction conditions can be derived determining d_{nhnknl} using the International Crystallographic tables).[61]

So we arrive at a very serious discrepancy between the predictions of the Hartman Perdok theory and reality, where flat faces (100), (110), (113) etc. are observed both on silicon and GaAs (see also the work of Lagally, Vvedensky et al. and van Loenen et al. with beautiful STM pictures of large flat areas of the (001) face of Si (this volume). Now using the Hartman Perdok theory, the morphology of crystals for almost all crystals is quite well predicted. So Si and GaAs show a very exceptional behavior.

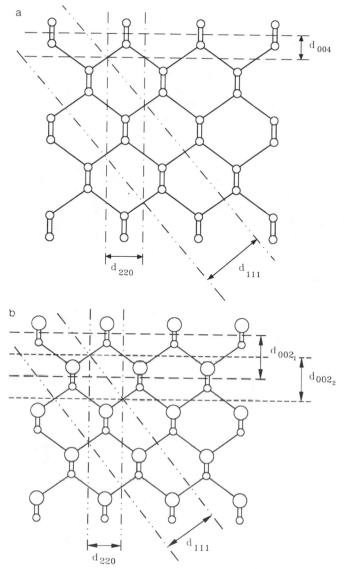

Fig. 6 a) Projection of the diamond structure along [110]. Each circle
represents a carbon atom. Adjacent PBC's are bonded to a
connected net in slice d_{111}, not in d_{220}. In layer d_{001} no
complete PBC occurs. So (220) and (110) are not connected nets.
b) Projection of the GaAs structure along [110]. ○ As, o Ga. Now
the slice thickness of (001) is doubled as compared to the slice
thickness of the diamond structure. But again only (111) is the
only connected (polar) net. Two alternative nets $(001)_{1,2}$ limited
above by As or Ga can be distinguished.

3.b   The Role of Surface Reconstruction

Now it is suggested from Molecular Dynamical Simulations, using
instead of only first-nearest-neighbor bond pair potentials also
Stillinger Weber (like) three point potentials, that the face (001) will

be reconstructed and the reconstruction will be a 2x1 reconstruction. (See van der Eerden et al.[62,63])

So obviously due to the occurrence of reconstruction, (001) becomes a kind of pseudo connected net which may have a roughening transition temperature. The beautiful STM pictures of large areas with flat surfaces strongly suggest this for a reconstructed surface. The problem now is how we can imagine a connected net.

A problem related to this is what the relation will be between the anisotropic Ising models and the blocks or cells used by Clark et al.[22,23] and van Loenen et al. (this volume) for a successful simulation of MBE experiments and growth of (pairs of) Si atoms. It is interesting to note that the Ising blocks have a height of a/4 in agreement with Fig. 6a. Comparing the unreconstructed Fig. 6a with the 2x1 reconstructed Fig. 7b it is not easy to imagine how a connected net will develop. But obviously due to "under earth bonds" (i.e. under the steps and the blocks with a height of a/4) a kind of connected net develops with a height of a/4, which may give effectively a connected net.

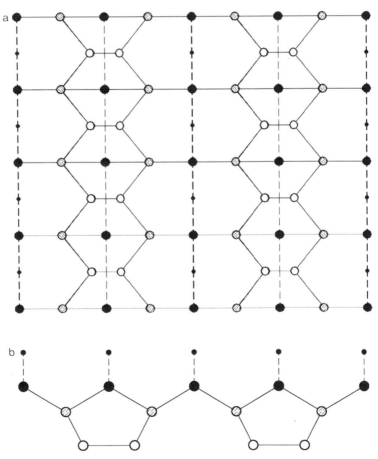

Fig. 7 a) Reconstructed surface of silicon, seen from above. b) Seen in the [110] direction.

In order to clarify this problem, more study is necessary. This Section 3.b was only meant to focus attention on the discrepancy between the results of the classical morphological theory of Hartman and Perdok and its modern extensions and the reconstructed surfaces. A logical integration of the concept of connected net and roughening transition on one hand and of a reconstructed surface on the other hand has to be developed, but as yet it seems that we have not reached this ideal.

We note that apart from real bonds of the bulk crystals, which are cut and replaced by sf bonds, obviously also new or pseudo bonds may be formed at the surface. First of all we have to consider reconstruction. But as shown by Lapujoulade et al. (this volume) due to all kind of interactions between trains of steps, pseudo bonds are as it were formed making pseudo connected nets which show a real roughening (or order-disorder) phase transition. Also impurities may form extra bonds and hence pseudo connected nets, showing a roughening transition (see Refs. 36,37).

Acknowledgements

The authors would like to thank Dr. G. H. Gilmer, Dr. H. J. F. Knops and Dr. J. P. van der Eerden for stimulating discussions on the background of Ising models.

REFERENCES

1. P. Bennema and G. H. Gilmer, in: Crystal Growth, an Introduction, P. Hartman, ed. North Holland 1973, p. 63-327.
2. P. Bennema and J. P. van der Eerden, in: Morphology of Crystals, Part, A, Ed. I. Sunagawa, Terra Scient. Publ. Comp. Tokyo D. Reidlel Publ. Comp. Dordrecht 1987, p. 1-75.
3. D. E. Temkin, in: Crystallization Processes Consult. Bureau, New York, p. 15 (1966).
4. G. H. Gilmer and P. Bennema, J. Appl. Phys. 43, 1347 (1972).
5. G. H. Gilmer, J. Crystal Growth 36, 15 (1976).
6. S. W. H. de Haan, V. J. A. Meeuwesen, B. P. Veltman, P. Bennema, C. van Leeuwen, and G. H. Gilmer, J. Crystal Growth 24, 491 (1974).
7. H. J. Leamy and G. H. Gilmer, J. Crystal Growth 24/25, 766 (1974).
8. H. van Beijeren, Phys. Rev. Lett. 38, 993 (1977).
9. R. Swendsen, Phys. Rev. B 17, 3710 (1978).
10. H. Müller-Krumbhaar, in: Current Topics in Material Science, Eds. E. Kaldis and H. J. Scheel, North Holland (1978), p. 1-46.
11. J. P. van der Eerden and H. F. J. Knops, Phys. Lett. 66A, 334 (1978).
12. J. D. Weeks and G. H. Gilmer, in: Adv. Chem. Phys. 40, 157 (1979).
13. H. J. Human, J. P. van der Eerden, L. A. M. J. Jetten and J. G. M. Odekerke, J. Crystal Growth 51, 589 (1981).
14. L. A. M. J. Jetten, H. J. Human, P. Bennema and J. P. van der Eerden, J. Crystal Growth 68, 503 (1984).
15. M. Elwenspoek and J. P. van der Eerden, J. Phys. A. Math. Gen. 20, 669 (1987).
16. M. Elwenspoek, P. Bennema and J. P. van der Eerden, J. Crystal Growth 83, 297 (1987).
17. A. A. Chernov and J. Lewis, J. Phys. Chem. Solids 23, 2185 (1967).
18. J. P. van der Eerden, P. Bennema and T. A. Cherepanova, Prog. Crystal Growth Charc. 1, 219 (1978).
19. Liu Guang Zhao, J. P. van der Eerden and P. Bennema, J. Crystal Growth 58, 131 (1982).
20. Liu Guang Zhao, J. Crystal Growth 89, 478 (1988).

21. Jenn-Shing Chen, Nai-Ben Ming and F. Rosenberger, J. Chem. Phys. $\underline{84}$, 2365 (1986).
22. S. Clarke, M. R. Wilby, D. D. Vvedensky and T. Kawamura, Phys. Rev. $\underline{B40}$, 1369 (1989).
23. S. Clarke, M. R. Wilby, D. D. Vvedensky and T. Kawamura, Appl. Phys. Lett. $\underline{54}$, 2417 (1989).
24. D. J. van Dijk, C. van Leeuwen and P. Bennema, J. Crystal Growth $\underline{23}$, 81 (1974).
25. J. P. van der Eerden, C. van Leeuwen, P. Bennema, W. L. van der Kruk and B. P. Th. Veltman, J. Appl. Phys. $\underline{48}$, 2124 (1977).
26. W. K. Burton, N. Cabrera and F. C. Frank, Phil. Trans. Roy. Soc. $\underline{A243}$, 299 (1951).
27. K. A. Jarkson, in: Liquid Metals and Solidification, Am. Soc. for Metals Cleveland (1958).
28. B. Mutaftschiev in: Adsorption et Croissance Cristalline (Centre Nationale de la Recherche Scientifique, 1965) p. 231.
29. V. V. Voronkov and A. A. Chernov, Structure of Crystal/Ideal Solution Interface, Proc. Int. Conf. on Crystal Growth, Boston, Ed. H. S. Peiser 593-597 (1966).
30. L. Onsager. Phys. Rev. $\underline{65}$, 117 (1944).
31. J. J. M. Rijpkema, H. J. F. Knops, P. Bennema and J. P. van der Eerden, J. Crystal Growth $\underline{61}$, 295 (1983).
32. L. P. Kadanoff and H. Ceva, Phys. Rev. $\underline{B\ 3}$, 3918 (1971).
33. T. A. Cherepanova, G. T. Didrihsons, P. Bennema and K. Tsukamoto, to be published in book on Oji Seminar (1985) ed. by I. Sunagawa (1990).
34. T. A. Cherepanova, P. Bennema, Yu. A. Yanson and K. Tsukamoto, to be published.
35. (a) P. Hartman and W. G. Perdok, Acta Cryst. $\underline{8}$, 49 (1955a).
    (b) P. Hartman and W. G. Perdok, Acta Cryst. $\underline{8}$, 521 (1955b).
    (c) P. Hartman and W. G. Perdok, Acta Cryst. $\underline{8}$, 525 (155c).
36. P. Hartman, Structure and Morphology In Crystal Growth: an Introduction, P. Hartman, ed. North Holland, Amsterdam 367 (1973).
37. P. Hartman, Modern PBC Theory in Morphology of Crystals, Part A, ed. I. Sunagawa, Terra Scient. Publ. Comp. Tokyo, D. Reidel Publishing Comp. Dordrecht 272 (1987).
38. P. Bennema, E. A. Giess and J. E. Weidenborner, J. Cryst. Growth $\underline{62}$, 41 (1983).
39. P. Bennema, in: Industrial Crystallization, ed. S. J. Jancic and E. J. de Jong Elsevier Science Publ. V. Amsterdam $\underline{84}$, 339 (1984).
40. Z. Berkovitch-Yellin, J. Am. Chem. Soc. $\underline{107}$, 8239 (1985).
41. R. A. Terpstra, P. Bennema, P. Hartman, C. F. Woensdregt, W. G. Perdok and M. L. Senechal, J. Crystal Growth $\underline{78}$, 468 (1986).
42. R. A. Terpstra, J. J. M. Rijpkema and P. Bennema, J. Crystal Growth $\underline{76}$, 494 (1986).
43. F. M. Smet, P. Bennema, J. P. van der Eerden and W. J. P. van Enckevort, J. Crystal Growth $\underline{97}$, 430 (1989).
44. M. P. C. Weijnen, G. M. van Rosmalen, P. Bennema and J. J. M. Rijpkema, J. Crystal Growth $\underline{82}$, 509 (1987).
45. R. Bolt, and P. Bennema, to be published in J. of Crystal Growth (1990).
46. L. E. C. van der Leemput, P. J. M. van Bentum, F. A. J. M. Driessen, J. W. Gerritsen, H. van Kempen, L. W. M. Schreurs and P. Bennema, J. of Microscopy $\underline{152}$, 103 (1988).
47. M. H. J. Hottenhuis, J. G. E. Gardeniers, L. A. M. J. Jetten and P. Bennema, J. Crystal Growth $\underline{92}$, 171 (1988).
48. P. Bennema, J. J. M. Rijpkema, Liu Xiang Yang, K. Lewtas, R. D. Tack and K. J. Roberts, to be published.
49. P. Bennema, S. de Jong, L. J. P. Vogels and K. J. Roberts, to be published.
50. R. A. Visser and P. Bennema, Neth. Milk Dairy J. $\underline{37}$, 109 (1983).
51. P. Hartman and P. Bennema, J. Crystal Growth $\underline{49}$, 145 (1980).

52. T. Janssen and A. Janner, Adv. in Phys. <u>36</u>, 519 (1987).
53. A. Janner, Th. Rasing, P. Bennema and W. H. van der Linden, Phys. Rev. Lett. <u>45</u>, 1700 (1980).
54. B. Dam and A. Janner, Z. Kristall. <u>165</u>, 247 (1983).
55. B. Dam, A. Janner, Th. Rasing, P. Bennema and W. H. van der Linden, Phys. Rev. Lett. <u>50</u>, 75 (1983).
56. B. Dam and A. Janner, Acta Cryst. <u>B42</u>, 69 (1986).
57. B. Dam and P. Bennema, Acta Cryst. <u>B43</u>, 64 (1987).
58. B. Dam, A. Janner and J. D. H. Donnay, Phys. Rev. Lett. <u>55</u>, 123 (1985).
59. A. Janner and B. Dam, Acta Cryst. <u>A45</u>, 115 (1989).
60. T. Janssen, A. Janner and P. Bennema, Phil. Magn. <u>B59</u>, 233 (1989).
61. International Tables for X-ray Crystallography, published for the International Union of Crystallography by the Kynoch Press Birmingham, England (1969).
62. J. P. van der Eerden, Liu Guang Zhao, F. de Jong and M. J. Anders, to be published in <u>J. Crystal Growth</u> (1990).
63. J. P. van der Eerden, A. Roos and J. M. van der Veer, to be published in <u>J. Crystal Growth</u> (1990).

STUDIES OF SURFACE DIFFUSION AND CRYSTAL GROWTH BY SEM AND STEM

J. A. Venables*+, T. Doust*#, J. S. Drucker+ and
M. Krishnamurthy+

* School of Mathematical and Physical Sciences,
University of Sussex, Brighton BN1 9QH, England
+ Department of Physics,
Arizona State University, Tempe, AZ 85287   USA
Present address: Department of Physics,
University of York, Heslington, York YO1 5DD, England

ABSTRACT

Surface diffusion, crystal nucleation and growth are being studied using UHV-scanning electron (SEM) and scanning transmission electron (STEM) microscopy-based techniques. A simple technique, biassed secondary electron imaging (b-SEI) has been demonstrated to be sensitive to intermediate layers at the 0.1ML level, and also to surface step topography. Auger electron spectroscopy (AES), on a lateral scale sufficient to probe the chemical composition of intermediate layers between islands, has been developed at the 0.05ML level. Crystallographic information in the form of reflection (RHEED) and transmission (THEED) electron diffraction is also available. These technique developments are briefly reviewed and examples given.

SEM studies of several Stranski-Krastanov growth systems have been made, and nucleation densities interpreted in terms of rate-equation models to derive atomic parameters. We have also used a shadow mask technique, in order to study surface diffusion in competition with evaporation (at high temperatures) and crystal nucleation, growth and annealing (at lower temperatures). Examples are given for Ag/Si(111).

Initial STEM studies have concentrated on cleaning electron transparent Si(100) samples in-situ and observing individual surface steps using b-SEI. A program of work on Ge/Si(100) has been started and preliminary results are shown.

1.   INTRODUCTION AND BACKGROUND INFORMATION

Surface diffusion, nucleation and growth processes are all important in determining the structure of thin films grown on surfaces. It is generally accepted that there are three growth modes possible in the simplest cases where interdiffusion does not take place. These are the island (or Volmer-Weber), layer (Frank-van der Merwe) and layer plus island (or Stranski-Krastanov) growth modes. The extensive experimental work, undertaken to determine and quantify the nucleation and growth processes in these modes, has been reviewed.[1,2]

Kinetics of Ordering and Growth at Surfaces
Edited by M. G. Lagally
Plenum Press, New York, 1990

By contrast, surface diffusion (in the context of thin films) has not been much studied. A useful review is available,[3] complementing a previous NATO meeting proceedings of a few years ago.[4] One of the problems of studying or modelling surface diffusion in detail is the existence of several diffusion mechanisms, and the possible dependence of diffusion coefficients on crystallographic face and direction, adsorbate concentration and structure, and on surface steps or other defects. Nucleation can also be, and often is, a heterogeneous process, so there is a potential difficulty in formulating simple theories which are testable.

Our approach to the theory has been to model these surface processes in terms of the minimum number of parameters, but to regard them as 'effective' parameters, which may be influenced by defects such as steps or impurities, or by complex cluster geometry.[5] If the diffusion is long range, and/or the nucleation density is very low, it is inevitable that such effects are involved. In some of the Stranski-Krastanov systems we have studied by SEM-based techniques the diffusion distance can be up to $100\mu m$; no-one to our knowledge has produced surfaces which are step - or impurity-free over such distances!

With the increasing spatial resolution of microscopic techniques it has become possible, at least in principle, to study the nucleation, growth and surface diffusion behavior of atoms and molecules on individual terraces, and to quantify their interaction with steps. This is seen dramatically in the recent Scanning Tunnelling Microscopy (STM) results featured at this workshop,[6] but also in the results of the Low Energy Electron Microscope (LEEM),[7] Transmission and Reflection Electron Microscopy (TEM and REM),[8] and the more restricted area studies based on Field Ion Microscopy (FIM).[9]

Our experimental approach has been to develop techniques based on secondary and Auger electrons in three directions. The first involves use of a simple new technique, biassed secondary electron imaging (b-SEI);[10] this is capable of visualizing sub-monolayer deposits on surfaces via changes in the surface electronic properties influencing the secondary electron yield, such as work function and/or band-bending. The second route is to provide microscopic chemical analysis via quantitative development of Auger Electron Spectroscopy (AES).[11,12] The third approach is to incorporate these surface sensitive techniques into a new high spatial resolution Scanning Transmission Electron Microscope (STEM),[13] while retaining the many other microanalytic facilities on a modern field emission STEM.

This paper is organized as follows. Section 2 briefly outlines the experimental techniques. Sections 3 and 4 review nucleation and surface diffusion work respectively, concentrating on one system Ag/Si(111). The extension to high spatial resolution using the STEM is described in section 5, where initial results for steps on Si(100) and for Ge/Si(100) are shown.

2.   EXPERIMENTAL TECHNIQUES AND EXAMPLES

The SEM experiments described were conducted in the Sussex UHV SEM.[14] The main techniques used were AES, RHEED and biassed-secondary electron imaging (b-SEI), as indicated schematically in Fig. 1. Of these techniques only the last needs some introduction. It is a very simple technique for observing sub-monolayer deposits via the change in secondary electron emission. The biassed images have been shown to give a lower detection limit of $\leq$ 0.1ML with a spatial resolution of

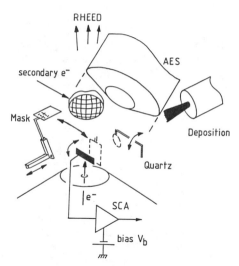

Fig. 1 Schematic diagram of the experimental chamber in the Sussex UHV SEM. The geometry of the holes can be seen on the mask which is shown in its 'flipped-back' position. The samples can be imaged using b-SEI or biased specimen current imaging. The latter technique requires the specimen current amplifier (SCA) illustrated.

$\leq$100nm.[10] The contrast for a 1ML Ag deposit on Si(111) at a bias $V_b \simeq$-500V is about 10%. A sequence of submonolayer line scans is shown in Fig. 2. In the room temperature (RT) deposit, essentially no diffusion is observed, the 20μm wide images being sharp at the 0.5 μm level, determined by the penumbra of the mask shadow. At higher temperatures, we see lateral diffusion, and also 'spikes' due to islands, which can be readily seen in detail in SEM images.[11,15] The experiments are performed in-situ as a function of deposition rate and substrate temperature, as described in sections 3 and 4.

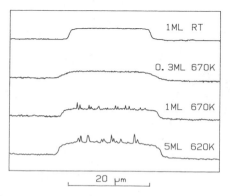

Fig. 2 Typical b-SEI line scans showing a range of contrast for different doses as well as the presence of islands at doses of 1 and 5 ML.

The STEM experiments described in section 5 are based on our development of the MIDAS instrument (A Microscope for Imaging, Diffraction and Analysis of Surfaces) at the NSF-ASU High Resolution Electron Microscopy Facility at Arizona State University.[13] As explained in more detail elsewhere,[16] the incorporation of secondary and Auger electron spectroscopy and imaging utilizes the principle of the 'magnetic parallelizer', in which the electrons spiral in a controlled magnetic field.

The MIDAS system has been 'stretched' beyond a standard STEM to allow us to detect these electrons, as illustrated in Fig. 3. There are two analysis chambers (AC1 and AC2), and the bores of the objective lens (O) contain lower and upper 'parallelizer' coils (LP and UP) terminated by adjustable magnetic apertures (arrows on Fig. 3). These apertures reduce the magnetic field abruptly to zero, causing the electrons to stop spiralling, and to proceed in straight lines away from the axis, within a narrow cone of order 6°.[13,14] The main problem is how to deflect the low energy electrons off the axis. We have chosen to do this with a small Wien ($\underline{E} \wedge \underline{B}$) filter followed by a sector of a gridless Cylindrical Mirror Analyser (CMA). The Wien filter (W) and CMA (C), plus secondary electron detectors (S) are shown in two projections in the analyser chambers on Fig. 3. In this way the cone of low energy electrons is transferred from the exit of the parallelizer to an intermediate focus (F) and then into an energy analyser (E), while the Wien filter keeps the 100 keV beam on-axis. Secondary electron detection (without analysis) is achieved simply by reversing the sign of the fields in the Wien filter.

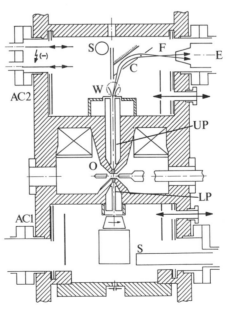

Fig. 3   Arrangement for analysis of low energy electrons in 'MIDAS'.[16]
See text for discussion of the symbols.

In initial studies, without the Auger analyzer in place, we have shown that biassed secondary electron images can very readily be observed from both sides of a thin sample.[17] However, it is clear that the mechanical mask method shown in Fig. 1 cannot be used to create a concentration gradient which is sharp on the 10nm scale. Consequently, we are experimenting with ridged samples, as shown in Fig. 4. A simple geometric calculation involving d and l shows that the definition, Δx, can be made $\approx$ 10nm for s = 1μm, d = 0.2μm and l = 20cm; the value of the maximum diffusion length x_m can be increased, with Δx, by decreasing the glancing angle θ.

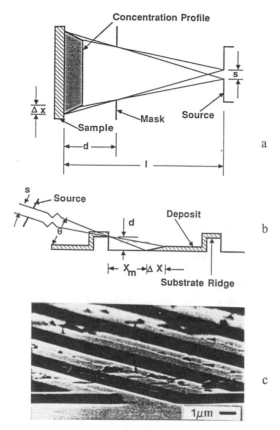

Fig. 4   Mask deposition geometries.   a) width of penumbra Δx as a function of instrumental variables s, d and l; b) use of ridged samples at glancing angle θ for high-resolution studies; c) SEM micrograph of such a sample, with d = 0.2μm mesas, 3μm wide, repeated every 10 μm.

3.   NUCLEATION AND GROWTH EXAMPLES

Most experiments to test nucleation ideas have been done on the island growth mode.[12] We have extended these ideas to form a simple picture of the Stranski-Krastanov mode.[5] Since, in the early stages, this is a strictly two-dimensional (2D) model, it is also appropriate for the first layer in a layer growth system. But in this case the model needs further extension to cope with the simultaneous nucleation and hole filling which occurs at several levels.[1,18]

The basis of these models, in which we make the simplest possible assumptions, is illustrated in Fig. 5. It is assumed that the intermediate, or Stranski, layer forms first, and that the nucleation events take place on top of this completed layer. Because S-K growth is very close to layer growth, we assume that the critical nuclei are in a 2D form with conversion to 3D islands occurring at a later stage. A simplified set of rate equations, which derives from work in the early 1970's,[1] is used to formulate the expected nucleation density, n_x, of stable clusters in terms of the experimental variables (deposition rate R, substrate temperature T) and the material parameters E_a, E_d and E_i, the energy of the critical cluster of size i. In the most important complete condensation regime, this results in a predicted nucleation density of the form

$$(n_x/N_0) \sim (R/N_0\nu)^p \exp(-E/kT) , \qquad (1)$$

where $p = i/(i+2)$ and $E = (iE_d + E_i)/(i + 2)$, N_0 is the atomic site density, and ν a characteristic vibration frequency. Only minor changes results if 3D islands are considered, with 2 replaced by 2.5 in these expressions.[1]

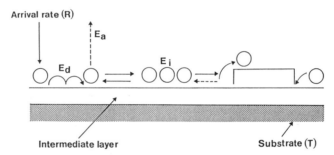

Fig. 5   Schematic illustration of the processes occurring in the layer-plus-island, or Stranski-Krastanov growth mode. The independent variables are the arrival rate (R) and substrate temperature (T). The activation energies for adsorption (E_a) and diffusion (E_d), and the binding energy E_i of critical clusters, which contain i atoms, are shown. The dotted lines indicate the less important processes.[5]

Further progress cannot be made unless the model is made highly specific; in particular, we need to know the binding energy of all clusters of size j, E_j, in order to calculate the critical cluster size i, which is the most unstable cluster. We have chosen to make the model a simple 2-parameter fit by constructing $E_j(j)$ for quasi-hexagonal 2D clusters in terms of nearest-neighbor bonds of strength E_b; at the highest temperatures, re-evaporation becomes important and E_a is also involved. Contact with the equilibrium vapor pressure also requires a rudimentary treatment of atomic vibrations, which has been done within the Einstein model.[5]

Examples of such calculations have been given previously[5] to compare with SEM-based deposition experiments on Ag/Mo(100),[19] Ag/W(110),[20] and Ag/Si(100) and (111).[11] The last system Ag/Si(111) was found to be very sensitive to surface defects, with an 'effective' value of the surface diffusion energy E_d which reduced from 0.55 to $\sim$ 0.33 eV as surface preparation procedures were improved. This system has been re-examined more recently[15] and the lower value confirmed as illustrated in Fig. 6. These data were taken in a relatively narrow range of temperatures 650<T<800K at several deposition rates. The calculations are for E_d=0.30eV and E_b = 0.08eV; this gives critical sizes (i) in the region of 100-400 due to the small pair binding energy, the breaks in the curves corresponding to changes in the value of i. The value of E_a, 2.55eV in this calculation, only enters at the highest temperatures and lowest deposition rates. Note that a density $N = 10^4$ cm^{-2} corresponds to islands ⌄ 100μm apart, and that diffusion distances are therefore of this order.

The parameter values deduced for the above Stranski-Krastanov systems are shown in Table I. The values are quoted to the nearest 0.05eV and are probably accurate to about 0.1 eV (E_a), 0.05 eV (E_d) and 0.03 eV (E_b). The sensitivity to various pre-exponential and parameter combinations has been explored;[5] predictions generally fall within these error bars, although it must be remembered that they constitute 'effective' values in the sense described in the introduction.

Table I

Substrate	N_0 (10^{15}cm^{-2})	Layer Thickness	Island Orientation	E_b (eV)	E_d (eV)	E_a (eV)
W(110) Ref. 20	1.38	2-3	(111)	0.3	0.1	2.1
Mo(100) Ref. 19	1.20	1-2	(100)	0.15	0.45	2.5
Si(100) Refs. 5 and 11	0.69	1/4	(100) and (210)	0.1	0.7	2.6
Si(111) Refs. 11 and 15	0.79	2/3	(111)	0.1	0.35	2.45

Some clear trends are apparent from Table I. The low-E_d and high-E_b values for Ag/W(110) are associated with the intermediate layer being very close to the (111) plane of bulk Ag, both in structure and lattice parameter. Nonetheless the value of $E_b \simeq 0.3$ eV is low, and $E_a \simeq 2.1$ eV is high, which clearly reflects the non-pair additive nature of metallic bonding. It is interesting that Kolaczkiewicz and Bauer[21] have found similar values for E_b from a thermodynamic analysis of this system. The strong lack of pair additivity is also seen in the binding energy of isolated Ag$_2$ and Ag$_3$ molecules, 1.65 and 2.62 eV, respectively.[22]

The values for Ag/Mo(100) imply higher $E_d \simeq 0.45$ eV and lower $E_b \simeq$ 0.15 eV. This may be rationalized in terms of the more open structure of the (100) surface, the slightly larger lattice parameter (lower N_0), and the fact that the (100)-oriented intermediate layer will inhibit the formation of compact hexagonal clusters. In fact the islands grow in a (100) orientation also, so the value of E_b is only an effective value, which is a suitable average of nearest and next-nearest neighbor bond strengths. In order to get condensation at all at the highest temperatures we must have $E_a > 2.5$ eV.

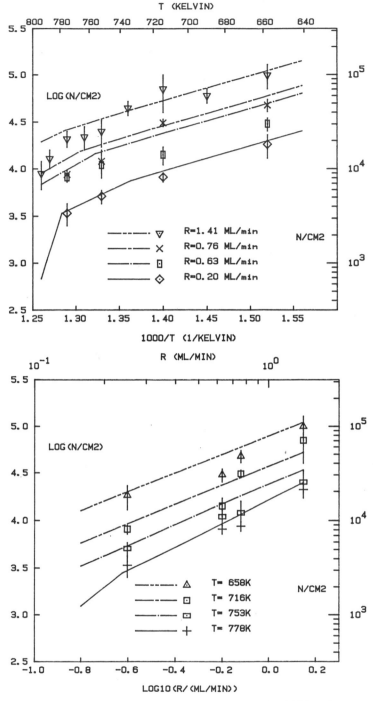

Fig. 6 Nucleation density (log scale) for Ag/Si(111) plotted (a) as a function of T^{-1} for several values of R; (b) as a function of log R for several values of T. The curves are calculated for $E_d = 0.30$ eV, $E_b = 0.08$ eV and $E_a = 2.55$ eV. See text for discussion 15.

The Ag/Si(100) and Ag/Si(111) cases have very low E_b values ≤ 0.1 eV and E_d values ranging upwards from ~ 0.3 eV for well-prepared (111) surfaces to ~ 0.7 eV for (100). The monolayer density is only about half that of bulk silver, and this, in addition to the surface reconstructions, must be very effective in keeping the diffusing Ag atoms apart. This leads to large critical nucleus sizes. On the (100) surface there is clearly competition between at least two epitaxial orientations for the islands, and the diffusion coefficient may well be anisotropic due to the 2 x 1 reconstruction. On the (111) surface the values are those appropriate to diffusion on top of the $\sqrt{3}$ x $\sqrt{3}$ R30° reconstruction, as discussed in the next section. The adsorption energy E_a has to be high (~ 2.5 eV) for both (100) and (111) surfaces.

4. A SURFACE DIFFUSION EXAMPLE: Ag/Si(111)

Surface diffusion has been observed directly, using the capability of b-SEI to visualize monolayer and multilayer deposits. Detailed work has been performed so far on Ag/Si(111)[15] and Ag/W(110);[23] here we concentrate exclusively on the first system, in which we have demonstrated that diffusion within the $\sqrt{3}$ x $\sqrt{3}$ R30° intermediate (hereafter $\sqrt{3}$) layer is very slow.[24] In this case, diffusion across the layer is by the so-called rolling carpet mechanism, as illustrated in Fig. 7.

The single atom concentration $n_1(\underline{r}, t)$ on top of this layer is governed by a diffusion equation

$$\frac{\partial n_1(\underline{r}, t)}{\partial t} = g(\underline{r}, t) - \frac{n_1(\underline{r}, t)}{\tau} + \nabla \cdot [D(\underline{r}, t) \, \nabla n_1(\underline{r}, t)] . \tag{2}$$

In this equation, g is a source term, which is equal to R inside the mask area for the duration of deposition, and zero otherwise; D is the adatom diffusion coefficient. The sink term $-n_1/\tau$ reflects the loss of adatoms by various processes. At high temperatures, the characteristic time $\tau = \tau_a$, the evaporation stay time. However, as developed in nucleation theory[1,5] this approach can be generalized, such that τ^{-1} is the sum of the inverse characteristic times for all competing loss processes. Thus when capture by stable clusters (τ_c) takes place we can write

$$\tau^{-1} = \tau_a^{-1} + \tau_c^{-1} , \tag{3}$$

where τ_c is written in terms of a capture number σ_x and the stable cluster density n_x as

$$\tau_c^{-1} = \sigma_x D n_x . \tag{4}$$

The ratio $r = \tau_a/\tau_c = \sigma_x D \tau_a n_x$ is useful to distinguish 'incomplete condensation' ($r \ll 1$) from 'complete condensation' ($r \geq 1$).[5] In Eq. (4) n_x will be a function of position which varies in a non-linear manner with n_1; so for $r \ll 1$ Eq. (2) is linear, whereas for $r \geq 1$ it is not. The capture number σ_x varies slowly with cluster size,[1] and a constant ~ 5 is a good first approximation.

The diffusion coefficient can be allowed without difficulty to be a function of position also. At low concentrations $D = D_1$ in Eq. (2), but when the second layer is not sufficiently dilute, the chemical diffusion coefficient will contain effects due to the creation and dissociation of

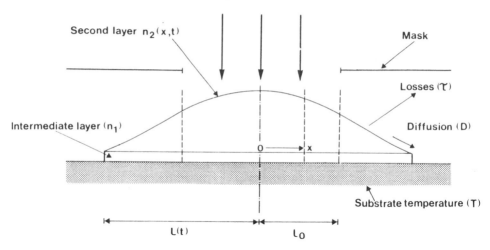

Fig. 7   Schematic diagram of one-dimensional diffusion model, which
          shows the sources and sinks for single adatoms on the
          intermediate layer and defines the measured patch widths l_o and
          $l(t)$.

sub-critical clusters. The effective, concentration dependent, diffusion
coefficient of single adatoms in the presence of clusters can be written
as[25]

$$D_{eff}(\underline{r}) = \frac{\Sigma_j j^2 n_j(\underline{r}) D_j}{\Sigma_j j^2 n_j(\underline{r})} \quad , \tag{5}$$

where n_j is the concentration of j-sized clusters and D_j is the diffusion
coefficient of such a cluster; this formulation enables diffusion 'in the
presence of clustering, ripening or annealing to be investigated[26].

With suitable boundary conditions, Eq. (2) can be solved in one
dimension and using more computer time in 2D. In the 1D case the rate of
patch broadening, $\partial l(t)/\partial t$, is given by

$$J = -D_1 (\partial n_1/\partial x)_{x=1} = n_0 (\partial l/\partial t) \quad , \tag{6}$$

where n_0^{-1} is the area per Ag atom in the $\sqrt{3}$ layer. The simplest
solutions to Eqs. (2) and (6) are the steady state solutions, with
$\partial n_1/\partial t = 0$ in Eq. (2) and $\partial l/\partial t$ small in Eq. (6). In this case the
behaviour is governed by a single diffusion length

$$l_d = (D_1 \tau)^{1/2} \quad , \tag{7}$$

with $\tau = \tau_a$ at the highest temperatures. Under these conditions the
patch width predicted is only a function of dose, $\theta = Rt$, and temperature
via l_d, and does not depend explicitly on R. The steady state model can
be extended to lower temperatures, but it is less accurate since n_x is a
function of position in eqn. (4). Nonetheless, the use of eqn. (1) to
estimate n_x, and hence l_d via Eq. (7) represents a quick

446

semiquantitative check on more detailed explicit solutions of the non-linear diffusion equation.[15]

Experimental results for the Ag/Si(111) have confirmed some of these predictions and enabled the lumped parameters $(E_a - E_d)$ and $(E_d + 3E_b)$ to be extracted, as indicated in Fig. 8.[15] The high temperature values (patch width vs T^{-1}) are determined by reevaporation, so that Eqn. (7) yields an activation energy $(E_a - E_d)/2$; we have shown that the dose θ is the relevant variable, and that l_d is independent of crystallographic orientation, indicating that steps are not important in this T range, 770<T<850K. No stable crystals are observed on such patches, and there is no effect of annealing for times up to t ∿ 20 min. We find $(E_a - E_d) \simeq 2.0$ eV.

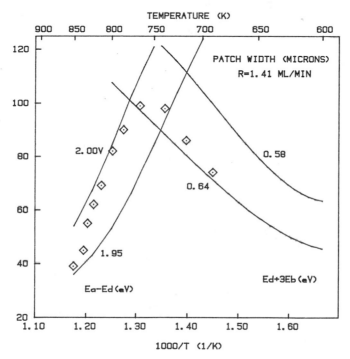

Fig. 8   Patch width as a function of substrate temperature during deposition, for a Ag dose of 5 ML onto Si(111) at R = 1.41 ML/min. Model calculations for the indicated parameter values are superimposed.

At the lower temperatures we see the effect of diffusion in competition with nucleation, and the curves shown indicate that the patch width values are satisfied by $(E_d + 3E_b) \simeq 0.6$ eV; this value is close to the value 0.54 eV used for nucleation predictions in Fig. 5. Near the peak of the curves in Fig. 8, extra atoms are present, even though no nuclei are seen, and these atoms anneal out to broaden the √3 patch in times t ∿ 5 min. This and other evidence based on AES amplitudes as a function of T,[27] argues that substantial extra atoms may be present in (or on) the √3 layer at elevated temperatures, and that diffusion may be slowed down in this region according to Eq. (5), especially when the stable cluster density is so low. We feel that such observations may go a long way to explaining the continuing controversy about the coverage and structure of the $\downarrow$3 layer.[15,27,28]

5. TOWARDS HIGHER SPATIAL RESOLUTION: Ge/Si(100)

In this section we describe our work on extending similar studies to higher spatial resolution using the UHV-STEM 'MIDAS' in Arizona,[13] where we have chosen initially to study Ge/Si(100). Any attempt to study crystal growth on surfaces must begin with well defined and repeatable methods of obtaining clean surfaces. In our case, these methods include the need to prepare electron transparent samples for transmission microscopy. To this end, we have expended considerable effort in the development of techniques for preparing both bulk (SEM) and thin (TEM) samples. A number of parallel experiments in an off-line UHV chamber have proven useful. These experiments were instrumental in developing surface preparation techniques, characterizing this film growth system and assessing the effects of a negative specimen bias and Ge deposition on the secondary electron yield. It is intended that this offline work will provide a suitable base for interpretation of microscope based experiments.

The offline experimental work was performed in a UHV chamber with a base pressure of 5×10^{-10} mbar; surface cleanliness and quality was assessed by AES and RHEED. Following a procedure which has been shown by STM to minimize surface defects,[29] we are able to prepare surfaces routinely with sharp 2 x 1 RHEED patterns and contaminant levels below the Auger detection limit.

A similar technique was used to prepare the Si surfaces in the microscope preparation chamber. The samples are 3mm discs which are ultrasonically cut from larger wafers. TEM samples must undergo a sequence of mechanical dimple grinding and acid etching to achieve electron transparency. We have elected to eliminate the traditional ion milling step in TEM sample preparation since we believe that this will implant impurities and amorphize the surface. Bulk 3 mm disk SEM samples were placed directly into the Mo cups and Ta carriers without any other preparation. These samples were then cleaned by electron bombardment heating in a process identical to that in the offline UHV chamber. The only difference was that the sample contained in its Mo cup was removed and the Ta carrier outgassed to 1450°C prior to the thermal processing.

Sample cleanliness has been assessed by inspection in the microscope. Contaminants show up as bright cubic or hexagonal particles in the secondary electron images. For SEM samples, contaminant particle densities as low as 1 particle per 100 μm^2 are routinely attained. TEM samples are at a developmental stage but we are able to get clean, electron transparent, regions of at least $1\mu m.^2$ When the Auger spectrometers are installed on the microscope column, we will be able to assess the sample cleanliness more quantitatively.

We have characterized the initial stages of film growth in the Ge/Si(100) epitaxial system using RHEED and AES. We confirm that this is a Stranski-Krastanov system with an intermediate layer thickness $\sim$3-4 ML.[30] This is evident in Fig. 9 where we show Auger peak height ratios and RHEED patterns for films of various thickness both before and after annealing. Typically, Ge deposition was done onto a room temperature substrate at R = 1 ML/min, followed by an anneal at 700°C for 2 minutes. The length of the anneal seems to have no effect on the film for thicknesses less than the intermediate layer thickness. However, once

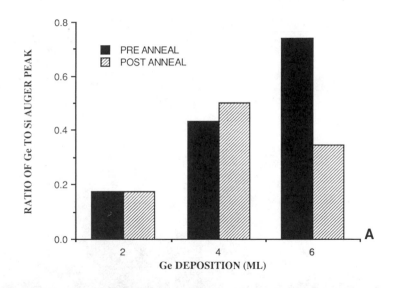

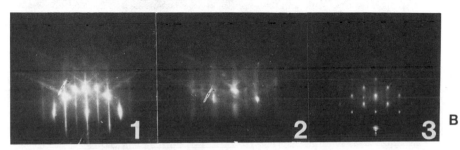

Fig. 9   (a) Auger peak height ratios for Ge(1147eV)/Si(1620eV) for
         Ge/Si(100) at coverages of 2, 4 and 6ML before and after
         annealing at T = 700°C for 2 min.; (b) corresponding RHEED
         patterns; 1) clean Si(100) 2x1, 2) 2ML before, and 3) 6ML after
         annealing. Each 2 min. deposit was made onto the previously
         annealed surface, at a rate of 1ML/min.

islands begin to form on top of the intermediate layers, longer anneals
result in sharper transmission electron diffraction spots which seems to
indicate that the film is coarsening.[26] Work now in progress will
determine the growth mode of Ge deposited on vicinal Si(100) samples.

Biassed secondary electron imaging (b-SEI) is a simple technique for
increasing the secondary electron collection efficiency by applying a
negative bias to the sample. Our initial experiments in MIDAS showed
that we were able to observe both sides of a thin sample and to correlate
these images with other STEM signals.[15] This facility is illustrated here
in Fig. 10, which shows b-SE images of the entrance (a) and exit (b)
surfaces of a thin reactively ion etched Si(100) sample composed of 1μm
wide, 0.7μm tall mesas separated by 2μm wide valleys. This is the type
of sample which will be used in conjunction with oblique deposition to
fabricate very sharp composition profiles for high resolution surface

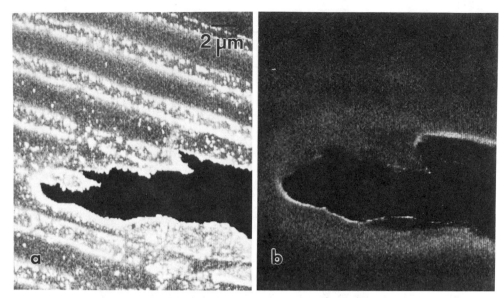

Fig. 10 Visualization of both surfaces of a reactively ion etched
Si(100) sample using b-SEI in MIDAS. a) entrance surface showing
the ridges, decorated by contaminant particle; b) exit surface
free of contaminants. Note the almost complete absence of
cross-talk between these two images.

diffusion measurements as indicated in Fig. 4. The bright contrast in
Fig. 10(a) and missing in 10(b) is due to contamination, probably SiC,
and is evidently located on the entrance surface. Note also that the
mesa structure etched in the entrance surface is not apparent in the exit
image.

This imaging mode is also able to visualize surface steps. Figure
11 shows a plan view b-SE micrograph of steps on a Si(100) surface
misoriented 1° toward the <110> direction. Since the interstep spacing
corresponds well with that expected for this surface and annealing
history, we tentatively identify these as bi-atomic ($a_o/2$ = 2.72Å high)
steps.[31] Further work, including REM and microdiffraction, is in
progress to identify the step height positively. The ability to image
single atomic steps in plan view will be extremely useful for studying
the atomistic processes leading to epitaxy on single terraces as well as
the interaction of adatoms and clusters with steps. Given the ability of
STM,[5] LEEM,[6] TEM and REM[7] to detect these surface steps it is reassuring
that we can also do so.[32]

The biassed secondary signal is very sensitive to sub-monolayer
deposits of Ge on Si(100), as illustrated here in Fig. 12. Room
temperature deposits in the 0-2ML range increase both the height and
breadth of the secondary electron spectrum. The secondary electron yield
increases by around a factor of two for the 1.8ML deposit shown.
Preliminary work on imaging this system in MIDAS has shown very strong
b-SEI contrast associated with the islands of Ge, which form both on
annealing low-temperature deposits of thickness ≥ 3ML and after hot
deposition. This work will be published elsewhere.[33]

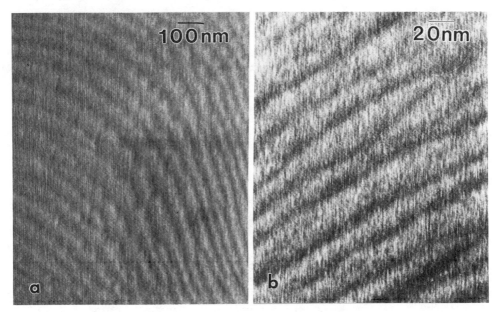

Fig. 11 Biased secondary electron images (entrance face) of steps on a vicinal Si(100) surface after cleaning by heating as described in the text: a) curved steps showing kinks; b) detail of relatively regular steps.

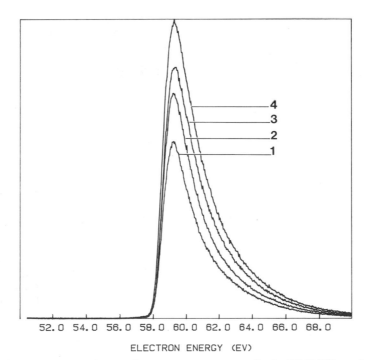

ELECTRON ENERGY (EV)

Fig. 12 Biassed secondary electron spectra of Ge/Si(100), showing the increase in secondary electron yield after Ge deposition at room temperature 1) clean Si(100; 2) 0.6ML; 3) 1.2 ML; 4) 1.8 ML Ge. (1ML = 0.141 nm).

6. CONCLUSIONS

Progress in the development of UHV-SEM and -STEM instrumentation for studies of surface diffusion and crystal growth has been reviewed. We have highlighted a simple technique, biassed secondary electron imaging (b-SEI), which is sensitive to sub-monolayer deposits and to individual surface steps.

SEM studies of nucleation densities have been made on a range of Ag-based deposition systems, and atomic parameters extracted from a comparison with nucleation theory calculations. In one case Ag/Si(111), the values have been validated by direct measurement of surface diffusion using b-SEI. STEM studies, which are in their infancy, have demonstrated images of surface steps on Si(100). Comparison of b-SE and Auger spectra suggests that monolayer sensitive studies of the system Ge/Si(100) can be pursued with high spatial resolution.

Acknowledgements

We acknowledge the continued support of the SERC for the Sussex work described. The Arizona work is supported by the Office of Naval Research (award #N-00014-84-G-0203), the National Science Foundation (grant #DMR-8500659), a grant from the Shell Oil Company Foundation and by Arizona State University. We are grateful to several colleagues at Sussex and Arizona for their participation in the programs described in this paper.

REFERENCES

1. J. A. Venables, G. D. T. Spiller and M. Hanbücken, Rept. Progr. Phys. 47, 399 (1984); J. A. Venables, Phil. Mag. 27, 697 (1973).
2. J. L. Robins, Appl. Surface Sci. 33/34, 379 (1988).
3. A. G. Naumovets and Yu. S. Vedula, Surface Sci. Rep. 4, 365 (1985).
4. Surface Mobilities on Solid Materials, NATO-ASI vol. B86, edited by Vu Thien Binh (Plenum, New York, 1982).
5. J. A. Venables, Phys. Rev. B36, 4153 (1987); J. Vac. Sci. Tech. B4, 870 (1986).
6. See, e.g., papers by E. J. van Loenen et al., M. G. Lagally et al, and O. Jusko et al., in this volume, also R. J. Hamers, U. K. Köhler and J. E. Demuth, Ultramicroscopy 31, 10 (1989), and refs. quoted.
7. E. Bauer, M. Mundschau, W. Swiech and W. Telieps, Ultramicroscopy 31, 49 (1989) and refs. quoted.
8. K. Yagi, in: Advances in Optical and Electron Microscopy (Academic Press, N.Y.) 11, 57 (1989) and refs. quoted; Y. Tanashiro and K. Takayanagi, Ultramicroscopy 31, 20 (1989).
9. H. W. Fink and G. Ehrlich, Surface Sci. 173, 128 (1986) and refs. quoted; see also articles by D. W. Bassett in Ref. [4], p. 63,83.
10. M. Futamoto, M. Hanbücken, C. J. Harland, G. W. Jones and J. A. Venables, Surface Sci. 150, 430 (1985); C. J. Harland, G. W. Jones, T. Doust and J. A. Venables, Proc. 5th Pfefferkorn Conference, Scanning Microscopy Suppl. 1, 109 (1987).
11. M. Hanbücken, M. Futamoto, and J. A. Venables, Surface Sci. 147, 433 (1984).
12. D. R. Batchelor, P. Rez, D. J. Fathers and J. A. Venables, Surface and Interface Analysis 13, 193 (1988); D. R. Batchelor, H. E. Bishop and J. A. Venables, 14, 700, 709 (1989).

13. J. A. Venables, H. S. von Harrach and J. M. Cowley, Proc. Inst. Phys. Conf. Ser. 90 (EMAG '87), 85 (1987); J. S. Drucker, M. Krishnamurthy, Luo Chuan Hong, G. G. Hembree and J. A. Venables, Proc. Inst. Phys. Conf. Ser. 98 (EMAG '89), 303 (1989); Proc. 47th EMSA meeting (1989) p. 208.

14. J. A. Venables, A. P. Janssen, P. Akhter, J. Derrien and C. J. Harland, J. Microscopy 118, 351 (1980); J. A. Venables, G. D. T. Spiller, D. J. Fathers, C. J. Harland and M. Hanbücken, Ultramicroscopy 11, 149 (1983); C. J. Harland and J. A. Venables, Ultramicroscopy 17, 9 (1985).

15. T. Doust, F. L. Metcalfe and J. A. Venables, Ultramicroscopy 31, 116 (1989); J. A. Venables, T. Doust and R. Kariotis, Mat. Res. Soc. Symp. 94, 3 (1987); to be published.

16. P. Kruit and J.A. Venables, Ultramicroscopy 25, 183 (1988); J. A. Venables, P. S. Flora, C. J. Harland, Luo Chuan Hung and G. G. Hembree, Proc. Inst. Phys. Conf. Ser. 98 (EMAG '89), 289; J. A. Venables and P. A. Bennett, Proc. NATO-ARW on Evaluation of Advanced Semiconductor Materials by Electron Microscopy (D. Cherns ed., Plenum 1989) B203, 306 (1989).

17. G. G. Hembree, P. A. Crozier, J. S. Drucker, M. Krishnamurthy and J. M. Cowley, Ultramicroscopy 31, 111 (1989).

18. D. Kashchiev, J. Cryst. Growth 40, 29 (1977); G. H. Gilmer, J. Cryst. Growth 49, 465 (1980); J. A. Nieminen and K. Kaski, Phys. Rev. A40, 2088, 2096 (1989); R. Kariotis, J. Phys. A22, 2781 (1989).

19. K. Hartig, A. P. Janssen and J. A. Venables, Surface Sci. 74, 69 (1978).

20. G. D. T. Spiller, P. Akhter and J. A. Venables, Surface Sci. 131, 517 (1983).

21. J. Kolaczkiewicz and E. Bauer, Surface Sci. 151, 333 (1985).

22. K. A. Gringerich, I. Shim, S. K. Gupta and J.E. Kingcade, Surface Sci. 156, 495 (1985), Table 5.

23. G. W. Jones and J.A. Venables, Ultramicroscopy 18, 439 (1985); to be published.

24. M. Hanbücken, T. Doust, O. Osasona, G. LeLay and J. A. Venables, Surface Sci. 168, 133 (1986).

25. C. P. Flynn, Phys. Rev. 134, A241 (1964).

26. M. Zinke-Allmang and L. C. Feldman, paper in this volume, and references quoted.

27. G. Raynerd, M. Hardiman and J. A. Venables, to be published.

28. For a recent bibliography, see paper by G. LeLay et al. in this volume.

29. See, e.g., M. G. Lagally et al., this volume; M. G. Lagally, R. Kariotis, B. S. Swartzentruber and Y-W. Mo, Ultramicroscopy 31, 87 (1989).

30. T. Sakamoto, this volume and refs. quoted; J. E. MacDonald et al, this volume.

31. J. E. Griffith, J. A. Kubby, P. E. Wieringa, R. S. Becker and J. S. Vickers, J. Vac. Sci. Technol, A6 493 (1988).

32. See also A. Bleloch, A. Howie and R. H. Milne, Ultramicroscopy 31, 99 (1989) and refs. quoted.

33. M. Krishnamurthy, J. S. Drucker and J. A. Venables, Mat. Res. Soc. Symp. (1990) in press, and to be published.

453

THREE-DIMENSIONAL CLUSTERING ON SURFACES: OVERLAYERS ON Si

M. Zinke-Allmang

Institute of Thin Film and Ion Technology
KFA Jülich
D-5170 Jülich, Federal Republic of Germany

L.C. Feldman and S. Nakahara

AT&T Bell Laboratories
600 Mountain Ave
Murray Hill, N.J. 07974, USA

ABSTRACT

Heteroepitaxy, the formation of crystalline, layered structures of different materials, is the driving force for a large part of current solid state science. Most heteroepitaxial applications require and assume a uniform two-dimensional planar structure, based on the simplest film growth concepts. Nevertheless, the propensity for three-dimensional nucleation is a strong element in film formation. In this paper we describe our recent work in studies of clustering for overlayers on Si. Our studies consider the set of adsorbates, Ge, Sn and Ga on substrates of Si, GaAs and As terminated Si surfaces. The use of the As terminated surface permits an interesting transition between the two primary semiconductors. We show that the kinetics of cluster growth is well described by ripening processes, which can, in turn, be thought of as special cases of scaling laws predicted by self-similarity concepts.

INTRODUCTION

The growth of new thin film structures by modern deposition techniques is a driving force for much of current-day solid-state and interface science. New optical and electronic properties arise from precise epitaxial composition modulations of single layers in lattice matched and lattice mismatched systems.[1] The propensity, however, for three-dimensional nucleation is a strong element in film formation. In particular, the equilibrium configuration of a lattice mismatched structure is a clustered configuration.[2,3] Since most applications depend upon a planar geometry, a fundamental understanding of cluster nucleation and growth is critical for heteroepitaxy applications. In this paper we describe recent studies of the cluster growth limits which reveal the clustering mechanism. Earlier reports of parts of this work can be found in Ref. 4.

Kinetics of Ordering and Growth at Surfaces
Edited by M. G. Lagally
Plenum Press, New York, 1990

GENERAL CONCEPTS OF CLUSTER FORMATION AND GROWTH

The growth of heteroepitaxial film systems, i.e. the crystalline growth of a different material on a clean substrate surface at elevated temperature, usually involves two materials with a lattice constant mismatch. In many cases match of the film lattice to the substrate is desired leading to strained layer epitaxy. Such systems are of technological and scientific interest since new semiconductor properties may be achieved.

The earliest understanding of the stability of such films originates from the pioneering work of van der Merwe[5] and Mathews.[6] They considered the balance associated with the energy of the strained film to that for the relaxed film containing defects and misfit dislocations. Because each energy component has a different dependence on film thickness a critical thickness for epitaxy, t_c, can be defined. Films thinner than this thickness may be totally strained and the film will realize the same lattice constant as the substrate in the direction parallel to the growing surface. Films thicker than t_c may contain extended defects as misfit dislocations and are therefore partially relaxed and unstrained.

Another important concept for the understanding of the stability of these epitaxial films is illustrated by the molecular dynamics simulations of Gilmer and Grabow[2] who showed that a strained epitaxial film is always unstable against clustering, for thicknesses greater than and less than t_c.

The following simplified calculation illustrates this general result concerning clustering.[3] Consider the ratio R of the energy of the strained film to the energy of the corresponding clustered film when the latter consists of m relaxed cubic clusters of linear dimension x. The ratio can be written as

$$R = \frac{A\gamma + An\epsilon_{str}}{A\gamma + 4mx^2\gamma} \, , \tag{1}$$

where A is the surface area of the substrate, γ is the surface energy density and n is the number of monolayers in the strained film. ϵ_{str} is the strain energy density for one monolayer given by

$$\epsilon_{str} = e^2 \, 2\sigma \left(\frac{1 + \nu}{1 - \nu} \right) \, , \tag{2}$$

where e is the strain, σ is the shear modulus, ν is Poisson's ratio and h is the thickness for one monolayer. In this rough calculation we have taken the surface energy of the substrate and overlayer as equal and the interfacial energy as zero. Relating the cluster density to the number of monolayers by $mNx^3 = nn_sA$ (mass conservation) we have,

$$R = \frac{1 + n(\epsilon_{str}/\gamma)}{1 + (n/x) \, (4n_s/N)} \, , \tag{3}$$

where N is the number of atoms/volume of the film material and n_s is the number of atoms/area of the surface. The energies are equal (R = 1), when $x_c = (4n_s/N) / (\epsilon_{str}/\gamma)$ where x_c may be thought of as a critical cluster size. For the case of Ge on Si (4.2% mismatch), $\epsilon_{str} \approx 0.05$ eV/atom and $\gamma \approx 2.0$ eV/atom (for both Ge and Si) so that R > 1 for x > 200 Å. In this case the clustered structure represents a lower energy state when the

456

(cubic) clusters are larger than 200 Å. This simple calculation shows that x_c is inversely proportional to the square of the strain so that small mismatches are inherently more stable (or easier to grow) than large mismatched systems. This has been demonstrated experimentally by Bean et al. for the system of Ge_xSi_{1-x} on Si(100).[7]

The fact that clusters on a surface are energetically favoured over a coherent film is equivalent to the thermodynamic expression that the system is brought into a two phase coexistence area of the surface phase diagram. There are several models describing possible ways for clusters to appear under these circumstances, e.g. random nucleation as introduced by Becker and Döring or spinodal decomposition. A good review of these nucleation processes is given by Binder.[8] In a recent study Bruinsma and Zangwill discussed these models and showed, for example, that the actual process depends on the temperature relative to the roughening temperature.[9]

Small clusters, initially appearing at a high density on the substrate surface, are not the equilibrium state of the system, rather further cluster growth occurs. Cluster growth involves mainly Ostwald ripening, growth of larger clusters at the expense of small clusters, and coalescence, the motion of clusters and their combining upon contact. The theoretical description of the latter process originates from the work of von Smoluchowski.[10] For Ostwald ripening, the process emphasized in this paper, the first analytical treatment came from Lifshitz and Slyozov.[11] Wagner[12] extended their work, including different processes which may limit the mass transport between the clusters. Chakraverty[13] applied this concept first to cluster growth on surfaces. We summarize these latter calculations[13] for the later discussion of the experimental data. For simplification we confine our discussion to hemispherical clusters thus avoiding additional geometrical factors.[13]

The driving force for the growth of clusters on a surface can be understood[14] from the pressure difference across a curved surface. Assume a growing bubble in a solution. The resistance to expansion is the increasing surface area of the bubble and the increased surface energy. At equilibrium the work $\Delta P\, dV$ must be equal to the work done at the surface, γdS. Using dV and dS for a special cluster we get $\Delta P = \gamma\,(dS/dV) = \gamma\,(8\pi r\, dr)/(4\pi r^2 dr) = 2\gamma/r$. This causes an increased vapor pressure or solubility at a highly curved surface. The second law of thermodynamics relates the change of surface energy to the isothermal osmotic work by:

$$2\gamma\, V_M/r = kT\, \ln\,(p/p_o) \,, \tag{4}$$

with p_o the pressure over a flat surface and V_M the molecular volume. Using $c(r)$ for the solubility of a cluster of radius r and $c_\infty = c(r = \infty)$, the equilibrium solubility of an infinite large cluster, we rewrite the Gibbs-Thomson equation (4):[11,12]

$$c(r) = c_\infty \exp\left\{\frac{2\gamma V_M}{kT}\frac{1}{r}\right\} . \tag{5a}$$

For large clusters the exponent on the right hand side vanishes and $c(r)$ approaches c_∞. This equation describes the excess concentration of a cluster of radius r in equilibrium with a surface. This dependence is sketched and its implication on the behaviour of two clusters in diffusive contact is highlighted in Fig. 1. Note that the differences in equilibrium concentrations are in general small, e.g. of the order of less than 1% for clusters larger than 100 Å, therefore the exponential term in Eq. (5a) can be replaced by the first two terms of a Taylor

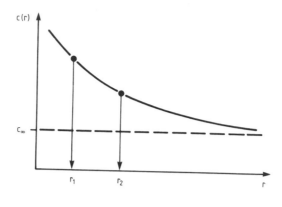

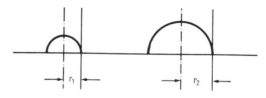

Fig. 1   The Gibbs-Thomson effect illustrated for the equilibrium
         concentration of a cluster as a function of cluster radius.   Two
         clusters and their respective equilibrium concentration are
         indicated.

series for cluster radii larger than 30 Å:[12]

$$c(r) = c_\infty \left[1 + \frac{2\gamma V_M}{kT} \frac{1}{r} \right] .$$

(5b)

An analytical treatment of the cluster growth requires an equation
connecting the time and radial dependence of the cluster size
distribution, f(r,t), to the Gibbs-Thomson effect. An implicit
representation follows from the mass conservation condition. Combined
with the mass transport mechanism these relations allow the calculation
of the explicit cluster size distribution.

The conservation of mass is included within the equation of
continuity.[15] Since we exclude at this stage coalescence phenomena, i.e.,
the combining of entire clusters, the number of clusters with radius
larger than r is conserved (since clusters disappear only in the vicinity
of r = 0). Writing f(r,t) for the cluster size distribution and F(r,t)
for the total number of clusters larger than r

$$F(r,t) = \int_{r=r'}^{\infty} f(r',t)dr' ,$$

(6a)

we get for the conservation of mass

458

$$\frac{\partial F}{\partial t} \, dt + \frac{\partial F}{\partial r} \, dr = 0 \qquad\qquad (6b)$$

and this corresponds to

$$\frac{\partial f(r,t)}{\partial t} = - \frac{\partial(f(r,t) \, (dr/dt))}{\partial r} . \qquad\qquad (6c)$$

Eq. (6c) can be solved if dr/dt is known. As originally discussed by Chakraverty[13] this is equivalent to finding a proper representation of the mass transport, i.e. a combination of surface diffusion and the passing of the cluster surface barrier. Schematics of several examples of the latter are shown in Fig. 2.

(i)   The surface diffusion mechanism has to be treated carefully in two dimensions since Fick's second law of diffusion in cylindrical coordinates does not have a steady state solution reaching a finite value far from the origin.[16] The steady state solution is applicable when the diffusion length Λ is larger than the inter-cluster distance on the surface d_{ic}, $\Lambda \gg d_{ic}$. The diffusion

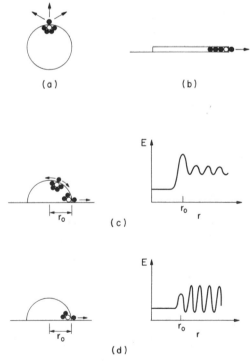

Fig. 2   Schematics of the mass transfer process for (a) three-dimensional clusters in solution, (b) two-dimensional islands on a surface, (c) three-dimensional clusters on a surface with contributions from the entire cluster surface and (d) three-dimensional clusters on a surface with contributions only from the interfacial area.

length is given by $K = 2(Dt)^{1/2}$ with D the diffusion coefficient. Then a quasi steady state solution can be given by introducing a screening length factor l_{sc} for which the free concentration is assumed to reach a constant value at the distance $l_{sc}r$. The screening length allows for a boundary condition at finite distance from the cluster, thus avoiding the divergence in the logarithmic solution of the diffusion equation. This yields for the change of the amount of material in a cluster of radius r, dn/dt:

$$dn/dt = - S_D \frac{D}{r\ lg(l_{sc})} (c'(r) - c_{free}) , \qquad (7a)$$

with the concentration gradient between c'(r), the actual concentration at the cluster surface and c_{free}, the average concentration between the clusters on the surface. S_D is the surface of the cluster active in the diffusive mass transport (index D), usually $S_D = 2\pi r$.

(ii) The change in the amount of material passing the energy barrier at the cluster surface is given by:

$$dn/dt = - S_R \kappa (c(r) - c'(r)) \qquad (7b)$$

with the driving force due to the concentration difference between the equilibrium concentration at the cluster surface c(r), as given by the Gibbs-Thomson equation, and the actual concentration at the cluster surface c'(r). The term S_R represents the cluster surface active in the mass transfer of atoms out of the cluster (index R). This interface transfer can either be through the full cluster surface, $S_R = \pi r^2$, as discussed by Chakraverty,[13] or only through the collar of the cluster at the substrate surface, $S_R = 2\pi r$, assuming a lower energy barrier for cluster atoms at this specific surface site (Fig. 2). Note that κ corresponds to a probability factor (dimension sec^{-1}) in the first case which we designate as κ_a with index a for areal, and to a speed constant for mass transfer (dimension cm/sec) in the second case designated as κ_r with index r for radial.

The change in radius of the cluster is related to the mass transfer by

$$S_C (dr/dt) = V_M (dn/dt) , \qquad (8)$$

with S_C the total surface of the cluster (index C). Rewriting Eqs. (7a) and (7b) using this relation and eliminating c'(r) by combining both equations we get

$$dr/dt = - \frac{V_M}{r^3\pi} \frac{S_R S_D \kappa D/lg(l_{sc})}{S_R\kappa + S_D D/(r lg(l_{sc}))} (c(r) - c_{free}) , \qquad (9a)$$

where $S_D = 2\pi r$ and $\kappa S_R = \pi r^2 \kappa_a$ or $\kappa S_R = 2\kappa r\kappa_r$ depending on the barrier passing mechanism. This yields three different limits for Eq. (9a) (see also Fig. 2c and 2d):

(a) $\kappa_r r > D$ or $\kappa_a r^2 > D$, i.e., the mass transport is limited by surface diffusion. In this limit

$$dr/dt = - V_M \frac{D}{r^2} (c(r) - c_{free}) . \tag{9b}$$

(b) $S_R = 2\pi r$ and $\kappa r < D$, i.e., the mass transport is limited by surface barrier passing with the mass transfer restricted to the contact line of the cluster to the substrate:

$$dr/dt = - V_M \frac{\kappa r}{r} (c(r) - c_{free}) . \tag{9c}$$

(c) $S_R = \pi r^2$ and $\kappa r < D$, i.e., the mass transport is limited by surface barrier passing with the mass transfer through the full cluster surface:

$$dr/dt = - V_M \kappa_a (c(r) - c_{free}) . \tag{9d}$$

Note that the first limit will eventually dominate since the radii of the clusters grow during ripening.

Each of the clusters of the distribution $f(r,t)$ tries to develop an equilibrium concentration according to Eq. (5b). Since we assume any initial supersaturation to be gone and no further material to be deposited (Eq. (6c)), this results in a free concentration, c_{free}, which is in equilibrium with some average sized cluster of the actual cluster size distribution. Small clusters decompose since their equilibrium concentration exceeds this free concentration, clusters with sufficiently large radii grow. Growth and decomposition rates are limited by Eqs. (9b-d). Of special importance therefore is the radius of those clusters which are just in equilibrium with the free concentration, which we define as the "critical radius" r_c

$$c(r_c) = c_{free} . \tag{10}$$

Note that r_c is a function of time in the case of ripening. Based on Eqs. (5b),(6c) and (9b-d) we can determine the time dependence of the cluster distribution $f(r,t)$. Introducing the Gibbs-Thomson equation in Eq. (9c) and using Eq. (10) we get

$$dr/dt = - \frac{\beta}{r} \left[\frac{1}{r} - \frac{1}{r_c} \right] , \tag{11}$$

with the abbreviation $\beta = 2\kappa V_M^2 c_\infty \gamma/kT$. Using the equation of continuity, Eq. (6c) we find the corresponding implicit time dependence of the cluster size distribution:

$$\frac{\partial f(r,t)}{\partial t} = \beta \frac{\partial}{\partial r} \left[\frac{f(r,t)}{r} \left(\frac{1}{r} - \frac{1}{r_c} \right) \right] . \tag{12}$$

Analogous equations can be given using the limits of Eqs. (9a) or (9c) instead of Eq. (9b). Note that the major difference is the power of the r dependence in the denominator on the right hand side of Eqs. (11) and (12). This difference persists throughout the following calculations to obtain $f(r,t)$ explicitly.[13] The first result of these calculations is the time dependence of the critical cluster radius for different mass-transport limits. If the mass transport is limited by surface diffusion we get

$$r_c(t) = r_c(0) \ (1 + t/\tau)^{1/4} \ , \tag{13a}$$

if the mass transport is limited by the passing of the surface barrier with the mass transfer through the collar of the cluster only:

$$r_c(t) = r_c(0) \ (1 + t/\tau)^{1/3} \ , \tag{13b}$$

and if the mass transport is limited by passing the surface barrier with the mass transfer through the full cluster surface:

$$r_c(t) = r_c(0) \ (1 + t/\tau)^{1/2} \ . \tag{13c}$$

The corresponding cluster size distributions are shown in Fig. 3. Note that the probability to find a cluster of a given size is only a function of a single dimensionless parameter ρ, $\rho = r/r_c(t)$. The observation that the cluster size distribution can be written without explicit time or radius terms is an important finding: As a consequence these distributions are indistinguishable for any time during the cluster growth process, i.e. they obey the hypothesis of statistical self-similarity.

The hypothesis of self-similarity is quantitatively defined as follows: Assume that a given cluster size distribution $f(r,t)$ is viewed at the later time λt.[17] Statistical self-similarity holds if the new cluster size distribution is indistinguishable from the distribution resulting from a magnification λ of all dimensions in the initial distribution. The mean cluster size is then increased by the same factor

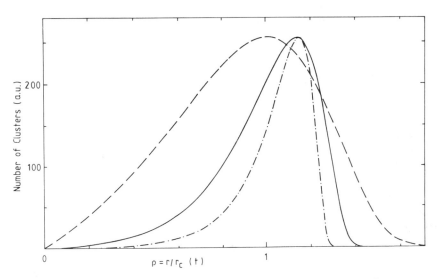

Fig. 3   Theoretical cluster size distributions for ripening processes with different power laws for the growth of the critical cluster with time: $t^{1/2}$ (dashed line), $t^{1/3}$ (solid line) and $t^{1/4}$ (dashed-dotted line).

$$f(r,t) = f\left[\frac{r_c(\lambda t)}{r_c(t)} r, \lambda t\right] .$$ (15)

For a system following self-similar behaviour the prefactor of r can be written as λ^n for $r_c(\lambda t) \gg r_c(t)$[17] with n to be determined by the dimensionality of the specific problems. For two dimensional diffusion and three dimensional clusters n is obtained from dimensional considerations based on the mass conservation law, the Gibbs-Thomson equation and the mass transport equation and is shown to be n = 1/4 in agreement with the thermodynamical model.

An important limitation of the approach discussed above has been discussed by Gunton et al.,[18] Kawasaki et al.[19] and Huse[20] who consider finite-size effects. The fact that the clusters, as the minority phase, cover a finite area of the surface does not alter the power law dependence but can broaden the cluster size distribution. This effect is observable from a lower coverage limit on. These calculations[18-20] are only discussed for two-dimensional islands on surfaces or three-dimensional clusters in a solution while we study three-dimensional clusters on a two-dimensional surface. Nevertheless we infer from that calculations that finite-size effects are not important for the systems investigated in this paper.

EXPERIMENTAL

All samples were prepared in an ultrahigh vacuum system containing Knudsen cells for Ga, Sn, As and Ge deposition and an electron beam gun evaporator for growth of Si buffer layers. Substrates were usually sputtered and annealed Si(100) and Si(111) surfaces. Before deposition all surfaces were atomically clean, as indicated by Auger electron spectroscopy, and well ordered as indicated by sharp 7x7 (for Si(111)) and 2x1 (for Si(100)) LEED patterns.

Auger spectroscopy allows detection of clustering for very small cluster sizes, i.e. at early stages of the growth process. This is demonstrated in Fig. 4 showing the attenuation of the Si(LVV) line at 92 eV as a function of Sn coverage on a Si(111) sample at different substrate temperatures. Note the small escape length of $\sim$ 10 Å at this electron energy. Deviations from an exponential decay sensitively indicate unregularities (non-uniformity) in the thickness of an overlayer (e.g. at 300K and 470K). Only the deposition at 150K is unaffected indicating a uniform amorphous growth. At the higher temperatures a uniform deposition occurs only to 1.7 monolayer forming an initial Stranski-Krastanov layer,[21] followed by immediate nucleation of three dimensional clusters even at room temperature.

Rutherford backscattering techniques (RBS) were used to determine the height of clusters on the surface.[22] Figure 5 shows a typical spectrum obtained using 1.0 MeV He$^+$ ions on a Si(100) sample covered with Ga clusters. For illustration the hypothetical spectrum of a single hemisperical cluster of radius r is sketched in Fig. 6 together with the principle of the RBS technique: Ions from the ion beam with a much larger diameter than the cluster are backscattered from atoms of the cluster at any depth between the surface of the cluster and the substrate interface with a yield proportional to the cluster material at the respective depth. The actual spectrum (Fig. 5) results by convolution with two other contributions: The real cluster size distribution and the experimental depth resolution. As long as the cluster size distribution does not

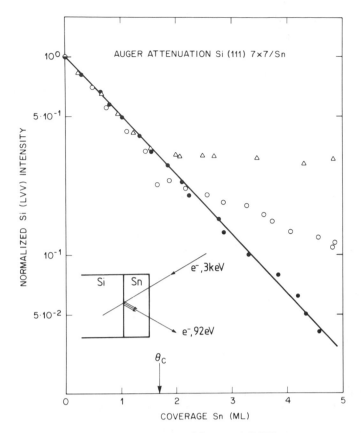

Fig. 4 Normalized intensity for the 92 eV Si(LVV) Auger transition as a
        function of Sn coverage for deposition on Si(111)-7x7 at 150K
        (full circles), 300K (open circles) and 470K (triangles). THe
        solid line is the result of an exponential least squares fit to
        the 150K data[32].

broaden extensively the height of a mean cluster can be extracted. This
was confirmed by ex-situ scanning electron microscopy (SEM) and
transmission electron microscopy (TEM) directly imaging the clusters.

In Fig. 7 we show cluster growth data comparing the third power (a)
and the fourth power (b) dependence of the cluster height vs. time. Both
dependencies are possible based on different cases summarized in the
theoretical section. The data are for clusters grown from an equivalent
coverage of 3.8 ML Sn on Si(100) and heat treated at 795 K. Figure 8
shows an equivalent plot of the fourth power of the cluster height vs
time for 2.9 ML Sn on Si(111) annealed at 525 K. In all cases the best
fit to the data consists of a combination of a third-power dependence at
short times, i.e., while the clusters are relatively small, and a
fourth-power dependence for the growth of larger clusters.

Activation energies for clustering follow from the temperature
dependence of the growth rates $\Delta r_c^4/\Delta t$: Fig. 9 shows an Arrhenius
representation of growth rates for Ga on Si(111) and Si(100). The solid

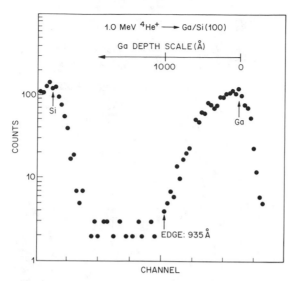

Fig. 5   Ion scattering spectrum from Si(100)-2x1 after deposition of 6
ML equivalent coverage of Ga and annealing at 475°C. The Ga and
Si edges are indicated. A depth scale for scattering in Ga is
given at the top.

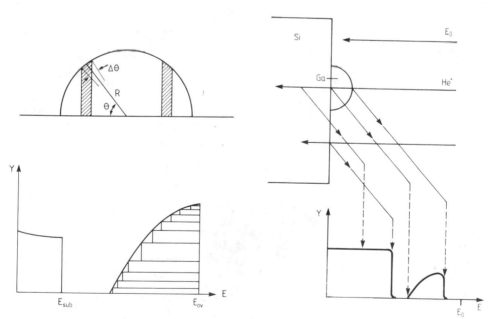

Fig. 6   Schematics of the energy distribution observed in RBS for
scattering from hemispherical clusters. Left: hypothetical
spectrum, right, principle of RBS.

465

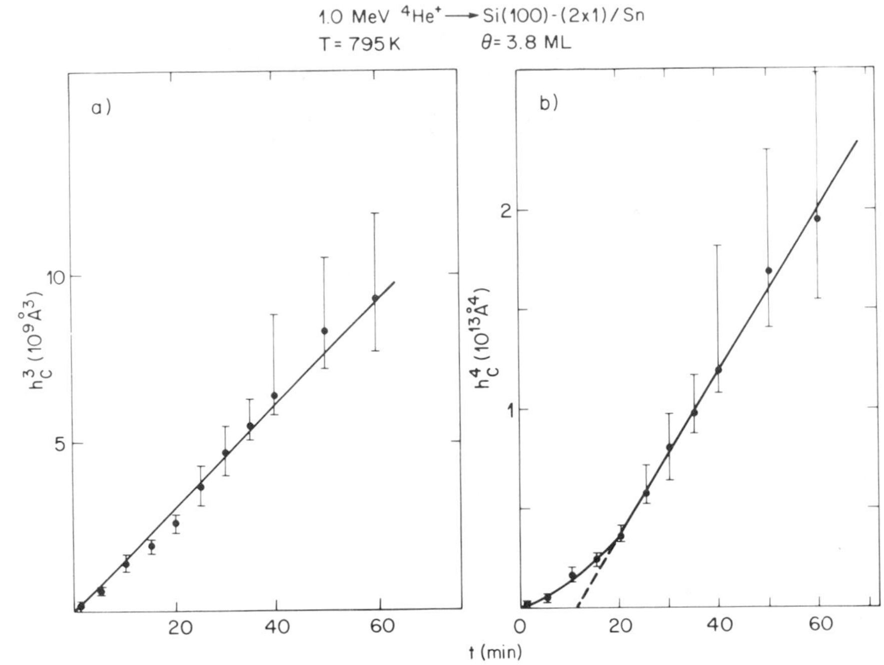

1.0 MeV ^{4}He$^+$ ⟶ Si(100)-(2x1)/Sn

T = 795 K θ = 3.8 ML

Fig. 7   Power law dependence of the height of clusters as a function of
          time for 3.8 ML equivalent coverage of Sn on Si(100)
          post-deposit annealed to 795K (a) using the cube of the cluster
          height and (b) using the fourth power of the cluster height.
          The dashed line in (b) extrapolates the late-stage growth
          linearly while the solid line illustrates a transition from
          third to fourth-power behavior.

lines fit the data with activation energies of Q = 0.80 ± 0.07 eV on
Si(100) and Q = 0.49 ± 0.05 eV on Si(111). We have systematically
investigated these parameters on Si and GaAs surfaces with different
adatoms. The results are summarized in Table I.

We now interpret these results based on the model and calculations
described by Chakraverty[13].

The linear fit to the data of Figs. 7 and 8 using the fourth power
of the cluster height strongly suggests surface diffusion as the mass
transport limiting process for the late stage of the cluster growth. This
regime corresponds to cluster radii of 1000 Å and larger. This conclusion
is in agreement with the conditions discussed above comparing the two
limiting processes of surface diffusion and mass transfer through the
surface energy barrier. Since speed constants for mass transfer are
unavailable[12] we test the condition $\kappa r^2 > D$ using the probability rate of
passing the surface. The term κ for the probability to pass the cluster
surface can be written as $\kappa = \kappa_0 \exp(-Q/kT)$ with κ_0 a frequency factor,
typically 10^{12} sec^{-1}, and Q the energy of the barrier to pass. Venables[23]
suggests estimating Q from the sublimation energy with two contributions,
the desorption from the free surface between the clusters and the energy
of a single chemical bond times a coordination number. This number can be

466

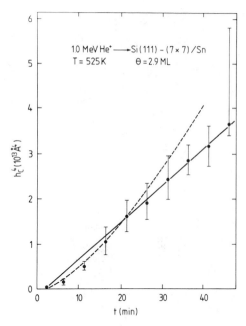

Fig. 8   RBS measurement of the fourth power of the cluster height as a function of time for 2.9 ML equivalent coverage of Sn on Si(111) deposited at room temperature and held at 525K.

achieved by assuming a nearest neighbour bond model. Unfortunately, the energies required for this estimate are not known. We therefore use a more crude estimate using the known energy of the strength of a Sn-Sn bond from bond dissociation energies[24]. This gives a lower limit for the term κr^2. The bond dissociation energy for Sn is 2.1 eV and hence $\kappa \approx 10^{-1}$ sec^{-1}. Therefore κr^2 is of the order of 10^{-11} cm^2/sec for cluster radii $r \geq 1000$ Å. A transition from the surface barrier passing limit ($t^{1/3}$ power law) to the diffusion limit ($t^{1/4}$ power law) in the region of this radius is consistent with our data for Sn on Si(100) (Figs. 7 and 8). The activation energies listed in Table I are from the later part of the growth curves and are assumed to be surface diffusion coefficients.

In addition, microscopic measurements support the interpretation of the data based on surface diffusion as the mass transport limiting process: Table I summarizes activation energies for Ga and Sn on Si(111) and Si(100), respectively. In both cases reflection electron microscopy (REM) showed that the cluster shape was independent of the underlaying surface structure. This implies that the surface tensions, which define the contact angle of clusters on surfaces, are the same for both surfaces. Still, the activation energy for clustering as well as the absolute rate of growth at a given temperature differ greatly. This result would not be expected for the passing of a surface barrier limiting the mass transfer, but can be understood for surface diffusion limited transport. In summary, there are three arguments which indicate that surface diffusion is the limiting process: (1) agreement with $t^{1/4}$ law, (2) an order of magnitude estimate which shows that $\kappa r^2 > D$ for large r and (3) dependence on the surface structure.

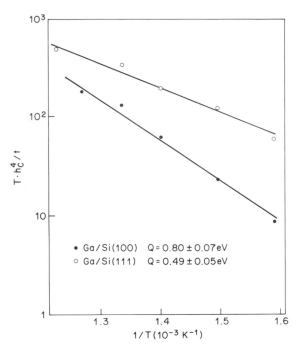

Fig. 9 Arrhenius plot of growth rates of Ga clusters on Si(100) and Si(111) to extract activation energies for the clustering process. The additional temperature factor on the ordinate is due to the Ostwald ripening model which predicts an $\exp(-E_a/kT)/T$ dependence for the ripening.

SPECIAL CASE: GROWTH ON As-TERMINATED SILICON

The use of the As terminated Si surface permits an interesting transition between the two primary semiconductors Si and GaAs. The arsenic deposition was carried out with the source at 280°C, and the sample at about 700°C; during exposure to the As source, the sample temperature was reduced to 400°C and held at this temperature for about 3 min in the As beam. Then the shutter of the As source was closed and the substrate cooled to room temperature. The resulting films displayed a sharp 1x1 LEED pattern.[25]

The deposition procedure for As on Si(111) yields a saturation coverage of 0.93 ± 0.03 ML.[26] Heating the sample gently (500°C for less than 5 min) reduces the coverage to 0.85 ± 0.04 ML indicating a small, loosely bound component. At this coverage the As/Si(111) system continues to display a sharp 1x1 LEED pattern.[27] This latter coverage corresponds closely to the density of second-layer sites in the DAS-model of the clean Si(111) reconstruction.[28] However, a simple substitution with As on these sites should yield a 7x7 LEED pattern. From X-ray standing wave experiments we know that As atoms occupy exclusively the top half of the Si(111) double plane and lie at 0.17 Å above the unrelaxed bulk terminated Si(111) plane.[29] Ion scattering experiments are in general agreement with this picture, showing a significant, but not complete, reordering of the Si surface upon As deposition.[27]

Table I

Activation Energies and Preexponential Factors for
Clustering From Cluster Growth Experiments.

System	Q (eV)	D_o (cm^2/s)	D(750 K) (cm^2/s)	Method
Sn/Si(111)	0.32 ± 0.04	5×10^{-8}	3×10^{-10}	RBS/SEM
Sn/Si(100)	1.0 ± 0.2	1×10^{-4}	2×10^{-11}	RBS/SEM
Ge/Si(100)	1.0 ± 0.1			RBS/SEM/TEM
Ga/Si(111)	0.49 ± 0.05	8×10^{-9}	4×10^{-12}	RBS/REM
Ga/Si(100)	0.80 ± 0.07	1×10^{-6}	6×10^{-12}	RBS/REM
Ga/Si(100) 4° miscut	0.80 ± 0.07	4×10^{-6}	2×10^{-11}	RBS/REM
Ga/As/Si(111)	1.23 ± 0.05			RBS/REM
Ga/GaAs(100)	1.15 ± 0.20	1×10^{-4}	2×10^{-12}	RBS/REM/SEM
GaAs/Si(100)a	0.7 ± 0.4			TEM/SEM
GaAs/Si(100)b	1.0 ± 0.1			TEM
Ga/GaAs(100)c	1.3 ± 0.1			RHEED
Sn/GaAs(100)d	1.8 ± 0.3			TEM

[a]based on D. K. Biegelsen, F. A. Ponce, A. J. Smith and J. C. Tramontana, J. Appl. Phys. 61, 1856 (1987).
[b]D. K. Biegelsen, F. A. Ponce, B. S. Krusor, J. C. Tramontana and R. D. Yingling, Appl. Phys. Lett. 52, 1779 (1988).
[c]J. H. Neave, P. J. Dobson, B.A. Joyce and Jing Zhang, Appl. Phys. Lett. 47, 101 (1985).
[d]based on J. J. Harris, B. A. Joyce, J. P. Gowers and J. H. Neave, Appl. Phys. A28, 63 (1982).

Figure 10 shows the cluster growth of 3 ML equivalent coverage of Ga deposited on the As terminated Si(111) surface and annealed at 715 K. The observed power law dependence is the same as that for Ga and Sn on the bare Si substrate (•, Fig. 10). Thus the same mechanism can be assumed for Ga cluster growth on Si(100), Si(111), As/Si(111) and GaAs(100) substrates. Note that additional As deposition during the Ga cluster growth terminates the ripening process (O, Fig. 10).[30] The diffusion coefficients for these systems are also given in Table I. The activation energies on the As terminated surface and GaAs(100) are higher, indicating an energetically rougher surface or a situation where the diffusion is governed by the strong Ga/As interaction.

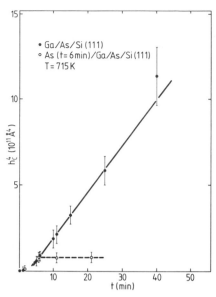

Fig. 10 RBS measurement of the fourth power of the cluster height for about 3 ML equivalent coverage on Ga on As terminated Si(111) post-deposit annealed to 715 K(●). In a second experiment on the same surface (O) an additional As deposition after 6 minutes terminated the Ga growth[33].

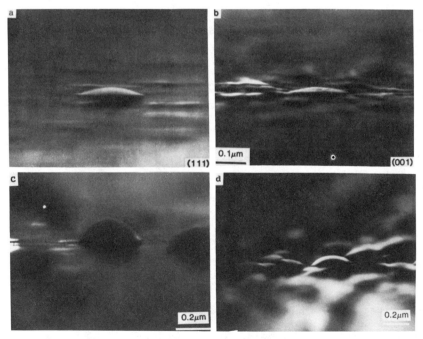

Fig. 11 Reflection electron microscopy pictures of Ga clusters on: (a) Si(111), at 750K for 100 min, (b) Si(100), at 715K for 60 min, (c) As terminated Si(111), at 715K for 40 min and (d) GaAs(100), at 825K for 10 min.

Reflection Electron Microscopy pictures of Ga clusters on these four different surfaces are shown in Fig. 11. While the contact angle on both Si surfaces is very low ($\sim 15°$) a dramatic change is observed on the As terminated surface displaying nearly hemispherical clusters. The GaAs(100) surface lies between these extreme cases. The contact angle, θ, can be related to the surface energy of the substrate, γ_s, the surface energy of the cluster material, γ_f, and the interfacial energy, γ_{sf}, through[31]

$$\cos \theta = (\gamma_s - \gamma_{sf})/\gamma_f . \tag{16}$$

Thus in the case of the As-terminated surface ($\theta \approx 90°$) the surface energy and interfacial energy are of equal magnitude while for the Ga/Si interaction ($\theta \approx 15°$) the interfacial energy is small as Si and Ga have comparable surface energies.[31] The high Ga/As interfacial energy may be correlated with the high activation energy for diffusion from the ripening data.

CONCLUSION

We have shown that the growth of clusters on surfaces can be described by an Ostwald ripening mechanism, predicting a transition from the third to the fourth power of the growth of the mean cluster radius with time. The regime described by the fourth power is shown to be diffusion limited growth. Surface diffusion activation energies and pre-exponential factors have been extracted. Data on Ga cluster growth and cluster shapes on Si(111), Si(100), As terminated Si(111) and GaAs(100) have been compared and show a correlation between interfacial energies and diffusion parameters.

Acknowledgements

We thank B. A. Davidson, G. J. Fisanick, H.-J. Gossmann and B. E. Weir for helpful discussions and technical support.

REFERENCES

1. T. P. Pearsall, J. Bevk, L. C. Feldman, J. M. Bonar, J. P. Mannaerts and A. Ourmazd, Phys. Rev. Lett. 58, 729 (1987).
2. M. H. Grabow and G. H. Gilmer, in Semiconductor-based Heterostructures: Interfacial Structure and Stability, L. Green, J. E. E. Baglin, G. Y. Chin, H. W. Deckman, W. Mayo and D. Narasinham eds., (Metallurgical Society, Warrendale, PA, 1986), p. 3.
3. L. C. Feldman, J. Bevk, B. A. Davidson, H.-J. Gossmann, A. Ourmazd, T. P. Pearsall and M. Zinke-Allmang, Mat. Res. Soc. Symp. Proc. 102, 405 (1988).
4. M. Zinke-Allmang, L. C. Feldman and S. Nakahara, Appl. Phys. Lett. 51 975 (1987); M. Zinke-Allmang, L. C. Feldman, S. Nakahara and B. A. Davidson, Phys. Rev. B39, 7848 (1989).
5. J. H. van der Merwe, in Single Crystal Films, M. H. Francombe and H. Sato eds., (Pergamon, Oxford, 1964), p. 139.
6. J. W. Mathews, in Epitaxial Growth, Part 2, J. W. Mathews ed., (Academic Press, New York, 1975), p. 559.
7. J. C. Bean, T. T. Sheng, L. C. Feldman, A. T. Fiory and R. T. Lynch, Appl. Phys. Lett. 44, 102 (1984).
8. K. Binder and D. Stauffer, Adv. Phys. 25, 343 (1976).

9. R. Bruinsma and A. Zangwill, Europhys. Lett. $\underline{4}$, 729 (1987).

10. M. von Smoluchowski, Z. Phys. Chem. $\underline{92}$, 129 (1918); see also J. T. G. Overbeek, in Colloid Science, vol. 1, H. R. Kruyt ed., (Elsevier, London, 1952), p. 278.

11. I. M. Lifshitz and V. V. Slyozov, Sov. Phys. JETP $\underline{35}$, 331 (1959); J. Phys. Chem. Solids $\underline{19}$, 35 (1961).

12. C. Wagner, Z. Elektrochem. $\underline{65}$, 581 (1961).

13. B. K. Chakraverty, J. Phys. Chem. Solids $\underline{28}$, 2401 (1967).

14. W. D. Kingery, H. K. Bowen and D. R. Uhlmann, Introduction to Ceramics (Wiley, New York), p. 185.

15. C. S. Lialikov, V. N. Piscounova, J. P. Chipilov and C. V. Cerdycev, Proc. 9th Int. Conf. on Photographic Science and Applications (Paris, 1935), p. 277.

16. J. Crank, in The Mathematics of Diffusion (Clarendon, Oxford, 1975).

17. W. W. Mullins, J. Appl. Phys. $\underline{59}$, 1341 (1986); W. W. Mullins and J. Vinals, Acta Metall. $\underline{37}$, 991 (1989).

18. J. D. Gunton, in Kinetics of Interface Reactions, Vol. 8, M. Grunze and H. J. Kreuzer eds., (Springer Series in Surf. Sci., Berlin, 1987), p. 238.

19. Y. Enomoto, M. Tokuyama and K. Kawasaki, Acta Metall. $\underline{34}$, 2119 (1986), and references herein.

20. D. Huse, Phys. Rev. $\underline{B34}$ 7845 (1986).

21. E. Bauer and H. Poppa, Thin Solid Films $\underline{12}$ 167 (1972).

22. L. C. Feldman and J. W. Mayer, in Fundamentals of Surface and Thin Film Analysis (Elsevier, New York, 1986).

23. J. A. Venables, J. Vac. Sci. Technol. $\underline{B4}$, 870 (1986).

24. J. A. Kerr and A. F. Trotman-Dickenson, in Handbook of Chemistry and Physics, 57th Edition, R. C. Weast ed., (CRC Press, Cleveland, 1976), p. F-219.

25. M. A. Olmstead, R. D. Bringans, R. I. G. Uhrberg and R. Z. Bachrach, Phys. Rev. $\underline{B34}$, 6041 (1986); R. I. G. Uhrberg, R. D. Bringans, M. A. Olmstead, R. Z. Bachrach and J. E. Northrup, Phys. Rev. $\underline{B35}$ 3945 (1987).

26. M. Zinke-Allmang, L. C. Feldman, J. R. Patel and J. C. Tully, Surf. Sci. $\underline{197}$, 1 (1988).

27. R. L. Headrick and W. R. Graham, Phys. Rev. $\underline{B37}$ 1051 (1988); J. Vac. Sci. Technol. $\underline{A6}$ 637 (1988); M. Copel and R. M. Tromp, Phys. Rev. $\underline{B37}$ 2766 (1988).

28. K. Takayanagi, Y. Tanishiro, M. Takahashi and S. Takahashi, J. Vac. Sci. Technol. $\underline{A3}$, 1502 (1985).

29. J. R. Patel, J. A. Golovchenko, P. E. Freeland and H.-J. Gossmann, Phys. Rev. $\underline{B36}$, 7715 (1987).

30. M. Zinke-Allmang, L. C. Feldman and S. Nakahara, Appl. Phys. Lett. $\underline{52}$, 144 (1988).

31. L. E. Murr in Interfacial Phenomena in Metals and Alloys (Addison-Wesley, Reading MA, 1975), p. 101ff.

32. M. Zinke-Allmang, H.-J. Gossmann, L. C. Feldman and G. J. Fisanick, Mat. Res. Soc. Symp. Proc. $\underline{77}$, 703 (1987).

33. M. Zinke-Allmang, L. C. Feldman and S. Nakahara, J. Vac. Sci. Technol. $\underline{B6}$, 1137 (1988).

STRAIN RELAXATION IN Ge/Si(001) STUDIED USING X-RAY DIFFRACTION

J. E. Macdonald,[1] A. A. Williams,[1] R. van Silfhout,[2]
J. F. van der Veen,[2] M. S. Finney,[3] A. D. Johnson[3] and
C. Norris[3]

[1]Physics Department, University of Wales College of
Cardiff, P.O. Box 913
Cardiff CF1 3TH, U.K.
[2]FOM Institute for Atomic and Molecular Physics,
Kruislaan 407
1098 SJ Amsterdam, The Netherlands
[3]Physics Department, University of Leicester
Leicester LE1 7RH, U.K.

ABSTRACT

Grazing incidence x-ray diffraction has been utilized to give a
direct measure of the lateral strain distribution during in-situ MBE
deposition of Ge onto Si(001). Differences in growth conditions and
thermal treatment result in significantly different strain relaxation
behavior. The results demonstrate the gradual relaxation of strain,
which is incomplete at a coverage of 11 monolayers. At this coverage
the strain distribution exhibits two components, one of which is almost
fully relaxed and the other having a range of lattice spacings
intermediate between those for bulk Si and Ge. The results are discussed
in terms of current models for strain relaxation.

INTRODUCTION

The use of strained layer epitaxy for semiconductor materials has
expanded rapidly in recent years.[1] Interest in heteroepitaxial structures
based on Ge/Si has centered on the possible fabrication of artificial
crystals having a quasidirect bandgap, resulting from the effect of
strain on the electronic structure. One of the most important parameters
in the production of such structures is the critical layer thickness at
which the strain energy in the film is relieved by the generation of
misfit dislocations. Critical thicknesses have been calculated by
considering either the energy or mechanical forces required for the
propagation of existing dislocations[2,3] and also including the formation
of new dislocations.[4,5] These models lead to values of the equilibrium
critical thickness h_e, below which no strain relaxation should be
observed. However, experimental values of critical thicknesses are often
considerably larger than calculated values of h_e.[6] The extent of
relaxation has also been found to depend on growth temperature and
annealing processes,[7,8] pointing to a kinetic view of strain relaxation.
Consequently other models have been developed involving

Kinetics of Ordering and Growth at Surfaces
Edited by M. G. Lagally
Plenum Press, New York, 1990

thermally-activated plastic deformation of the epitaxial layer.[9] This leads to a second critical thickness h_m, above which significant relaxation is observed. At layer thicknesses intermediate between h_e and h_m, the layer is metastable. It should however be pointed out that lower critical thicknesses are likely to be observed using techniques which detect dislocations, such as photoluminescence, than with those which are sensitive to strain relaxation, such as x-ray diffraction and ion scattering.[10,11] Models of critical thicknesses have been reviewed recently by Tuppen et al.[12]

Most of the existing experimental data for critical thicknesses involve $Si_{1-x}Ge_x$ films with $0.0 < x < 0.5$. The relaxation behavior of Ge-rich films, for which the critical thickness is small, has not been studied so extensively. Ion scattering studies of Ge/Si(001) yield a critical coverage θ_c of 6 monolayers (ML), where $1ML = 1.41Å$.[13] Changes in the profile of RHEED oscillations have been identified with changes from 2-dimensional to 3-dimensional growth, leading to values of θ_c of between 3 ML and 5 ML for pure Ge on Si(001).[14,15] Scanning electron microscopy studies of Ge layers deposited onto Si(001) substrates at room temperature showed islanding for coverages greater than 3ML upon heating above 250°C[16]. Ion scattering and electron microscopy studies have been used to analyze the kinetics of clustering after a post-deposition anneal and during low flux growth.[17]

In this work we aim to address the following questions:

1. What is the critical thickness for strain relaxation in pure Ge epilayers on Si(001)?
2. Does the strain relax fully when the critical thickness is exceeded and is the strain uniformally distributed?
3. How does the strain relaxation depend on growth temperature and thermal treatment?
4. Is strain relief primarily related to islanding or to dislocation formation?

Grazing incidence x-ray diffraction (GIXRD) gives a direct measure of the lateral strain parallel to the surface. In this geometry the peak from an unrelaxed epilayer coincides with that from the substrate as a result of their identical in-plane lattices spacings. This is independent of any intermixing at the interface. This contrasts with the conventional diffraction geometry, in which the momentum transfer $Q = Q_{/\!/} + Q_{\perp}$ has a large component $Q_{\perp}$ normal to the surface, the lateral component $Q_{/\!/}$ being limited by shadowing of either incident or scattered beams by the flat crystal surface. Strain relaxation and intermixing are often difficult to distinguish with the conventional geometry since both effects alter the epilayer peak position. The conventional geometry is more suited to monitoring the strain normal to the surface, but off-axis strain may also be probed by employing an asymmetric scattering geometry. For studying thin epilayers GIXRD benefits from the fact that the intrinsic peak width is much narrower due to the much larger extent of the film parallel to than normal to the interface, thus providing a direct measure of the in-plane strain distribution. For symmetric peaks ($Q_{/\!/} = 0$) in the conventional geometry, the epilayer peak is obscured by the substrate crystal truncation rod for symmetric reflections in the conventional geometry.[18] Thus GIXRD with $Q_{\perp} \approx 0$ is more sensitive to the occurrence of strain relaxation in very thin epilayers whereas the $Q_{\perp}$-dependence of the scattering is most useful for the study of strain normal to the surface, particularly in thicker layers and superlattices.

Following a preliminary investigation,[19] three samples were investigated in order to study the effect of thermal treatment on the relaxation process. These are denoted:

I: Substrate held at 550°C during deposition and immediately cooled to room temperature before measurement. This procedure was repeated after deposition of each monolayer.
II: Substrate maintained at 520°C during deposition and subsequent measurements.
III: Deposition of 2ML at a substrate temperature of 320°C, as suggested to avoid intermixing at the interface detected using Raman scattering,[20] followed by an anneal at 520°C for 30 minutes. Measurements were then performed at room temperature and this procedure was repeated for each subsequent deposition of 1ML.

EXPERIMENTAL DETAILS

The experiments were performed at the Synchrotron Radiation Source in Daresbury (UK) using unfocussed radiation from the superconducting Wiggler beamline. The x-rays were monochromatized using the (111) reflection from a channel cut Si crystal. An incident beam wavelength of 0.8 was employed for sample I and of 1.38Å for samples II and III, leading to critical angles for total reflection in Si of 0.12° and 0.20° respectively. The incident and exit grazing angles were set to $\beta = 0.09°$ and $\beta' = 0.13°$ respectively for sample I and $\beta = 0.08$, $\beta' = 0.09°$ for samples II and III. The detector acceptance angle $\Delta\beta'$ was 0.10°. The use of grazing incidence geometry resulted in suppression of bulk thermal diffuse scattering around the substrate Bragg peak, increasing sensitivity to strain in the epilayer.

The equipment consists of a large 5-circle diffractometer, coupled to an ultrahigh-vacuum chamber having in-situ MBE growth facilities as well as standard surface science analytical techniques.[21] The Si(001) substrates were cleaned by light sputtering with 800eV Ar^+ ions for 60 seconds followed by an anneal for 3 min at 1060°C. This gave a sharp 2x1 diffraction pattern, observed with RHEED and x-ray diffraction. The full width at half maximum (FWHM) of the (3/2,0) and (0,3/2) fractional order reflections, arising from the double-domain 2x1 reconstruction, were ∿0.02° for the clean surface, corresponding to an average reconstructed domain size of ∿7000Å.

The sample physical surface normal was aligned by reflection of a laser beam, while the crystallographic alignment was determined from the position of four bulk reflections.[22] Thus the miscut of the surface relative to the (001) crystallographic axis was determined to be 0.04°. The real unit cell vectors, used to define the scattering vector $Q = h\underline{b}_1 + k\underline{b}_2 + l\underline{b}_3$ in reciprocal space may be related to the conventional bulk cubic real cell vectors by $\underline{a}_1 = (1,1,0)_{cubic}$, $\underline{a}_2 = (1,-1,0)_{cubic}$ and $\underline{a}_3 = (0,0,1/4)_{cubic}$.

Deposition of Ge was performed using a Knudsen effusion cell which gave a deposition rate of 1 monolayer per 18 min ± 10% , calibrated by Rutherford Backscattering. Radial scans (along h) through the (2,0) Bragg peak were performed at grazing incidence at each coverage in order to give the distribution of lattice spacings parallel to the interface. The resolution along h is 0.02 r.l.u. as determined by the detector aperture slits and the illuminated surface area. Additional transverse scans (along k) were employed to determine the lateral extent of correlations and rod scans (along l) to give the correlation length normal to the interface.

RESULTS AND DISCUSSION

The radial scans for the different samples are shown in Fig. 1-3. For sample I, the peak profile remains unchanged for $\theta \leq$ 3ML due to the homoepitaxial nature of the Ge layer. The wings of the peak are not substantially broadened indicating the high crystalline quality of the overlayer. At 4ML, a weak shoulder appears on the Bragg peak due to the onset of strain relaxation. Upon further deposition the shoulder develops as the layer relaxes further. At a coverage of 11ML the peak from the overlayer occurs at h = 1.925 reciprocal lattice units (r.l.u.), which is close to that expected for bulk Ge (h = 1.920 r.l.u.). This gradual shift shows that the strain is relaxed gradually after exceeding the critical thickness, which is 3-4ML in this case. Strain relief is incomplete even after deposition of 11ML of Ge. A further striking feature of the radial scan for sample I is the 'plateau' of scattering between the bulk values of Ge and Si. This indicates that the strain is not constant in the epilayer, but rather that there is a distribution of intermediate values of strain. The question arises whether the strain is distributed laterally across the layer or as a function of height above the interface. Consequently the scan was repeated with β = 0.07°, β' = 0.05° ensuring that the detected beam was entirely below the critical angle. The effective penetration depth was thus reduced from 100Å to 40Å and led to the scan profile denoted by triangles in Fig. 1. Here the scattering from the plateau in the region h = 1.95-2.00 r.l.u. suffers more severe attenuation than the peak at h = 1.925 r.l.u.. This indicates that the atomic layers closest to the interface are still strained while the topmost layers are almost fully relaxed. It should be noted that the values of the penetration depth quoted above are difficult to interpret in view of the islanding that occurs at this coverage, as discussed later. A significant fraction of the beam may be totally externally reflected from the tops of islands thus shadowing parts of the surface.

Information concerning the extent of correlations laterally across the surface, for both the fully-relaxed and partially-relaxed regions, may be gleaned from transverse scans (varying k) at different values of h (Fig. 4). Assuming isotropic correlations in two dimensions, the FWHM at h = 1.925 r.l.u. and at h = 1.970 r.l.u. result in correlation lengths of 80Å and 100Å for the relaxed and partially relaxed regions of the epilayer respectively, after correction for resolution effects. The detailed interpretation of these values of the correlation lengths in terms of parameters such as dislocation densities is unclear at present.

It should be noted that the intermediate distribution of lattice parameters detected in the range h = 1.94-2.00 could also arise from an alloyed region at the interface, as indicated by the large number of Ge-Si bonds detected using Raman scattering for material grown at $\sim$500°C.[20] While the x-ray data cannot be used to establish whether intermixing does occur, it is unlikely that such an alloyed region would relax at a coverage of $\sim$4ML as the critical thickness for alloys containing significant concentrations of Si would be much larger.[4] Analysis is currently underway of specular reflectivity data collected simultaneously. The results will provide an electron density profile normal to the interface, which should clarify the extent of intermixing at the interface.

The corresponding radial scans for sample II, for which deposition and measurements were performed at 520°C, are shown in Fig. 2.

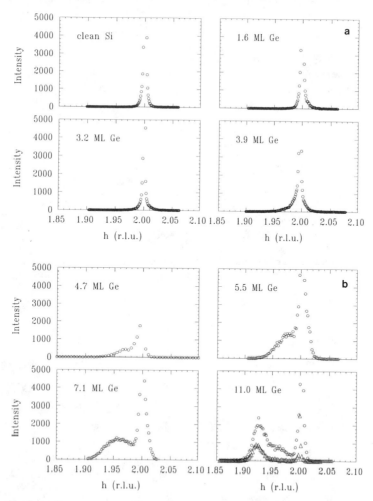

Fig. 1 Radial scans as a function of coverage for sample I. The intensity is plotted on an arbitrary scale. On this scale the Bragg peak intensity is $\sim 10^5$. For $\theta = 11ML$, circles denote data for $\beta = 0.09°$, $\beta' = 0.13°$ and triangles represent the same scan with $\beta = 0.07°$, $\beta' = 0.05°$, leading to a reduced penetration depth. a) Lower and b) higher coverages.

In this case, substantial relaxation is observed at $\theta = 3ML$. At $\theta = 4ML$, the peak from the relaxed Ge occurs at h = 1.95 r.l.u. and thus the strain in the overlayer is reduced to 1.5% whereas in sample I the extent of strain relief is minimal at this coverage. This provides further evidence that strain relaxation is indeed a thermally-activated process and is consistent with the clustering behavior observed by Zinke-Allmang et al.[17]

The radial scans for sample III, for which deposition was performed at a substrate temperature of 320°C followed by an anneal at 520°C, are shown in Fig. 3. At $\theta = 3 ML$, a weak peak is observed at h = 1.94 r.l.u. and this peak becomes more prominent at $\theta = 4 ML$. However the

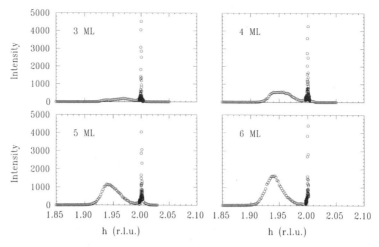

Fig. 2 Radial scans as a function of coverage for sample II.

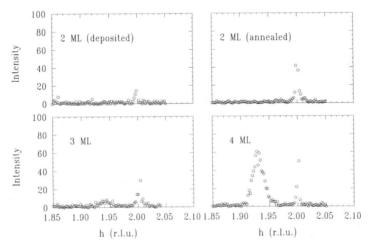

Fig. 3 Radial scans as a function of coverage for sample III.

scan profile differs from those for the other two samples in two ways. Firstly, the Ge peak is almost fully relaxed at very low coverages and there is no scattering at intermediate values of h indicating that there is no gradual transition of strain at the interface as was observed for sample I. Secondly, the scattered intensity is much weaker than that observed for samples I and II which suggests that the relaxation does not occur throughout the epilayer, but rather occurs in localized regions, possibly as bulk-like Ge islands nucleating at defects. The repeated thermal cycling involved in the measurement precludes drawing any firm conclusions about the effect of the low growth temperature on the relaxation behavior.

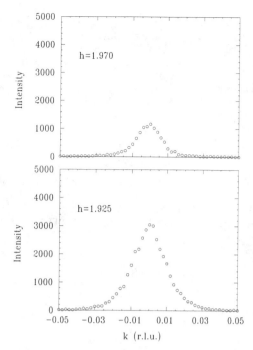

Fig. 4 Transverse scans at θ = 11ML for sample I. These are shown for the relaxed peak (h = 1.925 r.l.u.) and the intermediate strained region (h = 1.970 r.l.u.). After correction for resolution, these correspond to correlation lengths of 80Å and 100Å respectively.

CONCLUSIONS

The results demonstrate the sensitivity of GIXRD to lateral strain relaxation. Radial scans in reciprocal space provide a direct measure of the strain distribution and the depth dependence of the strain may be probed by varying the grazing angles of the incident and scattered beams. The results indicate that the critical thickness for strain relaxation of Ge epilayers on grown on Si(001) substrates is about 3 ML. This coincides with the coverage at which changes in the surface morphology are observed with RHEED,[14,15] which are most sensitive to surface morphology. As the Ge coverage is increased the layer relaxes gradually until almost fully-relaxed Ge is formed at $\theta \sim$ 11ML. The strain distribution is not constant, but intermediate lattice spacings between those of bulk Si and Ge are observed. By varying the grazing conditions, the intermediate strained material is shown to occur at the interface. This shows that a strained region of Ge persists at the interface even after relaxation of the surface region. Strain relaxation is enhanced by extended heating, evidencing that relaxation is thermally-activated in qualitative agreement with other work.[7,9] While it should be noted that this investigation was performed using very low deposition rates, recent results for samples prepared by more rapid deposition (0.5 Ås^{-1}) onto substrates at 400±50°C and capped with amorphous silicon (prepared by C. Gibbings, British Telecom Research Laboratories), showed strain distributions similar to those for sample I.

The onset of strain relaxation at 3-4 ML coincides with the coverage at which clustering is observed.[17] Such clustering, is also thermally activated by annealing at temperatures ≥500°C. This indicates that, for pure Ge epilayers on Si(001) surfaces, clustering is intimately related to strain relaxation. Further work is required to determine the role of islanding and dislocation formation, and to investigate in detail the effects of growth conditions and thermal treatment on the strain distribution in Si/Ge heterojunctions and multilayers.

Acknowledgements

We thank J. Flapper and Th. Michielsen from the Philips Research Laboratories (Eindhoven) for providing us with accurately oriented and polished Si(001) substrates. Discussions with S. Jain, C. Matthai and C. Tuppen are gratefully acknowledged. This work is supported by the U. K. Science and Engineering Research Council (SERC) and the Nederlandse Organisatie voor Wetenschappelijk Onderzoek (NWO).

REFERENCES

1. E. P. O'Reilly, Semicond. Sci. Technol. 4, 121 (1989).
2. J. H. van der Merwe, J. Appl. Phys. 34, 123 (1962).
3. J. W. Matthews and A. E. Blakeslee, J. Vac. Sci. Technol. 12, 126 (1975).
4. R. People and J. C. Bean, Appl. Phys. Lett. 47, 322 (1985) and Appl. Phys. Lett. 49, 229 (1986).
5. P. M. J. Marée, J. C. Barbour, J. F. van der Veen, K. L. Kavanagh, C. W. T. Bulle-Lieuwma and M. P. A. Wiegers, J. Appl. Phys. 62, 4413 (1987).
6. J. C. Bean, L. C. Feldman, A. T. Fiory, S. Nakahara and I. K. Robinson, J. Vac. Sci. Technol. A2, 436 (1984).
7. R. H. Miles, T. C. Gill, P. P. Chow, D. C. Johnson, R. J. Hauenstein, C. W. Nieh and M. D. Strathman, Appl. Phys. Lett. 52, 916 (1988).
8. D. J. Lockwood, J.-M. Baribeau and P. Y. Timbrell, J. Appl. Phys. 65, 3049 (1989).
9. J. Y. Tsao, B. W. Dodson, S. T. Picraux and D. M. Cornelison, Phys. Rev. Lett. 59, 2455 (1987).
10. I. J. Fritz, Appl. Phys. Lett. 51, 1080 (1987).
11. Y. Kohama, Y. Fukada and M. Seki, Appl. Phys. Lett. 52, 380 (1988).
12. C. G. Tuppen, C. J. Gibbings and M. Hockly, J. Cryst. Growth 94, 392 (1989).
13. J. Bevk, J. P. Mannaerts, L. C. Feldman, B. A. Davidson and A. Ourmazd, Appl. Phys. Lett. 49, 286 (1986).
14. S. S. Iyer, P. R. Pukite, J. C. Tsang and M. W. Copel, J. Cryst. Growth 95, 439 (1989).
15. K. Sakamoto, T. Sakamoto, S. Nagao, G. Hashiguchi, K. Kuniyoshi and Y. Bando, Jap. J. Appl. Phys. 26, 666 (1987).
16. H.-J. Gossmann and G. J. Fisanick, J. Vac. Sci. Technol. A6, 2037 (1988).
17. M. Zinke-Allmang, L. C. Feldman, S. Nakara and B. A. Davidson, Phys. Rev. B39, 7848 (1989).
18. I. K. Robinson, Phys. Rev. B33, 3830 (1986).
19. A. A. Williams, J. E. Macdonald, R. G. van Silfhout, J. F. van der Veen, A. D. Johnson and C. Norris, J. Phys.: Condens. Matter 1, SB273 (1989).
20. S. S. Iyer, J. C. Tsang, M. W. Copel, P. R. Pukite and R. M. Tromp, Appl. Phys. Lett. 54, 219 (1989).

21. E. Vlieg, A. van't Ent, A. P. de Jong, H. Neerings and J. F. van der Veen, Nucl. Instr. Methods A262, 522 (1987).
22. E. Vlieg, J. F. van der Veen, J. E. Macdonald and M. Miller, J. Appl. Cryst. 20, 330 (1987).

ON THE MAGNETIC PROPERTIES OF ULTRATHIN EPITAXIAL COBALT

FILMS AND SUPERLATTICES

J. J. de Miguel, A. Cebollada, J. M. Gallego and
R. Miranda

Departamento de Física de la Materia Condensada
Universidad Autónoma de Madrid, Cantoblanco
28049 Madrid, Spain

ABSTRACT

The growth and magnetic properties of films of fcc cobalt on Cu(100) substrates has been characterized by a multitechnique approach. The films are ferromagnetically ordered in-plane at temperatures below T_c. The Curie temperature of the films displays a linear dependence with the coverage reaching bulk-like behaviour at coverages of 5-6 monolayers. Spin-polarized photoemission shows that the band structure is already close to that of the bulk at 5 ML. Crystalline Co/Cu superlattices have also been grown on Cu(100) substrates. The magnetic ordering of Co slabs across Cu layers of varying thicknesses, as explored by neutron diffraction, has been found to be antiferromagnetic. When applying an external magnetic field in the plane of the layers, the magnetization shows a complex behaviour related to in-plane anisotropy.

INTRODUCTION

Substantial modifications of many physical (magnetic, transport, superconducting) properties are expected when the effective dimension of a system drops below a "critical" length (magnetic coherence, mean free path, coherence length) characteristic of the relevant phenomenon. The recently developed capability to grow in UHV epitaxial films of metals and artificially layered materials, while controlling the deposited amount in fractions of a monolayer (ML) has allowed us to explore experimentally this field.[1] In particular, an immense amount of work has been devoted to elucidate the magnetic properties of thin films[2] and multilayers.[3] In spite of these efforts, an accepted picture of the magnetic properties of thin heteroepitaxial films has not been reached so far.[4] Even worse, there are many examples in the literature of contradictory reports concerning the magnetic properties of the same experimental system.[5-18]

We are deeply convinced that the origin of the conflicting results lies in differences in the growth and characterization of the deposited adlayers among the different laboratories. Accordingly, we set out to study, in a selected model system, the influence of the growth conditions in the magnetic properties of thin films and superlattices. The purpose

Kinetics of Ordering and Growth at Surfaces
Edited by M. G. Lagally
Plenum Press, New York, 1990

of the present paper is to review the results of such an study for a particular system, fcc Co films and {Co/Cu} multilayers, epitaxially grown on single-crystal, Cu(100) substrates, which has attracted a considerable attention recently.[5-18] This study, carried out by means of a multitechnique approach, has allowed us to resolve some contradictions concerning the magnetic properties of ultrathin Co films[13,15] and has yielded a more detailed understanding of the magnetic properties of metallic superlattices than previously available.[17,18]

In summary, we have found that the structural parameters influencing the magnetic properties are the following: i) the surface perfection of the single-crystal substrate prior to deposition; ii) the calibration of the deposited coverage; iii) the nucleation and growth processes in the adlayer; iv) the mode of growth; v) the structural perfection of the film and vi) the interdiffusion conditions. This latter is specially important for multilayers.

EXPERIMENTAL METHODS

The data have been obtained in various experimental chambers which have been described in detail in the original publications.[6,15,18] All the Cu substrates employed were cut from a single-crystal bar (1.3 cm^2) and oriented to within 0.1° of the (001) direction by Laue diffraction. The substrate temperature, that could be changed from 150 K to 1200 K, was measured by Cr-Al thermocouples spot-welded to the back side of the samples. They were systematically cleaned by cycles of sputtering with Ar (600 eV, 1μA/cm^2) and annealing to 1000 K. The surface cleanliness was always checked with Auger Electron Spectroscopy (AES), while the average crystalline perfection of the substrate was characterized by means of Low Energy Electron Diffraction (LEED) or Thermal Energy Atom Scattering (TEAS). TEAS data indicate that the cleaning procedure employed results in surfaces containing flat terraces, 400 Å wide on the average, separated by monoatomic steps.[15] Scanning Tunneling Microscopy (STM) reveals that only substrates with terraces larger than 400 Å were suitable for appropriate Co growth to occur.[16] In particular, Cu(100) surfaces with smaller terraces (50-100 Å) do not show oscillations during Co evaporation (see below).

The Co (Cu) layers were evaporated onto the substrates from ovens heated by electron bombardment and equipped with mechanical shutters. The pressure in the chambers was typically in the low 10^{-9} Torr range during evaporation. The deposition rate was low (0.01 Å/sec) and the quality of the growth was controlled by measuring TEAS and/or Medium Energy Electron Scattering (MEED) intensity oscillations.[15] The observation of a (1x1) LEED pattern and Kikuchi lines while displacing the substrate in front of the LEED optics ensures the lateral crystallinity of the growing films.

GEOMETRIC STRUCTURE OF THE COBALT FILMS

Crucial points in thin film studies are the accurate determination of the deposited coverage, the mode of growth and the conditions for interdiffusion.

1. Mode of Growth

It is known that the experimental realization of an epitaxial metallic film of high structural perfection is a difficult task. In principle, simple thermodynamic arguments would predict that films of a high surface-energy metal (e.g., Co, H$_{vap}$ = 93 kcal/gatom) deposited on a lower surface energy metal surface (e.g. Cu, H_v = 73 kcal/g atom) would tend to form three-dimensional crystallites, rather than adopting a layer-by-layer mode of growth.

The standard method to determine the mode of growth (and calibrate the thickness) of a deposited layer is based on the time evolution of the Auger peaks of the substrate and the deposited material[19]. In this way, it was reported some time ago[6] that cobalt grows epitaxially on Cu(100) in a layer-by-layer fashion. Since then, this has been confirmed by means of AES,[8] LEED,[11] TEAS[15] and MEED[14] data. The precise determination of the coverage from AES uptake curves depends, however, on the detection of "breaks" and relies on the values of the inelastic mean free path (IMFP) for electrons.[20] An independent method to obtain the growth mode and the deposited coverage is, thus, desirable.

Figure 1 displays the specularly reflected intensity of He atoms scattered by the surface during growth of Co on Cu(100) at different temperatures. A number of equally-spaced oscillations are detected. These TEAS oscillations have monolayer periodicity and their persistence signals a layer-by-layer mode of growth,[21] in close analogy to earlier

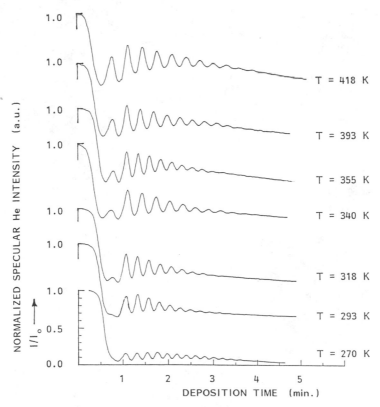

Fig. 1 Temperature dependence of the TEAS oscillations observed during deposition of Co on Cu(100).

observations of RHEED oscillations for semiconductors[22] and metals.[23] The microscopic origin of the oscillations is illustrated in Fig. 2. For systems growing in the layer-by-layer mode, Monte Carlo simulations[24] have revealed that nucleation and growth of two-dimensional islands on the terraces of the substrate provide the scenario oscillating with monolayer periodicity. Actually, many surface-sensitive quantities (work

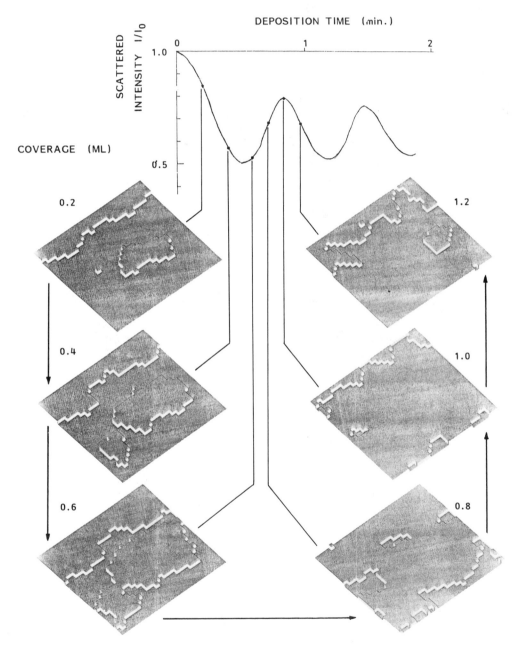

Fig. 2 Microscopic origin of the oscillations of the TEAS specular intensity during layer-by-layer growth. The images are the result of Monte Carlo simulations.[25]

486

function, sample current, diffracted intensities in LEED and RHEED) have been found to oscillate during layer growth. This also holds for the specularly reflected intensity in MEED, which oscillates periodically as a function of the evaporation time for almost every set of experimental parameters.[15] The period was independent of the azimuth and the particular diffracted spot being measured. By monitoring the time evolution of the AES signals of Cu and Co at 58 and 54 eV of kinetic energy simultaneously with the specularly reflected MEED intensity, we have demonstrated that the breaks detected in the AES uptake curve coincide nicely with the maxima in the MEED intensity.[16] Thus, the period of both MEED and TEAS oscillations corresponds to the growth of a single Co layer. This is very convenient because of the facility of implementing the MEED technique in any laboratory. The simple counting of the number of oscillations is a method that can be used with confidence to establish the Co coverage, specially for large coverages where the AES signal from the substrate is reduced and, accordingly, any AES calibration is prone to errors.

2.   Calibration of the Deposited Co Coverage

In Fig. 3 we show the ratio of the peak-to-peak intensities of the Auger peaks of Cu_{920} and Co_{716} as a function of the number of deposited Co layers, which was independently and simultaneously determined by our counting the number of MEED oscillations in the intensity of the specular beam, as described above. The continuous line is a fit to a model for layer-by-layer growth.[20] This calibration has been independently tested by chemical titration of the number of Cu atoms visible on the surface after deposition of a certain amount of Co by means of O_2[6] and CO[8] adsorption. These experiments indicate that at 1ML the Cu surface is completely covered by Co. The mean free paths deduced from the data in Fig. 3 are $\lambda_{Co}(716 \text{ eV}) = 9.4$ Å and $\lambda_{Cu}(920 \text{ eV}) = 13.2$ Å. These values are smaller than those calculated (14.9 and 17.2 Å, respectively) using theoretical expressions.[21] An early AES study reported inelastic mean free paths values of $\lambda_{Co}(95) = 5.9$ Å, $\lambda_{Cu}(105) = 6.5$ Å and $\lambda_{Co}(716) = 9.6$Å.[6] The AES data presented here are in excellent agreement with earlier calibrations.[6]

3.   Structure and Perfection of the Films

At temperatures below 690 K the stable phase of cobalt is hcp. An fcc phase exists at higher temperatures. However, we showed some years ago that the fcc phase of cobalt can be stabilized at 300 K by depositing Co in UHV on a Cu(001) substrate.[6] Despite the lateral mismatch of 1.9% ($a_{Co} = 3.548$ Å, $a_{Cu} = 3.615$ Å) the unit cell of the growing film and the substrate seems to be congruent as indicated by the sharp (1x1) LEED pattern observed during growth.[6,7,11] Earlier LEED measurements suggested that Co adsorbs in the four-fold hollow site to continue the fcc lattice of Cu. The Co-Cu vertical distance at the interface was d = 1.70 Å.[11] For subsequent cobalt layers grown at 300K the Co-Co vertical distance was found to be 1.74 Å[11] somewhat smaller than the bulk value. These observations seem to indicate that the lattice mismatch is accommodated by strain. In spite of the accumulation of strain energy in the growing overlayer, fcc films of Co of more than 20 ML have been grown with a (1x1) LEED pattern of reasonable quality,[6] suggesting pseudomorphic growth. The critical thickness of the cobalt film at which misfit dislocations develop to relief the strain, is not known.

The density of defects in the deposited layer can be estimated from the decrease of the specularly reflected TEAS intensity.[25] In the stationary state displayed in Fig. 1, they are of the order of a few percent (e.g. 3% at 418 K and 6% at 293 K). Figure 4 shows a three-

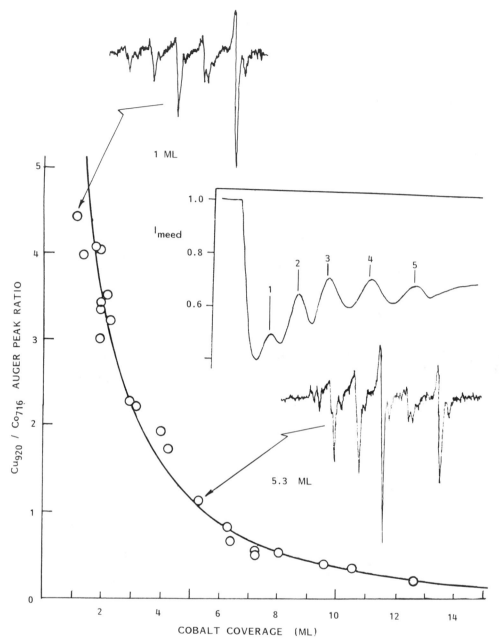

Fig. 3 Ratio of the peak-to-peak intensities of the Auger peaks of Cu at 920 eV and Co at 716 eV as a function of the number of Co layers deposited on a Cu(100) substrate. The inset shows the oscillations in the intensity of the specular beam in MEED, I_0, observed when 5.3 ML are evaporated on the substrate as well as the Auger spectrum measured afterwards, which is almost identical to the "1 ML" spectrum in Ref. 11. A spectrum of the high energy Auger peaks for a 1 ML film is also shown in the inset. The low energy Auger peaks of the Cu substrate are no longer visible with a coverage of 6.5 ML of Co.

dimensional plot of the specular intensity of TEAS as a function of both
the angle of incidence of the He beam and the evaporated Co coverage. For
constant angle of incidence, maxima in the reflected intensity appear at
completion of every Co layer. The amplitude of the oscillations is more
pronounced at angles corresponding to destructive interference.[25] For any
given coverage, there is a rocking curve containing information on the
interference conditions (e.g. vertical distances, density) of atomic
terraces. Figure 4 clearly demonstrates that these are, for the first
half-layer of Co in direct contact with the Cu substrate, different from
the rest.

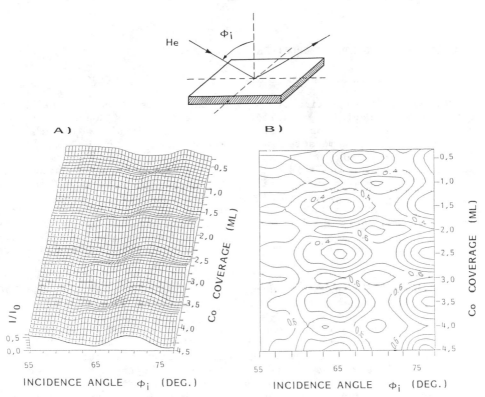

Fig. 4 a) Specular intensity of TEAS as a function of the angle of
incidence and the deposited coverage for Co/Cu(100).   b) Contour
plot of the data in a).

4.   Interdiffusion Conditions

In order to obtain meaningful magnetic measurements it is very
important to perform a careful investigation of the conditions for
interdiffusion. It has been shown previously how this problem affects
SMOKE data[15] for Co/Cu(100). In other cases, such as Fe/Cu(100),
interdiffusion can be the main reason of conflicting reports on the
magnetic properties.[26] In spite of the limited (9% at.) solubility of Co
in Cu seen in the bulk phase diagram, a noticeable degree of
interdiffusion may exist at the monolayer range, as indicated by the fact
that Co can be totally dissolved into the Cu substrate by heating to 1000
K. It is our experience that interdiffusion of Co into the Cu substrate

does depend basically on the temperature and the condition of the substrate (defects, thin film vs bulk). The interdiffusion is faster if the sample is held at a constant temperature during evaporation than if it is maintained at the same temperature _after_ a Co film has been deposited at lower temperature. At constant temperature the dependence with time is weak. The amount of Co that can be dissolved into Cu does so rapidly. All this points to the existence of an interfacial alloyed layer. This does not mean, however, that epitaxial layer of Co of satisfactory degree of perfection can not be obtained. For example, intermixing does not occur if the deposition is carried out at temperatures below 450 K. For films of Co deposited at 300 K, the low energy AES peaks reveal an appreciable interdiffusion only after annealing at 500 K during 30 min. Thus, an abrupt Co/Cu interface can be expected for processing temperatures below those mentioned.

ELECTRONIC AND MAGNETIC PROPERTIES OF COBALT FILMS

1.   Ground Magnetic State. Spontaneous Magnetization Versus Temperature

The magnetic properties of (single) thin films of Co on Cu(100) are controversial, specially with regard to the thickness dependence. Most authors agree that the cobalt films are magnetized in-plane, as indicated by hysteresis loops measured with Surface Magneto-Optic Kerr Effect (SMOKE).[27] No sign of perpendicular orientation of the magnetization has been found even for the thinnest films.[14] Some authors,[9,10,12] however, have claimed that fcc cobalt films are ferromagnetic down to the 1ML limit, with T_c well above room temperature.[13] In more detail, Beier et al[13] reported SMOKE measurements displaying a hysteresis loop at room temperature for a "single monolayer" of fcc Co epitaxially grown on Cu(100). No change in the magnetic properties was observed up to 450K.[13] These results seemed to support previous data by Pescia et al.[9] for the same system which indicated that T_c was well above 300K even for a single layer film. Furthermore, above 1 ML, T_c was reported to be independent of the Co coverage,[9] implying a surprising insensitivity of the magnetic properties with respect to the film thickness.

In contradiction with these results, we have found by longitudinal SMOKE measurements[15,16] performed in carefully characterized fcc Co layers that T_c for the films is drastically lower than the T_c value of bulk fcc Co (1388 K), as reproduced in Fig. 5. Furthermore, T_c shows a linear dependence with the Co coverage in the range accessible to experiments (the upper limit being determined by interdiffusion, the lower by the present experimental set-up). Extrapolation of the data indicates that T_c reaches bulk-like values at coverages of 5.4 ML. It is suggestive that the simple linear extrapolation indicates that the Curie point for a monolayer film would be _zero_. This would be an unexpected support of the Mermin-Wagner theorem[28] that states that in a 2D system with only short-range isotropic magnetic interaction, the magnetization is zero at any finite temperature. Strong dependences of T_c with the evaporated coverage has been reported also for nickel[29] and iron[30] films.

The reason for the apparent discrepancy with the data by Beier et al.[13] is the erroneous determination of the coverage in their films.[11] The experimental evidence suggests that the "1 ML" film of Beier et al.[13] is, actually, a 5 ML-thick film. This will explain the lack of dependence of T_c with Co coverage and the claim of "1 ML" having a T_c well above 300K.[9,13] We have plotted in Fig. 3 (see above) the high energy Auger peaks and the MEED oscillations for 5.3 ML of Co deposited at 450K on the Cu(100) surface. This Auger spectrum is identical to the one shown by Clarke et al. in Fig. 1 of Ref. 11 which, however, they assign to a

490

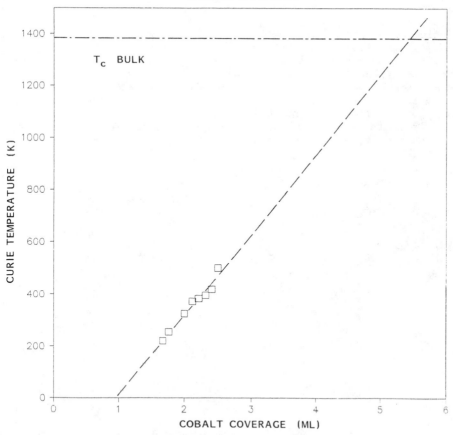

Co/Cu(100)

Fig. 5   Coverage dependence of the Curie temperature determined from
SMOKE hysteresis curves recorded as a function of the sample
temperature with the magnetizing field in-plane and normal to
the surface plane.   There was no component of the magnetization
perpendicular to the plane.

deposit of only 1ML. This discrepancy in the coverage calibration is due
to the erroneous identification made by Clarke et al.[11] of the first
break visible in their AES data (see Fig. 1 of Refs. 2,11) with 1ML
coverage. In this break the reported value of the Auger peak ratio
$(Co_{656}+Co_{716}/Cu_{845}+Cu_{920}+Co_{656}+Co_{716})$ is 0.53.[11] If the assignment of 1ML
to this point were correct the IMFP of electrons of high kinetic energy
would be unusually small, i.e. for the Cu_{920} Auger electrons it would
only be 5 Å.   As we have seen, this value would be completely
inconsistent with experimentally determined[6,14] and calculated[20] values
of the IMFP for electrons in this range of kinetic energy. The surprising
observation of the disappearance of the high energy Auger peaks of the
substrate after deposition of only 6 ML of Co[11] further indicates that
the calibration is inconsistent.   The lack of a fine-enough mesh of data
points and the low surface sensitivity of high energy Auger peaks could

be the reasons for the missing of additional breaks in the calibration of Clarke et al.[11]

2.   Electronic Structure

The Co/Cu system is a convenient scenario for two-dimensional effects to appear, as demonstrated by the photoelectron spectroscopy data reproduced in Fig. 6. In these angle-resolved UPS spectra, the evaporation of Co produces the development of the d band of Co near E_F, well separated in energy from the d bands of Cu, located 2-3 eV below E_F. Thus, the itinerant d-electrons of Co can be considered not to interact with those of Cu, whereby two dimensional effects can be expected to

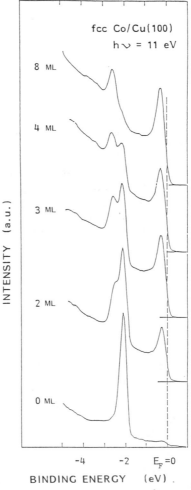

Fig. 6 Photoelectron spectra of the valence band for increasing coverages of Co deposited on Cu(100). The data have been taken at normal emission and 300 K, using a photon energy of 11 eV.

occur.[6] Angle-resolved UPS data, actually showed the two-dimensional character of the d-bands of a single deposited layer of cobalt.[7] In this connection, it is suggestive to recall the purely two-dimensional behaviour displayed by T_c. However, the transition of the electronic structure to the three dimensional behaviour takes place rapidly, as happens with the macroscopic magnetization. For Co thicknesses above 5ML, the electronic structure of the film is in excellent agreement with the calculated band structure of bulk fcc Co, with respect to exchange splitting, band dispersion, etc. as given by recent spin-resolved photoemission data.[14] Representative spectra are reproduced in Fig. 7. They clearly prove that the bands of Co, when spin-resolved, display the expected dispersion, with minority electrons dominating the spectra in the vicinity of E_F at higher photon energies (i.e. near the center of the Brillouin zone). The exchange splitting for the Δ_5 band has been found to be 1.4 eV, slightly smaller than the calculated value.[14]

Instead of changing the temperature, one can study the modifications produced in the photoemission spectra by crossing T_c by using the strong coverage dependence of T_c. The intensity spectra measured at 300 K for 1.8 ML (above T_c) and 2.1 ML-thick (below T_c) films do not show important differences. We conclude that the electronic structure does not change significantly upon crossing T_c, in particular the exchange splitting does not go to zero above T_c in agreement with previous reports.[31]

COBALT/COPPER SUPERLATTICES

We have used our experience with the Co/Cu system to explore the magnetic properties of layered metallic systems with an artificial periodicity. New phenomena are expected due to coupling effects between magnetic layers across non-magnetic materials.[4] The aim of this effort is the production of an ideal superlattice of X repeating bilayers, $[A_M/B_N]_X$, each bilayer consisting of M atomic planes of a magnetic metal A and N atomic planes of a non-magnetic material B (see Fig. 8). It is customary to use the term "superlattice" when the atomic planes are stacked coherently within each A and B layer. If only the periodicity normal to the surface, Λ_{SL}, is well defined, the material is labelled "multilayer". So far, most of the studies reported have been performed on multilayers.[4]

1. Mode of Growth

Crystalline metallic superlattices are more difficult to produce than those of semiconductors due to the much worse quality of metallic single crystal substrates as compared to semiconductors substrates. We have grown single-crystal superlattices by depositing alternating layers of Co and Cu on single-crystal Cu(100) substrates with mosaic spreads of less than 1°, as given by neutron diffraction.[18] In order to explore superlattice effects, we have chosen Co thicknesses in the range where the single-film data indicate bulk-like behaviour with respect to magnetic moment, electronic bands or T_c. The thickness of the evaporated layers was determined by our counting the number of oscillations in TEAS and cross-checked with a calibrated quartz balance. The samples were finally covered with 1000 Å of an epitaxial Cu buffer to protect the multilayer against exposure to air. It is known by Scanning AES Microscopy that oxygen does not penetrate more than 30 ML in these conditions.[31] Superlattices grown at 450 K, where optimum quality was obtained for Co deposition on bulk Cu(100), were found to be interdiffused. Thus, in spite of a slightly worse crystalline perfection, lower temperatures must be chosen in order to preserve the structure.

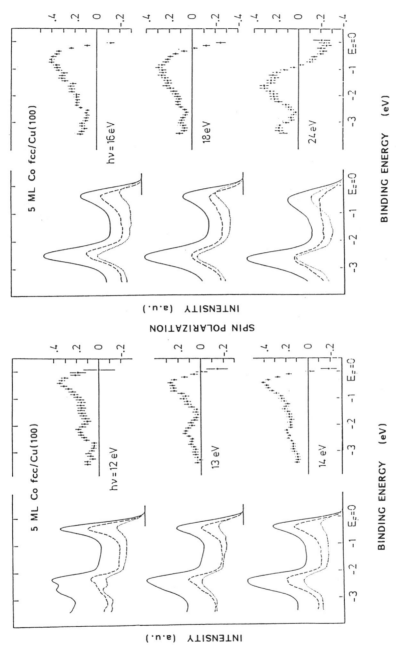

Fig. 7 Spin-resolved photoemission spectra at different photon energies for 5 ML of Co epitaxially grown on Cu(100). The data have been taken in the normal-incidence normal-emission geometry. Each panel shows the spin polarization (right) as a function of the binding energy, as well as the total (continuous line) majority (broken line), and minority (dotted line) electron energy distribution curves (left).

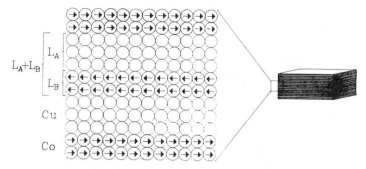

Fig. 8 Scheme of a metallic superlattice. The superlattice periodicity
is $L_A + L_B$, with L_A and L_B being the thickness of metals A and B
respectively. In our case A = Cu and B = Co. The arrows indicate
the magnetic moments of the cobalt atoms which are aligned
ferromagnetically <u>within</u> each period, but antiferromagnetically
from layer to layer (see text).

2.   Structural Characterization

Neutron diffraction with the scattering vector perpendicular to the
layer plane was used to verify the periodicity and crystalline character
of the multilayers. Figure 9 shows data corresponding to two selected
superlattices, [6 Co/8 Cu]$_{62}$ and [9 Co/5 Cu]$_{103}$, which have almost the
same periodicity and were both grown at 300K.[18] Low and high Q satellites
are observed in the data. The position of the satellites is related to
the chemical modulation, Λ_{SL}, by $Q = m \, 2\pi/\Lambda_{SL}$, (with m = 1,2,3..). The
periodicity of the superlattices, Λ_{SL}, is 26.4 Å and 26.6 Å respectively,
in excellent agreement with bilayer thicknesses estimated by the quartz
balance.

The presence of high Q superlattice satellites located around the
(002) Bragg reflection of the Cu substrate confirms the crystalline
character of the multilayer. The high Q satellites have an asymmetric
shape which can be ascribed to i) non-uniform d spacing, i.e. periodicity
fluctuation, or ii) strain in the film. The peak intensities for the
different satellites can be nicely reproduced by a simple step model
calculation[33] where the fluctuations in d-spacing are of the order of the
lattice mismatch (2%), indicating that the superlattices have atomically
sharp interfaces. The broadening of the high Q satellites indicates
coherence lengths of several hundred angstroms.

3.   Magnetic Properties

Extra satellites with half-integer index, m = 1/2,3/2...
corresponding to a doubling of the chemical bilayer periodicity, $2\Lambda_{SL}$,
are seen in both superlattices. The extra satellites have been assigned
to antiferromagnetic ordering of the Co layers by means of polarized
neutron diffraction, which has shown that the satellites have a purely
magnetic origin[18] and that they disappear under an external magnetic
field of 10 kOe applied parallel to the layers. This field is large
enough to saturate the magnetization of the sample at 4.2 K, as we will
show below.

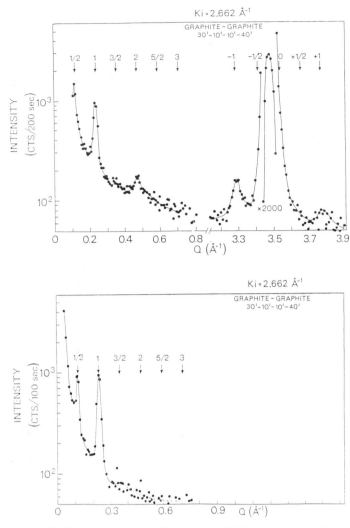

Fig. 9   Neutron diffraction data for two different superlattices in the low and high Q regions. Top: a) $[9 \text{ Co}/5 \text{ Cu}]_{103}$, Λ_{SL} = 26.6Å; bottom: $[6\text{Co}/8\text{Cu}]_{62}$, Λ_{SL} = 26.4Å. The data have been taken at 300 K with zero magnetic field. The incident neutron beam has k_i = 2.662Å^{-1}.

The magnetization of the multilayer films was measured by means of a 2-SQUID magnetometer. Representative data taken at 4K for low magnetic fields (≤ 1 kOe) applied parallel to the film plane of the $[6 \text{ Co/8 Cu}]_{62}$ sample are shown in Fig. 10. The inset displays the data for high magnetic fields (≥ 1 kOe) showing that the value of the saturation magnetization, M_S, is close to the one of bulk hcp Co (1.76 T). It is reached for fields of 7.5 kOe, well above the corresponding value for hcp Co along the easy axis (1.5 kOe). The magnetic moment deduced from these and other measurements[18] is 1.7 ± 0.1 μ_B, i.e. almost identical to the bulk fcc value, 1.75 μ_B, and in agreement with previous spin polarized neutron reflection measurements.[12]

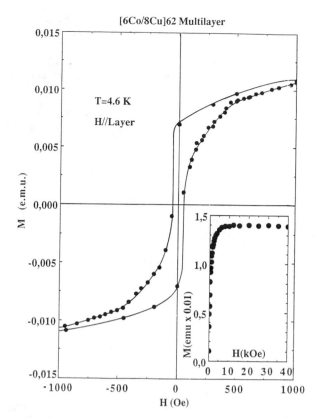

Fig. 10 Hysteresis loop for the [6 Co/8 Cu]$_{62}$ sample in the range up to
1 kOe at T = 4.6K. The magnetic field was applied parallel to
the layers. The inset shows the behavior of the magnetization
at higher magnetic field.

The saturation magnetization of these Co/Cu superlattices shows only
a very small decrease (7-8%) when changing the temperature from 4.2 K to
300 K (to avoid interdiffusion). This is in agreement with expectations
based on the single-film data indicating that T$_c$ will be well above 300K
for this range of thickness of the Co layers.

Data taken with H applied along the [110] and the [010] direction in
the plane, respectively, reveal different shapes in the hysteresis loops
indicating the existence of in-plane magneto-crystalline anisotropy.[34,35]
This in-plane anisotropy can be related to the difference between the
hard, [010], and the easy, [110], directions in the fcc structure of the
Co layers.

Our findings can be summarized as follows: At and below 300K each
group of 6 (9) ML of Co is ferromagnetically ordered in-plane in its
interior. The Co atoms have an average magnetic moment similar to the
bulk value. The magnetic ground state of the artificial film is
characterized by an antiferromagnetic ordering of the Co layers across

the Cu interlayers as revealed by polarized neutron diffraction. Accordingly, the total magnetization is zero and a new magnetic periodicity ($2\Lambda_{SL}$) appears. This situation is schematically depicted in Fig. 8. Upon increasing the field parallel to the surface plane, the magnetic moments of the Co layers begin to reverse their direction with a complex behaviour related to the in-plane anisotropy.[34,35] At 7.5 kOe all the Co layers are aligned along the field direction, the saturation magnetization is reached and the magnetic and chemical periodicities coincide as shown by the complete disappearance of the half-order satellites in neutron diffraction.[18]

The microscopic origin of the interaction between cobalt layers giving rise to the antiferromagnetic ground state of the superlattice is not yet known. Two extreme possibilities are i) long-range dipole interaction, i.e. the interaction responsible for domain formation in ferromagnets and ii) short-range exchange interaction, i.e. a coherent propagation of magnetic correlations across the non-magnetic Cu layers via an exchange coupling between Co moments and Cu valence electrons similar to the Ruderman-Kittel-Kasuda-Yosida (RKKY) interaction.[37] This second sort of coupling requires crystallographic perfection at the atomic level. A different dependence (monotonic or oscillatory, respectively) of the sign of the magnetic coupling with the thickness of the Cu intralayer could be used to distinguish experimentally between the two possibilities.

Acknowledgements

The authors are deeply indebted to many friends and colleagues who have contributed with discussions and various kinds of help with the experimental data to increase their understanding of this problem: J. Kirschner, C. M. Schneider, S. Ferrer, J. L. Martinez, G. Fillion, J. P. Rebouillat, J. Ferrón, C. Ocal and J. L. Vicent. We acknowledge the expert technical assistance from P.Flores during the measurements at the ILL. Work at the Universidad Autónoma de Madrid was supported by the CICyT under contract No. 3788.

REFERENCES

1. R. W. Vook, Int. Metals Rev. <u>27</u>, 209 (1982).
2. U. Gradmann, J. Mag. Magn. Mater. <u>54-57</u>, 733 (1986).
3. For a review of metallic multilayers see, e.g., I.K. Schuller in <u>Physics, Fabrication and Applications of Multilayered Structures</u>, P. Dhez and C. Weisbuch eds. (Springer, Berlin, 1989).
4. J. P. Renard and P. Beauvillain, Phys. Scripta <u>19</u>, 405 (1987).
5. U. Gradmann, Appl. Phys. <u>3</u>, 161 (1974).
6. L. González, R. Miranda, M. Salmerón, J. A. Vergés and F. Ynduráin, Phys. Rev. B <u>24</u>, 3245 (1981).
7. R. Miranda, D. Chandesris and J. Lecante, Surface Sci. <u>130</u>, 269 (1983).
8. F. Falo, I. Cano and M. Salmerón, Surface Sci. <u>143</u>, 303 (1984).
9. D. Pescia, G. Zampieri, M. Stampanoni, G. L. Bona, R. F. Willis and F. Meier, Phys, Rev. Lett. <u>58</u>, 933 (1987).
10. J. A. C. Bland, D. Pescia, and R. F. Willis, Phys. Scripta <u>19</u>, 413 (1987).
11. A. Clarke, G. Jennings, R. F. Willis, P. J. Rous and J. B. Pendry, Surface Sci. <u>187</u>, 327 (1987).
12. J. A. C. Bland, D. Pescia and R. F. Willis, Phys. Rev. Lett. <u>58</u>, 1244 (1987).
13. T. Beier, H. Jahrreis, D. Pescia, Th. Woike and W. Gudat, Phys. Rev. Lett. <u>61</u>, 1875 (1988).

14. C. M. Schneider, J. J. de Miguel, P. Bressler, J. Garbe, R. Miranda, S. Ferrer and J. Kirschner, J. Phys. (Paris) Suppl. C8, 1657 (1989).

15. J. J. de Miguel, A. Cebollada, J. M. Gallego, S. Ferrer, R. Miranda, C. M. Schneider, P. Bressler, J. Garbe, K. Bethke and J. Kirschner, Surface Sci. 211/212, 732 (1989).

16. C. M. Schneider, P. Bressler, P. Schuster, J. Kirschner, J. J. de Miguel, and R. Miranda (to be published).

17. C. Chappert and P. Bruno, J. Appl. Phys. 64, 5736 (1988).

18. A. Cebollada, J. L. Martínez, J. M. Gallego, J. J. de Miguel, R. Miranda, S. Ferrer, F. Batallán, G. Fillion and J. P. Rebouillat, Phys. Rev. B39, 9726 (1989).

19. J. P. Biberian and G. Somorjai, Appl. Surf. Sci. 2, 352 (1979). S. Ossicini, R. Memeo and F. Ciccacci, J. Vac. Sci. Technol. A3, 387 (1985).

20. M. P. Seah and W. A. Dench, Surf. Interface Anal. 1, 2 (1979).

21. L. J. Gómez, S. Bourgeal, J. Ibañez and M. Salmerón, Phys. Rev. B31, 2551 (1985).

22. J. H. Neave, B. A. Joyce, P. J. Dobson, and N. Norton, Appl. Phys. A31, 1 (1983).

23. S. T. Purcell, B. Heinrich, and A. S. Arrot, Phys. Rev. B35, 6458 (1987).

24. J. J. de Miguel, J. Ferrón, A. Cebollada, J. M. Gallego and S. Ferrer, J. Crystal Growth 91, 5915 (1988).

25. J. J. de Miguel, A. Cebollada, J. M. Gallego, J. Ferrón and S. Ferrer, J. Crystal Growth 88, 442 (1988).

26. D. A. Steigerwald and W. F. Egelhoff, Jr., Surface Sci. 192, L887 (1987).

27. S. D. Bader, E. R. Moog and P. Grunberg, J. Mag. Magn. Mat. 53, L295 (1986).

28. N. D. Mermin and H. Wagner, Phys. Rev. Lett. 17, 1133 (1966).

29. R. Bergholz and U. Gradmann, J. Mag. Magn. Mat. 45, 389 (1984).

30. O. Paul, M. Taborelli and M. Landolt, Surface Sci. 211/212, 724 (1989).

31. J. Kirschner, J. Appl. Phys. 64, 5915 (1988).

32. J. R. Dutcher, B. Heinrich, J. F. Cochran, D. A. Steigerwald and W. F. Egelhoff Jr., J. Appl. Phys. 63, 3464 (1988).

33. C. F. Majkrzak, J. D. Axe, and P. Boni, J. Appl. Phys. 57, 3657 (1985).

34. J. P. Rebouillat et al, to be published.

35. C. M. Schneider, J. J. de Miguel, P. Bressler, S. Ferrer, P. Schuster, R. Miranda and J. Kirschner, to be published.

36. J. P. Locquet, D. Neerinck, L. Stockman, Y. Bruynseraede and I. K. Schuller, Phys. Rev. B38, 3572 (1988).

37. Y. Yafet, J. Appl. Phys. 61, 4058 (1987).

RAMAN SCATTERING AND DISORDERED THIN-FILM GROWTH

PHENOMENA

J. S. Lannin

Department of Physics
Pennsylvania State University
University Park, PA 16802 U.S.A.

ABSTRACT

The use of Raman scattering as a probe of growth of thin films is reviewed. Enhancement of Raman signals to study early stages of growth and submonolayer films is described.

INTRODUCTION

The structural, physical and chemical properties of thin films in their various growth stages are of considerable current interest. A number of established methods such as LEED and TEM, as well as recent structural probes such as STM have provided important information on film structure, particularly the earliest submonolayer film growth stages of highly ordered systems. Given the quite diverse range of thin film systems of interest, which include more disordered noncrystalline, as well as partially ordered crystalline structures, considerable value exists in developing additional structural probes. Examples include spectroscopic probes, which in addition to providing physical property information, also indirectly yield structural data. Both vibrational, electronic and optical spectroscopies may be useful in this regard. Of these, vibrational Raman scattering provides a high resolution, sensitive probe of structural changes. While certain methods such as HREELS and FIRS provide information on surface phonons of ordered crystalline surfaces or adsorbed molecular vibrations, they have, as yet, generally not been applied to the surfaces of disordered crystalline or noncrystalline thin film systems. As discussed below, Raman scattering is particularly useful under appropriate conditions for obtaining information about whether the phase of an ultrathin film is crystalline or noncrystalline as well as information about the nature of that phase. For crystalline solids this also includes the presence of film strain or textural ordering.[1] A special capability, which is difficult to obtain by diffraction methods, is the potential of Raman scattering to observe small variations in network structural order within an amorphous phase. This is a consequence of the dependence of the phonon density of states of amorphous (a-) solids in such systems as elemental a-Ge and a-Si on short range order.[2,3] Raman scattering is also particularly useful in studying interfacial bonding in a number of thin film systems.

Kinetics of Ordering and Growth at Surfaces
Edited by M. G. Lagally
Plenum Press, New York, 1990

501

In this paper we focus on applications of Raman scattering to the study of the early stages of thin film growth phenomena of both selected polycrystalline and amorphous interfacial systems, as well as quite recent work in our laboratory on ultrathin films, which include submonolayer semiconductor island structures.[4] As Raman scattering is an optical method it probes both surface and bulk regions and differs from surface sensitive methods in its capabilities to study "buried" interfacial and near surface regions. The extended penetration depth of Raman scattering, that is of order of the inverse optical absorption coefficient, implies that surface sensitivity may be achieved for ultrathin films if substrate contributions are small or may be accurately subtracted. As Raman scattering is inherently a weak, higher order process, conventional methods do not generally yield detection capabilities at the monolayer level. To this end, we have utilized a combination of enhancement methods that provide for sufficient sensitivity for submonolayer detection of weak scattering amorphous semiconductors. In addition to signal level constraints, the obvious need for in situ UHV studies exists to avoid possible surface contamination effects in ultrathin films.

Previous in situ multichannel Raman scattering studies at submonolayer coverage have been restricted to high frequency scattering of molecules physisorbed on well polished metal surfaces.[5] In situ studies of crystalline solids utilizing multichannel detection have observed scattering from approximately 2-3 monolayers.[6] One possible exception to this is a conventional Raman scattering study of an approximate monolayer of Sb deposited on III-V compounds (GaAs and InP).[7]

While Raman scattering of thin films may involve scattering processes in which energy loss or gain occurs within the phonon, electron or magnon systems,[8] emphasis is placed here on structure sensitive phonon processes. In addition, more emphasis is given to noncrystalline, amorphous thin films. Substantial interest in the formation of these materials exists, especially in the early stages of film formation, given the often inherent low atom mobility and the resulting nonhomogeneous structure that may contain internal surfaces and under coordinated atoms. The absence of periodicity in amorphous solids implies in general rather weak signals and as such necessitates substantial enhancement for submonolayer detection. Recent success in this area provides a means of addressing questions of structural order within the amorphous state and variations in such order with film thickness.[9]

After briefly reviewing in an introductory manner major aspects of Raman scattering via phonons in ordered and disordered systems, including enhancement methods, we discuss examples of interfacial related problems in crystalline and noncrystalline systems. Emphasis is placed on Raman enhancement methods including multichannel Raman scattering (MCRS)[6] for increased signal detection and interference enhanced Raman scattering (IERS)[10] that yields an increased scattering due to local electric field enhancement in ultrathin films and near interfaces. Also discussed are conventional, nonenhanced Raman scattering studies of multilayer amorphous films which yield information on interfacial structure. A recent example of in situ UHV Raman scattering, in which the detection of equivalent coverage of submonolayer island structures of amorphous(a-) Ge by a combination of enhancement methods is then presented[4]. The observed changes in the Raman spectra yield information on the structure of ultrathin films in both cluster form, as well as in their coalesced state.

RAMAN SCATTERING: DISORDER AND SIZE EFFECTS

   Vibrational Raman scattering involves inelastic light scattering in which phonons are either created within the solid in the Stokes process or destroyed in the antiStokes process. Of usual interest are either one- or two-phonon processes. For crystalline solids without significant defect concentrations, the first ordered scattering involves zone center, $k \approx 0$ optic phonons of appropriate symmetry. The discrete, narrow $k \approx 0$ response is particularly useful for studying the effects of interfacial strain in thin films. This discrete spectrum contrasts with the first order scattering of disordered crystalline or amorphous solids which may often yield broad bands associated with the relaxation of the $k \approx 0$ selection rule due to modification or total loss of crystalline periodicity.[11,12] The broader peaks of the disorder induced first order scattering in highly defective crystals, particularly nonstoichiometric solids, yield a matrix element weighted phonon density of states. Information about the frequency dependence of crystalline matrix elements that modulate the induced scattering relative to the density of states is generally not known. For amorphous solids, a combination of Raman and inelastic neutron scattering in several elemental systems, has yielded information on the corresponding matrix element or coupling parameter.[13,14] Of interest here is that the frequency dependence of the coupling parameter in a-Ge or a-Si for high frequency optic-like modes is relatively independent of disorder in the amorphous network. As such, variations in the Raman scattering of thin film tetrahedral amorphous solids are primarily due to changes in the phonon density of states.[9] In addition to first order scattering, both ordered and disordered crystalline as well as amorphous systems exhibit a continuous second order scattering that may for crystals, under appropriate conditions, yield information about the phonon density of states.[15,16] For amorphous solids the absence of crystal momentum implies that the second order scattering is a matrix element weighted, self convolution of the density of states.[17]

   If single crystals are reduced in size below approximately 100Å, the first order allowed Raman scattering is modified by an admixture of other $k \neq 0$ Fourier components into the scattering.[18] The k range of these components is approximately of order $2\pi/D$, where D is the average particle size. In this case the Raman scattering becomes asymmetric in form and reduced somewhat in intensity. This implies the addition value of Raman scattering to differentiate between small microcrystallites, which exhibit perturbed optic-like scattering, and amorphous solids, which manifest continuous scattering extending to lower frequencies. In the latter, structural disorder often occurs on the scale of a few atoms as noted by appreciable fluctuations in geometrical parameters such as the bond angle or dihedral angle. In contrast, in microcrystalline solids, whose scale is usually greater than 25Å, the major disorder is currently believed to be at interfacial grain boundaries. One example of practical importance here involves carbon substrate films employed for the growth of island structures. As noted below, Raman scattering measurements indicate that either microcrystalline, amorphous or mixed phases may be present, depending on the deposition conditions.[19]

   In microcrystalline Si films, the Raman spectra indicate an asymmetric broadening to lower frequencies. This is a consequence of $k \approx 0$ optic phonons of Ge and Si occurring at the maximum of the optic branches.[20] In contrast, in semimetallic Bi the lowest $k \approx 0$ optic mode occurs at the minimum of its dispersion curves.[21] Recent thin films of Bi deposited on Si at $\sim 120°K$ of thickness <15Å, suggest asymmetric Raman spectral broadening of the Eg mode. The asymmetry to higher frequencies is consistent with finite size effects.[22] These measurements, as well as

Bi annealing studies, demonstrate the crystalline nature of these low temperature, sputter deposited films. This suggests that UPS studies[23] of ion bombarded Bi at 300°K do not reflect the valence states of amorphous Bi, but rather final state orientational effects. Angular dependent UPS spectra on films deposited at 120°K are also consistent with crystal formation and final state effects.[22]

Studies of microcrystalline materials, such as Si have yielded to date a minimum grain size of ∿30Å.[24] These measurements have been in relatively continuous high density films. In situ Raman studies of island films may allow smaller stable microcrystals to be formed. It is expected, however, that for very small microclusters that the geometry will change from that of crystalline phases.[25,26] Extensive theoretical work is currently in progress concerning the structure of such microclusters. A number of these studies suggest that small microclusters may have ring structures uncommon to crystalline phases, for example.

Raman scattering may also be utilized to determine if near surface bombardment related disorder results in formation of an amorphous component, given the characteristically distinct character of the scattering of this phase. As the amorphous state is itself variable, the detailed form of the Raman spectra may be utilized to determine the extent of the bond angle disorder. In a-Si and a-Ge, for example, the width of the high frequency TO bands have been shown to be approximately proportional to the width of the bond angle distribution about atoms in the amorphous work.[27,28] To date, Raman studies of bombarded systems have not been performed on ultrathin films, but have been confined to thin films[6] or ion implanted Si or Ge.[29] For crystalline Si and Ge high energy ion implantion at low temperatures yields Raman spectra that indicate substantial disorder in the amorphous phase. This bond angle related disorder may be reduced by annealing prior to crystallization.[29] Similar lower energy, shallow ion bombardment effects, expected for ultrathin films, are now feasible to study utilizing enhancement methods.

RAMAN SCATTERING ENHANCEMENT

Conventional Raman scattering from highly absorbing semiconductors or metals utilizes a back reflection geometry in which the focused laser beam penetrates to approximate depths often of order 200-1000Å. As signal levels are often weak, particularly for noncrystalline systems, a reduction of the thickness to increase surface sensitivity thus requires signal enhancement. One means of doing this utilizes a multilayer, interference enhancing film configuration, similar to that shown in Fig. 1 for the three uniform layers shown. The three (tri) layer structure is prepared to have a low reflectivity for the laser wavelength of interest. This antireflection coating configuration allows for constructive interference from light that does not undergo the phase shift of the elastic reflected beam.[10] With this geometry, the total scattering from the thin third layer may exceed that of a thick film by more than an order of magnitude. Typical thicknesses of top absorbing layers range from ∿40-200Å, while the intermediate transparent spacer layer, SiO_2 thickness is often between 500-1000Å. A bottom layer of Al of thickness typically greater than several hundred Angstroms is usually employed on smooth glass or polished Si substrates.

The IERS method has proved useful in the study of ultrathin film effects such as metal diffusion and silicide growth on thin film a-Si:H alloys,[30] annealing studies of thin (<50Å) a-Ge films,[31] and in the achievement of Raman scattering from amorphous metals.[32] One limitation of this method is that the thickness of the individual components is restricted for maximum enhancement to the conditions for a low reflectivity, typically ≲ 5%. The relative thickness of the transparent and top layers are obtained utilizing computer programs that assume homogeneous layers of known optical constants. For new materials, we employ spectroscopic ellipsometry measurements of thicker top layer films to estimate these parameters for the laser frequencies employed for Raman studies. As the top layer thickness is often small, the optical constants utilized may vary from those of the actual nonhomogeneous thin film. This requires some amount of trial and error in the fabrication process. For ultrathin films, a fourth layer may be deposited, as shown in Fig. 1 for island films, to form a high Raman scattering, quadlayer structure without substantially modifying the antireflection condition. At present the IERS method is limited to the study of polycrystalline or amorphous films, given the polycrystalline nature of the trilayer structures.

While IERS allows for a larger signal from the scatterer due to incident electric field enhancement, additional sensitivity is achieved by employed multichannel Raman scattering (MCRS) with an area sensitive detector. A detailed review of different detectors and their utilization for Raman scattering has recently been presented.[6] As MCRS provides a means of sampling the entire spectral range of interest continuously, it substantially reduces the time to obtain a given signal to noise. In addition, certain multichannel detectors have a quite low background of dark counts. Combining MCRS with IERS provides for an approximate two

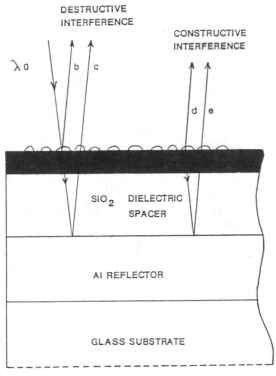

Fig. 1 Multilayer geometry for interference enhanced Raman scattering study of island films deposited on a trilayer structure.

505

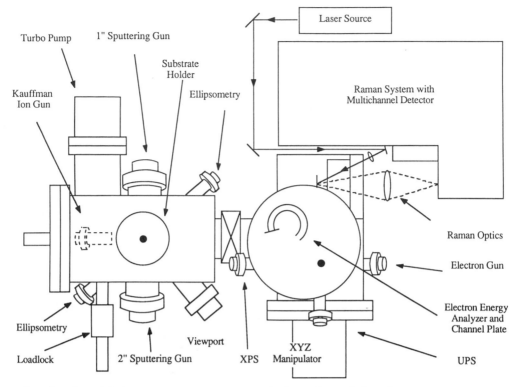

Fig. 2 Ultrahigh vacuum system utilized for film formation, Raman
scattering, UPS and ellipsometry.

orders of magnitude enhancement over wide band, conventional Raman
scattering. It should be noted that the monochromators employed for MCRS
have generally poorer stray light rejection than do corresponding
scanning systems with a narrow exit slit. As such, low frequency Raman
scattering in opaque materials requires more detailed estimates of the
stray light background due to surface roughness. Figure 2 schematically
illustrates the UHV Raman system employed in our laboratory for studies
of ultrathin island film phenomena. A preparation chamber contains two dc
magnetron sources, an evaporation source and a Kaufman ion source for
ion-assisted growth or substrate bombardment. Two optical ports with 70°
glancing angles to the substrate normal are utilized for real time,
continuous monitoring of the growth with visible ellipsometry. An
analysis chamber has an electron analyzer for Auger, UPS and XPS
analysis. The entire chamber system is mobile on tracks that allow
placement adjacent to the Raman system for MCRS utilizing f/1.2 optics.
The incident Raman exciting laser beam forms an angle of ~35°, providing
for improved polarization studies over glancing angle Raman studies.

RAMAN SCATTERING FROM SINGLE AND MULTILAYER INTERFACES

Raman scattering studies have been performed to obtain information
on interfacial structure and bonding. These studies have also noted
changes in the substrate Raman spectra due to local electric fields that
arise from band bending effects of Schottky barrier formation. In
addition, the formation of new bonds or phases has been studied. For
amorphous interfaces evidence has been obtained for additional structural
disorder within several bond distances of the interface. Examples of new

506

bond formation involve studies of the interface between Si and Ge in crystalline, as well as amorphous states. In bulk $Ge_{1-x}Si_x$ crystalline and thin film amorphous alloys one observes, in addition to Ge-Ge and Si-Si derived optic bands, new first and second order Raman bands associated with the formation of Ge-Si bonds.[33-35] The relative intensity of the Ge-Si band in Si/Ge layers has suggested an interfacial region of a few interatomic spacings. Annealing studies of the Ge-Si band in amorphous interfaces have been employed to obtain information on the activation energy for diffusion of Si into a-Ge.[36] Extensive Raman studies[6,37,38] have also been performed on crystalline strained layer $Ge_{1-x}Si_x$ superlattices.

Information about interfacial reactions during film deposition or post annealing has also been obtained for transition metal silicide films utilizing either IERS[30,39] or MCRS[6]. Annealing IERS studies of 50Å of Ni deposited on 100Å of a-Si indicate at 200°C modification of the a-Si:H intensity and the growth of a broad peak at $\sim$120cm^{-1}. These spectral changes were attributed to atomic interdiffusion and the formation of a disordered intermixed phase of unidentified origin. Conventional Raman studies of the amorphous transition metal alloy a-NiSi$_2$ indicate a broad band at $\sim$110cm.[40] This suggests that a-Si$_{1-x}$Ni$_x$ alloy formation occurs in the interdiffusion process. At temperatures of 250°C evidence for crystalline NiSi$_2$ formation is observed in the IERS studies. Similar silicide formation has been obtained for thin Pd and Pt deposited on a-Si and c-Si utilizing both conventional, IERS and in situ MCRS.[6,30]

Raman scattering studies in ultrahigh vacuum of Sb evaporated onto crystalline InP (110) exhibit a number of interesting transitions for low coverage.[7,41] For approximately one monolayer a quite weak, narrow feature has been attributed to the presence of Sb chains epitaxial to InP(100). For concentrations of $\sim$3 monolayers evidence of a broad a-Sb derived feature is observed, while at thicknesses exceeding $\sim$11 monolayers a transition to crystalline Sb is observed by two narrower peaks. The peak positions are shifted to higher frequency relative to c-Sb, a fact attributed to interfacial stress. The broadened width of the c-Sb line was associated with the presence of small crystallites which were estimated to be $\sim$100Å in size. Similar effects are observed for Sb on GaAs substrates. The asymmetry in the E_g mode of Sb suggests that finite size, k mixing effects may, however, also influence peak shape and position in microcrystalline Sb films.

Although current detection capabilities of Raman scattering are not comparable to UPS and Auger methods, single crystal substrate spectra may be highly sensitive to low electric field effects which modify Raman selection rules. It has been observed, for example, that coverages as low as 0.01 monolayers of Sb result in Raman scattering, under resonance conditions, from the InP substrate LO phonons that are normally Raman inactive.[41] Similar effects have been observed in a number of other III-V semiconductor systems, such as amorphous Ge deposited on GaAs, due to local field induced band bending effects.[42] Recent STM measurements have shown that Schottky barrier related band bending is apparently related to structural defects or edge atoms of the growing layer-like films.[43]

The sensitivity of the vibrational states to interfacial interactions has been studied in thick film amorphous multilayer structures utilizing conventional Raman scattering. These variable thickness films have been produced by glow discharge methods in which gas flow variations were employed.[44] TEM studies have shown relatively sharp amorphous interfaces.[45] Amorphous multilayer structures composed of a-Si:H/a-SiN$_x$:H or a-Si:H/a-SiO$_2$ exhibit Raman scattering spectra which indicate a strong dependence on a-Si:H thickness.[46] Examples of this are shown in Fig. 3 for films prepared by glow discharge deposition on anode and cathode surfaces. In these systems the transparent a-SiN$_x$ and a-SiO$_2$ layers yield insignificant Raman scattering. Of interest here is thus the observation that with decreasing a-Si:H thickness the Raman scattering exhibits an increased width and decreased relative intensity of the high frequency a-Si derived TO band at 475cm^{-1}. Figure 4 indicates the variation of this Raman band width for films deposited on anode and cathode surfaces.

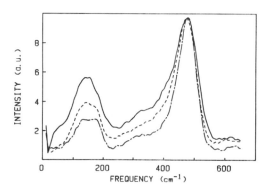

Fig. 3 Peak-normalized Raman scattering and of two a-Si:H/a-SiN$_x$:H multilayers with thickness 12Å (solid) and 24Å (dashed), and of a thick a-Si:H film (broken). From Ref. 46.

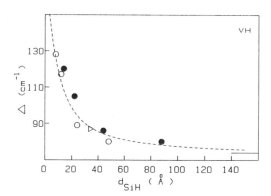

Fig. 4 Variation of the VH Raman width of the high frequency TO peak as a function of a-Si:H thickness of (o) anode and (•) cathode deposited a-Si:H/a-SiN$_x$ films and an a-Si:H/a-SiO$_2$ film (Δ). The dashed curve is a model fit based on inhomogeneous disorder of 6Å correlation length in a-Si:H layers. From Ref. 51.

In thick film amorphous solids, such as a-Si, a-Ge and their alloys with H, the width of the TO band and its relative intensity to the low frequency TA band have been shown to vary with conditions of film formation.[47,48] The origin of these changes are associated with variations in the short range bond angle order within the amorphous state.[49,50] The spectral variations of Fig. 3 and 4 exhibit increased disorder, for fixed a-SiN$_x$:H or a-SiO$_2$ thickness, thus indicating the greater influence of interfacial interactions in thin a-Si:H layers. An analysis of the decrease of the TO width with film thickness suggested that Si atoms of a-Si:H within ~ 2-3 bond lengths of the interface experience an increased disorder in their bond angle distribution. The model employed in this analysis assumed a growth of order from the interface with an exponential correlation length.[51] The dashed curve of Fig. 4 indicates the excellent agreement between this simple model and experiment. Relatively similar results are obtained by assuming a two layer model with interfacial and bulk disorder in the a-Si:H film.[52] The origin of the increased disorder may be generally related to the presence of new bonds at the interface that place additional constraints on the growing amorphous network. This contrasts with crystalline interfaces in which lattice mismatch results in strained layers, interface dislocations or new crystalline phases.

A number of other Raman measurements have been applied to study the nature of amorphous multilayer interfacial bonding, including a-Ge/a-Se and a-Ge/aSiO$_2$ films prepared by multisource evaporation.[53,54] For the latter case, additional high frequency modes were observed at $\sim 370 cm^{-1}$ that were attributed to the formation of a quite thin alloy of a-Ge$_y$(SiO$_x$)$_{1-y}$. Evidence for this was based on Raman spectra of thicker film alloys, which also suggested y $\simeq 0.35$. Modeling of the interface in terms of a 4 layer model, suggested that the observed broadening of the a-Ge TO peak was not, as in a-Si:H/SiO$_2$ due to disorder in a-Si:H, but rather to an alloy contribution whose spectrum overlapped the lower frequency a-Ge peak.

RAMAN SCATTERING OF ISLAND FILMS

The observation of interfacial disorder in amorphous multilayers of total optical penetration depth of order 1000A, suggests that the influence of single amorphous interfacial layers might be detected with Raman enhancement methods. Initial focus in our studies in this area have been on utilizing substrates which weakly interact with the growing film, allowing island formation.[4,55] Also in progress are studies of Stranski-Krastanov related growth of amorphous and crystalline systems. Of particular value for island studies are disordered C substrates, given the high nucleation density and primary strong graphitic-like bonding of these films. Such films have been extensively utilized for electron microscopy and photoemission studies of island growth of metals.[56] The use of the term "disordered C" (d-C) is employed here since Raman scattering measurements indicate a mixture of microcrystalline and amorphous C phases present in our low pressure rf sputtered films. Although various workers often refer to C films as amorphous, it is likely in many instances that these are either microcrystalline or mixtures of amorphous and microcrystalline, unless deposition temperature is maintained sufficiently low or films are relatively impure. Bombardment of these films at energies of $\leq 500 eV$ for purposes of surface cleaning is expected to enhance the near surface amorphous fraction. Evidence that weak interactions occur between a growing film and d-C substrates may be noted by the absence of new Raman scattering bands that are carbide-like, or the observation of no new photoemission valence band features. For the case of a-Ge on d-C these conditions are satisfied. Auger lineshapes support this conclusion.

Examples of in situ UHV Raman spectra of ultrathin a-Ge films are shown in Fig. 5 for the depolarization VH configuration (incident and scattered fields orthogonal). While the spectra indicate a broadening of the TO Peak that is qualitatively similar to the effect seen in Fig. 4 of reduced thickness in a-Si:H/a-SiN$_x$:H multilayers, qualitative differences occur. In particular, substantial, $\sim 10\%$ reductions in the position of the

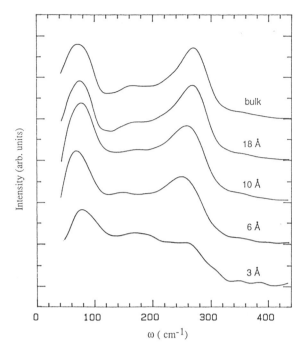

Fig. 5 Depolarized VH Raman spectra of ultrathin a-Ge films on carbon. From Ref. 4.

peak of the Ge TO band with decreasing coverage are observed. It has also been suggested that the observed increase in TO peak width with decreasing thickness is due to an increased scattering from near surface atoms having more distorted bond angles. These distortions are a consequence of growth constraints and local, near surface reconstructions to reduce system energy. In addition to this disorder, surfaces of a-Ge are expected to have undercoordinated atoms with dangling bonds. The influence of these bonds is to reduce Ge-Ge bond stretching force constants and shift the TO band peak to lower frequency. For thicknesses ≳ 10-12A, UPS substrate/film difference spectra suggest that substantial island coalescence has occurred. The observed changes in the Raman spectra for larger thickness imply that surface effects, including those of not completely coalesced ultrathin films, are significant.

Support for the idea that the size related peak shift in Fig. 5 is due to dangling, primarily 3-fold bonds is provided by theoretical studies of the local phonon density of states of 3-fold and 4-fold atoms in a-Ge random network models.[57] Strong experimental evidence is noted in the effect of chemisorbed H on the a-Ge clusters. Figure 6 contrasts the spectra, normalized to the TA peak, of an ∼10Å thick film before and after H chemisorption. With the addition of atomic hydrogen, produced with a hot filament, the TO peak is observed to shift to higher frequencies, independent to first order of the H concentration. While H reduces the width of the TO band, as it does in thick film alloys of a-Ge:H and a-Si:H,[58] it more gradually achieves its bulk value with increasing thickness. This effect, in which H decorating dangling bonds does not remove the TO broadening, is attributed to increased bond angle

510

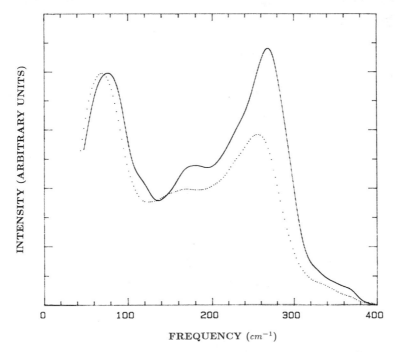

Fig. 6 Influence of atomic H chemisorbed (solid) on 10Å thick a-Ge (dotted) on VH Raman spectra. From Ref. 60.

disorder of atoms near the surface of the amorphous cluster. During the process of coalescence, as well as up to thickness of order 100Å (≈25ML), these surface atoms may influence the physical properties of thin films. For systems such as amorphous solids deposited at moderate temperatures, low surface mobility may result in the formation of nonhomogeneous films and the presence of micromorphological defects of low density, including small as well as large voids.[59]

Ultraviolet photoemission measurements (UPS) on ultrathin Ge films on d-C have been utilized to model film growth as well as note changes in the valence bands with cluster size or upon coalescence.[60] The reduction of the UPS substrate C signal has been modeled with good agreement with experiment. The results are consistent with approximate hemispherical islands with a Ge escape depth of approximately 7Å for 16.9eV NeI excitation. These results are consistent with similar escape depth estimates for Si.[61] The a-Ge difference UPS valence band spectra indicate (Fig. 7) an increase in binding energy of the Ge p band peak and a narrowing of this band as cluster size decreases. These effects are primarily attributed to a decrease in average cluster coordination due to surface atoms with dangling bonds. In addition, finite size effects may also influence the band width. As the islands have a distribution of sizes, this may yield an additional broadening of the valence band spectra with coverage. The chemisorption of H onto the a-Ge clusters results in the formation of a band at ∿4.8eV below the Fermi edge associated with Ge-H monohydride bonds.[54] The intensity of the Ge-H band is qualitatively consistent with the H primarily populating surface dangling bonds. The photoemission spectra are thus in accord with the

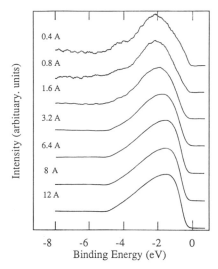

Fig. 7 UPS NeI difference spectra of a-Ge on C as a function of coverage. From Ref. 55.

Raman scattering peak shift arising from the presence of surface or near surface dangling bonds.

The success of combining MCRS and IERS to obtain submonolayer detection of a-Ge films clearly implies higher Raman sensitivity for microcrystalline island structures. Although broadening of the crystalline band will result in reduced peak intensity in such films, the scattering intensity is expected, nonetheless, to be substantially above that of the corresponding amorphous phase. Future studies of resonance Raman scattering in microcrystallites will thus be useful in extending the study of size related phenomena in this and other systems.

Acknowledgements

A number of graduate students and postdoctoral fellows have provided important contributions to aspects of our research at Penn State University presented here, including J. Fortner, R. Q. Yu, R. Fainchtein, N. Maley and N. Lustig. This work was supported by US NSF Grant DMR 8602391 and US DOE Grant DE-FG02-84ER45095.

REFERENCES

1. S. Nakashima, K. Mizoguchi, Y. Inoue, M. Miyauchi, A. Mitsuishi, T. Nishimura and Y. Akasaka, Jpn. J. Appl. Phys. 25, L222 (1986).
2. R. Alben, D. Weaire, J. E. Smith, Jr., and M. H. Brodsky, Phys. Rev. B11, 2271 (1975).
3. J. S. Lannin, Physics Today, 41, #7, 28 (1988).
4. J. Fortner, R. Q. Yu and J. S. Lannin, Proc. 13th Intern. Conf. Amorphous and Liquid Semiconductors, ed. M. Paesler and R. Zallen, J. Non-Cryst. Solids (in press).
5. A. Campion, J. K. Brown and V. M. Grizzle, Surf. Sci. 115, L153 (1982).
6. J. C. Tsang, in Light Scattering in Solids, V4, M. Cardona and G. Güntherodt ed. (Springer-Verlag), Berlin, 1989), p. 233 and references therein.

7. M. Hunermann, W. Pletschen, V. Resch, V. Rettweiler, W. Richter, J. Geurts and P. Lautenschläger, Surf. Sci. 189/190, 322 (1987).
8. W. Hayes and R. Loudon, in Scattering of Light by Crystals (Wiley, New York, 1978).
9. J. S. Lannin, Proc. 12th Int. Conf. Amorphous and Liquid Semiconductors, J. Noncryst. Solids 97 and 98, 39 (1987).
10. G. A. N. Connell, R. J. Nemanich, and C. C. Tsai, Appl. Phys. Lett. 36, 31 (1980).
11. J. M. Worlock and S. P. S. Porto, Phys. Rev. Lett. 15, 697 (1965).
12. R. Shuker and R. W. Gammon, Phys. Rev. Lett. 25, 222 (1970).
13. N. Maley and J. S. Lannin, Proc. 18th Intern. Conf. Phys. Semiconductors, Stockholm 1986, O. Engström ed. (World Scientific, Singapore, 1987), p. 1053.
14. F. Li and J. S. Lannin, Phys. Rev. B39, 6220 (1989).
15. P. A. Temple and C. E. Hathaway, Phys. Rev. B7, 3685 (1973).
16. B. Weinstein and M. Cardona, Phys. Rev. B7, 2545 (1973)
17. J. S. Lannin and P.J. Carroll, Phil. Mag. 45, 155 (1982).
18. R. J. Nemanich, S. A. Solin and R. M. Martin, Phys. Rev. B23, 635 (1981).
19. S. A. Solin and R. J. Kobliska, Proc. 5th Intern. Conf. Amorphous and Liquid Semicond., J. Stuke and W. Brenig eds. (Taylor and Francis, London, 1974), p. 1251.
20. G. Nelin and G. Nilsson, Phys. Rev. B5, 3151 (1972).
21. R. J. MacFarlane, in Physics of Semimetals and Narrow Gap Semiconductors, D. L. Carter and R. T. Bate ed. (Pergamon, New York, 1971), p. 289.
22. J. Fortner, R. Q. Yu and J. S. Lannin (to be published).
23. L. Ley, R. A. Pollak, S. P. Kowalczyk, R. McFeely and D. A. Shirley, Phys. Rev. B8, 641 (1973).
24. Z.Iqbal and S. Veprek, J. Phys. C15, 377 (1982).
25. W. L. Brown, R. R. Freeman, K. Raghavachari and M. Schlüter, Science 233, 860 (1987).
26. S. Saito and S. Ohnishi, in Microclusters, S. Sugano, S. Ohnishi and Y. Nishina, (Springer-Verlag, Berlin, 1987) ed. p. 263.
27. D. Beeman, R. Tsu and M. F. Thorpe, Phys. Rev. B32, 874 (1985).
28. N. Maley, D. Beeman and J. S. Lannin, Phys. Rev. B38, 10611 (1988).
29. J. Fortner and J. S. Lannin, Phys. Rev. B37 (1988).
30. R. J. Nemanich and C. M. Doland, J. Vac. Sci. Tech. B3, 1142 (1985).
31. J. E. Yehoda and J. S. Lannin, J. Vac. Sci. Tech. A1, 392 (1983).
32. N. Lustig, R. Fainchtein and J. S. Lannin, Phys. Rev. Lett. 55, 1775 (1985).
33. M. A. Renucci, J. B. Renucci and M. Cardona, Proc. 2nd Intern. Conf. Light Scattering in Solids, M. Balkanski ed. (Flammarion, Paris, 1971), p. 326.
34. J. S. Lannin, Phys. Rev. B16, 1510 (1977).
35. J. S. Lannin, Proc. 5th Int. Conf. on Amorphous and Liquid Semiconductors, Garmisch-Partenkirchen, 1973 J. Stuke and W. Brenig, eds. (Taylor and Francis, London, 1974), p. 1245.
36. P. Persans, in Amorphous Silicon and Related Materials, H. Fritzsche, ed. (Word Scientific, Singapore, 1988), p. 1045.
37. F. Cerdeira, A. Pinczuk, J. C. Bean, B. Batlogg and B. A. Wilson, Appl. Phys. Lett. 45, 1138 (1984).
38. G. Abstreiter, Festkörperprobleme 26, 41 (1986).
39. R. J. Nemanich, M. J. Thompson, W. B. Jackson, C. C. Tsai and B. L. Stafford, J. Non-Cryst. Solids 50 and 60, 513 (1983).
40. L. Koudelka, N. Lustig and J. S. Lannin, Solid State Commun. 63, 163 (1987).
41. W. Pletschen, N. Esser, J. Geurts, W. Richter, A. Tulke, M. Mattern-Klosson and H. Lüth, Proc. 18th Intern. Conf. Phys. Semicond., O. Engström ed. (World Scientific, Singapore, 1987), p. 367.

42. H. Brugger, F. Schäffler and G. Abstreiter, Phys. Rev. Lett. 52, 141 (1984).
43. R. Feenstra, Bull. Amer. Phys. Soc. 34, 820 (1989).
44. H. Munekata and H. Kukimoto, Jpn. J. Appl. Phys. 12, 213 (1982); B. Abeles and T. Tiedje, Phys. Rev. Lett. 51, 2003 (1983).
45. H. Deckman, J. H. Dunsmuir and B. Abeles, Appl. Phys. Lett. 46, 592 (1985).
46. N. Maley and J. S. Lannin, Phys. Rev. Rapid Commun. 31, 5577 (1985).
47. J. S. Lannin, in Amorphous Hydrogenated Silicon, J. Pankove, ed.; Vol. 21B of Semiconductors and Semimetals, Beer and Willardson eds., p. 159 (1984); S. T. Kshirsagar and J. S. Lannin, J. de Phys. 42, 54 (1981).
48. R. Tsu, J. G. Hernandez and F. H. Pollak, J. Non-Cryst. Solids 66, 109 (1984).
49. D. Beeman, R. Tsu and M. F. Thorpe, Phys. Rev. B32, 874 (1985).
50. N. Maley, D. Beeman and J. S. Lannin, Phys. Rev. B38, 10611 (1988).
51. N. Maley, J. S. Lannin and H. Ugar, J. Non-Cryst. Solids 77 & 78, 1073 (1985).
52. C. Roxlo, B. Abeles and P. D. Persans, J. Vac. Sci. Techn. B4, 1430 (1986).
53. H. J. Trodahl, M. W. Wright and A. Bittar, Solid State Commun. 59, 699 (1986).
54. G. V. M. Williams, A. Bittar and H. J. Trodahl, J. Appl. Phys. 60, 5148 (1988).
55. R. Q. Yu, J. Fortner, and J. S. Lannin, (to be published).
56. R. C. Baetzold and J. F. Hamilton, Prog. Solid State Chem. 15, 1 (1983) and references therein.
57. R. Biswas (private communication).
58. N. Maley and J. S. Lannin, Phys. Rev. B36, 1146 (1987).
59. R. W. Collins, in Amorphous Silicon and Related Materials, H. Fritzsche ed. (World Scientific, Singapore, 1988), p. 1003.
60. R. Q. Yu, J. Fortner and J. S. Lannin, J. Vac. Sci. Techn. (in press).
61. C. M. Garner, I. Lindau, C. Y. Su, P. Pianetta and W. Spicer, Phys. Rev. B19, 3944 (1979).